MW00353766

U.S. Customary unit		Times conversion factor		Equals SI unit	
		Accurate	Practical		
Moment of inertia (area)					
inch to fourth power	in.^{4}	416,231	416,000	millimeter to fourth power	mm^4
inch to fourth power	in.^{4}	0.416231×10^{-6}	0.416×10^{-6}	meter to fourth power	m^4
Moment of inertia (mass)					
slug foot squared	slug-ft^2	1.35582	1.36	kilogram meter squared	kg·m^2
Power					
foot-pound per second	ft-lb/s	1.35582	1.36	watt (J/s or N·m/s)	W
foot-pound per minute	ft-lb/min	0.0225970	0.0226	watt	W
horsepower (550 ft-lb/s)	hp	745.701	746	watt	W
Pressure; stress					
pound per square foot	psf	47.8803	47.9	pascal (N/m^2)	Pa
pound per square inch	psi	6894.76	6890	pascal	Pa
kip per square foot	ksf	47.8803	47.9	kilopascal	kPa
kip per square inch	ksi	6.89476	6.89	megapascal	MPa
Section modulus					
inch to third power	in.^{3}	16,387.1	16,400	millimeter to third power	mm^3
inch to third power	in.^{3}	16.3871×10^{-6}	16.4×10^{-6}	meter to third power	m^3
Velocity (linear)					
foot per second	ft/s	0.3048*	0.305	meter per second	m/s
inch per second	in./s	0.0254*	0.0254	meter per second	m/s
mile per hour	mph	0.44704*	0.447	meter per second	m/s
mile per hour	mph	1.609344*	1.61	kilometer per hour	km/h
Volume					
cubic foot	ft^3	0.0283168	0.0283	cubic meter	m^3
cubic inch	in.^{3}	16.3871×10^{-6}	16.4×10^{-6}	cubic meter	m^3
cubic inch	in.^{3}	16.3871	16.4	cubic centimeter (cc)	cm^3
gallon (231 in.^{3})	gal.	3.78541	3.79	liter	L
gallon (231 in.^{3})	gal.	0.00378541	0.00379	cubic meter	m^3

*An asterisk denotes an *exact* conversion factor

Note: To convert from SI units to USCS units, *divide* by the conversion factor

Temperature Conversion Formulas

$$T(°C) = \frac{5}{9}[T(°F) - 32] = T(K) - 273.15$$

$$T(K) = \frac{5}{9}[T(°F) - 32] + 273.15 = T(°C) + 273.15$$

$$T(°F) = \frac{9}{5}T(°C) + 32 = \frac{9}{5}T(K) - 459.67$$

Mechanics of Materials

BRIEF EDITION

James M. Gere
Late Professor Emeritus, Stanford University

Barry J. Goodno
Georgia Institute of Technology

CENGAGE
Learning™

Australia • Brazil • Japan • Korea • Mexico • Singapore • Spain • United Kingdom • United States

CENGAGE
Learning™

Mechanics of Materials, Brief Edition
James M. Gere and Barry J. Goodno

Publisher, Global Engineering:
   Christopher M. Shortt

Acquisitions Editor: Randall Adams

Senior Developmental Editor: Hilda Gowans

Editorial Assistant: Tanya Altieri

Team Assistant: Carly Rizzo

Marketing Manager: Lauren Betsos

Media Editor: Chris Valentine

Content Project Manager: Jennifer Ziegler

Production Service: RPK Editorial Services

Copyeditor: Shelly Gerger-Knechtl

Proofreader: Martha McMaster

Indexer: Shelly Gerger-Knechtl

Compositor: Integra Software Solutions

Senior Art Director: Michelle Kunkler

Internal Designer: Peter Papayanakis

Cover Designer: Andrew Adams/4065042
   Canada Inc.

Cover Images: © Baloncici/Shutterstock;
   © Carlos Neto/Shutterstock

Text and Image Permissions Researcher:
   Kristiina Paul

First Print Buyer: Arethea L. Thomas

For product information and technology assistance, contact us at **Cengage Learning Customer & Sales Support, 1-800-354-9706.**
For permission to use material from this text or product, submit all requests online at **www.cengage.com/permissions.** Further permissions questions can be emailed to **permissionrequest@cengage.com.**

Library of Congress Control Number: 2010932704

ISBN-13: 978-1-111-13602-4

ISBN-10: 1-111-13602-5

Cengage Learning
200 First Stamford Place, Suite 400
Stamford, CT 06902
USA

Cengage Learning is a leading provider of customized learning solutions with office locations around the globe, including Singapore, the United Kingdom, Australia, Mexico, Brazil, and Japan. Locate your local office at: **international.cengage.com/region.**

Cengage Learning products are represented in Canada by Nelson Education Ltd.

For your course and learning solutions, visit **www.cengage.com/engineering.**

Purchase any of our products at your local college store or at our preferred online store **www.cengagebrain.com.**

Printed in Canada
1 2 3 4 5 6 7 13 12 11 10

Contents

James Monroe Gere
1925–2008

James Monroe Gere, Professor Emeritus of Civil Engineering at Stanford University, died in Portola Valley, CA, on January 30, 2008. Jim Gere was born on June 14, 1925, in Syracuse, NY. He joined the U.S. Army Air Corps at age 17 in 1942, serving in England, France and Germany. After the war, he earned undergraduate and master's degrees in Civil Engineering from the Rensselaer Polytechnic Institute in 1949 and 1951, respectively. He worked as an instructor and later as a Research Associate for Rensselaer between 1949 and 1952. He was awarded one of the first NSF Fellowships, and chose to study at Stanford. He received his Ph.D. in 1954 and was offered a faculty position in Civil Engineering, beginning a 34-year career of engaging his students in challenging topics in mechanics, and structural and earthquake engineering. He served as Department Chair and Associate Dean of Engineering and in 1974 co-founded the John A. Blume Earthquake Engineering Center at Stanford. In 1980, Jim Gere also became the founding head of the Stanford Committee on Earthquake Preparedness, which urged campus members to brace and strengthen office equipment, furniture, and other contents items that could pose a life safety hazard in the event of an earthquake. That same year, he was invited as one of the first foreigners to study the earthquake-devastated city of Tangshan, China. Jim retired from Stanford in 1988 but continued to be a most valuable member of the Stanford community as he gave freely of his time to advise students and to guide them on various field trips to the California earthquake country.

Jim Gere was known for his outgoing manner, his cheerful personality and wonderful smile, his athleticism, and his skill as an educator in Civil Engineering. He authored nine textbooks on various engineering subjects starting in 1972 with *Mechanics of Materials*, a text that was inspired by his teacher and mentor Stephan P. Timoshenko. His other well-known textbooks, used in engineering courses around the world, include: *Theory of Elastic Stability*, co-authored with S. Timoshenko; *Matrix Analysis of Framed Structures* and *Matrix Algebra for Engineers*, both co-authored with W. Weaver; *Moment Distribution*; *Earthquake Tables: Structural and Construction Design Manual*, co-authored with H. Krawinkler; and *Terra Non Firma: Understanding and Preparing for Earthquakes*, co-authored with H. Shah.

Jim Gere in the Timoshenko Library at Stanford holding a copy of the 2nd edition of this text (photo courtesy of Richard Weingardt Consultants, Inc.)

Respected and admired by students, faculty, and staff at Stanford University, Professor Gere always felt that the opportunity to work with and be of service to young people both inside and outside the classroom was one of his great joys. He hiked frequently and regularly visited Yosemite and the Grand Canyon national parks. He made over 20 ascents of Half Dome in Yosemite as well as "John Muir hikes" of up to 50 miles in a day. In 1986 he hiked to the base camp of Mount Everest, saving the life of a companion on the trip. James was an active runner and completed the Boston Marathon at age 48, in a time of 3:13.

James Gere will be long remembered by all who knew him as a considerate and loving man whose upbeat good humor made aspects of daily life or work easier to bear. His last project (in progress and now being continued by his daughter Susan of Palo Alto) was a book based on the written memoirs of his great-grandfather, a Colonel (122d NY) in the Civil War.

Preface

Mechanics of materials is a basic engineering subject that, along with statics, must be understood by anyone concerned with the strength and physical performance of structures, whether those structures are man-made or natural. At the college level, mechanics of materials is usually taught during the sophomore and junior years. The subject is required for most students majoring in mechanical, structural, civil, biomedical, petroleum, aeronautical, and aerospace engineering. Furthermore, many students from such diverse fields as materials science, industrial engineering, architecture, and agricultural engineering also find it useful to study this subject.

About the Brief Edition

In many university engineering programs today, both statics and mechanics of materials are now taught in large sections comprised of students from the variety of engineering disciplines listed above. Instructors for the various parallel sections must cover the same material, and all of the major topics must be presented so that students are well prepared for the more advanced and follow-on courses required by their specific degree programs. There is little time for advanced or specialty topics because fundamental concepts such as stress and strain, deformations and displacements, flexure and torsion, shear and stability must be covered before the term ends. As a result, there has been increased interest in a more streamlined, or brief, text on mechanics of materials that is focused on the essential topics that can and must be covered in the first undergraduate course. This text has been designed to meet this need.

The main topics covered in this book are the analysis and design of structural members subjected to tension, compression, torsion, and bending, including the fundamental concepts mentioned above. Other important topics are the transformations of stress and strain, combined loadings and combined stress, deflections of beams, and stability of columns. Unfortunately, it is no longer possible in most programs to cover a number of specialized subtopics which were removed to produce this "brief" edition. This streamlined text is based on the review comments of many instructors who asked for a text specifically tailored to the needs of their semester length course, with advanced material removed. The resulting brief text, based upon and derived from the full 7th edition of this text book, covers the essential topics in the full text with the same level of detail and rigor.

Some of the specialized topics no longer covered here include the following: stress concentrations, dynamic and impact loadings, nonprismatic members, shear centers, bending of unsymmetric beams, maximum stresses in beams, energy based approaches for computing deflections of beams, and statically indeterminate beams. A discussion of beams of two materials, or composite beams, was retained but moved to the end of the chapter on stresses in beams. Review material on centroids and moments of inertia was also removed from the text but was placed online so is still available to the student. Finally, Appendices A-H, as well as References and Historical Notes, were moved online to shorten the text while retaining a comprehensive discussion of major topics.

As an aid to the student reader, each chapter begins with a *Chapter Overview* which highlights the major topics to be covered in that chapter, and closes with a *Chapter Summary & Review* in which the key points as well as major mathematical formulas presented in the chapter are listed for quick review (in preparation for examinations on the material). Each chapter also opens with a photograph of a component or structure which illustrates the key concepts to be discussed in that chapter.

Considerable effort has been spent in checking and proofreading the text so as to eliminate errors, but if you happen to find one, no matter how trivial, please notify me by e-mail (*bgoodno@ce.gatech.edu*). We will correct any errors in the next printing of the book.

Examples

Examples are presented throughout the book to illustrate the theoretical concepts and show how those concepts may be used in practical situations. In some cases, photographs have been added showing actual engineering structures or components to reinforce the tie between theory and application. Many instructors discuss lessons learned from engineering failures to motivate student interest in the subject matter and to illustrate basic concepts. In both lecture and text examples, it is appropriate to begin with simplified analytical models of the structure or component and the associated free-body diagram(s) to aid the student in understanding and applying the relevant theory in engineering analysis of the system. The text examples vary in length from one to four pages, depending upon the complexity of the material to be illustrated. When the emphasis is on concepts, the examples are worked out in symbolic terms so as to better illustrate the ideas, and when the emphasis is on problem-solving, the examples are numerical in character. In selected examples throughout the text, graphical display of results (e.g., stresses in beams) has been added to enhance the student's understanding of the problem results.

Problems

In all mechanics courses, solving problems is an important part of the learning process. This textbook offers more than 700 problems for homework assignments and classroom discussions. The problems are placed at the end of each chapter so that they are easy to find and don't break up the presentation of the main subject matter. Also, problems are generally arranged in order of increasing difficulty thus alerting students to the time necessary for solution. Answers to all problems are listed near the back of the book. An Instructor Solution Manual (ISM) is available to registered instructors at the publisher's web site.

In addition to the end of chapter problems, a new appendix has been added to this brief edition containing more than 100 FE Exam type problems. Many students take the *Fundamentals of Engineering Examination* upon graduation, the first step on their path to registration as a Professional Engineer. These problems cover all of the major topics presented in the text and are thought to be representative of those likely to appear on an FE exam. Most of these problems are in SI units which is the system of units used on the FE Exam itself, and require use of an engineering calculator to carry out the solution. Each of the problems is presented in the FE Exam format. The student must select from 4 available answers (A, B, C or D), only one of which is the correct answer. The correct answer choices are listed in the Answers section at the back of this text, and the detailed solution for each problem is available on the student website. It is expected that careful review of these problems will serve as a useful guide to the student in preparing for this important examination.

Units

Both the International System of Units (SI) and the U.S. Customary System (USCS) are used in the examples and problems. Discussions of both systems and a table of conversion factors are given in online Appendix A. For problems involving numerical solutions, odd-numbered problems are in USCS units and even-numbered problems are in SI units. This convention makes it easy to know in advance which system of units is being used in any particular problem. In addition, tables containing properties of structural-steel shapes in both USCS and SI units may be found in online Appendix E so that solution of beam analysis and design examples and end-of-chapter problems can be carried out in either USCS or SI units.

DIGITAL SUPPLEMENTS

Instructor Resources Web site

As noted above, an Instructor Solution Manual (ISM) is available to registered instructors at the publisher's web site. This web site also includes a full set of PowerPoint slides containing all graphical images in the text for use by instructors during lecture or review sessions. Finally, to reduce the length of the printed book, Chapter 10 on *Review of Centroids and Moments of Inertia* has also been moved to the instructor web site, as have *Appendices A-H* (see listing below) and the *References and Historical Notes* sections from the full seventh edition text.

For reference:

Free Student Companion Web site

A free student companion web site is available for student users of the brief edition. The web site contains Chapter 10 on *Review of Centroids and Moments of Inertia*, as well as *Appendices A-H (see above)* and the *References and Historical Notes* sections from the full seventh edition text. Lastly, solutions to all FE Exam type problems presented in the appendix of this text are listed so the student can check not only answers but also detailed solutions in preparation for the FE Exam.

CourseMate Premium Web site

CourseMate from Cengage Learning offers students book-specific interactive learning tools at and incredible value. Each CourseMate website includes an e-book and interactive learning tools. To access additional course materials (including CourseMate), please visit www.cengagebrain.com. At the CengageBrain.com home page, search for the ISBN of your title (from the back cover of your book) using the search box at the top of the page. This will take you to the product page where these resources can be found.

S. P. Timoshenko (1878–1972) and J. M. Gere (1925–2008)

Many readers of this book will recognize the name of Stephen P. Timoshenko–probably the most famous name in the field of applied mechanics. Timoshenko is generally recognized as the world's most outstanding pioneer in applied mechanics. He contributed many new ideas and concepts and became famous for both his scholarship and his teaching.

Through his numerous textbooks he made a profound change in the teaching of mechanics not only in this country but wherever mechanics is taught. Timoshenko was both teacher and mentor to James Gere and provided the motivation for the first edition of this text, authored by James M. Gere and published in 1972; the second and each subsequent edition of this book were written by James Gere over the course of his long and distinguished tenure as author, educator and researcher at Stanford University. James Gere started as a doctoral student at Stanford in 1952 and retired from Stanford as a professor in 1988 having authored this and eight other well known and respected text books on mechanics, and structural and earthquake engineering. He remained active at Stanford as Professor Emeritus until his death in January of 2008.

A brief biography of Timoshenko appears in the first reference in the online *References and Historical Notes* section, and also in an August 2007 *STRUCTURE magazine* article entitled *"Stephen P. Timoshenko: Father of Engineering Mechanics in the U.S."* by Richard G. Weingardt, P.E. This article provides an excellent historical perspective on this and the many other engineering mechanics textbooks written by each of these authors.

Acknowledgments

To acknowledge everyone who contributed to this book in some manner is clearly impossible, but I owe a major debt to my former Stanford teachers, especially my mentor and friend, and lead author, James M. Gere.

I am grateful to my many colleagues teaching Mechanics of Materials at various institutions throughout the world who have provided feedback and constructive criticism about the text; for all those anonymous reviews, my thanks. With each new edition, their advice has resulted in significant improvements in both content and pedagogy.

My appreciation and thanks also go to the reviewers who provided specific comments for this Brief Edition:

Hank Christiansen, Brigham Young University
Paul R. Heyliger, Colorado State University
Richard Johnson, Montana Tech, University of Montana
Ronald E. Smelser, University of North Carolina at Charlotte
Candace S. Sulzbach, Colorado School of Mines

I wish to also acknowledge my Structural Engineering and Mechanics colleagues at the Georgia Institute of Technology, many of whom provided valuable advice on various aspects of the revisions and additions leading to the current edition. It is a privilege to work with all of these educators and to learn from them in almost daily interactions and discussions about structural engineering and mechanics in the context of research and higher education. Finally, I wish to extend my thanks to my many current and former students who have helped to shape this text in its various editions.

The editing and production aspects of the book were always in skillful and experienced hands, thanks to the talented and knowledgeable personnel of Cengage Learning (formerly Thomson Learning). Their goal was the same as mine–to produce the best possible brief edition of this text, never compromising on any aspect of the book.

The people with whom I have had personal contact at Cengage Learning are Christopher Carson, Executive Director, Global Publishing Program, Christopher Shortt, Publisher, Global Engineering Program, Randall Adams and Swati Meherishi, Senior Acquisitions Editors, who provided guidance throughout the project; Hilda Gowans, Senior Developmental Editor, Engineering, who was always available to provide information and encouragement; Nicola Winstanley who managed all aspects of new photo selection; Andrew Adams who created the cover design for the book; and Lauren Betsos, Global Marketing Manager, who developed promotional material in support of the text. I would like to especially acknowledge the work of Rose Kernan of RPK Editorial Services, who edited the manuscript and designed the pages. To each of these individuals I express my heartfelt thanks not only for a job well done but also for the friendly and considerate way in which it was handled.

I am deeply appreciative of the patience and encouragement provided by my family, especially my wife, Lana, throughout this project.

Finally, I am very pleased to be involved in this endeavor, at the invitation of my mentor and friend of thirty eight years, Jim Gere, which extends this textbook toward the forty year mark. I am committed to the continued excellence of this text and welcome all comments and suggestions. Please feel free to provide me with your critical input at *bgoodno@ce.gatech.edu.*

BARRY J. GOODNO
Atlanta, Georgia

Symbols

A	area
A_f, A_w	area of flange; area of web
a, b, c	dimensions, distances
C	centroid, compressive force, constant of integration
c	distance from neutral axis to outer surface of a beam
D	diameter
d	diameter, dimension, distance
E	modulus of elasticity
E_r, E_t	reduced modulus of elasticity; tangent modulus of elasticity
e	eccentricity, dimension, distance, unit volume change (dilatation)
F	force
f	shear flow, shape factor for plastic bending, flexibility, frequency (Hz)
f_T	torsional flexibility of a bar
G	modulus of elasticity in shear
g	acceleration of gravity
H	height, distance, horizontal force or reaction, horsepower
h	height, dimensions
I	moment of inertia (or second moment) of a plane area
I_x, I_y, I_z	moments of inertia with respect to x, y, and z axes
I_{x1}, I_{y1}	moments of inertia with respect to x_1 and y_1 axes (rotated axes)
I_{xy}	product of inertia with respect to xy axes
I_{x1y1}	product of inertia with respect to x_1y_1 axes (rotated axes)
I_P	polar moment of inertia
I_1, I_2	principal moments of inertia
J	torsion constant
K	effective length factor for a column

k	spring constant, stiffness, symbol for $\sqrt{P/EI}$
k_T	torsional stiffness of a bar
L	length, distance
L_E	effective length of a column
ln, log	natural logarithm (base e); common logarithm (base 10)
M	bending moment, couple, mass
m	moment per unit length, mass per unit length
N	axial force
n	factor of safety, integer, revolutions per minute (rpm)
O	origin of coordinates
O'	center of curvature
P	force, concentrated load, power
P_{allow}	allowable load (or working load)
P_{cr}	critical load for a column
p	pressure (force per unit area)
Q	force, concentrated load, first moment of a plane area
q	intensity of distributed load (force per unit distance)
R	reaction, radius
r	radius, radius of gyration ($r = \sqrt{I/A}$)
S	section modulus of the cross section of a beam, shear center
s	distance, distance along a curve
T	tensile force, twisting couple or torque, temperature
t	thickness, time, intensity of torque (torque per unit distance)
t_f, t_w	thickness of flange; thickness of web
u_r, u_t	modulus of resistance; modulus of toughness
V	shear force, volume, vertical force or reaction
v	deflection of a beam, velocity
v', v'', etc.	dv/dx, d^2v/dx^2, etc.
W	force, weight, work
w	load per unit of area (force per unit area)
x, y, z	rectangular axes (origin at point O)
x_c, y_c, z_c	rectangular axes (origin at centroid C)
$\bar{x}, \bar{y}, \bar{z}$	coordinates of centroid
α	angle, coefficient of thermal expansion, nondimensional ratio
β	angle, nondimensional ratio, spring constant, stiffness
β_R	rotational stiffness of a spring
γ	shear strain, weight density (weight per unit volume)
$\gamma_{xy}, \gamma_{yz}, \gamma_{zx}$	shear strains in xy, yz, and zx planes

$\gamma_{x_1y_1}$	shear strain with respect to x_1y_1 axes (rotated axes)
γ_θ	shear strain for inclined axes
δ	deflection of a beam, displacement, elongation of a bar or spring
ΔT	temperature differential
ϵ	normal strain
$\epsilon_x, \epsilon_y, \epsilon_z$	normal strains in x, y, and z directions
$\epsilon_{x_1}, \epsilon_{y_1}$	normal strains in x_1 and y_1 directions (rotated axes)
ϵ_θ	normal strain for inclined axes
$\epsilon_1, \epsilon_2, \epsilon_3$	principal normal strains
ϵ'	lateral strain in uniaxial stress
ϵ_T	thermal strain
ϵ_Y	yield strain
θ	angle, angle of rotation of beam axis, rate of twist of a bar in torsion (angle of twist per unit length)
θ_p	angle to a principal plane or to a principal axis
θ_s	angle to a plane of maximum shear stress
κ	curvature ($\kappa = 1/\rho$)
λ	distance, curvature shortening
ν	Poisson's ratio
ρ	radius, radius of curvature ($\rho = 1/\kappa$), radial distance in polar coordinates, mass density (mass per unit volume)
σ	normal stress
$\sigma_x, \sigma_y, \sigma_z$	normal stresses on planes perpendicular to x, y, and z axes
$\sigma_{x_1}, \sigma_{y_1}$	normal stresses on planes perpendicular to x_1y_1 axes (rotated axes)
σ_θ	normal stress on an inclined plane
$\sigma_1, \sigma_2, \sigma_3$	principal normal stresses
σ_{allow}	allowable stress (or working stress)
σ_{cr}	critical stress for a column ($\sigma_{cr} = P_{cr}/A$)
σ_{pl}	proportional-limit stress
σ_r	residual stress
σ_T	thermal stress
σ_U, σ_Y	ultimate stress; yield stress
τ	shear stress
$\tau_{xy}, \tau_{yz}, \tau_{zx}$	shear stresses on planes perpendicular to the x, y, and z axes and acting parallel to the y, z, and x axes
$\tau_{x_1y_1}$	shear stress on a plane perpendicular to the x_1 axis and acting parallel to the y_1 axis (rotated axes)
τ_θ	shear stress on an inclined plane
τ_{allow}	allowable stress (or working stress) in shear

τ_U, τ_Y ultimate stress in shear; yield stress in shear

ϕ angle, angle of twist of a bar in torsion

ψ angle, angle of rotation

ω angular velocity, angular frequency ($\omega = 2\pi f$)

★A star attached to a section number indicates a specialized or advanced topic. One or more stars attached to a problem number indicate an increasing level of difficulty in the solution.

Greek Alphabet

A	α	Alpha	N	ν	Nu
B	β	Beta	Ξ	ξ	Xi
Γ	γ	Gamma	O	o	Omicron
Δ	δ	Delta	Π	π	Pi
E	ϵ	Epsilon	P	ρ	Rho
Z	ζ	Zeta	Σ	σ	Sigma
H	η	Eta	T	τ	Tau
Θ	θ	Theta	Υ	υ	Upsilon
I	ι	Iota	Φ	ϕ	Phi
K	κ	Kappa	X	χ	Chi
Λ	λ	Lambda	Ψ	ψ	Psi
M	μ	Mu	Ω	ω	Omega

Mechanics of
Materials

This telecommunications tower is an assemblage of many members that act primarily in tension or compression. (Photo by Bryan Tokarczyk, PE/KPFF Tower Engineers)

Tension, Compression, and Shear

CHAPTER OVERVIEW

In Chapter 1, we are introduced to mechanics of materials, which examines the **stresses**, **strains**, and **displacements** in bars of various materials acted on by axial loads applied at the centroids of their cross sections. We will learn about **normal stress** (σ) and **normal strain** (ε) in materials used for structural applications, then identify key properties of various materials, such as the modulus of elasticity (E) and yield (σ_y) and ultimate (σ_u) stresses, from plots of stress (σ) versus strain (ε). We will also plot shear stress (τ) versus shear strain (γ) and identify the shearing modulus of elasticity (G). If these materials perform only in the linear range, stress and strain are related by Hooke's Law for normal stress and strain ($\sigma = E \cdot \epsilon$) and also for shear stress and strain ($\tau = G \cdot \gamma$). We will see that changes in lateral dimensions and volume depend upon Poisson's ratio (v). Material properties E, G, and v, in fact, are directly related to one another and are not independent properties of the material.

Assemblage of bars to form structures (such as trusses) leads to consideration of average shear (τ) and bearing (σ_b) stresses in their connections as well as normal stresses acting on the net area of the cross section (if in tension) or on the full cross-sectional area (if in compression). If we restrict maximum stresses at any point to **allowable** values by use of factors of safety, we can identify allowable levels of axial loads for simple systems, such as cables and bars. **Factors of safety** relate actual to required strength of structural members and account for a variety of uncertainties, such as variations in material properties and probability of accidental overload. Lastly, we will consider **design**: the iterative process by which the appropriate size of structural members is determined to meet a variety of both **strength** and **stiffness requirements** for a particular structure subjected to a variety of different loadings.

Chapter 1 is organized as follows:

1.1 INTRODUCTION TO MECHANICS OF MATERIALS

Mechanics of materials is a branch of applied mechanics that deals with the behavior of solid bodies subjected to various types of loading. Other names for this field of study are *strength of materials* and *mechanics of deformable bodies*. The solid bodies considered in this book include bars with axial loads, shafts in torsion, beams in bending, and columns in compression.

The principal objective of mechanics of materials is to determine the stresses, strains, and displacements in structures and their components due to the loads acting on them. If we can find these quantities for all values of the loads up to the loads that cause failure, we will have a complete picture of the mechanical behavior of these structures.

An understanding of mechanical behavior is essential for the safe design of all types of structures, whether airplanes and antennas, buildings and bridges, machines and motors, or ships and spacecraft. That is why mechanics of materials is a basic subject in so many engineering fields. Statics and dynamics are also essential, but those subjects deal primarily with the forces and motions associated with particles and rigid bodies. In mechanics of materials we go one step further by examining the stresses and strains inside real bodies, that is, bodies of finite dimensions that deform under loads. To determine the stresses and strains, we use the physical properties of the materials as well as numerous theoretical laws and concepts.

Theoretical analyses and experimental results have equally important roles in mechanics of materials. We use theories to derive formulas and equations for predicting mechanical behavior, but these expressions cannot be used in practical design unless the physical properties of the materials are known. Such properties are available only after careful experiments have been carried out in the laboratory. Furthermore, not all practical problems are amenable to theoretical analysis alone, and in such cases physical testing is a necessity.

The historical development of mechanics of materials is a fascinating blend of both theory and experiment—theory has pointed the way to useful results in some instances, and experiment has done so in others. Such famous persons as Leonardo da Vinci (1452–1519) and Galileo Galilei (1564–1642) performed experiments to determine the strength of wires, bars, and beams, although they did not develop adequate theories (by today's standards) to explain their test results. By contrast, the famous mathematician Leonhard Euler (1707–1783) developed the mathematical theory of columns and calculated the critical load of a column in 1744, long before any experimental evidence existed to show the significance of his results. Without appropriate tests to back up his theories, Euler's results remained unused for over a hundred years, although today they are the basis for the design and analysis of most columns.[*]

---

[*]The history of mechanics of materials, beginning with Leonardo and Galileo, is given in Refs. 1-1, 1-2, and 1-3 (a list of references is available online).

Problems

When studying mechanics of materials, you will find that your efforts are divided naturally into two parts: first, understanding the logical development of the concepts, and second, applying those concepts to practical situations. The former is accomplished by studying the derivations, discussions, and examples that appear in each chapter, and the latter is accomplished by solving the problems at the ends of the chapters. Some of the problems are numerical in character, and others are symbolic (or algebraic).

An advantage of *numerical problems* is that the magnitudes of all quantities are evident at every stage of the calculations, thus providing an opportunity to judge whether the values are reasonable or not. The principal advantage of *symbolic problems* is that they lead to general-purpose formulas. A formula displays the variables that affect the final results; for instance, a quantity may actually cancel out of the solution, a fact that would not be evident from a numerical solution. Also, an algebraic solution shows the manner in which each variable affects the results, as when one variable appears in the numerator and another appears in the denominator. Furthermore, a symbolic solution provides the opportunity to check the dimensions at every stage of the work.

Finally, the most important reason for solving algebraically is to obtain a general formula that can be used for many different problems. In contrast, a numerical solution applies to only one set of circumstances. Because engineers must be adept at both kinds of solutions, you will find a mixture of numeric and symbolic problems throughout this book.

Numerical problems require that you work with specific units of measurement. In keeping with current engineering practice, this book utilizes both the International System of Units (SI) and the U.S. Customary System (USCS). A discussion of both systems appears in Appendix B (available online), where you will also find many useful tables, including a table of conversion factors.

All problems appear at the ends of the chapters, with the problem numbers and subheadings identifying the sections to which they belong. In the case of problems requiring numerical solutions, odd-numbered problems are in USCS units and even-numbered problems are in SI units.

The techniques for solving problems are discussed in detail in Appendix C (available online). In addition to a list of sound engineering procedures, Appendix C includes sections on dimensional homogeneity and significant digits. These topics are especially important, because every equation must be dimensionally homogeneous and every numerical result must be expressed with the proper number of significant digits. In this book, final numerical results are usually presented with three significant digits when a number begins with the digits 2 through 9, and with four significant digits when a number begins with the digit 1. Intermediate values are often recorded with additional digits to avoid losing numerical accuracy due to rounding of numbers.

1.2 NORMAL STRESS AND STRAIN

The most fundamental concepts in mechanics of materials are **stress** and **strain**. These concepts can be illustrated in their most elementary form by considering a prismatic bar subjected to axial forces. A **prismatic bar** is a straight structural member having the same cross section throughout its length, and an **axial force** is a load directed along the axis of the member, resulting in either tension or compression in the bar. Examples are shown in Fig. 1-1, where the tow bar is a prismatic member in tension and the landing gear strut is a member in compression. Other examples are the members of a bridge truss, connecting rods in automobile engines, spokes of bicycle wheels, columns in buildings, and wing struts in small airplanes.

For discussion purposes, we will consider the tow bar of Fig. 1-1 and isolate a segment of it as a free body (Fig. 1-2a). When drawing this free-body diagram, we disregard the weight of the bar itself and assume that the only active forces are the axial forces P at the ends. Next we consider two views of the bar, the first showing the same bar *before* the loads are applied (Fig. 1-2b) and the second showing it *after* the loads are applied (Fig. 1-2c). Note that the original length of the bar is denoted by the letter L, and the increase in length due to the loads is denoted by the Greek letter δ (delta).

The internal actions in the bar are exposed if we make an imaginary cut through the bar at section *mn* (Fig. 1-2c). Because this section is taken perpendicular to the longitudinal axis of the bar, it is called a **cross section**.

We now isolate the part of the bar to the left of cross section *mn* as a free body (Fig. 1-2d). At the right-hand end of this free body (section *mn*) we show the action of the removed part of the bar (that is, the part to the right of section *mn*) upon the part that remains. This action consists of continuously distributed *stresses* acting over the entire cross section, and the axial force P acting at the cross section is the *resultant* of those stresses. (The resultant force is shown with a dashed line in Fig. 1-2d.)

Stress has units of force per unit area and is denoted by the Greek letter σ (sigma). In general, the stresses σ acting on a plane surface may be uniform throughout the area or may vary in intensity from one point to another. Let us assume that the stresses acting on cross section *mn*

FIG. 1-1 Structural members subjected to axial loads. (The tow bar is in tension and the landing gear strut is in compression.)

Landing gear strut

Tow bar

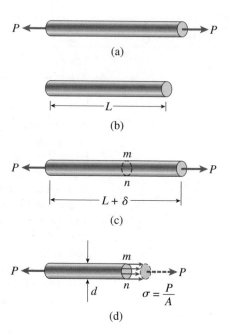

FIG. 1-2 Prismatic bar in tension:
(a) free-body diagram of a segment of
the bar, (b) segment of the bar before
loading, (c) segment of the bar after
loading, and (d) normal stresses in the
bar

(Fig. 1-2d) are *uniformly distributed* over the area. Then the resultant of
those stresses must be equal to the magnitude of the stress times the
cross-sectional area A of the bar, that is, $P = \sigma A$. Therefore, we obtain
the following expression for the magnitude of the stresses:

$$\sigma = \frac{P}{A} \tag{1-1}$$

This equation gives the intensity of uniform stress in an axially loaded,
prismatic bar of arbitrary cross-sectional shape.

When the bar is stretched by the forces P, the stresses are **tensile
stresses**; if the forces are reversed in direction, causing the bar to be
compressed, we obtain **compressive stresses**. Inasmuch as the stresses
act in a direction perpendicular to the cut surface, they are called **normal
stresses**. Thus, normal stresses may be either tensile or compressive.
Later, in Section 1.6, we will encounter another type of stress, called
shear stress, that acts parallel to the surface.

When a **sign convention** for normal stresses is required, it is
customary to define tensile stresses as positive and compressive stresses
as negative.

Because the normal stress σ is obtained by dividing the axial force
by the cross-sectional area, it has **units** of force per unit of area. When
USCS units are used, stress is customarily expressed in pounds per
square inch (psi) or kips per square inch (ksi).[*] For instance, suppose

---

[*]One kip, or kilopound, equals 1000 lb.

that the bar of Fig. 1-2 has a diameter d of 2.0 inches and the load P has a magnitude of 6 kips. Then the stress in the bar is

$$\sigma = \frac{P}{A} = \frac{P}{\pi d^2/4} = \frac{6\,\text{k}}{\pi(2.0\,\text{in.})^2/4} = 1.91\,\text{ksi (or 1910 psi)}$$

In this example the stress is tensile, or positive.

When SI units are used, force is expressed in newtons (N) and area in square meters (m^2). Consequently, stress has units of newtons per square meter (N/m^2), that is, pascals (Pa). However, the pascal is such a small unit of stress that it is necessary to work with large multiples, usually the megapascal (MPa).

To demonstrate that a pascal is indeed small, we have only to note that it takes almost 7000 pascals to make 1 psi.[*] As an illustration, the stress in the bar described in the preceding example (1.91 ksi) converts to 13.2 MPa, which is 13.2×10^6 pascals. Although it is not recommended in SI, you will sometimes find stress given in newtons per square millimeter (N/mm^2), which is a unit equal to the megapascal (MPa).

Limitations

The equation $\sigma = P/A$ is valid only if the stress is uniformly distributed over the cross section of the bar. This condition is realized if the axial force P acts through the centroid of the cross-sectional area, as demonstrated later in this section. When the load P does not act at the centroid, bending of the bar will result, and a more complicated analysis is necessary (see Sections 5.12 and 11.5). However, in this book (as in common practice) it is understood that axial forces are applied at the centroids of the cross sections unless specifically stated otherwise.

The uniform stress condition pictured in Fig. 1-2d exists throughout the length of the bar except near the ends. The stress distribution at the end of a bar depends upon how the load P is transmitted to the bar. If the load happens to be distributed uniformly over the end, then the stress pattern at the end will be the same as everywhere else. However, it is more likely that the load is transmitted through a pin or a bolt, producing high localized stresses called **stress concentrations**.

One possibility is illustrated by the eyebar shown in Fig. 1-3. In this instance the loads P are transmitted to the bar by pins that pass through the holes (or eyes) at the ends of the bar. Thus, the forces shown in the figure are actually the resultants of bearing pressures between the pins and the eyebar, and the stress distribution around the holes is quite complex. However, as we move away from the ends and toward the middle of the bar, the stress distribution gradually approaches the uniform distribution pictured in Fig. 1-2d.

As a practical rule, the formula $\sigma = P/A$ may be used with good accuracy at any point within a prismatic bar that is at least as far away

FIG. 1-3 Steel eyebar subjected to tensile loads P

---

[*]Conversion factors between USCS units and SI units are listed in Table B-5, Appendix B (available online).

from the stress concentration as the largest lateral dimension of the bar. In other words, the stress distribution in the steel eyebar of Fig. 1-3 is uniform at distances b or greater from the enlarged ends, where b is the width of the bar, and the stress distribution in the prismatic bar of Fig. 1-2 is uniform at distances d or greater from the ends, where d is the diameter of the bar (Fig. 1-2d).

Of course, even when the stress is *not* uniformly distributed, the equation $\sigma = P/A$ may still be useful because it gives the *average* normal stress on the cross section.

Normal Strain

As already observed, a straight bar will change in length when loaded axially, becoming longer when in tension and shorter when in compression. For instance, consider again the prismatic bar of Fig. 1-2. The elongation δ of this bar (Fig. 1-2c) is the cumulative result of the stretching of all elements of the material throughout the volume of the bar. Let us assume that the material is the same everywhere in the bar. Then, if we consider half of the bar (length $L/2$), it will have an elongation equal to $\delta/2$, and if we consider one-fourth of the bar, it will have an elongation equal to $\delta/4$.

In general, the elongation of a segment is equal to its length divided by the total length L and multiplied by the total elongation δ. Therefore, a unit length of the bar will have an elongation equal to $1/L$ times δ. This quantity is called the *elongation per unit length*, or **strain**, and is denoted by the Greek letter ϵ (epsilon). We see that strain is given by the equation

$$\epsilon = \frac{\delta}{L} \tag{1-2}$$

If the bar is in tension, the strain is called a **tensile strain**, representing an elongation or stretching of the material. If the bar is in compression, the strain is a **compressive strain** and the bar shortens. Tensile strain is usually taken as positive and compressive strain as negative. The strain ϵ is called a **normal strain** because it is associated with normal stresses.

Because normal strain is the ratio of two lengths, it is a **dimensionless quantity**, that is, it has no units. Therefore, strain is expressed simply as a number, independent of any system of units. Numerical values of strain are usually very small, because bars made of structural materials undergo only small changes in length when loaded.

As an example, consider a steel bar having length L equal to 2.0 m. When heavily loaded in tension, this bar might elongate by 1.4 mm, which means that the strain is

$$\epsilon = \frac{\delta}{L} = \frac{1.4 \text{ mm}}{2.0 \text{ m}} = 0.0007 = 700 \times 10^{-6}$$

In practice, the original units of δ and L are sometimes attached to the strain itself, and then the strain is recorded in forms such as mm/m, μm/m, and in./in. For instance, the strain ϵ in the preceding illustration could be given as 700 μm/m or 700×10^{-6} in./in. Also, strain is sometimes expressed as a percent, especially when the strains are large. (In the preceding example, the strain is 0.07%.)

Uniaxial Stress and Strain

The definitions of normal stress and normal strain are based upon purely static and geometric considerations, which means that Eqs. (1-1) and (1-2) can be used for loads of any magnitude and for any material. The principal requirement is that the deformation of the bar be uniform throughout its volume, which in turn requires that the bar be prismatic, the loads act through the centroids of the cross sections, and the material be **homogeneous** (that is, the same throughout all parts of the bar). The resulting state of stress and strain is called **uniaxial stress and strain**.

Further discussions of uniaxial stress, including stresses in directions other than the longitudinal direction of the bar, are given later in Section 2.6. We will also analyze more complicated stress states, such as biaxial stress and plane stress, in Chapter 6.

Line of Action of the Axial Forces for a Uniform Stress Distribution

Throughout the preceding discussion of stress and strain in a prismatic bar, we assumed that the normal stress σ was distributed uniformly over the cross section. Now we will demonstrate that this condition is met if the line of action of the axial forces is through the centroid of the cross-sectional area.

Consider a prismatic bar of arbitrary cross-sectional shape subjected to axial forces P that produce uniformly distributed stresses σ (Fig. 1-4a). Also, let p_1 represent the point in the cross section where the line of action of the forces intersects the cross section (Fig. 1-4b). We construct a set of xy axes in the plane of the cross section and denote the coordinates of point p_1 by $\bar{x}$ and $\bar{y}$. To determine these coordinates, we observe that the moments M_x and M_y of the force P about the x and y axes, respectively, must be equal to the corresponding moments of the uniformly distributed stresses.

The moments of the force P are

$$M_x = P\bar{y} \qquad M_y = -P\bar{x} \qquad \text{(a,b)}$$

in which a moment is considered positive when its vector (using the right-hand rule) acts in the positive direction of the corresponding axis.[*]

---

[*]To visualize the right-hand rule, imagine that you grasp an axis of coordinates with your right hand so that your fingers fold around the axis and your thumb points in the positive direction of the axis. Then a moment is positive if it acts about the axis in the same direction as your fingers.

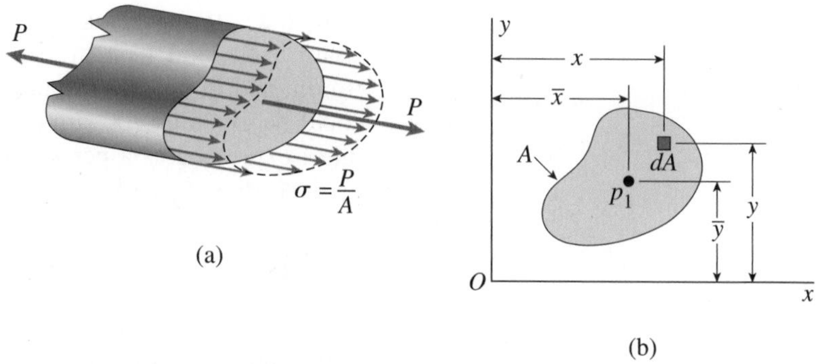

FIG. 1-4 Uniform stress distribution in a prismatic bar: (a) axial forces P, and (b) cross section of the bar

The moments of the distributed stresses are obtained by integrating over the cross-sectional area A. The differential force acting on an element of area dA (Fig. 1-4b) is equal to σdA. The moments of this elemental force about the x and y axes are $\sigma y dA$ and $-\sigma x dA$, respectively, in which x and y denote the coordinates of the element dA. The total moments are obtained by integrating over the cross-sectional area:

$$M_x = \int \sigma y \, dA \qquad M_y = -\int \sigma x \, dA \qquad \text{(c,d)}$$

These expressions give the moments produced by the stresses σ.

Next, we equate the moments M_x and M_y as obtained from the force P (Eqs. a and b) to the moments obtained from the distributed stresses (Eqs. c and d):

$$P\bar{y} = \int \sigma y \, dA \qquad P\bar{x} = \int \sigma x \, dA$$

Because the stresses σ are uniformly distributed, we know that they are constant over the cross-sectional area A and can be placed outside the integral signs. Also, we know that σ is equal to P/A. Therefore, we obtain the following formulas for the coordinates of point p_1:

$$\bar{y} = \frac{\int y \, dA}{A} \qquad \bar{x} = \frac{\int x \, dA}{A} \qquad \text{(1-3a,b)}$$

These equations are the same as the equations defining the coordinates of the centroid of an area (see Eqs. 10-3a and b in Chapter 10 available online). Therefore, we have now arrived at an important conclusion: *In order to have uniform tension or compression in a prismatic bar, the axial force must act through the centroid of the cross-sectional area.* As explained previously, we always assume that these conditions are met unless it is specifically stated otherwise.

The following examples illustrate the calculation of stresses and strains in prismatic bars. In the first example we disregard the weight of the bar and in the second we include it. (It is customary when solving textbook problems to omit the weight of the structure unless specifically instructed to include it.)

Example 1-1

A short post constructed from a hollow circular tube of aluminum supports a compressive load of 26 kips (Fig. 1-5). The inner and outer diameters of the tube are $d_1 = 4.0$ in. and $d_2 = 4.5$ in., respectively, and its length is 16 in. The shortening of the post due to the load is measured as 0.012 in.

Determine the compressive stress and strain in the post. (Disregard the weight of the post itself, and assume that the post does not buckle under the load.)

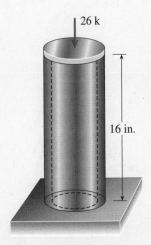

FIG. 1-5 Example 1-1. Hollow aluminum post in compression

Solution

Assuming that the compressive load acts at the center of the hollow tube, we can use the equation $\sigma = P/A$ (Eq. 1-1) to calculate the normal stress. The force P equals 26 k (or 26,000 lb), and the cross-sectional area A is

$$A = \frac{\pi}{4}\left(d_2^2 - d_1^2\right) = \frac{\pi}{4}\left[(4.5 \text{ in.})^2 - (4.0 \text{ in.})^2\right] = 3.338 \text{ in.}^2$$

Therefore, the compressive stress in the post is

$$\sigma = \frac{P}{A} = \frac{26,000 \text{ lb}}{3.338 \text{ in.}^2} = 7790 \text{ psi}$$

The compressive strain (from Eq. 1-2) is

$$\epsilon = \frac{\delta}{L} = \frac{0.012 \text{ in.}}{16 \text{ in.}} = 750 \times 10^{-6}$$

Thus, the stress and strain in the post have been calculated.

Note: As explained earlier, strain is a dimensionless quantity and no units are needed. For clarity, however, units are often given. In this example, ϵ could be written as 750×10^{-6} in./in. or 750 μin./in.

Example 1-2

A circular steel rod of length L and diameter d hangs in a mine shaft and holds an ore bucket of weight W at its lower end (Fig. 1-6).

(a) Obtain a formula for the maximum stress σ_{max} in the rod, taking into account the weight of the rod itself.

(b) Calculate the maximum stress if $L = 40$ m, $d = 8$ mm, and $W = 1.5$ kN.

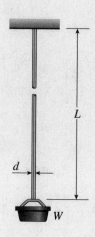

FIG. 1-6 Example 1-2. Steel rod supporting a weight W

Solution

(a) The maximum axial force F_{max} in the rod occurs at the upper end and is equal to the weight W of the ore bucket plus the weight W_0 of the rod itself. The latter is equal to the weight density γ of the steel times the volume V of the rod, or

$$W_0 = \gamma V = \gamma A L \qquad (1\text{-}4)$$

in which A is the cross-sectional area of the rod. Therefore, the formula for the maximum stress (from Eq. 1-1) becomes

$$\sigma_{max} = \frac{F_{max}}{A} = \frac{W + \gamma A L}{A} = \frac{W}{A} + \gamma L \qquad (1\text{-}5) \quad \Longleftarrow$$

(b) To calculate the maximum stress, we substitute numerical values into the preceding equation. The cross-sectional area A equals $\pi d^2/4$, where $d = 8$ mm, and the weight density γ of steel is 77.0 kN/m³ (from Table I-1 in Appendix I available online). Thus,

$$\sigma_{max} = \frac{1.5\,\text{kN}}{\pi(8\,\text{mm})^2/4} + (77.0\,\text{kN/m}^3)(40\,\text{m})$$

$$= 29.8\,\text{MPa} + 3.1\,\text{MPa} = 32.9\,\text{MPa} \quad \Longleftarrow$$

In this example, the weight of the rod contributes noticeably to the maximum stress and should not be disregarded.

1.3 MECHANICAL PROPERTIES OF MATERIALS

The design of machines and structures so that they will function properly requires that we understand the **mechanical behavior** of the materials being used. Ordinarily, the only way to determine how materials behave when they are subjected to loads is to perform experiments in the laboratory. The usual procedure is to place small specimens of the material in testing machines, apply the loads, and then measure the resulting deformations (such as changes in length and changes in diameter). Most materials-testing laboratories are equipped with machines capable of loading specimens in a variety of ways, including both static and dynamic loading in tension and compression.

A typical **tensile-test machine** is shown in Fig. 1-7. The test specimen is installed between the two large grips of the testing machine and then loaded in tension. Measuring devices record the deformations, and the automatic control and data-processing systems (at the left in the photo) tabulate and graph the results.

A more detailed view of a **tensile-test specimen** is shown in Fig. 1-8 on the next page. The ends of the circular specimen are enlarged where they fit in the grips so that failure will not occur near the grips themselves. A failure at the ends would not produce the desired information about the material, because the stress distribution near the grips is not uniform, as explained in Section 1.2. In a properly designed specimen, failure will occur in the prismatic portion of the specimen where the stress distribution is uniform and the bar is subjected only to pure tension. This situation is shown in Fig. 1-8, where the steel specimen has just fractured under load. The device at the left, which is attached by

FIG. 1-7 Tensile-test machine with automatic data-processing system (Courtesy of MTS Systems Corporation)

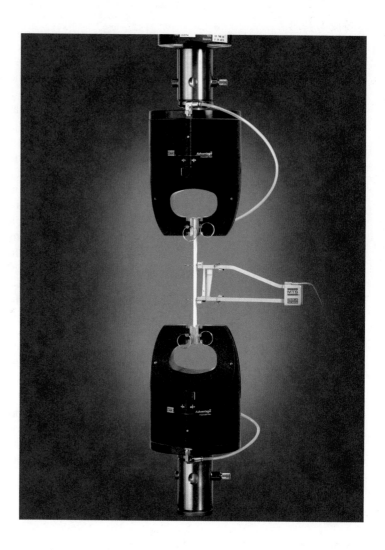

FIG. 1-8 Typical tensile-test specimen with extensometer attached; the specimen has just fractured in tension (Courtesy of MTS Systems Corporation)

two arms to the specimen, is an **extensometer** that measures the elongation during loading.

In order that test results will be comparable, the dimensions of test specimens and the methods of applying loads must be standardized. One of the major standards organizations in the United States is the American Society for Testing and Materials (ASTM), a technical society that publishes specifications and standards for materials and testing. Other standardizing organizations are the American Standards Association (ASA) and the National Institute of Standards and Technology (NIST). Similar organizations exist in other countries.

The ASTM standard tension specimen has a diameter of 0.505 in. and a **gage length** of 2.0 in. between the gage marks, which are the points where the extensometer arms are attached to the specimen (see Fig. 1-8). As the specimen is pulled, the axial load is measured and recorded, either automatically or by reading from a dial. The elongation over the gage length is measured simultaneously, either by mechanical gages of the kind shown in Fig. 1-8 or by electrical-resistance strain gages.

In a **static test**, the load is applied slowly and the precise *rate* of loading is not of interest because it does not affect the behavior of the specimen. However, in a **dynamic test** the load is applied rapidly and sometimes in a cyclical manner. Since the nature of a dynamic load affects the properties of the materials, the rate of loading must also be measured.

Compression tests of metals are customarily made on small specimens in the shape of cubes or circular cylinders. For instance, cubes may be 2.0 in. on a side, and cylinders may have diameters of 1 in. and lengths from 1 to 12 in. Both the load applied by the machine and the shortening of the specimen may be measured. The shortening should be measured over a gage length that is less than the total length of the specimen in order to eliminate end effects.

Concrete is tested in compression on important construction projects to ensure that the required strength has been obtained. One type of concrete test specimen is 6 in. in diameter, 12 in. in length, and 28 days old (the age of concrete is important because concrete gains strength as it cures). Similar but somewhat smaller specimens are used when performing compression tests of rock (Fig. 1-9, on the next page).

Stress-Strain Diagrams

Test results generally depend upon the dimensions of the specimen being tested. Since it is unlikely that we will be designing a structure having parts that are the same size as the test specimens, we need to express the test results in a form that can be applied to members of any size. A simple way to achieve this objective is to convert the test results to stresses and strains.

The axial stress σ in a test specimen is calculated by dividing the axial load P by the cross-sectional area A (Eq. 1-1). When the initial area of the specimen is used in the calculation, the stress is called the **nominal stress** (other names are *conventional stress* and *engineering stress*). A more exact value of the axial stress, called the **true stress**, can be calculated by using the actual area of the bar at the cross section where failure occurs. Since the actual area in a tension test is always less than the initial area (as illustrated in Fig. 1-8), the true stress is larger than the nominal stress.

The average axial strain ϵ in the test specimen is found by dividing the measured elongation δ between the gage marks by the gage length L (see Fig. 1-8 and Eq. 1-2). If the initial gage length is used in the calculation (for instance, 2.0 in.), then the **nominal strain** is obtained. Since the distance between the gage marks increases as the tensile load is applied, we can calculate the **true strain** (or *natural strain*) at any value of the load by using the actual distance between the gage marks. In tension, true strain is always smaller than nominal strain. However, for

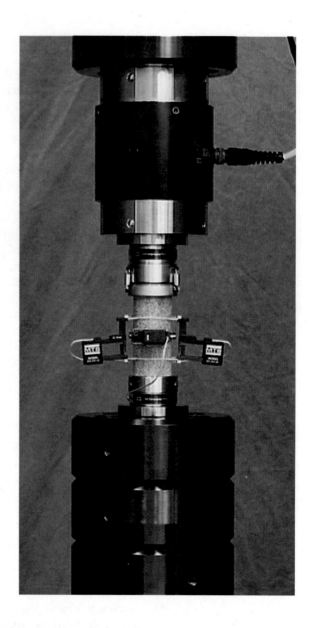

FIG. 1-9 Rock sample being tested in compression to obtain compressive strength, elastic modulus and Poisson's ratio (Courtesy of MTS Systems Corporation)

most engineering purposes, nominal stress and nominal strain are adequate, as explained later in this section.

After performing a tension or compression test and determining the stress and strain at various magnitudes of the load, we can plot a diagram of stress versus strain. Such a **stress-strain diagram** is a characteristic of the particular material being tested and conveys important information about the mechanical properties and type of behavior.[*]

---

[*]Stress-strain diagrams were originated by Jacob Bernoulli (1654–1705) and J. V. Poncelet (1788–1867); see Ref. 1-4 (available online).

The first material we will discuss is **structural steel**, also known as *mild steel* or *low-carbon steel*. Structural steel is one of the most widely used metals and is found in buildings, bridges, cranes, ships, towers, vehicles, and many other types of construction. A stress-strain diagram for a typical structural steel in tension is shown in Fig. 1-10. Strains are plotted on the horizontal axis and stresses on the vertical axis. (In order to display all of the important features of this material, the strain axis in Fig. 1-10 is not drawn to scale.)

The diagram begins with a straight line from the origin O to point A, which means that the relationship between stress and strain in this initial region is not only *linear* but also *proportional*.[*] Beyond point A, the proportionality between stress and strain no longer exists; hence the stress at A is called the **proportional limit**. For low-carbon steels, this limit is in the range 30 to 50 ksi (210 to 350 MPa), but high-strength steels (with higher carbon content plus other alloys) can have proportional limits of more than 80 ksi (550 MPa). The slope of the straight line from O to A is called the **modulus of elasticity**. Because the slope has units of stress divided by strain, modulus of elasticity has the same units as stress. (Modulus of elasticity is discussed later in Section 1.5.)

With an increase in stress beyond the proportional limit, the strain begins to increase more rapidly for each increment in stress. Consequently, the stress-strain curve has a smaller and smaller slope, until, at point B, the curve becomes horizontal (see Fig. 1-10). Beginning at this point, considerable elongation of the test specimen occurs with no

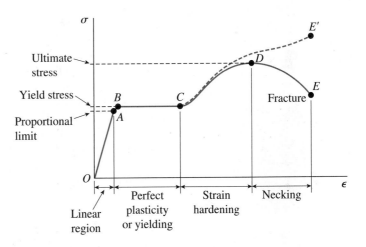

FIG. 1-10 Stress-strain diagram for a typical structural steel in tension (not to scale)

[*]Two variables are said to be *proportional* if their ratio remains constant. Therefore, a proportional relationship may be represented by a straight line through the origin. However, a proportional relationship is not the same as a *linear* relationship. Although a proportional relationship is linear, the converse is not necessarily true, because a relationship represented by a straight line that does *not* pass through the origin is linear but not proportional. The often-used expression "directly proportional" is synonymous with "proportional" (Ref. 1-5; a list of references is available online).

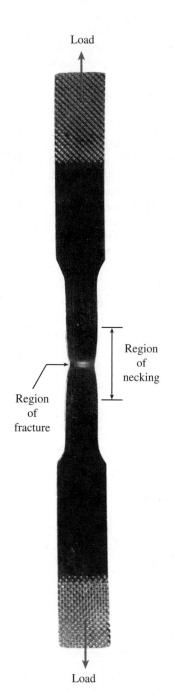

Load

Region
of
necking

Region
of
fracture

Load

noticeable increase in the tensile force (from *B* to *C*). This phenomenon is known as **yielding** of the material, and point *B* is called the **yield point**. The corresponding stress is known as the **yield stress** of the steel.

In the region from *B* to *C* (Fig. 1-10), the material becomes **perfectly plastic**, which means that it deforms without an increase in the applied load. The elongation of a mild-steel specimen in the perfectly plastic region is typically 10 to 15 times the elongation that occurs in the linear region (between the onset of loading and the proportional limit). The presence of very large strains in the plastic region (and beyond) is the reason for not plotting this diagram to scale.

After undergoing the large strains that occur during yielding in the region *BC*, the steel begins to **strain harden**. During strain hardening, the material undergoes changes in its crystalline structure, resulting in increased resistance of the material to further deformation. Elongation of the test specimen in this region requires an increase in the tensile load, and therefore the stress-strain diagram has a positive slope from *C* to *D*. The load eventually reaches its maximum value, and the corresponding stress (at point *D*) is called the **ultimate stress**. Further stretching of the bar is actually accompanied by a reduction in the load, and fracture finally occurs at a point such as *E* in Fig. 1-10.

The yield stress and ultimate stress of a material are also called the **yield strength** and **ultimate strength**, respectively. **Strength** is a general term that refers to the capacity of a structure to resist loads. For instance, the yield strength of a beam is the magnitude of the load required to cause yielding in the beam, and the ultimate strength of a truss is the maximum load it can support, that is, the failure load. However, when conducting a tension test of a particular material, we define load-carrying capacity by the stresses in the specimen rather than by the total loads acting on the specimen. As a result, the strength of a material is usually stated as a stress.

When a test specimen is stretched, **lateral contraction** occurs, as previously mentioned. The resulting decrease in cross-sectional area is too small to have a noticeable effect on the calculated values of the stresses up to about point *C* in Fig. 1-10, but beyond that point the reduction in area begins to alter the shape of the curve. In the vicinity of the ultimate stress, the reduction in area of the bar becomes clearly visible and a pronounced **necking** of the bar occurs (see Figs. 1-8 and 1-11).

If the actual cross-sectional area at the narrow part of the neck is used to calculate the stress, the **true stress-strain curve** (the dashed line *CE'* in Fig. 1-10) is obtained. The total load the bar can carry does indeed diminish after the ultimate stress is reached (as shown by curve *DE*), but this reduction is due to the decrease in area of the bar and not to a loss in strength of the material itself. In reality, the material withstands an increase in true stress up to failure (point *E'*). Because most structures are expected to function at stresses below the proportional limit, the **conventional stress-strain curve** *OABCDE*, which is based upon the original cross-sectional area of the specimen and is easy to determine, provides satisfactory information for use in engineering design.

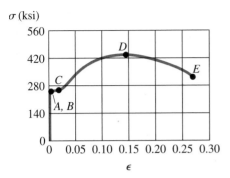

FIG. 1-12 Stress-strain diagram for a typical structural steel in tension (drawn to scale)

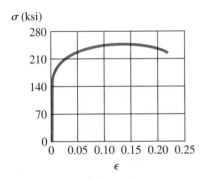

FIG. 1-13 Typical stress-strain diagram for an aluminum alloy

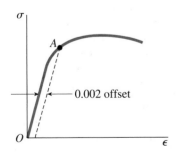

FIG. 1-14 Arbitrary yield stress determined by the offset method

The diagram of Fig. 1-10 shows the general characteristics of the stress-strain curve for mild steel, but its proportions are not realistic because, as already mentioned, the strain that occurs from B to C may be more than ten times the strain occurring from O to A. Furthermore, the strains from C to E are many times greater than those from B to C. The correct relationships are portrayed in Fig. 1-12, which shows a stress-strain diagram for mild steel drawn to scale. In this figure, the strains from the zero point to point A are so small in comparison to the strains from point A to point E that they cannot be seen, and the initial part of the diagram appears to be a vertical line.

The presence of a clearly defined yield point followed by large plastic strains is an important characteristic of structural steel that is sometimes utilized in practical design. Metals such as structural steel that undergo large *permanent* strains before failure are classified as **ductile**. For instance, ductility is the property that enables a bar of steel to be bent into a circular arc or drawn into a wire without breaking. A desirable feature of ductile materials is that visible distortions occur if the loads become too large, thus providing an opportunity to take remedial action before an actual fracture occurs. Also, materials exhibiting ductile behavior are capable of absorbing large amounts of strain energy prior to fracture.

Structural steel is an alloy of iron containing about 0.2% carbon, and therefore it is classified as a low-carbon steel. With increasing carbon content, steel becomes less ductile but stronger (higher yield stress and higher ultimate stress). The physical properties of steel are also affected by heat treatment, the presence of other metals, and manufacturing processes such as rolling. Other materials that behave in a ductile manner (under certain conditions) include aluminum, copper, magnesium, lead, molybdenum, nickel, brass, bronze, monel metal, nylon, and teflon.

Although they may have considerable ductility, **aluminum alloys** typically do not have a clearly definable yield point, as shown by the stress-strain diagram of Fig. 1-13. However, they do have an initial linear region with a recognizable proportional limit. Alloys produced for structural purposes have proportional limits in the range 10 to 60 ksi (70 to 410 MPa) and ultimate stresses in the range 20 to 80 ksi (140 to 550 MPa).

When a material such as aluminum does not have an obvious yield point and yet undergoes large strains after the proportional limit is exceeded, an *arbitrary* yield stress may be determined by the **offset method**. A straight line is drawn on the stress-strain diagram parallel to the initial linear part of the curve (Fig. 1-14) but offset by some standard strain, such as 0.002 (or 0.2%). The intersection of the offset line and the stress-strain curve (point A in the figure) defines the yield stress. Because this stress is determined by an arbitrary rule and is not an inherent physical property of the material, it should be distinguished from a true yield stress by referring to it as the **offset yield stress**. For a

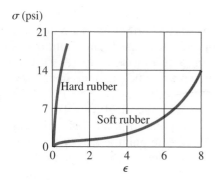

FIG. 1-15 Stress-strain curves for two kinds of rubber in tension

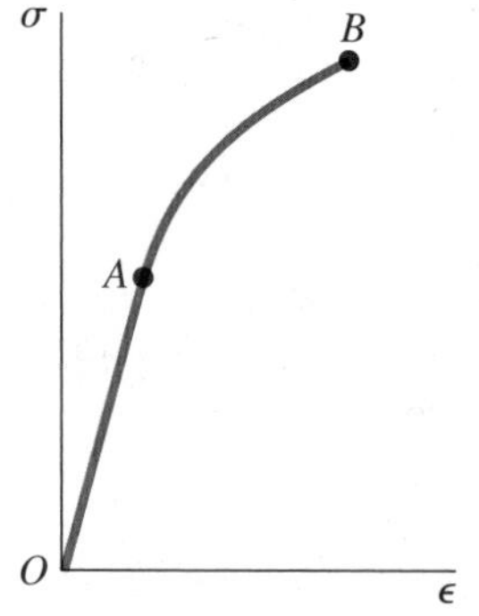

material such as aluminum, the offset yield stress is slightly above the proportional limit. In the case of structural steel, with its abrupt transition from the linear region to the region of plastic stretching, the offset stress is essentially the same as both the yield stress and the proportional limit.

Rubber maintains a linear relationship between stress and strain up to relatively large strains (as compared to metals). The strain at the proportional limit may be as high as 0.1 or 0.2 (10% or 20%). Beyond the proportional limit, the behavior depends upon the type of rubber (Fig. 1-15). Some kinds of soft rubber will stretch enormously without failure, reaching lengths several times their original lengths. The material eventually offers increasing resistance to the load, and the stress-strain curve turns markedly upward. You can easily sense this characteristic behavior by stretching a rubber band with your hands. (Note that although rubber exhibits very large strains, it is not a ductile material because the strains are not permanent. It is, of course, an elastic material; see Section 1.4.)

The ductility of a material in tension can be characterized by its elongation and by the reduction in area at the cross section where fracture occurs. The **percent elongation** is defined as follows:

$$\text{Percent elongation} = \frac{L_1 - L_0}{L_0}(100) \qquad (1\text{-}6)$$

in which L_0 is the original gage length and L_1 is the distance between the gage marks at fracture. Because the elongation is not uniform over the length of the specimen but is concentrated in the region of necking, the percent elongation depends upon the gage length. Therefore, when stating the percent elongation, the gage length should always be given. For a 2 in. gage length, steel may have an elongation in the range from 3% to 40%, depending upon composition; in the case of structural steel, values of 20% or 30% are common. The elongation of aluminum alloys varies from 1% to 45%, depending upon composition and treatment.

The **percent reduction in area** measures the amount of necking that occurs and is defined as follows:

$$\text{Percent reduction in area} = \frac{A_0 - A_1}{A_0}(100) \qquad (1\text{-}7)$$

in which A_0 is the original cross-sectional area and A_1 is the final area at the fracture section. For ductile steels, the reduction is about 50%.

Materials that fail in tension at relatively low values of strain are classified as **brittle**. Examples are concrete, stone, cast iron, glass, ceramics, and a variety of metallic alloys. Brittle materials fail with only little elongation after the proportional limit (the stress at point A in Fig. 1-16) is exceeded. Furthermore, the reduction in area is insignificant, and so the nominal fracture stress (point B) is the same as the true ultimate stress. High-carbon steels have very high yield stresses—over

FIG. 1-16 Typical stress-strain diagram for a brittle material showing the proportional limit (point A) and fracture stress (point B)

100 ksi (700 MPa) in some cases—but they behave in a brittle manner and fracture occurs at an elongation of only a few percent.

Ordinary **glass** is a nearly ideal brittle material, because it exhibits almost no ductility. The stress-strain curve for glass in tension is essentially a straight line, with failure occurring before any yielding takes place. The ultimate stress is about 10,000 psi (70 MPa) for certain kinds of plate glass, but great variations exist, depending upon the type of glass, the size of the specimen, and the presence of microscopic defects. **Glass fibers** can develop enormous strengths, and ultimate stresses over 1,000,000 psi (7 GPa) have been attained.

Many types of **plastics** are used for structural purposes because of their light weight, resistance to corrosion, and good electrical insulation properties. Their mechanical properties vary tremendously, with some plastics being brittle and others ductile. When designing with plastics it is important to realize that their properties are greatly affected by both temperature changes and the passage of time. For instance, the ultimate tensile stress of some plastics is cut in half merely by raising the temperature from 50° F to 120° F. Also, a loaded plastic may stretch gradually over time until it is no longer serviceable. For example, a bar of polyvinyl chloride subjected to a tensile load that initially produces a strain of 0.005 may have that strain doubled after one week, even though the load remains constant. (This phenomenon, known as *creep*, is discussed in the next section.)

Ultimate tensile stresses for plastics are generally in the range 2 to 50 ksi (14 to 350 MPa) and weight densities vary from 50 to 90 lb/ft^3 (8 to 14 kN/m^3). One type of nylon has an ultimate stress of 12 ksi (80 MPa) and weighs only 70 lb/ft^3 (11 kN/m^3), which is only 12% heavier than water. Because of its light weight, the strength-to-weight ratio for nylon is about the same as for structural steel (see Prob. 1.3-4).

A **filament-reinforced material** consists of a base material (or *matrix*) in which high-strength filaments, fibers, or whiskers are embedded. The resulting composite material has much greater strength than the base material. As an example, the use of glass fibers can more than double the strength of a plastic matrix. Composites are widely used in aircraft, boats, rockets, and space vehicles where high strength and light weight are needed.

Compression

Stress-strain curves for materials in compression differ from those in tension. Ductile metals such as steel, aluminum, and copper have proportional limits in compression very close to those in tension, and the initial regions of their compressive and tensile stress-strain diagrams are about the same. However, after yielding begins, the behavior is quite different. In a tension test, the specimen is stretched, necking may occur, and fracture ultimately takes place. When the material is compressed, it bulges outward on the sides and becomes barrel shaped, because friction between the specimen and the end plates prevents lateral expansion. With increasing

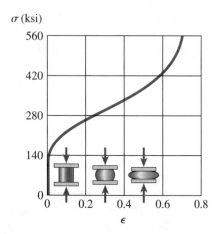

σ (ksi)

FIG. 1-17 Stress-strain diagram for copper in compression

load, the specimen is flattened out and offers greatly increased resistance to further shortening (which means that the stress-strain curve becomes very steep). These characteristics are illustrated in Fig. 1-17, which shows a compressive stress-strain diagram for copper. Since the actual cross-sectional area of a specimen tested in compression is larger than the initial area, the true stress in a compression test is smaller than the nominal stress.

Brittle materials loaded in compression typically have an initial linear region followed by a region in which the shortening increases at a slightly higher rate than does the load. The stress-strain curves for compression and tension often have similar shapes, but the ultimate stresses in compression are much higher than those in tension. Also, unlike ductile materials, which flatten out when compressed, brittle materials actually break at the maximum load.

Tables of Mechanical Properties

Properties of materials are listed in the tables of Appendix I (available online). The data in the tables are typical of the materials and are suitable for solving problems in this book. However, properties of materials and stress-strain curves vary greatly, even for the same material, because of different manufacturing processes, chemical composition, internal defects, temperature, and many other factors.

For these reasons, data obtained from Appendix I (or other tables of a similar nature) should not be used for specific engineering or design purposes. Instead, the manufacturers or materials suppliers should be consulted for information about a particular product.

1.4 ELASTICITY, PLASTICITY, AND CREEP

Stress-strain diagrams portray the behavior of engineering materials when the materials are loaded in tension or compression, as described in the preceding section. To go one step further, let us now consider what happens when the load is removed and the material is *unloaded.*

Assume, for instance, that we apply a load to a tensile specimen so that the stress and strain go from the origin O to point A on the stress-strain curve of Fig. 1-18a. Suppose further that when the load is removed, the material follows exactly the same curve back to the origin O. This property of a material, by which it returns to its original dimensions during unloading, is called **elasticity**, and the material itself is said to be *elastic*. Note that the stress-strain curve from O to A need not be linear in order for the material to be elastic.

Now suppose that we load this same material to a higher level, so that point B is reached on the stress-strain curve (Fig. 1-18b). When unloading occurs from point B, the material follows line BC on the diagram. This unloading line is parallel to the initial portion of the loading curve; that is, line BC is parallel to a tangent to the stress-strain curve at the origin. When point C is reached, the load has been entirely removed, but a **residual strain**, or *permanent strain*, represented by line OC, remains in

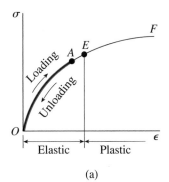

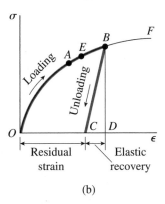

FIG. 1-18 Stress-strain diagrams illustrating (a) elastic behavior, and (b) partially elastic behavior

the material. As a consequence, the bar being tested is longer than it was before loading. This residual elongation of the bar is called the **permanent set**. Of the total strain *OD* developed during loading from *O* to *B*, the strain *CD* has been recovered elastically and the strain *OC* remains as a permanent strain. Thus, during unloading the bar returns partially to its original shape, and so the material is said to be **partially elastic**.

Between points *A* and *B* on the stress-strain curve (Fig. 1-18b), there must be a point before which the material is elastic and beyond which the material is partially elastic. To find this point, we load the material to some selected value of stress and then remove the load. If there is no permanent set (that is, if the elongation of the bar returns to zero), then the material is fully elastic up to the selected value of the stress.

The process of loading and unloading can be repeated for successively higher values of stress. Eventually, a stress will be reached such that not all the strain is recovered during unloading. By this procedure, it is possible to determine the stress at the upper limit of the elastic region, for instance, the stress at point *E* in Figs. 1-18a and b. The stress at this point is known as the **elastic limit** of the material.

Many materials, including most metals, have linear regions at the beginning of their stress-strain curves (for example, see Figs. 1-10 and 1-13). The stress at the upper limit of this linear region is the proportional limit, as explained in the preceding section. The elastic limit is usually the same as, or slightly above, the proportional limit. Hence, for many materials the two limits are assigned the same numerical value. In the case of mild steel, the yield stress is also very close to the proportional limit, so that for practical purposes the yield stress, the elastic limit, and the proportional limit are assumed to be equal. Of course, this situation does not hold for all materials. Rubber is an outstanding example of a material that is elastic far beyond the proportional limit.

The characteristic of a material by which it undergoes inelastic strains beyond the strain at the elastic limit is known as **plasticity**. Thus, on the stress-strain curve of Fig. 1-18a, we have an elastic region followed by a plastic region. When large deformations occur in a ductile material loaded into the plastic region, the material is said to undergo **plastic flow**.

Reloading of a Material

If the material remains within the elastic range, it can be loaded, unloaded, and loaded again without significantly changing the behavior. However, when loaded into the plastic range, the internal structure of the material is altered and its properties change. For instance, we have already observed that a permanent strain exists in the specimen after unloading from the plastic region (Fig. 1-18b). Now suppose that the material is **reloaded** after such an unloading (Fig. 1-19). The new loading begins at point *C* on the diagram and continues upward to point *B*, the point at which unloading began during the first loading cycle. The material then follows the original stress-strain curve toward point *F*. Thus, for the second loading, we can imagine that we have a new stress-strain diagram with its origin at point *C*.

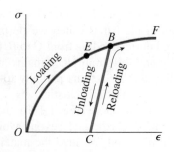

FIG. 1-19 Reloading of a material and raising of the elastic and proportional limits

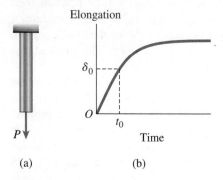

FIG. 1-20 Creep in a bar under constant load

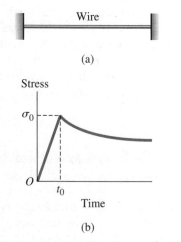

FIG. 1-21 Relaxation of stress in a wire under constant strain

During the second loading, the material behaves in a linearly elastic manner from C to B, with the slope of line CB being the same as the slope of the tangent to the original loading curve at the origin O. The proportional limit is now at point B, which is at a higher stress than the original elastic limit (point E). Thus, by stretching a material such as steel or aluminum into the inelastic or plastic range, the *properties of the material are changed*—the linearly elastic region is increased, the proportional limit is raised, and the elastic limit is raised. However, the ductility is reduced because in the "new material" the amount of yielding beyond the elastic limit (from B to F) is less than in the original material (from E to F).[*]

Creep

The stress-strain diagrams described previously were obtained from tension tests involving static loading and unloading of the specimens, and the passage of time did not enter our discussions. However, when loaded for long periods of time, some materials develop additional strains and are said to **creep**.

This phenomenon can manifest itself in a variety of ways. For instance, suppose that a vertical bar (Fig. 1-20a) is loaded slowly by a force P, producing an elongation equal to δ_0. Let us assume that the loading and corresponding elongation take place during a time interval of duration t_0 (Fig. 1-20b). Subsequent to time t_0, the load remains constant. However, due to creep, the bar may gradually lengthen, as shown in Fig. 1-20b, even though the load does not change. This behavior occurs with many materials, although sometimes the change is too small to be of concern.

As another manifestation of creep, consider a wire that is stretched between two immovable supports so that it has an initial tensile stress σ_0 (Fig. 1-21). Again, we will denote the time during which the wire is initially stretched as t_0. With the elapse of time, the stress in the wire gradually diminishes, eventually reaching a constant value, even though the supports at the ends of the wire do not move. This process, is called **relaxation** of the material.

Creep is usually more important at high temperatures than at ordinary temperatures, and therefore it should always be considered in the design of engines, furnaces, and other structures that operate at elevated temperatures for long periods of time. However, materials such as steel, concrete, and wood will creep slightly even at atmospheric temperatures. For example, creep of concrete over long periods of time can create undulations in bridge decks because of sagging between the supports. (One remedy is to construct the deck with an upward **camber**, which is an initial displacement above the horizontal, so that when creep occurs, the spans lower to the level position.)

---

[*]The study of material behavior under various environmental and loading conditions is an important branch of applied mechanics. For more detailed engineering information about materials, consult a textbook devoted solely to this subject.

1.5 LINEAR ELASTICITY, HOOKE'S LAW, AND POISSON'S RATIO

Many structural materials, including most metals, wood, plastics, and ceramics, behave both elastically and linearly when first loaded. Consequently, their stress-strain curves begin with a straight line passing through the origin. An example is the stress-strain curve for structural steel (Fig. 1-10), where the region from the origin O to the proportional limit (point A) is both linear and elastic. Other examples are the regions below *both* the proportional limits and the elastic limits on the diagrams for aluminum (Fig. 1-13), brittle materials (Fig. 1-16), and copper (Fig. 1-17).

When a material behaves elastically and also exhibits a linear relationship between stress and strain, it is said to be **linearly elastic**. This type of behavior is extremely important in engineering for an obvious reason—by designing structures and machines to function in this region, we avoid permanent deformations due to yielding.

Hooke's Law

The linear relationship between stress and strain for a bar in simple tension or compression is expressed by the equation

$$\sigma = E\epsilon \tag{1-8}$$

in which σ is the axial stress, ϵ is the axial strain, and E is a constant of proportionality known as the **modulus of elasticity** for the material. The modulus of elasticity is the slope of the stress-strain diagram in the linearly elastic region, as mentioned previously in Section 1.3. Since strain is dimensionless, the units of E are the same as the units of stress. Typical units of E are psi or ksi in USCS units and pascals (or multiples thereof) in SI units.

The equation $\sigma = E\epsilon$ is commonly known as **Hooke's law**, named for the famous English scientist Robert Hooke (1635–1703). Hooke was the first person to investigate scientifically the elastic properties of materials, and he tested such diverse materials as metal, wood, stone, bone, and sinew. He measured the stretching of long wires supporting weights and observed that the elongations "always bear the same proportions one to the other that the weights do that made them" (Ref. 1-6 available online). Thus, Hooke established the linear relationship between the applied loads and the resulting elongations.

Equation (1-8) is actually a very limited version of Hooke's law because it relates only to the longitudinal stresses and strains developed in simple tension or compression of a bar (*uniaxial stress*). To deal with more complicated states of stress, such as those found in most structures and machines, we must use more extensive equations of Hooke's law (see Sections 6.5 and 6.6).

The modulus of elasticity has relatively large values for materials that are very stiff, such as structural metals. Steel has a modulus of approximately 30,000 ksi (210 GPa); for aluminum, values around 10,600 ksi (73 GPa) are typical. More flexible materials have a lower modulus—values for plastics range from 100 to 2,000 ksi (0.7 to 14 GPa). Some representative values of E are listed in Table I-2, Appendix I (available online). For most materials, the value of E in compression is nearly the same as in tension.

Modulus of elasticity is often called **Young's modulus**, after another English scientist, Thomas Young (1773–1829). In connection with an investigation of tension and compression of prismatic bars, Young introduced the idea of a "modulus of the elasticity." However, his modulus was not the same as the one in use today, because it involved properties of the bar as well as of the material (Ref. 1-7 available online).

Poisson's Ratio

When a prismatic bar is loaded in tension, the axial elongation is accompanied by **lateral contraction** (that is, contraction normal to the direction of the applied load). This change in shape is pictured in Fig. 1-22, where part (a) shows the bar before loading and part (b) shows it after loading. In part (b), the dashed lines represent the shape of the bar prior to loading.

Lateral contraction is easily seen by stretching a rubber band, but in metals the changes in lateral dimensions (in the linearly elastic region) are usually too small to be visible. However, they can be detected with sensitive measuring devices.

The **lateral strain** ϵ' at any point in a bar is proportional to the axial strain ϵ at that same point if the material is linearly elastic. The ratio of these strains is a property of the material known as **Poisson's ratio**. This dimensionless ratio, usually denoted by the Greek letter ν (nu), can be expressed by the equation

$$\nu = -\frac{\text{lateral strain}}{\text{axial strain}} = -\frac{\epsilon'}{\epsilon} \tag{1-9}$$

The minus sign is inserted in the equation to compensate for the fact that the lateral and axial strains normally have opposite signs. For instance, the axial strain in a bar in tension is positive and the lateral strain is negative (because the width of the bar decreases). For compression we have the opposite situation, with the bar becoming shorter (negative axial strain) and wider (positive lateral strain). Therefore, for ordinary materials Poisson's ratio will have a positive value.

When Poisson's ratio for a material is known, we can obtain the lateral strain from the axial strain as follows:

$$\epsilon' = -\nu\epsilon \tag{1-10}$$

When using Eqs. (1-9) and (1-10), we must always keep in mind that they apply only to a bar in uniaxial stress, that is, a bar for which the only stress is the normal stress σ in the axial direction.

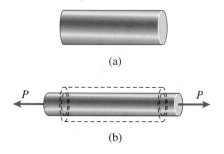

(a)

P ← → P

(b)

FIG. 1-22 Axial elongation and lateral contraction of a prismatic bar in tension: (a) bar before loading, and (b) bar after loading. (The deformations of the bar are highly exaggerated.)

Poisson's ratio is named for the famous French mathematician Siméon Denis Poisson (1781–1840), who attempted to calculate this ratio by a molecular theory of materials (Ref. 1-8 available online). For isotropic materials, Poisson found $\nu = 1/4$. More recent calculations based upon better models of atomic structure give $\nu = 1/3$. Both of these values are close to actual measured values, which are in the range 0.25 to 0.35 for most metals and many other materials. Materials with an extremely low value of Poisson's ratio include cork, for which ν is practically zero, and concrete, for which ν is about 0.1 or 0.2. A theoretical upper limit for Poisson's ratio is 0.5, as explained later in Section 6.5. Rubber comes close to this limiting value.

A table of Poisson's ratios for various materials in the linearly elastic range is given in Appendix I (see Table I-2 available online). For most purposes, Poisson's ratio is assumed to be the same in both tension and compression.

When the strains in a material become large, Poisson's ratio changes. For instance, in the case of structural steel the ratio becomes almost 0.5 when plastic yielding occurs. Thus, Poisson's ratio remains constant only in the linearly elastic range. When the material behavior is nonlinear, the ratio of lateral strain to axial strain is often called the *contraction ratio*. Of course, in the special case of linearly elastic behavior, the contraction ratio is the same as Poisson's ratio.

Limitations

For a particular material, Poisson's ratio remains constant throughout the linearly elastic range, as explained previously. Therefore, at any given point in the prismatic bar of Fig. 1-22, the lateral strain remains proportional to the axial strain as the load increases or decreases. However, for a given value of the load (which means that the axial strain is constant throughout the bar), additional conditions must be met if the lateral strains are to be the same throughout the entire bar.

First, the material must be **homogeneous**, that is, it must have the same composition (and hence the same elastic properties) at every point. However, having a homogeneous material does not mean that the elastic properties at a particular point are the same in all *directions*. For instance, the modulus of elasticity could be different in the axial and lateral directions, as in the case of a wood pole. Therefore, a second condition for uniformity in the lateral strains is that the elastic properties must be the same in all directions *perpendicular* to the longitudinal axis. When the preceding conditions are met, as is often the case with metals, the lateral strains in a prismatic bar subjected to uniform tension will be the same at every point in the bar and the same in all lateral directions.

Materials having the same properties in all directions (whether axial, lateral, or any other direction) are said to be **isotropic**. If the properties differ in various directions, the material is **anisotropic** (or **aeolotropic**).

In this book, all examples and problems are solved with the assumption that the material is linearly elastic, homogeneous, and isotropic, unless a specific statement is made to the contrary.

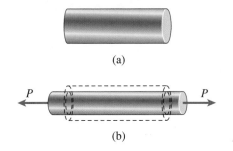

(a)

P ⟵   ⟶ P

(b)

FIG. 1-22 (Repeated)

Example 1-3

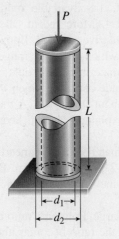

FIG. 1-23 Example 1-3. Steel pipe in compression

A steel pipe of length $L = 4.0$ ft, outside diameter $d_2 = 6.0$ in., and inside diameter $d_1 = 4.5$ in. is compressed by an axial force $P = 140$ k (Fig. 1-23). The material has modulus of elasticity $E = 30{,}000$ ksi and Poisson's ratio $\nu = 0.30$.

Determine the following quantities for the pipe: (a) the shortening δ, (b) the lateral strain ϵ', (c) the increase Δd_2 in the outer diameter and the increase Δd_1 in the inner diameter, and (d) the increase Δt in the wall thickness.

Solution

The cross-sectional area A and longitudinal stress σ are determined as follows:

$$A = \frac{\pi}{4}\left(d_2^2 - d_1^2\right) = \frac{\pi}{4}\left[(6.0 \text{ in.})^2 - (4.5 \text{ in.})^2\right] = 12.37 \text{ in.}^2$$

$$\sigma = -\frac{P}{A} = -\frac{140 \text{ k}}{12.37 \text{ in.}^2} = -11.32 \text{ ksi (compression)}$$

Because the stress is well below the yield stress (see Table I-3, Appendix I available online), the material behaves linearly elastically and the axial strain may be found from Hooke's law:

$$\epsilon = \frac{\sigma}{E} = \frac{-11.32 \text{ ksi}}{30{,}000 \text{ ksi}} = -377.3 \times 10^{-6}$$

The minus sign for the strain indicates that the pipe shortens.

(a) Knowing the axial strain, we can now find the change in length of the pipe (see Eq. 1-2):

$$\delta = \epsilon L = (-377.3 \times 10^{-6})(4.0 \text{ ft})(12 \text{ in./ft}) = -0.018 \text{ in.}$$

The negative sign again indicates a shortening of the pipe.

(b) The lateral strain is obtained from Poisson's ratio (see Eq. 1-10):

$$\epsilon' = -\nu\epsilon = -(0.30)(-377.3 \times 10^{-6}) = 113.2 \times 10^{-6}$$

The positive sign for ϵ' indicates an increase in the lateral dimensions, as expected for compression.

(c) The increase in outer diameter equals the lateral strain times the diameter:

$$\Delta d_2 = \epsilon' d_2 = (113.2 \times 10^{-6})(6.0 \text{ in.}) = 0.000679 \text{ in.}$$

Similarly, the increase in inner diameter is

$$\Delta d_1 = \epsilon' d_1 = (113.2 \times 10^{-6})(4.5 \text{ in.}) = 0.000509 \text{ in.}$$

(d) The increase in wall thickness is found in the same manner as the increases in the diameters; thus,

$$\Delta t = \epsilon' t = (113.2 \times 10^{-6})(0.75 \text{ in.}) = 0.000085 \text{ in.}$$

This result can be verified by noting that the increase in wall thickness is equal to half the difference of the increases in diameters:

$$\Delta t = \frac{\Delta d_2 - \Delta d_1}{2} = \frac{1}{2}(0.000679 \text{ in.} - 0.000509 \text{ in.}) = 0.000085 \text{ in.}$$

as expected. Note that under compression, all three quantities increase (outer diameter, inner diameter, and thickness).

Note: The numerical results obtained in this example illustrate that the dimensional changes in structural materials under normal loading conditions are extremely small. In spite of their smallness, changes in dimensions can be important in certain kinds of analysis (such as the analysis of statically indeterminate structures) and in the experimental determination of stresses and strains.

1.6 SHEAR STRESS AND STRAIN

Diagonal bracing for an elevated walkway showing a clevis and a pin in double shear (© Barry Goodno)

In the preceding sections we discussed the effects of normal stresses produced by axial loads acting on straight bars. These stresses are called "normal stresses" because they act in directions *perpendicular* to the surface of the material. Now we will consider another kind of stress, called a **shear stress**, that acts *tangential* to the surface of the material.

As an illustration of the action of shear stresses, consider the bolted connection shown in Fig. 1-24a. This connection consists of a flat bar *A*, a clevis *C*, and a bolt *B* that passes through holes in the bar and clevis. Under the action of the tensile loads *P*, the bar and clevis will press against the bolt in **bearing**, and contact stresses, called **bearing stresses**, will be developed. In addition, the bar and clevis tend to *shear* the bolt, that is, cut through it, and this tendency is resisted by shear stresses in the bolt. As an example, consider the bracing for an elevated pedestrian walkway shown in the photograph.

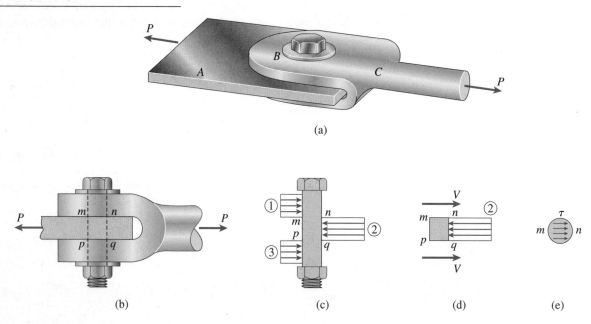

(a)

(b)　　　(c)　　　(d)　　　(e)

FIG. 1-24 Bolted connection in which the bolt is loaded in double shear

To show more clearly the actions of the bearing and shear stresses, let us look at this type of connection in a schematic side view (Fig. 1-24b). With this view in mind, we draw a free-body diagram of the bolt (Fig. 1-24c). The bearing stresses exerted by the clevis against the bolt appear on the left-hand side of the free-body diagram and are labeled 1 and 3. The stresses from the bar appear on the right-hand side and are labeled 2. The actual distribution of the bearing stresses is difficult to determine, so it is customary to assume that the stresses are uniformly distributed. Based upon the assumption of uniform distribution, we can calculate an **average bearing stress** σ_b by dividing the total bearing force F_b by the bearing area A_b:

$$\sigma_b = \frac{F_b}{A_b} \tag{1-11}$$

The **bearing area** is defined as the projected area of the curved bearing surface. For instance, consider the bearing stresses labeled 1. The projected area A_b on which they act is a rectangle having a height equal to the thickness of the clevis and a width equal to the diameter of the bolt. Also, the bearing force F_b represented by the stresses labeled 1 is equal to $P/2$. The same area and the same force apply to the stresses labeled 3.

Now consider the bearing stresses between the flat bar and the bolt (the stresses labeled 2). For these stresses, the bearing area A_b is a rectangle with height equal to the thickness of the flat bar and width equal to the bolt diameter. The corresponding bearing force F_b is equal to the load P.

The free-body diagram of Fig. 1-24c shows that there is a tendency to shear the bolt along cross sections mn and pq. From a free-body diagram of the portion $mnpq$ of the bolt (see Fig. 1-24d), we see that shear forces V act over the cut surfaces of the bolt. In this particular example there are two planes of shear (mn and pq), and so the bolt is said to be in **double shear**. In double shear, each of the shear forces is equal to one-half of the total load transmitted by the bolt, that is, $V = P/2$.

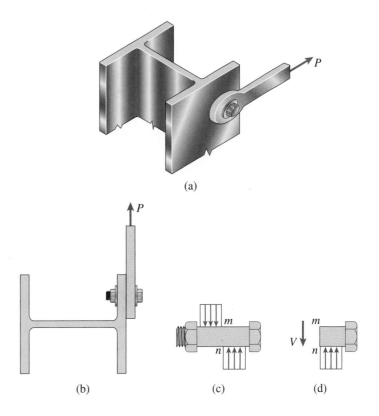

FIG. 1-25 Bolted connection in which the bolt is loaded in single shear

The shear forces V are the resultants of the shear stresses distributed over the cross-sectional area of the bolt. For instance, the shear stresses acting on cross section mn are shown in Fig. 1-24e. These stresses act parallel to the cut surface. The exact distribution of the stresses is not known, but they are highest near the center and become zero at certain locations on the edges. As indicated in Fig. 1-24e, shear stresses are customarily denoted by the Greek letter τ (tau).

A bolted connection in **single shear** is shown in Fig. 1-25a, where the axial force P in the metal bar is transmitted to the flange of the steel column through a bolt. A cross-sectional view of the column (Fig. 1-25b) shows the connection in more detail. Also, a sketch of the bolt (Fig. 1-25c) shows the assumed distribution of the bearing stresses acting on the bolt. As mentioned earlier, the actual distribution of these bearing stresses is much more complex than shown in the figure. Furthermore, bearing stresses are also developed against the inside surfaces of the bolt head and nut. Thus, Fig. 1-25c is *not* a free-body diagram—only the idealized bearing stresses acting on the shank of the bolt are shown in the figure.

By cutting through the bolt at section mn we obtain the diagram shown in Fig. 1-25d. This diagram includes the shear force V (equal to the load P) acting on the cross section of the bolt. As already pointed out, this shear force is the resultant of the shear stresses that act over the cross-sectional area of the bolt.

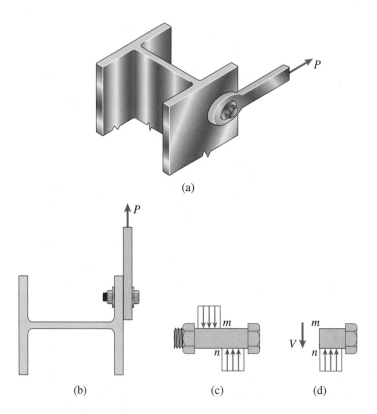

FIG. 1-25 (Repeated)

(a)

(b)   (c)   (d)

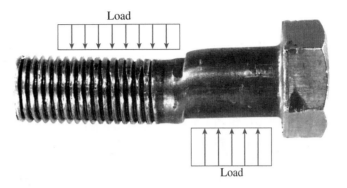

Load

Load

FIG. 1-26 Failure of a bolt in single shear

The deformation of a bolt loaded almost to fracture in single shear is shown in Fig. 1-26 (compare with Fig. 1-25c).

In the preceding discussions of bolted connections we disregarded **friction** (produced by tightening of the bolts) between the connecting elements. The presence of friction means that part of the load is carried by friction forces, thereby reducing the loads on the bolts. Since friction forces are unreliable and difficult to estimate, it is common practice to err on the conservative side and omit them from the calculations.

The **average shear stress** on the cross section of a bolt is obtained by dividing the total shear force V by the area A of the cross section on which it acts, as follows:

$$\tau_{\text{aver}} = \frac{V}{A} \tag{1-12}$$

In the example of Fig. 1-25, which shows a bolt in single shear, the shear force V is equal to the load P and the area A is the cross-sectional area of the bolt. However, in the example of Fig. 1-24, where the bolt is in double shear, the shear force V equals $P/2$.

From Eq. (1-12) we see that shear stresses, like normal stresses, represent intensity of force, or force per unit of area. Thus, the **units** of shear stress are the same as those for normal stress, namely, psi or ksi in USCS units and pascals or multiples thereof in SI units.

The loading arrangements shown in Figs. 1-24 and 1-25 are examples of **direct shear** (or *simple shear*) in which the shear stresses are created by the direct action of the forces in trying to cut through the material. Direct shear arises in the design of bolts, pins, rivets, keys, welds, and glued joints.

Shear stresses also arise in an indirect manner when members are subjected to tension, torsion, and bending, as discussed later in Sections 2.6, 3.3, and 5.7, respectively.

Equality of Shear Stresses on Perpendicular Planes

To obtain a more complete picture of the action of shear stresses, let us consider a small element of material in the form of a rectangular parallelepiped having sides of lengths a, b, and c in the x, y, and z directions, respectively (Fig. 1-27).[*] The front and rear faces of this element are free of stress.

Now assume that a shear stress τ_1 is distributed uniformly over the right-hand face, which has area bc. In order for the element to be in equilibrium in the y direction, the total shear force $\tau_1 bc$ acting on the right-hand face must be balanced by an equal but oppositely directed shear force on the left-hand face. Since the areas of these two faces are equal, it follows that the shear stresses on the two faces must be equal.

The forces $\tau_1 bc$ acting on the left- and right-hand side faces (Fig. 1-27) form a couple having a moment about the z axis of magnitude $\tau_1 abc$, acting counterclockwise in the figure.[**] Equilibrium of the element requires that this moment be balanced by an equal and opposite moment resulting from shear stresses acting on the top and bottom faces of the element. Denoting the stresses on the top and bottom faces as τ_2, we see that the corresponding horizontal shear forces equal $\tau_2 ac$. These forces form a clockwise couple of moment $\tau_2 abc$. From moment equilibrium of the element about the z axis, we see that $\tau_1 abc$ equals $\tau_2 abc$, or

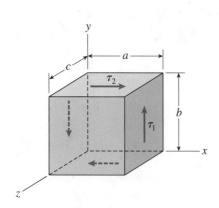

FIG. 1-27 Small element of material subjected to shear stresses

$$\tau_1 = \tau_2 \qquad (1\text{-}13)$$

Therefore, the magnitudes of the four shear stresses acting on the element are equal, as shown in Fig. 1-28a.

In summary, we have arrived at the following general observations regarding shear stresses acting on a rectangular element:

1. Shear stresses on opposite (and parallel) faces of an element are equal in magnitude and opposite in direction.
2. Shear stresses on adjacent (and perpendicular) faces of an element are equal in magnitude and have directions such that both stresses point toward, or both point away from, the line of intersection of the faces.

These observations were obtained for an element subjected only to shear stresses (no normal stresses), as pictured in Figs. 1-27 and 1-28. This state of stress is called **pure shear** and is discussed later in greater detail (Section 3.5).

For most purposes, the preceding conclusions remain valid even when normal stresses act on the faces of the element. The reason is that the normal stresses on opposite faces of a small element usually are equal in magnitude and opposite in direction; hence they do not alter the equilibrium equations used in reaching the preceding conclusions.

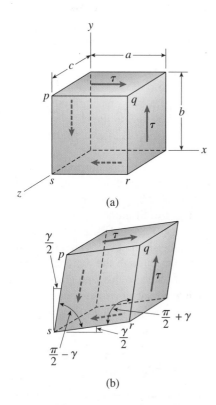

FIG. 1-28 Element of material subjected to shear stresses and strains

---

[*]A **parallelepiped** is a prism whose bases are parallelograms; thus, a parallelepiped has six faces, each of which is a parallelogram. Opposite faces are parallel and identical parallelograms. A **rectangular parallelepiped** has all faces in the form of rectangles.

[**]A **couple** consists of two parallel forces that are equal in magnitude and opposite in direction.

Shear Strain

Shear stresses acting on an element of material (Fig. 1-28a) are accompanied by *shear strains*. As an aid in visualizing these strains, we note that the shear stresses have no tendency to elongate or shorten the element in the x, y, and z directions—in other words, the lengths of the sides of the element do not change. Instead, the shear stresses produce a change in the *shape* of the element (Fig. 1-28b). The original element, which is a rectangular parallelepiped, is deformed into an oblique parallelepiped, and the front and rear faces become rhomboids.[*]

Because of this deformation, the angles between the side faces change. For instance, the angles at points q and s, which were $\pi/2$ before deformation, are reduced by a small angle γ to $\pi/2 - \gamma$ (Fig. 1-28b). At the same time, the angles at points p and r are increased to $\pi/2 + \gamma$. The angle γ is a measure of the **distortion**, or change in shape, of the element and is called the **shear strain**. Because shear strain is an angle, it is usually measured in degrees or radians.

Sign Conventions for Shear Stresses and Strains

As an aid in establishing sign conventions for shear stresses and strains, we need a scheme for identifying the various faces of a stress element (Fig. 1-28a). Henceforth, we will refer to the faces oriented toward the positive directions of the axes as the positive faces of the element. In other words, a positive face has its outward normal directed in the positive direction of a coordinate axis. The opposite faces are negative faces. Thus, in Fig. 1-28a, the right-hand, top, and front faces are the positive x, y, and z faces, respectively, and the opposite faces are the negative x, y, and z faces.

Using the terminology described in the preceding paragraph, we may state the sign convention for shear stresses in the following manner:

> *A shear stress acting on a positive face of an element is positive if it acts in the positive direction of one of the coordinate axes and negative if it acts in the negative direction of an axis. A shear stress acting on a negative face of an element is positive if it acts in the negative direction of an axis and negative if it acts in a positive direction.*

Thus, all shear stresses shown in Fig. 1-28a are positive.

The sign convention for shear strains is as follows:

> *Shear strain in an element is positive when the angle between two positive faces (or two negative faces) is reduced. The strain is negative when the angle between two positive (or two negative) faces is increased.*

---

[*]An **oblique angle** can be either acute or obtuse, but it is *not* a right angle. A **rhomboid** is a parallelogram with oblique angles and adjacent sides *not* equal. (A *rhombus* is a parallelogram with oblique angles and all four sides equal, sometimes called a *diamond-shaped figure*.)

Thus, the strains shown in Fig. 1-28b are positive, and we see that positive shear stresses are accompanied by positive shear strains.

Hooke's Law in Shear

The properties of a material in shear can be determined experimentally from direct-shear tests or from torsion tests. The latter tests are performed by twisting hollow, circular tubes, thereby producing a state of pure shear, as explained later in Section 3.5. From the results of these tests, we can plot **shear stress-strain diagrams** (that is, diagrams of shear stress τ versus shear strain γ). These diagrams are similar in shape to tension-test diagrams (σ versus ϵ) for the same materials, although they differ in magnitudes.

From shear stress-strain diagrams, we can obtain material properties such as the proportional limit, modulus of elasticity, yield stress, and ultimate stress. These properties in shear are usually about half as large as those in tension. For instance, the yield stress for structural steel in shear is 0.5 to 0.6 times the yield stress in tension.

For many materials, the initial part of the shear stress-strain diagram is a straight line through the origin, just as it is in tension. For this linearly elastic region, the shear stress and shear strain are proportional, and therefore we have the following equation for **Hooke's law in shear**:

$$\tau = G\gamma \tag{1-14}$$

in which G is the **shear modulus of elasticity** (also called the *modulus of rigidity*).

The shear modulus G has the same **units** as the tension modulus E, namely, psi or ksi in USCS units and pascals (or multiples thereof) in SI units. For mild steel, typical values of G are 11,000 ksi or 75 GPa; for aluminum alloys, typical values are 4000 ksi or 28 GPa. Additional values are listed in Table I-2, Appendix I (available online).

The moduli of elasticity in tension and shear are related by the following equation:

$$G = \frac{E}{2(1 + \nu)} \tag{1-15}$$

in which ν is Poisson's ratio. This relationship, which is derived later in Section 3.6, shows that E, G, and ν are not independent elastic properties of the material. Because the value of Poisson's ratio for ordinary materials is between zero and one-half, we see from Eq. (1-15) that G must be from one-third to one-half of E.

The following examples illustrate some typical analyses involving the effects of shear. Example 1-4 is concerned with shear stresses in a plate, Example 1-5 deals with bearing and shear stresses in pins and bolts, and Example 1-6 involves finding shear stresses and shear strains in an elastomeric bearing pad subjected to a horizontal shear force.

Example 1-4

A punch for making holes in steel plates is shown in Fig. 1-29a. Assume that a punch having diameter $d = 20$ mm is used to punch a hole in an 8-mm plate, as shown in the cross-sectional view (Fig. 1-29b).

If a force $P = 110$ kN is required to create the hole, what is the average shear stress in the plate and the average compressive stress in the punch?

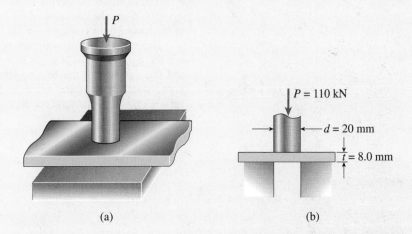

FIG. 1-29 Example 1-4. Punching a hole in a steel plate

(a)   (b)

Solution

The average shear stress in the plate is obtained by dividing the force P by the shear area of the plate. The shear area A_s is equal to the circumference of the hole times the thickness of the plate, or

$$A_s = \pi dt = \pi(20 \text{ mm})(8.0 \text{ mm}) = 502.7 \text{ mm}^2$$

in which d is the diameter of the punch and t is the thickness of the plate. Therefore, the average shear stress in the plate is

$$\tau_{\text{aver}} = \frac{P}{A_s} = \frac{110 \text{ kN}}{502.7 \text{ mm}^2} = 219 \text{ MPa}$$

The average compressive stress in the punch is

$$\sigma_c = \frac{P}{A_{\text{punch}}} = \frac{P}{\pi d^2/4} = \frac{110 \text{ kN}}{\pi(20 \text{ mm})^2/4} = 350 \text{ MPa}$$

in which A_{punch} is the cross-sectional area of the punch.

Note: This analysis is highly idealized because we are disregarding impact effects that occur when a punch is rammed through a plate. (The inclusion of such effects requires advanced methods of analysis that are beyond the scope of mechanics of materials.)

Example 1-5

A steel strut S serving as a brace for a boat hoist transmits a compressive force $P = 12$ k to the deck of a pier (Fig. 1-30a). The strut has a hollow square cross section with wall thickness $t = 0.375$ in. (Fig. 1-30b), and the angle θ between the strut and the horizontal is 40°. A pin through the strut transmits the compressive force from the strut to two gussets G that are welded to the base plate B. Four anchor bolts fasten the base plate to the deck.

The diameter of the pin is $d_{pin} = 0.75$ in., the thickness of the gussets is $t_G = 0.625$ in., the thickness of the base plate is $t_B = 0.375$ in., and the diameter of the anchor bolts is $d_{bolt} = 0.50$ in.

Determine the following stresses: (a) the bearing stress between the strut and the pin, (b) the shear stress in the pin, (c) the bearing stress between the pin and the gussets, (d) the bearing stress between the anchor bolts and the base plate, and (e) the shear stress in the anchor bolts. (Disregard any friction between the base plate and the deck.)

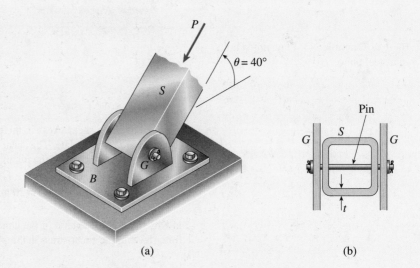

FIG. 1-30 Example 1-5. (a) Pin connection between strut S and base plate B (b) Cross section through the strut S

(a)   (b)

Solution

(a) *Bearing stress between strut and pin.* The average value of the bearing stress between the strut and the pin is found by dividing the force in the strut by the total bearing area of the strut against the pin. The latter is equal to twice the thickness of the strut (because bearing occurs at two locations) times the diameter of the pin (see Fig. 1-30b). Thus, the bearing stress is

$$\sigma_{b1} = \frac{P}{2td_{pin}} = \frac{12 \text{ k}}{2(0.375 \text{ in.})(0.75 \text{ in.})} = 21.3 \text{ ksi} \quad \longleftarrow$$

This bearing stress is not excessive for a strut made of structural steel.

(b) *Shear stress in pin.* As can be seen from Fig. 1-30b, the pin tends to shear on two planes, namely, the planes between the strut and the gussets. Therefore, the average shear stress in the pin (which is in double shear) is equal to the total load applied to the pin divided by twice its cross-sectional area:

$$\tau_{pin} = \frac{P}{2\pi d_{pin}^2/4} = \frac{12\text{ k}}{2\pi(0.75\text{ in.})^2/4} = 13.6\text{ ksi}$$

The pin would normally be made of high-strength steel (tensile yield stress greater than 50 ksi) and could easily withstand this shear stress (the yield stress in shear is usually at least 50% of the yield stress in tension).

(c) *Bearing stress between pin and gussets.* The pin bears against the gussets at two locations, so the bearing area is twice the thickness of the gussets times the pin diameter; thus,

$$\sigma_{b2} = \frac{P}{2t_G d_{pin}} = \frac{12\text{ k}}{2(0.625\text{ in.})(0.75\text{ in.})} = 12.8\text{ ksi}$$

which is less than the bearing stress between the strut and the pin (21.3 ksi).

(d) *Bearing stress between anchor bolts and base plate.* The vertical component of the force P (see Fig. 1-30a) is transmitted to the pier by direct bearing between the base plate and the pier. The horizontal component, however, is transmitted through the anchor bolts. The average bearing stress between the base plate and the anchor bolts is equal to the horizontal component of the force P divided by the bearing area of four bolts. The bearing area for one bolt is equal to the thickness of the base plate times the bolt diameter. Consequently, the bearing stress is

$$\sigma_{b3} = \frac{P\cos 40°}{4t_B d_{bolt}} = \frac{(12\text{ k})(\cos 40°)}{4(0.375\text{ in.})(0.50\text{ in.})} = 12.3\text{ ksi}$$

(e) *Shear stress in anchor bolts.* The average shear stress in the anchor bolts is equal to the horizontal component of the force P divided by the total cross-sectional area of four bolts (note that each bolt is in single shear). Therefore,

$$\tau_{bolt} = \frac{P\cos 40°}{4\pi d_{bolt}^2/4} = \frac{(12\text{ k})(\cos 40°)}{4\pi(0.50\text{ in.})^2/4} = 11.7\text{ ksi}$$

Any friction between the base plate and the pier would reduce the load on the anchor bolts.

Example 1-6

A bearing pad of the kind used to support machines and bridge girders consists of a linearly elastic material (usually an elastomer, such as rubber) capped by a steel plate (Fig. 1-31a). Assume that the thickness of the elastomer is h, the dimensions of the plate are $a \times b$, and the pad is subjected to a horizontal shear force V.

Obtain formulas for the average shear stress τ_{aver} in the elastomer and the horizontal displacement d of the plate (Fig. 1-31b).

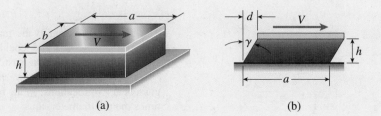

FIG. 1-31 Example 1-6. Bearing pad in shear

(a)　　　　　　　　　　　　　　(b)

Solution

Assume that the shear stresses in the elastomer are uniformly distributed throughout its entire volume. Then the shear stress on any horizontal plane through the elastomer equals the shear force V divided by the area ab of the plane (Fig. 1-31a):

$$\tau_{\text{aver}} = \frac{V}{ab} \qquad (1\text{-}16)$$

The corresponding shear strain (from Hooke's law in shear; Eq. 1-14) is

$$\gamma = \frac{\tau_{\text{aver}}}{G_e} = \frac{V}{abG_e} \qquad (1\text{-}17)$$

in which G_e is the shear modulus of the elastomeric material. Finally, the horizontal displacement d is equal to $h \tan \gamma$ (from Fig. 1-31b):

$$d = h \tan \gamma = h \tan \left(\frac{V}{abG_e}\right) \qquad (1\text{-}18)$$

In most practical situations the shear strain γ is a small angle, and in such cases we may replace $\tan \gamma$ by γ and obtain

$$d = h\gamma = \frac{hV}{abG_e} \qquad (1\text{-}19)$$

Equations (1-18) and (1-19) give approximate results for the horizontal displacement of the plate because they are based upon the assumption that the shear stress and strain are constant throughout the volume of the elastomeric material. In reality the shear stress is zero at the edges of the material (because there are no shear stresses on the free vertical faces), and therefore the deformation of the material is more complex than pictured in Fig. 1-31b. However, if the length a of the plate is large compared with the thickness h of the elastomer, the preceding results are satisfactory for design purposes.

1.7 ALLOWABLE STRESSES AND ALLOWABLE LOADS

Engineering has been aptly described as the *application of science to the common purposes of life*. In fulfilling that mission, engineers design a seemingly endless variety of objects to serve the basic needs of society. These needs include housing, agriculture, transportation, communication, and many other aspects of modern life. Factors to be considered in design include functionality, strength, appearance, economics, and environmental effects. However, when studying mechanics of materials, our principal design interest is **strength**, that is, *the capacity of the object to support or transmit loads*. Objects that must sustain loads include buildings, machines, containers, trucks, aircraft, ships, and the like. For simplicity, we will refer to all such objects as **structures**; thus, *a structure is any object that must support or transmit loads*.

Factors of Safety

If structural failure is to be avoided, the loads that a structure is capable of supporting must be greater than the loads it will be subjected to when in service. Since *strength* is the ability of a structure to resist loads, the preceding criterion can be restated as follows: *The actual strength of a structure must exceed the required strength*. The ratio of the actual strength to the required strength is called the **factor of safety** n:

$$\text{Factor of safety } n = \frac{\text{Actual strength}}{\text{Required strength}} \qquad (1\text{-}20)$$

Of course, the factor of safety must be greater than 1.0 if failure is to be avoided. Depending upon the circumstances, factors of safety from slightly above 1.0 to as much as 10 are used.

The incorporation of factors of safety into design is not a simple matter, because both strength and failure have many different meanings. Strength may be measured by the load-carrying capacity of a structure, or it may be measured by the stress in the material. Failure may mean the fracture and complete collapse of a structure, or it may mean that the deformations have become so large that the structure can no longer perform its intended functions. The latter kind of failure may occur at loads much smaller than those that cause actual collapse.

The determination of a factor of safety must also take into account such matters as the following: probability of accidental overloading of the structure by loads that exceed the design loads; types of loads (static or dynamic); whether the loads are applied once or are repeated; how accurately the loads are known; possibilities for fatigue failure; inaccuracies in construction; variability in the quality of workmanship; variations in properties of materials; deterioration due to corrosion or other environmental effects; accuracy of the methods of analysis; whether failure is gradual (ample warning) or sudden (no warning); consequences of failure (minor damage or major catastrophe); and other such considerations. If the factor of safety is too low, the likelihood of failure will be high and the structure will be unacceptable; if the factor is too large, the structure will be wasteful of materials and perhaps unsuitable for its function (for instance, it may be too heavy).

Because of these complexities and uncertainties, factors of safety must be determined on a probabilistic basis. They usually are established by groups of experienced engineers who write the codes and specifications used by other designers, and sometimes they are even enacted into law. The provisions of codes and specifications are intended to provide reasonable levels of safety without unreasonable costs.

In aircraft design it is customary to speak of the **margin of safety** rather than the factor of safety. The margin of safety is defined as the factor of safety minus one:

$$\text{Margin of safety} = n - 1 \tag{1-21}$$

Margin of safety is often expressed as a percent, in which case the value given above is multiplied by 100. Thus, a structure having an actual strength that is 1.75 times the required strength has a factor of safety of 1.75 and a margin of safety of 0.75 (or 75%). When the margin of safety is reduced to zero or less, the structure (presumably) will fail.

Allowable Stresses

Factors of safety are defined and implemented in various ways. For many structures, it is important that the material remain within the linearly elastic range in order to avoid permanent deformations when the loads are removed. Under these conditions, the factor of safety is established with respect to yielding of the structure. Yielding begins when the yield stress is reached at *any* point within the structure. Therefore, by applying a factor of safety with respect to the yield stress (or yield strength), we obtain an **allowable stress** (or *working stress*) that must not be exceeded anywhere in the structure. Thus,

$$\text{Allowable stress} = \frac{\text{Yield strength}}{\text{Factor of safety}} \tag{1-22}$$

or, for tension and shear, respectively,

$$\sigma_{\text{allow}} = \frac{\sigma_Y}{n_1} \quad \text{and} \quad \tau_{\text{allow}} = \frac{\tau_Y}{n_2} \tag{1-23a,b}$$

in which σ_Y and τ_Y are the yield stresses and n_1 and n_2 are the corresponding factors of safety. In building design, a typical factor of safety with respect to yielding in tension is 1.67; thus, a mild steel having a yield stress of 36 ksi has an allowable stress of 21.6 ksi.

Sometimes the factor of safety is applied to the **ultimate stress** instead of the yield stress. This method is suitable for brittle materials, such as concrete and some plastics, and for materials without a clearly defined yield stress, such as wood and high-strength steels. In these cases the allowable stresses in tension and shear are

$$\sigma_{\text{allow}} = \frac{\sigma_U}{n_3} \quad \text{and} \quad \tau_{\text{allow}} = \frac{\tau_U}{n_4} \tag{1-24a,b}$$

in which σ_U and τ_U are the ultimate stresses (or ultimate strengths). Factors of safety with respect to the ultimate strength of a material are usually larger than those based upon yield strength. In the case of mild steel, a factor of safety of 1.67 with respect to yielding corresponds to a factor of approximately 2.8 with respect to the ultimate strength.

Allowable Loads

After the allowable stress has been established for a particular material and structure, the **allowable load** on that structure can be determined. The relationship between the allowable load and the allowable stress depends upon the type of structure. In this chapter we are concerned only with the most elementary kinds of structures, namely, bars in tension or compression and pins (or bolts) in direct shear and bearing.

In these kinds of structures the stresses are uniformly distributed (or at least *assumed* to be uniformly distributed) over an area. For instance, in the case of a bar in tension, the stress is uniformly distributed over the cross-sectional area provided the resultant axial force acts through the centroid of the cross section. The same is true of a bar in compression provided the bar is not subject to buckling. In the case of a pin subjected to shear, we consider only the average shear stress on the cross section, which is equivalent to assuming that the shear stress is uniformly distributed. Similarly, we consider only an average value of the bearing stress acting on the projected area of the pin.

Therefore, in all four of the preceding cases the **allowable load** (also called the *permissible load* or the *safe load*) is equal to the allowable stress times the area over which it acts:

$$\text{Allowable load} = (\text{Allowable stress})(\text{Area}) \qquad \text{(1-25)}$$

For bars in direct *tension* and *compression* (no buckling), this equation becomes

$$P_{\text{allow}} = \sigma_{\text{allow}}\, A \qquad \text{(1-26)}$$

in which σ_{allow} is the permissible normal stress and A is the cross-sectional area of the bar. If the bar has a hole through it, the *net area* is normally used when the bar is in tension. The **net area** is the gross cross-sectional area minus the area removed by the hole. For compression, the gross area may be used if the hole is filled by a bolt or pin that can transmit the compressive stresses.

For pins in *direct shear*, Eq. (1-25) becomes

$$P_{\text{allow}} = \tau_{\text{allow}}\, A \qquad \text{(1-27)}$$

in which τ_{allow} is the permissible shear stress and A is the area over which the shear stresses act. If the pin is in single shear, the area is the cross-sectional area of the pin; in double shear, it is twice the cross-sectional area.

Finally, the permissible load based upon *bearing* is

$$P_{\text{allow}} = \sigma_b A_b \qquad \text{(1-28)}$$

in which σ_b is the allowable bearing stress and A_b is the projected area of the pin or other surface over which the bearing stresses act.

The following example illustrates how allowable loads are determined when the allowable stresses for the material are known.

Example 1-7

A steel bar serving as a vertical hanger to support heavy machinery in a factory is attached to a support by the bolted connection shown in Fig. 1-32. The main part of the hanger has a rectangular cross section with width $b_1 = 1.5$ in. and thickness $t = 0.5$ in. At the connection the hanger is enlarged to a width $b_2 = 3.0$ in. The bolt, which transfers the load from the hanger to the two gussets, has diameter $d = 1.0$ in.

Determine the allowable value of the tensile load P in the hanger based upon the following four considerations:

(a) The allowable tensile stress in the main part of the hanger is 16,000 psi.

(b) The allowable tensile stress in the hanger at its cross section through the bolt hole is 11,000 psi. (The permissible stress at this section is lower because of the stress concentrations around the hole.)

(c) The allowable bearing stress between the hanger and the bolt is 26,000 psi.

(d) The allowable shear stress in the bolt is 6,500 psi.

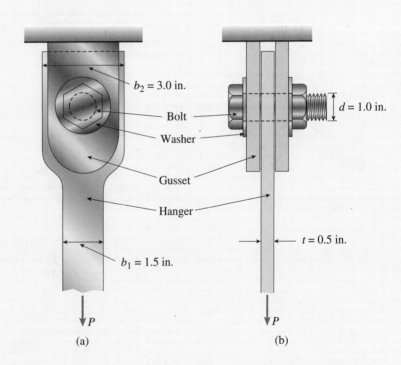

FIG. 1-32 Example 1-7. Vertical hanger subjected to a tensile load P: (a) front view of bolted connection, and (b) side view of connection

Solution

(a) The allowable load P_1 based upon the stress in the main part of the hanger is equal to the allowable stress in tension times the cross-sectional area of the hanger (Eq. 1-26):

$$P_1 = \sigma_{\text{allow}}A = \sigma_{\text{allow}}b_1t = (16{,}000 \text{ psi})(1.5 \text{ in.} \times 0.5 \text{ in.}) = 12{,}000 \text{ lb}$$

continued

A load greater than this value will overstress the main part of the hanger, that is, the actual stress will exceed the allowable stress, thereby reducing the factor of safety.

(b) At the cross section of the hanger through the bolt hole, we must make a similar calculation but with a different allowable stress and a different area. The net cross-sectional area, that is, the area that remains after the hole is drilled through the bar, is equal to the net width times the thickness. The net width is equal to the gross width b_2 minus the diameter d of the hole. Thus, the equation for the allowable load P_2 at this section is

$$P_2 = \sigma_{allow}A = \sigma_{allow}(b_2 - d)t = (11{,}000 \text{ psi})(3.0 \text{ in.} - 1.0 \text{ in.})(0.5 \text{ in.})$$
$$= 11{,}000 \text{ lb}$$

(c) The allowable load based upon bearing between the hanger and the bolt is equal to the allowable bearing stress times the bearing area. The bearing area is the projection of the actual contact area, which is equal to the bolt diameter times the thickness of the hanger. Therefore, the allowable load (Eq. 1-28) is

$$P_3 = \sigma_b A = \sigma_b dt = (26{,}000 \text{ psi})(1.0 \text{ in.})(0.5 \text{ in.}) = 13{,}000 \text{ lb}$$

(d) Finally, the allowable load P_4 based upon shear in the bolt is equal to the allowable shear stress times the shear area (Eq. 1-27). The shear area is twice the area of the bolt because the bolt is in double shear; thus:

$$P_4 = \tau_{allow}A = \tau_{allow}(2)(\pi d^2/4) = (6{,}500 \text{ psi})(2)(\pi)(1.0 \text{ in.})^2/4 = 10{,}200 \text{ lb}$$

We have now found the allowable tensile loads in the hanger based upon all four of the given conditions.

Comparing the four preceding results, we see that the smallest value of the load is

$$P_{allow} = 10{,}200 \text{ lb}$$

This load, which is based upon shear in the bolt, is the allowable tensile load in the hanger.

1.8 DESIGN FOR AXIAL LOADS AND DIRECT SHEAR

In the preceding section we discussed the determination of allowable loads for simple structures, and in earlier sections we saw how to find the stresses, strains, and deformations of bars. The determination of such quantities is known as **analysis**. In the context of mechanics of materials, analysis consists of determining the *response* of a structure to loads, temperature changes, and other physical actions. By the response of a structure, we mean the stresses, strains, and deformations produced by the loads.

Response also refers to the load-carrying capacity of a structure; for instance, the allowable load on a structure is a form of response.

A structure is said to be *known* (or *given*) when we have a complete physical description of the structure, that is, when we know all of its *properties*. The properties of a structure include the types of members and how they are arranged, the dimensions of all members, the types of supports and where they are located, the materials used, and the properties of the materials. Thus, when analyzing a structure, *the properties are given and the response is to be determined*.

The inverse process is called **design**. When designing a structure, *we must determine the properties of the structure in order that the structure will support the loads and perform its intended functions*. For instance, a common design problem in engineering is to determine the size of a member to support given loads. Designing a structure is usually a much lengthier and more difficult process than analyzing it—indeed, analyzing a structure, often more than once, is typically part of the design process.

In this section we will deal with design in its most elementary form by calculating the required sizes of simple tension and compression members as well as pins and bolts loaded in shear. In these cases the design process is quite straightforward. Knowing the loads to be transmitted and the allowable stresses in the materials, we can calculate the required areas of members from the following general relationship (compare with Eq. 1-25):

$$\text{Required area} = \frac{\text{Load to be transmitted}}{\text{Allowable stress}} \qquad (1\text{-}29)$$

This equation can be applied to any structure in which the stresses are uniformly distributed over the area. (The use of this equation for finding the size of a bar in tension and the size of a pin in shear is illustrated in Example 1-8, which follows.)

In addition to **strength** considerations, as exemplified by Eq. (1-29), the design of a structure is likely to involve **stiffness** and **stability**. Stiffness refers to the ability of the structure to resist changes in shape (for instance, to resist stretching, bending, or twisting), and stability refers to the ability of the structure to resist buckling under compressive stresses. Limitations on stiffness are sometimes necessary to prevent excessive deformations, such as large deflections of a beam that might interfere with its performance. Buckling is the principal consideration in the design of columns, which are slender compression members (Chapter 9).

Another part of the design process is **optimization**, which is the task of designing the best structure to meet a particular goal, such as minimum weight. For instance, there may be many structures that will support a given load, but in some circumstances the best structure will be the lightest one. Of course, a goal such as minimum weight usually must be balanced against more general considerations, including the aesthetic, economic, environmental, political, and technical aspects of the particular design project.

When analyzing or designing a structure, we refer to the forces that act on it as either **loads** or **reactions**. Loads are *active forces* that are applied to the structure by some external cause, such as gravity, water pressure, wind, amd earthquake ground motion. Reactions are *passive forces* that are induced at the supports of the structure—their magnitudes and directions are determined by the nature of the structure itself. Thus, reactions must be calculated as part of the analysis, whereas loads are known in advance.

Example 1-8, on the following pages, begins with a review of **free-body diagrams** and elementary statics and concludes with the design of a bar in tension and a pin in direct shear.

When drawing free-body diagrams, it is helpful to distinguish reactions from loads or other applied forces. A common scheme is to place a slash, or slanted line, across the arrow when it represents a reactive force, as illustrated in Fig. 1-34 of the following example.

Example 1-8

The two-bar truss *ABC* shown in Fig. 1-33 has pin supports at points *A* and *C*, which are 2.0 m apart. Members *AB* and *BC* are steel bars, pin connected at joint *B*. The length of bar *BC* is 3.0 m. A sign weighing 5.4 kN is suspended from bar *BC* at points *D* and *E*, which are located 0.8 m and 0.4 m, respectively, from the ends of the bar.

Determine the required cross-sectional area of bar *AB* and the required diameter of the pin at support *C* if the allowable stresses in tension and shear are 125 MPa and 45 MPa, respectively. (*Note:* The pins at the supports are in double shear. Also, disregard the weights of members *AB* and *BC*.)

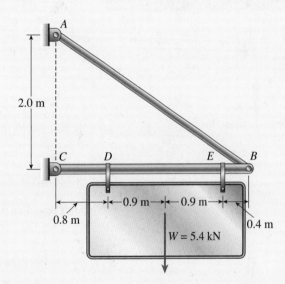

FIG. 1-33 Example 1-8. Two-bar truss *ABC* supporting a sign of weight *W*

Solution

The objectives of this example are to determine the required sizes of bar *AB* and the pin at support *C*. As a preliminary matter, we must determine the tensile force in the bar and the shear force acting on the pin. These quantities are found from free-body diagrams and equations of equilibrium.

Reactions. We begin with a free-body diagram of the entire truss (Fig. 1-34a). On this diagram we show all forces acting on the truss—namely, the loads from the weight of the sign and the reactive forces exerted by the pin supports at *A* and *C*. Each reaction is shown by its horizontal and vertical components, with the resultant reaction shown by a dashed line. (Note the use of slashes across the arrows to distinguish reactions from loads.)

The horizontal component R_{AH} of the reaction at support *A* is obtained by summing moments about point *C*, as follows (counterclockwise moments are positive):

$$\sum M_C = 0 \quad R_{AH}(2.0 \text{ m}) - (2.7 \text{ kN})(0.8 \text{ m}) - (2.7 \text{ kN})(2.6 \text{ m}) = 0$$

continued

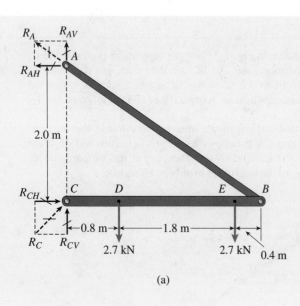

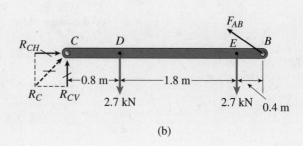

(a)   (b)

FIG. 1-34 Free-body diagrams for
Example 1-8

Solving this equation, we get

$$R_{AH} = 4.590 \text{ kN}$$

Next, we sum forces in the horizontal direction and obtain

$$\sum F_{\text{horiz}} = 0 \qquad R_{CH} = R_{AH} = 4.590 \text{ kN}$$

To obtain the vertical component of the reaction at support C, we may use
a free-body diagram of member BC, as shown in Fig. 1-34b. Summing moments
about joint B gives the desired reaction component:

$$\sum M_B = 0 \qquad -R_{CV}(3.0 \text{ m}) + (2.7 \text{ kN})(2.2 \text{ m}) + (2.7 \text{ kN})(0.4 \text{ m}) = 0$$

$$R_{CV} = 2.340 \text{ kN}$$

Now we return to the free-body diagram of the entire truss (Fig. 1-34a) and
sum forces in the vertical direction to obtain the vertical component R_{AV} of the
reaction at A:

$$\sum F_{\text{vert}} = 0 \qquad R_{AV} + R_{CV} - 2.7 \text{ kN} - 2.7 \text{ kN} = 0$$

$$R_{AV} = 3.060 \text{ kN}$$

As a partial check on these results, we note that the ratio R_{AV}/R_{AH} of the forces acting at point A is equal to the ratio of the vertical and horizontal components of line AB, namely, 2.0 m/3.0 m, or 2/3.

Knowing the horizontal and vertical components of the reaction at A, we can find the reaction itself (Fig. 1-34a):

$$R_A = \sqrt{(R_{AH})^2 + (R_{AV})^2} = 5.516 \text{ kN}$$

Similarly, the reaction at point C is obtained from its componets R_{CH} and R_{CV}, as follows:

$$R_C = \sqrt{(R_{CH})^2 + (R_{CV})^2} = 5.152 \text{ kN}$$

Tensile force in bar AB. Because we are disregarding the weight of bar AB, the tensile force F_{AB} in this bar is equal to the reaction at A (see Fig.1-34):

$$F_{AB} = R_A = 5.516 \text{ kN}$$

Shear force acting on the pin at C. This shear force is equal to the reaction R_C (see Fig. 1-34); therefore,

$$V_C = R_C = 5.152 \text{ kN}$$

Thus, we have now found the tensile force F_{AB} in bar AB and the shear force V_C acting on the pin at C.

Required area of bar. The required cross-sectional area of bar AB is calculated by dividing the tensile force by the allowable stress, inasmuch as the stress is uniformly distributed over the cross section (see Eq. 1-29):

$$A_{AB} = \frac{F_{AB}}{\sigma_{\text{allow}}} = \frac{5.516 \text{ kN}}{125 \text{ MPa}} = 44.1 \text{ mm}^2 \qquad \Longleftarrow$$

Bar AB must be designed with a cross-sectional area equal to or greater than 44.1 mm^2 in order to support the weight of the sign, which is the only load we considered. When other loads are included in the calculations, the required area will be larger.

Required diameter of pin. The required cross-sectional area of the pin at C, which is in double shear, is

$$A_{\text{pin}} = \frac{V_C}{2\tau_{\text{allow}}} = \frac{5.152 \text{ kN}}{2(45 \text{ MPa})} = 57.2 \text{ mm}^2 \qquad \Longleftarrow$$

from which we can calculate the required diameter:

$$d_{\text{pin}} = \sqrt{4A_{\text{pin}}/\pi} = 8.54 \text{ mm} \qquad \Longleftarrow$$

continued

A pin of at least this diameter is needed to support the weight of the sign without exceeding the allowable shear stress.

Notes: In this example we intentionally omitted the weight of the truss from the calculations. However, once the sizes of the members are known, their weights can be calculated and included in the free-body diagrams of Fig. 1-34.

When the weights of the bars are included, the design of member *AB* becomes more complicated, because it is no longer a bar in simple tension. Instead, it is a beam subjected to bending as well as tension. An analogous situation exists for member *BC*. Not only because of its own weight but also because of the weight of the sign, member *BC* is subjected to both bending and compression. The design of such members must wait until we study stresses in beams (Chapter 5).

In practice, other loads besides the weights of the truss and sign would have to be considered before making a final decision about the sizes of the bars and pins. Loads that could be important include wind loads, earthquake loads, and the weights of objects that might have to be supported temporarily by the truss and sign.

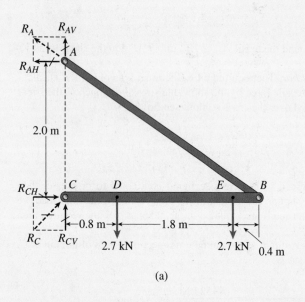

(a)

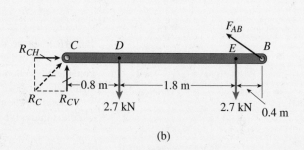

(b)

FIG. 1-34 (Repeated)

CHAPTER SUMMARY & REVIEW

In Chapter 1 we learned about mechanical properties of construction materials. We computed normal stresses and strains in bars loaded by centroidal axial loads, and also shear stresses and strains (as well as bearing stresses) in pin connections used to assemble simple structures, such as trusses. We also defined allowable levels of stress from appropriate factors of safety and used these values to set allowable loads that could be applied to the structure.

Some of the major concepts presented in this chapter are as follows.

1. The principal objective of mechanics of materials is to determine the **stresses**, **strains, and displacements** in structures and their components due to the loads acting on them. These components include bars with axial loads, shafts in torsion, beams in bending, and colums in compression.

2. Prismatic bars subjected to tensile or compressive loads acting through the centroid of their cross section (to avoid bending) experience **normal stress** (σ) **and strain** (ϵ)

$$\sigma = \frac{P}{A}$$

$$\epsilon = \frac{\delta}{L}$$

   and either extension or contraction proportional to their lengths. These stresses and strains are **uniform** except near points of load application where high localized stresses, or **stress-concentrations**, occur.

3. We investigated the **mechanical behavior** of various materials and plotted the resulting stress-strain diagram, which conveys important information about the material. **Ductile** materials (such as mild steel) have an initial linear relationship between normal stress and strain (up to the **proportional limit**) and are said to be **linearly elastic** with stress and strain related by **Hooke's law**

$$\sigma = E\epsilon$$

   they also have a well-defined yield point. Other ductile materials (such as aluminum alloys) typically do not have a clearly definable yield point, so an arbitrary yield stress may be determined by using the **offset method**.

4. Materials that fail in tension at relatively low values of strain (such as concrete, stone, cast iron, glass ceramics and a variety of metallic alloys) are classified as **brittle**. Brittle materials fail with only little elongation after the proportional limit.

5. If the material remains within the elastic range, it can be loaded, unloaded, and loaded again without significantly changing the behavior. However when loaded into the plastic range, the internal structure of the material is altered and its properties change. Loading and unloading behavior of materials depends on the **elasticity** and **plasticity** properties of the material, such as the **elastic limit** and possibility of **permanent set** (residual strain) in the material. Sustained loading over time may lead to creep and **relaxation**.

6. Axial elongation of bars loaded in tension is accompanied by lateral contraction; the ratio of lateral strain to normal strain is known as **Poisson's ratio** (ν).

$$\nu = -\frac{\text{lateral strain}}{\text{axial strain}} = -\frac{\epsilon'}{\epsilon}$$

continued

Poisson's ratio remains constant throughout the linearly elastic range, provided the material is homogeneous and isotropic. Most of the examples and problems in the text are solved with the assumption that the material is linearly elastic, homogeneous, and isotropic.

7. **Normal** stresses (σ) act perpendicular to the surface of the material and **shear stresses** (τ) act tangential to the surface. We investigated bolted connections between plates in which the bolts were subjected to either average single or double shear (τ_{aver}) where

$$\tau_{aver} = \frac{V}{A}$$

as well as average **bearing** stresses (σ_b). The bearing stresses act on the rectangular projected area (A_b) of the actual curved contact surface between a bolt and plate.

$$\sigma_b = \frac{F_b}{A_b}$$

8. We looked at an element of material acted on by shear stresses and strains to study a state of stress referred to as **pure shear**. We saw that shear strain (γ) is a measure of the distortion or change in shape of the element in pure shear. We looked at Hooke's law in shear in which shear stress (τ) is related to shear strain by the shearing modulus of elasticity G.

$$\tau = G\gamma$$

We noted that E and G are related and therefore are not independent elastic properties of the material.

$$G = \frac{E}{2(1 + \nu)}$$

9. **Strength** is the capacity of a structure or component to support or transmit loads. **Factors of safety** relate actual to required strength of structural members and account for a variety of uncertainties, such as variations in material properties, uncertain magnitudes or distributions of loadings, probability of accidental overload, and so on. Because of these uncertainties, factors of safety (n_1, n_2, n_3, n_4) must be determined using probabilistic methods.

10. Yield or ultimate level stresses can be divided by factors of safety to produce allowable values for use in design. For **ductile** materials,

$$\sigma_{allow} = \frac{\sigma_Y}{n_1}, \quad \tau_{allow} = \frac{\tau_Y}{n_2}$$

while for **brittle** materials,

$$\sigma_{allow} = \frac{\sigma_U}{n_3}, \quad \tau_{allow} = \frac{\tau_U}{n_4}.$$

A typical value of n_1 and n_2 is 1.67 while n_3 and n_4 might be 2.8.

For a pin-connected member in axial **tension**, the allowable load depends on the allowable stress times the appropriate area (e.g., net cross-sectional area for bars acted on by centroidal tensile loads, cross-sectional area of pin for pins in shear, and projected area for bolts in bearing). If the bar is in **compression**, net cross-sectional area need not be used, but buckling may be an important consideration.

11. Lastly, we considered **design**, the iterative process by which the appropriate size of structural members is determined to meet a variety of both **strength and stiffness requirements** for a particular structure subjected to a variety of different loadings. However, incorporation of factors of safety into design is not a simple matter, because both strength and failure have many different meanings.

PROBLEMS CHAPTER 1

Normal Stress and Strain

1.2-1 A hollow circular post ABC (see figure) supports a load $P_1 = 1700$ lb acting at the top. A second load P_2 is uniformly distributed around the cap plate at B. The diameters and thicknesses of the upper and lower parts of the post are $d_{AB} = 1.25$ in., $t_{AB} = 0.5$ in., $d_{BC} = 2.25$ in., and $t_{BC} = 0.375$ in., respectively.

(a) Calculate the normal stress σ_{AB} in the upper part of the post.

(b) If it is desired that the lower part of the post have the same compressive stress as the upper part, what should be the magnitude of the load P_2?

(c) If P_1 remains at 1700 lb and P_2 is now set at 2260 lb, what new thickness of BC will result in the same compressive stress in both parts?

1.2-2 A force P of 70 N is applied by a rider to the front hand brake of a bicycle (P is the resultant of an evenly distributed pressure). As the hand brake pivots at A, a tension T develops in the 460-mm long brake cable ($A_e = 1.075$ mm^2) which elongates by $\delta = 0.214$ mm. Find normal stress σ and strain ε in the brake cable.

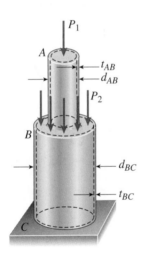

PROB. 1.2-1

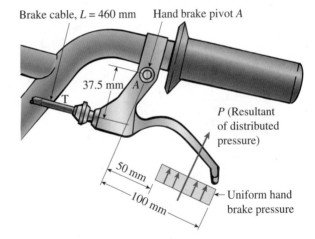

PROB. 1.2-2

1.2-3 A bicycle rider would like to compare the effectiveness of cantilever hand brakes [see figure part (a)] versus V brakes [figure part (b)].

(a) Calculate the braking force R_B at the wheel rims for each of the bicycle brake systems shown. Assume that all forces act in the plane of the figure and that cable tension $T = 45$ lbs. Also, what is the average compressive normal stress σ_c on the brake pad ($A = 0.625$ in^2)?

(b) For each braking system, what is the stress in the brake cable (assume effective cross-sectional area of 0.00167 in^2)?

(*HINT*: Because of symmetry, you only need to use the right half of each figure in your analysis.)

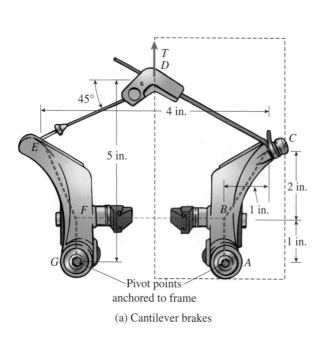

(a) Cantilever brakes

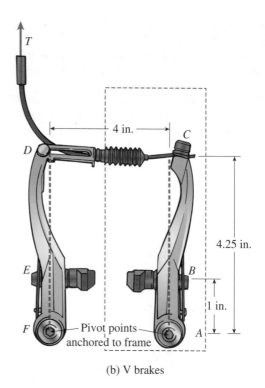

(b) V brakes

PROB. 1.2-3

1.2-4 A circular aluminum tube of length $L = 400$ mm is loaded in compression by forces P (see figure). The outside and inside diameters are 60 mm and 50 mm, respectively. A strain gage is placed on the outside of the bar to measure normal strains in the longitudinal direction.

(a) If the measured strain is $\epsilon = 550 \times 10^{-6}$, what is the shortening δ of the bar?

(b) If the compressive stress in the bar is intended to be 40 MPa, what should be the load P?

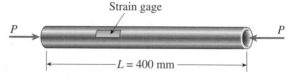

PROB. 1.2-4

1.2-5 The cross section of a concrete corner column that is loaded uniformly in compression is shown in the figure.

(a) Determine the average compressive stress σ_c in the concrete if the load is equal to 3200 k.

(b) Determine the coordinates x_c and y_c of the point where the resultant load must act in order to produce uniform normal stress in the column.

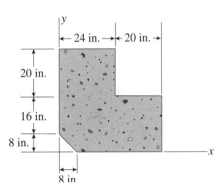

PROB. 1.2-5

1.2-6 A car weighing 130 kN when fully loaded is pulled slowly up a steep inclined track by a steel cable (see figure). The cable has an effective cross-sectional area of 490 mm², and the angle α of the incline is 30°.

Calculate the tensile stress σ_t in the cable.

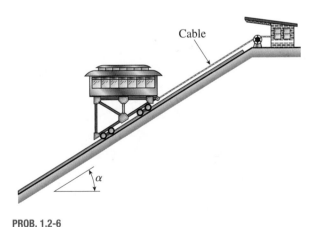

PROB. 1.2-6

1.2-7 Two steel wires support a moveable overhead camera weighing $W = 25$ lb (see figure) used for close-up viewing of field action at sporting events. At some instant, wire 1 is at an angle $\alpha = 20°$ to the horizontal and wire 2 is at an angle

$\beta = 48°$. Both wires have a diameter of 30 mils. (Wire diameters are often expressed in mils; one mil equals 0.001 in.)

Determine the tensile stresses σ_1 and σ_2 in the two wires.

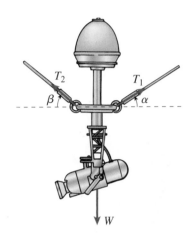

PROB. 1.2-7

1.2-8 A long retaining wall is braced by wood shores set at an angle of 30° and supported by concrete thrust blocks, as shown in the first part of the figure. The shores are evenly spaced, 3 m apart.

For analysis purposes, the wall and shores are idealized as shown in the second part of the figure. Note that the base of the wall and both ends of the shores are assumed to be pinned. The pressure of the soil against the wall is assumed to be triangularly distributed, and the resultant force acting on a 3-meter length of the wall is $F = 190$ kN.

If each shore has a 150 mm × 150 mm square cross section, what is the compressive stress σ_c in the shores?

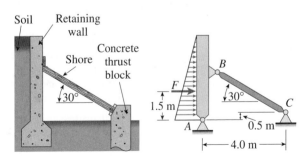

PROB. 1.2-8

1.2-9 A pickup truck tailgate supports a crate ($W_C = 150$ lb), as shown in the figure. The tailgate weighs $W_T = 60$ lb and is supported by two cables (only one is shown in the figure). Each cable has an effective cross-sectional area $A_e = 0.017$ in^2.

(a) Find the tensile force T and normal stress σ in each cable.

(b) If each cable elongates $\delta = 0.01$ in. due to the weight of both the crate and the tailgate, what is the average strain in the cable?

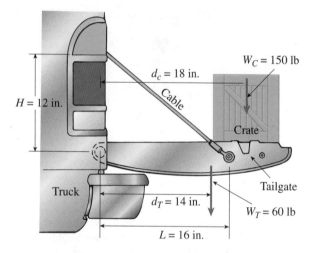

(© Barry Goodno)

PROBS. 1.2-9 and 1.2-10

1.2-10 Solve the preceding problem if the mass of the tailgate is $M_T = 27$ kg and that of the crate is $M_C = 68$ kg. Use dimensions $H = 305$ mm, $L = 406$ mm, $d_C = 460$ mm, and $d_T = 350$ mm. The cable cross-sectional area is $A_e = 11.0$ mm^2.

(a) Find the tensile force T and normal stress σ in each cable.

(b) If each cable elongates $\delta = 0.25$ mm due to the weight of both the crate and the tailgate, what is the average strain in the cable?

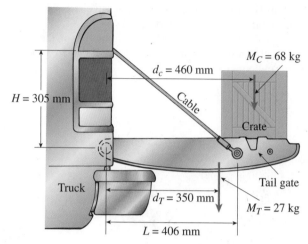

PROB. 1.2-10

1.2-11 An L-shaped reinforced concrete slab 12 ft $\times$ 12 ft (but with a 6 ft $\times$ 6 ft cutout) and thickness $t = 9.0$ in, is lifted by three cables attached at O, B and D, as shown in the figure. The cables are combined at point Q, which is 7.0 ft above the top of the slab and directly above the center of mass at C. Each cable has an effective cross-sectional area of $A_e = 0.12$ in^2.

(a) Find the tensile force T_i ($i = 1, 2, 3$) in each cable due to the weight W of the concrete slab (ignore weight of cables).

(b) Find the average stress σ_i in each cable. (See Table I-1 in Appendix I, available online, for the weight density of reinforced concrete.)

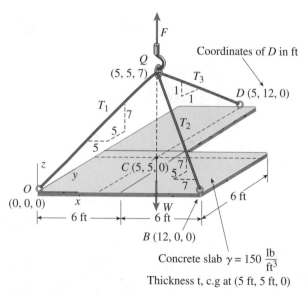

PROB. 1.2-11

1.2-12 A round bar ACB of length $2L$ (see figure) rotates about an axis through the midpoint C with constant angular speed ω (radians per second). The material of the bar has weight density γ.

(a) Derive a formula for the tensile stress σ_x in the bar as a function of the distance x from the midpoint C.

(b) What is the maximum tensile stress σ_{max}?

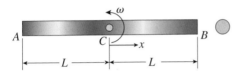

PROB. 1.2-12

1.2-13 Two gondolas on a ski lift are locked in the position shown in the figure while repairs are being made elsewhere. The distance between support towers is $L = 100$ ft. The length of each cable segment under gondola weights $W_B = 450$ lb and $W_C = 650$ lb are $D_{AB} = 12$ ft, $D_{BC} = 70$ ft, and $D_{CD} = 20$ ft. The cable sag at B is $\Delta_B = 3.9$ ft and that at $C(\Delta_C)$ is 7.1 ft. The effective cross-sectional area of the cable is $A_e = 0.12$ in.^{2}.

(a) Find the tension force in each cable segment; neglect the mass of the cable.

(b) Find the average stress (σ) in each cable segment.

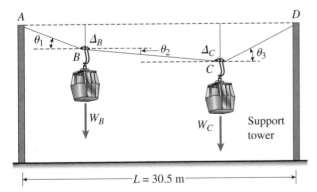

PROB. 1.2-13

1.2-14 A crane boom of mass 450 kg with its center of mass at C is stabilized by two cables AQ and BQ ($A_e = 304$ mm^2 for each cable) as shown in the figure. A load $P = 20$ kN is supported at point D. The crane boom lies in the y–z plane.

(a) Find the tension forces in each cable: T_{AQ} and T_{BQ} (kN); neglect the mass of the cables, but include the mass of the boom in addition to load P.

(b) Find the average stress (σ) in each cable.

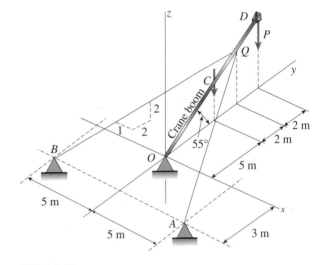

PROB. 1.2-14

Mechanical Properties and Stress-Strain Diagrams

1.3-1 Imagine that a long steel wire hangs vertically from a high-altitude balloon.

(a) What is the greatest length (feet) it can have without yielding if the steel yields at 40 ksi?

(b) If the same wire hangs from a ship at sea, what is the greatest length? (Obtain the weight densities of steel and sea water from Table I-1, Appendix I available online.)

1.3-2 Imagine that a long wire of tungsten hangs vertically from a high-altitude balloon.

(a) What is the greatest length (meters) it can have without breaking if the ultimate strength (or breaking strength) is 1500 MPa?

(b) If the same wire hangs from a ship at sea, what is the greatest length? (Obtain the weight densities of tungsten and sea water from Table I-1, Appendix I available online.)

1.3-3 Three different materials, designated *A, B,* and *C,* are tested in tension using test specimens having diameters of 0.505 in. and gage lengths of 2.0 in. (see figure). At failure, the distances between the gage marks are found to be 2.13, 2.48, and 2.78 in., respectively. Also, at the failure cross sections the diameters are found to be 0.484, 0.398, and 0.253 in., respectively.

Determine the percent elongation and percent reduction in area of each specimen, and then, using your own judgment, classify each material as brittle or ductile.

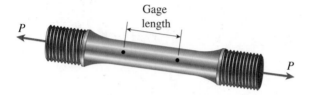

PROB. 1.3-3

1.3-4 The *strength-to-weight ratio* of a structural material is defined as its load-carrying capacity divided by its weight. For materials in tension, we may use a characteristic tensile stress (as obtained from a stress-strain curve) as a measure of strength. For instance, either the yield stress or the ultimate stress could be used, depending upon the particular application. Thus, the strength-to-weight ratio $R_{S/W}$ for a material in tension is defined as

$$R_{S/W} = \frac{\sigma}{\gamma}$$

in which σ is the characteristic stress and γ is the weight density. Note that the ratio has units of length.

Using the ultimate stress σ_U as the strength parameter, calculate the strength-to-weight ratio (in units of meters) for each of the following materials: aluminum alloy 6061-T6, Douglas fir (in bending), nylon, structural steel ASTM-A572, and a titanium alloy. (Obtain the material properties from Tables I-1 and I-3 of Appendix I available online. When a range of values is given in a table, use the average value.)

1.3-5 A symmetrical framework consisting of three pin-connected bars is loaded by a force *P* (see figure). The angle between the inclined bars and the horizontal is $\alpha = 48°$. The axial strain in the middle bar is measured as 0.0713.

Determine the tensile stress in the outer bars if they are constructed of aluminum alloy having the stress-strain diagram shown in Fig. 1-13. (Express the stress in USCS units.)

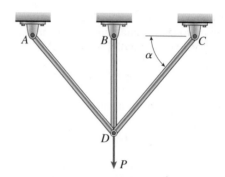

PROB. 1.3-5

1.3-6 A specimen of a methacrylate plastic is tested in tension at room temperature (see figure), producing the stress-strain data listed in the accompanying table (see the next page).

Plot the stress-strain curve and determine the proportional limit, modulus of elasticity (i.e., the slope of the initial part of the stress-strain curve), and yield stress at 0.2% offset. Is the material ductile or brittle?

PROB. 1.3-6

STRESS-STRAIN DATA FOR PROBLEM 1.3-6	
Stress (MPa)	Strain
8.0	0.0032
17.5	0.0073
25.6	0.0111
31.1	0.0129
39.8	0.0163
44.0	0.0184
48.2	0.0209
53.9	0.0260
58.1	0.0331
62.0	0.0429
62.1	Fracture

1.3-7 The data shown in the accompanying table were obtained from a tensile test of high-strength steel. The test specimen had a diameter of 0.505 in. and a gage length of 2.00 in. (see figure for Prob. 1.3-3). At fracture, the elongation between the gage marks was 0.12 in. and the minimum diameter was 0.42 in.

Plot the conventional stress-strain curve for the steel and determine the proportional limit, modulus of elasticity (i.e., the slope of the initial part of the stress-strain curve), yield stress at 0.1% offset, ultimate stress, percent elongation in 2.00 in., and percent reduction in area.

TENSILE-TEST DATA FOR PROBLEM 1.3-7

Load (lb)	Elongation (in.)
1,000	0.0002
2,000	0.0006
6,000	0.0019
10,000	0.0033
12,000	0.0039
12,900	0.0043
13,400	0.0047
13,600	0.0054
13,800	0.0063
14,000	0.0090
14,400	0.0102
15,200	0.0130
16,800	0.0230
18,400	0.0336
20,000	0.0507
22,400	0.1108
22,600	Fracture

Elasticity and Plasticity

1.4-1 A bar made of structural steel having the stress-strain diagram shown in the figure has a length of 48 in. The yield stress of the steel is 42 ksi and the slope of the initial linear part of the stress-strain curve (modulus of elasticity) is 30×10^3 ksi. The bar is loaded axially until it elongates 0.20 in., and then the load is removed.

How does the final length of the bar compare with its original length? (*Hint:* Use the concepts illustrated in Fig. 1-18b.)

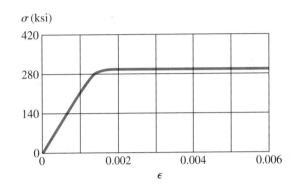

PROB. 1.4-1

1.4-2 A bar of length 2.0 m is made of a structural steel having the stress-strain diagram shown in the figure. The yield stress of the steel is 250 MPa and the slope of the initial linear part of the stress-strain curve (modulus of elasticity) is 200 GPa. The bar is loaded axially until it elongates 6.5 mm, and then the load is removed.

How does the final length of the bar compare with its original length? (*Hint:* Use the concepts illustrated in Fig. 1-18b.)

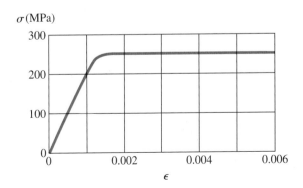

PROB. 1.4-2

1.4-3 An aluminum bar has length $L = 5$ ft and diameter $d = 1.25$ in. The stress-strain curve for the aluminum is shown in Fig. 1-13 of Section 1.3. The initial straight-line part of the curve has a slope (modulus of elasticity) of 10×10^6 psi. The bar is loaded by tensile forces $P = 39$ k and then unloaded.

(a) What is the permanent set of the bar?

(b) If the bar is reloaded, what is the proportional limit? (*Hint*: Use the concepts illustrated in Figs. 1-18b and 1-19.)

1.4-4 A circular bar of magnesium alloy is 750 mm long. The stress-strain diagram for the material is shown in the figure. The bar is loaded in tension to an elongation of 6.0 mm, and then the load is removed.

(a) What is the permanent set of the bar?

(b) If the bar is reloaded, what is the proportional limit? (*Hint*: Use the concepts illustrated in Figs. 1-18b and 1-19.)

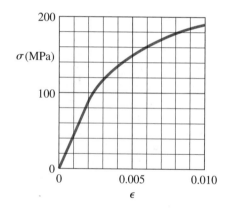

PROBS. 1.4-3 and 1.4-4

1.4-5 A wire of length $L = 4$ ft and diameter $d = 0.125$ in. is stretched by tensile forces $P = 600$ lb. The wire is made of a copper alloy having a stress-strain relationship that may be described mathematically by the following equation:

$$\sigma = \frac{18,000\epsilon}{1 + 300\epsilon} \quad 0 \leq \epsilon \leq 0.03 \quad (\sigma = \text{ksi})$$

in which ϵ is nondimensional and σ has units of kips per square inch (ksi).

(a) Construct a stress-strain diagram for the material.

(b) Determine the elongation of the wire due to the forces P.

(c) If the forces are removed, what is the permanent set of the bar?

(d) If the forces are applied again, what is the proportional limit?

Hooke's Law and Poisson's Ratio

When solving the problems for Section 1.5, assume that the material behaves linearly elastically.

1.5-1 A high-strength steel bar used in a large crane has diameter $d = 2.00$ in. (see figure). The steel has modulus of elasticity $E = 29 \times 10^6$ psi and Poisson's ratio $\nu = 0.29$. Because of clearance requirements, the diameter of the bar is limited to 2.001 in. when it is compressed by axial forces.

What is the largest compressive load P_{max} that is permitted?

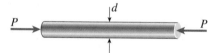

PROB. 1.5-1

1.5-2 A round bar of 10 mm diameter is made of aluminum alloy 7075-T6 (see figure). When the bar is stretched by axial forces P, its diameter decreases by 0.016 mm.

Find the magnitude of the load P. (Obtain the material properties from Appendix I available online.)

PROB. 1.5-2

1.5-3 A polyethylene bar having diameter $d_1 = 4.0$ in. is placed inside a steel tube having inner diameter $d_2 = 4.01$ in. (see figure). The polyethylene bar is then compressed by an axial force P.

At what value of the force P will the space between the polyethylene bar and the steel tube be closed? (For polyethylene, assume $E = 200$ ksi and $v = 0.4$.)

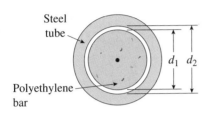

Steel tube

Polyethylene bar

d_1 d_2

PROB. 1.5-3

1.5-4 A prismatic bar with a circular cross section is loaded by tensile forces $P = 65$ kN (see figure). The bar has length $L = 1.75$ m and diameter $d = 32$ mm. It is made of aluminum alloy with modulus of elasticity $E = 75$ GPa and Poisson's ratio $v = 1/3$.

Find the increase in length of the bar and the percent decrease in its cross-sectional area.

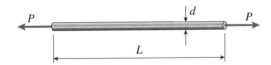

P d P

L

PROBS. 1.5-4 and 1.5-5

1.5-5 A bar of monel metal as in the figure (length $L = 9$ in., diameter $d = 0.225$ in.) is loaded axially by a tensile force P. If the bar elongates by 0.0195 in., what is the decrease in diameter d? What is the magnitude of the load P? Use the data in Table I-2, Appendix I available online.

1.5-6 A tensile test is peformed on a brass specimen 10 mm in diameter using a gage length of 50 mm (see figure). When the tensile load P reaches a value of 20 kN, the distance between the gage marks has increased by 0.122 mm.

(a) What is the modulus of elasticity E of the brass?
(b) If the diameter decreases by 0.00830 mm, what is Poisson's ratio?

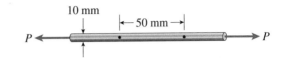

10 mm   50 mm

P P

PROB. 1.5-6

1.5-7 A hollow, brass circular pipe ABC (see figure) supports a load $P_1 = 26.5$ kips acting at the top. A second load $P_2 = 22.0$ kips is uniformly distributed around the cap plate at B. The diameters and thicknesses of the upper and lower parts of the pipe are $d_{AB} = 1.25$ in., $t_{AB} = 0.5$ in., $d_{BC} = 2.25$ in., and $t_{BC} = 0.375$ in., respectively. The modulus of elasticity is 14,000 ksi. When both loads are fully applied, the wall thickness of pipe BC increases by 200×10^{-6} in.

(a) Find the increase in the inner diameter of pipe segment BC.
(b) Find Poisson's ratio for the brass.
(c) Find the increase in the wall thickness of pipe segment AB and the increase in the inner diameter of AB.

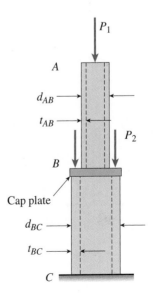

P_1

A

d_{AB}

t_{AB}

P_2

B

Cap plate

d_{BC}

t_{BC}

C

PROB. 1.5-7

1.5-8 A brass bar of length 2.25 m with a square cross section of 90 mm on each side is subjected to an axial tensile force of 1500 kN (see figure). Assume that $E = 110$ GPa and $\nu = 0.34$.

Determine the increase in volume of the bar.

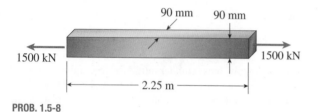

PROB. 1.5-8

Shear Stress and Strain

1.6-1 An angle bracket having thickness $t = 0.75$ in. is attached to the flange of a column by two 5/8-inch diameter bolts (see figure). A uniformly distributed load from a floor joist acts on the top face of the bracket with a pressure $p = 275$ psi. The top face of the bracket has length $L = 8$ in. and width $b = 3.0$ in.

Determine the average bearing pressure σ_b between the angle bracket and the bolts and the average shear stress τ_{aver} in the bolts. (Disregard friction between the bracket and the column.)

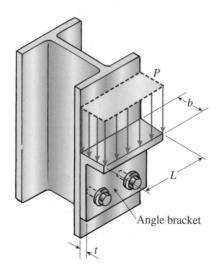

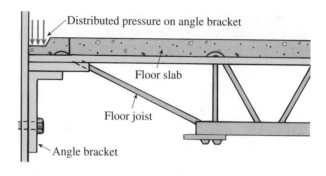

PROB. 1.6-1

1.6-2 Truss members supporting a roof are connected to a 26-mm-thick gusset plate by a 22 mm diameter pin as shown in the figure and photo. The two end plates on the truss members are each 14 mm thick.

(a) If the load $P = 80$ kN, what is the largest bearing stress acting on the pin?

(b) If the ultimate shear stress for the pin is 190 MPa, what force P_{ult} is required to cause the pin to fail in shear? (Disregard friction between the plates.)

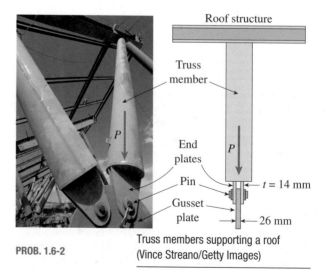

PROB. 1.6-2

Truss members supporting a roof
(Vince Streano/Getty Images)

1.6-3 The upper deck of a football stadium is supported by braces each of which transfers a load $P = 160$ kips to the base of a column [see figure part (a)]. A cap plate at the bottom of the brace distributes the load P to four flange plates ($t_f = 1$ in.) through a pin ($d_p = 2$ in.) to two gusset plates ($t_g = 1.5$ in.) [see figure parts (b) and (c)].

Determine the following quantities.
(a) The average shear stress τ_{aver} in the pin.
(b) The average bearing stress between the flange plates and the pin (σ_{bf}), and also between the gusset plates and the pin (σ_{bg}).
(Disregard friction between the plates.)

(b) Detail at bottom of brace
(© Barry Goodno)

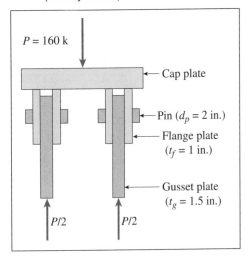

(c) Section through bottom of brace
(© Barry Goodno)

(a) Stadium brace
PROB. 1.6-3 (© Barry Goodno)

1.6-4 The inclined ladder AB supports a house painter (82 kg) at C and the self weight ($q = 36$ N/m) of the ladder itself. Each ladder rail ($t_r = 4$ mm) is supported by a shoe ($t_s = 5$ mm) which is attached to the ladder rail by a bolt of diameter $d_p = 8$ mm.

(a) Find support reactions at A and B.

(b) Find the resultant force in the shoe bolt at A.

(c) Find maximum average shear (τ) and bearing (σ_b) stresses in the shoe bolt at A.

Use dimensions shown in the figure. Neglect the weight of the brake system.

(a) Find the average shear stress τ_{aver} in the pivot pin where it is anchored to the bicycle frame at B.

(b) Find the average bearing stress $\sigma_{b,aver}$ in the pivot pin over segment AB.

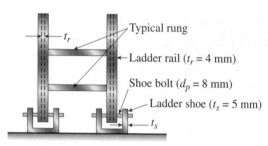

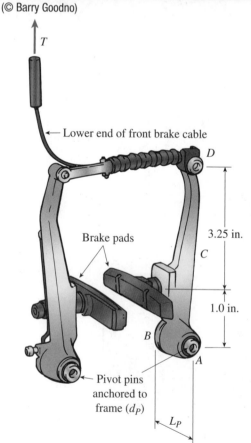

(© Barry Goodno)

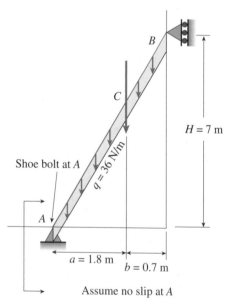

PROB. 1.6-4

1.6-5 The force in the brake cable of the V-brake system shown in the figure is $T = 45$ lb. The pivot pin at A has diameter $d_p = 0.25$ in. and length $L_p = 5/8$ in.

PROB. 1.6-5

1.6-6 A steel plate of dimensions $2.5 \times 1.5 \times 0.08$ m and weighing 23.1 kN is hoisted by steel cables with lengths $L_1 = 3.2$ m and $L_2 = 3.9$ m that are each attached to the plate by a clevis and pin (see figure). The pins through the clevises are 18 mm in diameter and are located 2.0 m apart. The orientation angles are measured to be $\theta = 94.4°$ and $\alpha = 54.9°$.

For these conditions, first determine the cable forces T_1 and T_2, then find the average shear stress τ_{aver} in both pin 1 and pin 2, and then the average bearing stress σ_b between the steel plate and each pin. Ignore the mass of the cables.

(c) Determine the average shear stress τ_{aver} in the nut and also in the steel plate.

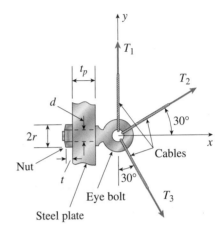

PROB. 1.6-7

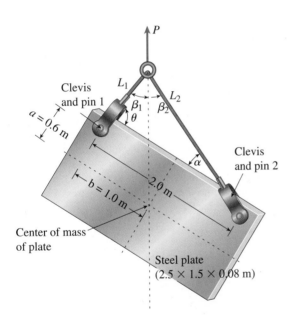

PROB. 1.6-6

1.6-8 An elastomeric bearing pad consisting of two steel plates bonded to a chloroprene elastomer (an artificial rubber) is subjected to a shear force V during a static loading test (see figure). The pad has dimensions $a = 125$ mm and $b = 240$ mm, and the elastomer has thickness $t = 50$ mm. When the force V equals 12 kN, the top plate is found to have displaced laterally 8.0 mm with respect to the bottom plate.

What is the shear modulus of elasticity G of the chloroprene?

1.6-7 A special-purpose eye bolt of shank diameter $d = 0.50$ in. passes through a hole in a steel plate of thickness $t_p = 0.75$ in. (see figure) and is secured by a nut with thickness $t = 0.25$ in. The hexagonal nut bears directly against the steel plate. The radius of the circumscribed circle for the hexagon is $r = 0.40$ in. (which means that each side of the hexagon has length 0.40 in.). The tensile forces in three cables attached to the eye bolt are $T_1 = 800$ lb., $T_2 = 550$ lb., and $T_3 = 1241$ lb.

(a) Find the resultant force acting on the eye bolt.

(b) Determine the average bearing stress σ_b between the hexagonal nut on the eye bolt and the plate.

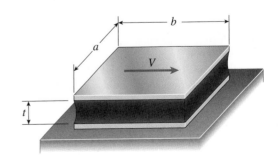

PROB. 1.6-8

1.6-9 A joint between two concrete slabs A and B is filled with a flexible epoxy that bonds securely to the concrete (see figure). The height of the joint is $h = 4.0$ in., its length is $L = 40$ in., and its thickness is $t = 0.5$ in. Under the action of shear forces V, the slabs displace vertically through the distance $d = 0.002$ in. relative to each other.

(a) What is the average shear strain γ_{aver} in the epoxy?

(b) What is the magnitude of the forces V if the shear modulus of elasticity G for the epoxy is 140 ksi?

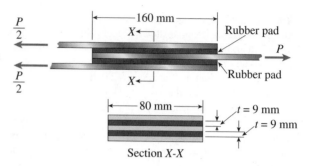

Section X-X

PROB. 1.6-10

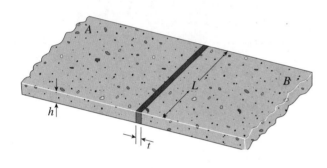

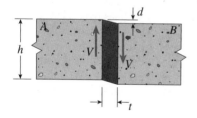

PROB. 1.6-9

1.6-10 A flexible connection consisting of rubber pads (thickness $t = 9$ mm) bonded to steel plates is shown in the figure. The pads are 160 mm long and 80 mm wide.

(a) Find the average shear strain γ_{aver} in the rubber if the force $P = 16$ kN and the shear modulus for the rubber is $G = 1250$ kPa.

(b) Find the relative horizontal displacement δ between the interior plate and the outer plates.

1.6-11 A spherical fiberglass buoy used in an underwater experiment is anchored in shallow water by a chain [see part (a) of the figure]. Because the buoy is positioned just below the surface of the water, it is not expected to collapse from the water pressure. The chain is attached to the buoy by a shackle and pin [see part (b) of the figure]. The diameter of the pin is 0.5 in. and the thickness of the shackle is 0.25 in. The buoy has a diameter of 60 in. and weighs 1800 lb on land (not including the weight of the chain).

(a) Determine the average shear stress τ_{aver} in the pin.

(b) Determine the average bearing stress σ_b between the pin and the shackle.

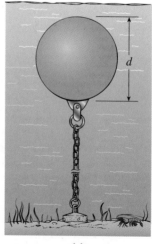

(a)

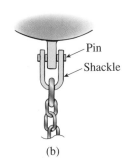

(b)

PROB. 1.6-11

1.6-12 The clamp shown in the figure is used to support a load hanging from the lower flange of a steel beam. The clamp consists of two arms (A and B) joined by a pin at C. The pin has diameter $d = 12$ mm. Because arm B straddles arm A, the pin is in double shear.

Line 1 in the figure defines the line of action of the resultant horizontal force H acting between the lower flange of the beam and arm B. The vertical distance from this line to the pin is $h = 250$ mm. Line 2 defines the line of action of the resultant vertical force V acting between the flange and arm B. The horizontal distance from this line to the centerline of the beam is $c = 100$ mm. The force conditions between arm A and the lower flange are symmetrical with those given for arm B.

Determine the average shear stress in the pin at C when the load $P = 18$ kN.

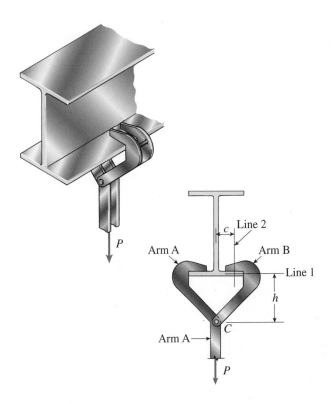

PROB. 1.6-12

1.6-13 A hitch-mounted bicycle rack is designed to carry up to four 30-lb. bikes mounted on and strapped to two arms GH [see bike loads in the figure part (a)]. The rack is attached to the vehicle at A and is assumed to be like a cantilever beam $ABCDGH$ [figure part (b)]. The weight of fixed segment AB is $W_1 = 10$ lb, centered 9 in. from A [see the figure part (b)] and the rest of the rack weighs $W_2 = 40$ lb, centered 19 in. from A. Segment $ABCDG$ is a steel tube, 2×2 in., of thickness $t = 1/8$ in. Segment $BCDGH$ pivots about a bolt at B of diameter $d_B = 0.25$ in. to allow access to the rear of the vehicle without removing the hitch rack. When in use, the rack is secured in an upright position by a pin at C (diameter of pin $d_p = 5/16$ in.) [see photo and figure part (c)]. The overturning effect of the bikes on the rack is equivalent to a force couple $F \cdot h$ at BC.

(a) Find the support reactions at A for the fully loaded rack.

(b) Find forces in the bolt at B and the pin at C.

(c) Find average shear stresses τ_{aver} in both the bolt at B and the pin at C.

(d) Find average bearing stresses σ_b in the bolt at B and the pin at C.

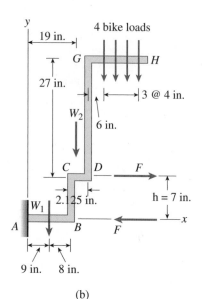

(b)

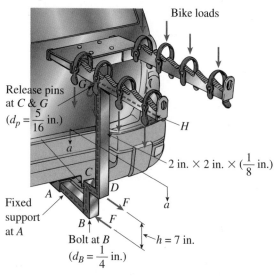

(a)

(© Barry Goodno)

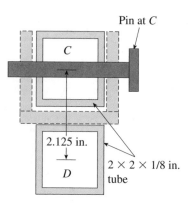

(c) Section a–a

PROB. 1.6-13

1.6-14 A bicycle chain consists of a series of small links, each 12 mm long between the centers of the pins (see figure). You might wish to examine a bicycle chain and observe its construction. Note particularly the pins, which we will assume to have a diameter of 2.5 mm.

In order to solve this problem, you must now make two measurements on a bicycle (see figure): (1) the length L of the crank arm from main axle to pedal axle, and (2) the radius R of the sprocket (the toothed wheel, sometimes called the chainring).

(a) Using your measured dimensions, calculate the tensile force T in the chain due to a force $F = 800$ N applied to one of the pedals.

(b) Calculate the average shear stress τ_{aver} in the pins.

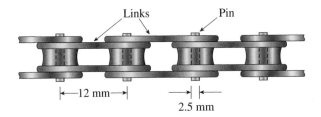

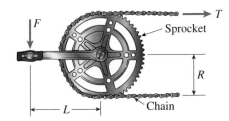

PROB. 1.6-14

1.6-15 A shock mount constructed as shown in the figure is used to support a delicate instrument. The mount consists of an outer steel tube with inside diameter b, a central steel bar of diameter d that supports the load P, and a hollow rubber cylinder (height h) bonded to the tube and bar.

(a) Obtain a formula for the shear stress τ in the rubber at a radial distance r from the center of the shock mount.

(b) Obtain a formula for the downward displacement δ of the central bar due to the load P, assuming that G is the shear modulus of elasticity of the rubber and that the steel tube and bar are rigid.

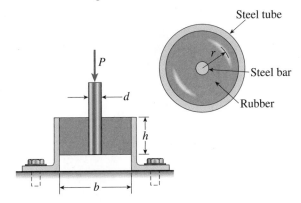

PROB. 1.6-15

1.6-16 The steel plane truss shown in the figure is loaded by three forces P, each of which is 490 kN. The truss members each have a cross-sectional area of 3900 mm^2 and are connected by pins each with a diameter of $d_p = 18$ mm. Members AC and BC each consist of one bar with thickness of $t_{AC} = t_{BC} = 19$ mm. Member AB is composed of two bars [see figure part (b)] each having thickness

$t_{AB}/2 = 10$ mm and length $L = 3$ m. The roller support at B, is made up of two support plates, each having thickness $t_{sp}/2 = 12$ mm.

(a) Find support reactions at joints A and B and forces in members AB, BC, and AB.

(b) Calculate the largest average shear stress $\tau_{p,\max}$ in the pin at joint B, disregarding friction between the members; see figures parts (b) and (c) for sectional views of the joint.

(c) Calculate the largest average bearing stress $\sigma_{b,\max}$ acting against the pin at joint B.

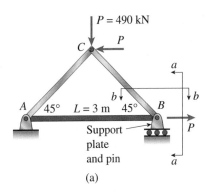

(a)

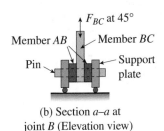

(b) Section a–a at joint B (Elevation view)

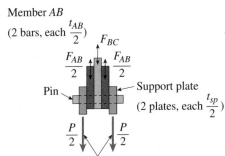

(c) Section b–b at joint B (Plan view)

PROB. 1.6-16

1.6-17 A spray nozzle for a garden hose requires a force $F = 5$ lb. to open the spring-loaded spray chamber AB. The nozzle hand grip pivots about a pin through a flange at O. Each of the two flanges has thickness $t = 1/16$ in., and the pin has diameter $d_p = 1/8$ in. [see figure part (a)]. The spray nozzle is attached to the garden hose with a quick release fitting at B [see figure part (b)]. Three brass balls (diameter $d_b = 3/16$ in.) hold the spray head in place under water pressure force $f_p = 30$ lb. at C [see figure part (c)]. Use dimensions given in figure part (a).

(a) Find the force in the pin at O due to applied force F.

(b) Find average shear stress τ_{aver} and bearing stress σ_b in the pin at O.

(c) Find the average shear stress τ_{aver} in the brass retaining balls at C due to water pressure force f_p.

1.6-18 A single steel strut AB with diameter $d_s = 8$ mm. supports the vehicle engine hood of mass 20 kg which pivots about hinges at C and D [see figures (a) and (b)]. The strut is bent into a loop at its end and then attached to a bolt at A with diameter $d_b = 10$ mm. Strut AB lies in a vertical plane.

(a) Find the strut force F_s and average normal stress σ in the strut.

(b) Find the average shear stress τ_{aver} in the bolt at A.

(c) Find the average bearing stress σ_b on the bolt at A.

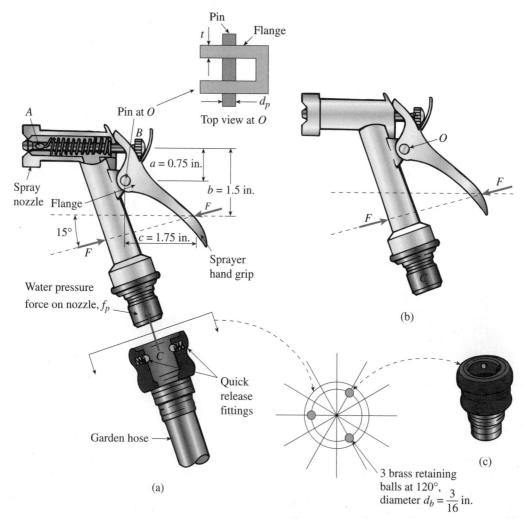

(a)

(b)

(c)

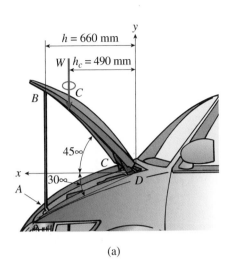

(a)

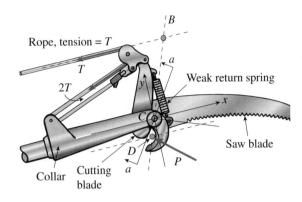

(a) Top part of pole saw

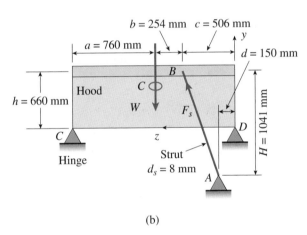

(b)

PROB. 1.6-18

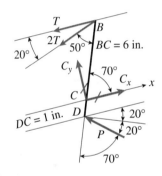

(b) Free-body diagram

1.6-19 The top portion of a pole saw used to trim small branches from trees is shown in the figure part (a). The cutting blade BCD [see figure parts (a) and (c)] applies a force P at point D. Ignore the effect of the weak return spring attached to the cutting blade below B. Use properties and dimensions given in the figure.

(a) Find the force P on the cutting blade at D if the tension force in the rope is $T = 25$ lb [(see free body diagram in part (b)].

(b) Find force in the pin at C.

(c) Find average shear stress τ_{aver} and bearing stress σ_b in the support pin at C [see Section a–a through cutting blade in figure part (c)].

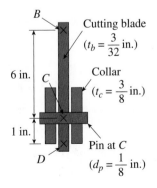

(c) Section a–a

PROB. 1.6-19

Allowable Loads

1.7-1 A bar of solid circular cross section is loaded in tension by forces P (see figure). The bar has length $L = 16.0$ in. and diameter $d = 0.50$ in. The material is a magnesium alloy having modulus of elasticity $E = 6.4 \times 10^6$ psi. The allowable stress in tension is $\sigma_{allow} = 17,000$ psi, and the elongation of the bar must not exceed 0.04 in.

What is the allowable value of the forces P?

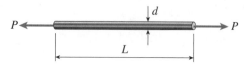

PROB. 1.7-1

1.7-2 A torque T_0 is transmitted between two flanged shafts by means of ten 20-mm bolts (see figure and photo). The diameter of the bolt circle is $d = 250$ mm.

If the allowable shear stress in the bolts is 85 MPa, what is the maximum permissible torque? (Disregard friction between the flanges.)

Drive shaft coupling on a ship propulsion motor
(Courtesy of American Superconductor)

PROB. 1.7-2

1.7-3 A tie-down on the deck of a sailboat consists of a bent bar bolted at both ends, as shown in the figure. The diameter d_B of the bar is 1/4 in., the diameter d_W of the washers is 7/8 in., and the thickness t of the fiberglass deck is 3/8 in.

If the allowable shear stress in the fiberglass is 300 psi, and the allowable bearing pressure between the washer and the fiberglass is 550 psi, what is the allowable load P_{allow} on the tie-down?

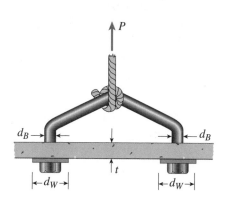

PROB. 1.7-3

1.7-4 Two steel tubes are joined at B by four pins ($d_p = 11$ mm), as shown in the cross section $a–a$ in the figure. The outer diameters of the tubes are $d_{AB} = 40$ mm and $d_{BC} = 28$ mm. The wall thicknesses are $t_{AB} = 6$ mm and $t_{BC} = 7$ mm. The yield stress in tension for the steel is $\sigma_Y = 200$ MPa and the ultimate stress in *tension* is $\sigma_U = 340$ MPa. The corresponding yield and ultimate values in *shear* for the pin are 80 MPa and 140 MPa, respectively. Finally, the yield and ultimate values in *bearing* between the pins and the tubes are 260 MPa and 450 MPa, respectively. Assume that the factors of safety with respect to yield stress and ultimate stress are 4 and 5, respectively.

(a) Calculate the allowable tensile force P_{allow} considering tension in the tubes.

(b) Recompute P_{allow} for shear in the pins.

(c) Finally, recompute P_{allow} for bearing between the pins and the tubes. Which is the controlling value of P?

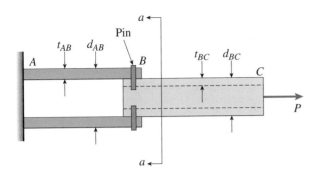

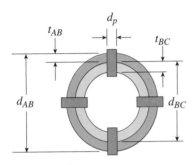

Section a–a

PROB. 1.7-4

1.7-5 A steel pad supporting heavy machinery rests on four short, hollow, cast iron piers (see figure). The ultimate strength of the cast iron in compression is 50 ksi. The outer diameter of the piers is $d = 4.5$ in. and the wall thickness is $t = 0.40$ in.

Using a factor of safety of 3.5 with respect to the ultimate strength, determine the total load P that may be supported by the pad.

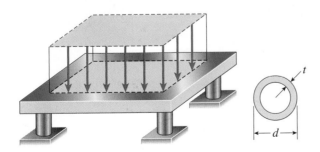

PROB. 1.7-5

1.7-6 The rear hatch of a van [$BDCF$ in figure part (a)] is supported by two hinges at B_1 and B_2 and by two struts A_1B_1 and A_2B_2 (diameter $d_s = 10$ mm) as shown in figure part (b). The struts are supported at A_1 and A_2 by pins, each with diameter $d_p = 9$ mm and passing through an eyelet of thickness $t = 8$ mm at the end of the strut [figure part (b)]. If a closing force $P = 50$ N is applied at G and the mass of the hatch $M_h = 43$ kg is concentrated at C:

(a) What is the force F in each strut? [Use the free-body diagram of one half of the hatch in the figure part (c)]

(b) What is the maximum permissible force in the strut, F_{allow}, if the allowable stresses are as follows: compressive stress in the strut, 70 MPa; shear stress in the pin, 45 MPa; and bearing stress between the pin and the end of the strut, 110 MPa.

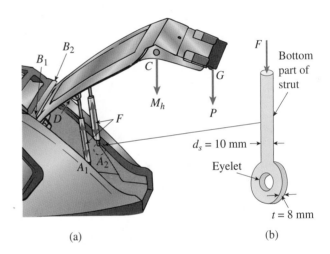

(a)   (b)

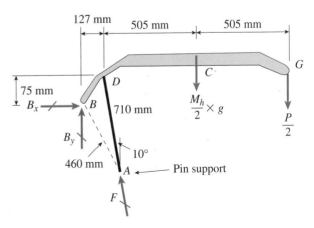

PROB. 1.7-6

1.7-7 A lifeboat hangs from two ships' davits, as shown in the figure. A pin of diameter $d = 0.80$ in. passes through each davit and supports two pulleys, one on each side of the davit.

Cables attached to the lifeboat pass over the pulleys and wind around winches that raise and lower the lifeboat. The lower parts of the cables are vertical and the upper parts make an angle $\alpha = 15°$ with the horizontal. The allowable tensile force in each cable is 1800 lb, and the allowable shear stress in the pins is 4000 psi.

If the lifeboat weighs 1500 lb, what is the maximum weight that should be carried in the lifeboat?

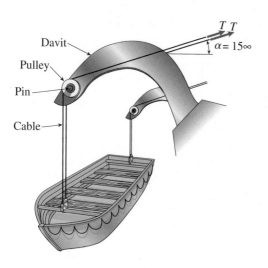

PROB. 1.7-7

1.7-8 A cable and pulley system in figure part (a) supports a cage of mass 300 kg at B. Assume that this includes the mass of the cables as well. The thickness of each the three steel pulleys is $t = 40$ mm. The pin diameters are $d_{pA} = 25$ mm, $d_{pB} = 30$ mm and $d_{pC} = 22$ mm [see figure, parts (a) and part (b)].

(a) Find expressions for the resultant forces acting on the pulleys at A, B, and C in terms of cable tension T.

(b) What is the maximum weight W that can be added to the cage at B based on the following allowable stresses? Shear stress in the pins is 50 MPa; bearing stress between the pin and the pulley is 110 MPa.

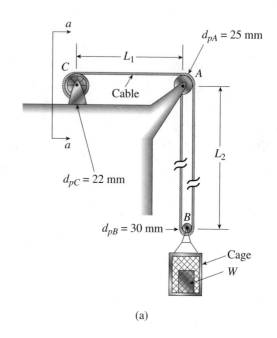

(a)

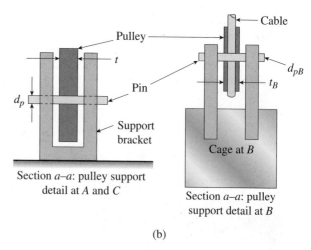

Section a–a: pulley support detail at A and C

Cage at B

Section a–a: pulley support detail at B

(b)

PROB. 1.7-8

1.7-9 A ship's spar is attached at the base of a mast by a pin connection (see figure). The spar is a steel tube of outer diameter $d_2 = 3.5$ in. and inner diameter $d_1 = 2.8$ in. The steel pin has diameter $d = 1$ in., and the two plates connecting the spar to the pin have thickness $t = 0.5$ in. The allowable stresses are as follows: compressive stress in the spar, 10 ksi; shear stress in the pin, 6.5 ksi; and bearing stress between the pin and the connecting plates, 16 ksi.

Determine the allowable compressive force P_{allow} in the spar.

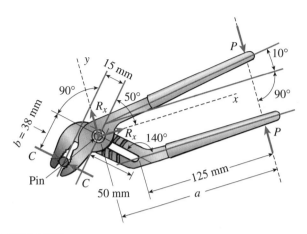

PROB. 1.7-10

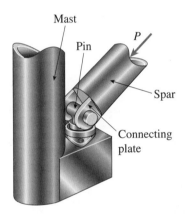

PROB. 1.7-9

1.7-11 A metal bar AB of weight W is suspended by a system of steel wires arranged as shown in the figure. The diameter of the wires is 5/64 in., and the yield stress of the steel is 65 ksi.

Determine the maximum permissible weight W_{max} for a factor of safety of 1.9 with respect to yielding.

1.7-10 What is the maximum possible value of the clamping force C in the jaws of the pliers shown in the figure if the ultimate shear stress in the 5-mm diameter pin is 340 MPa?

What is the maximum permissible value of the applied load P if a factor of safety of 3.0 with respect to failure of the pin is to be maintained?

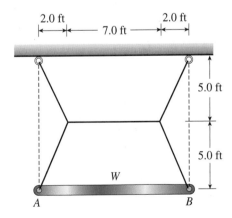

PROB. 1.7-11

1.7-12 A plane truss is subjected to loads $2P$ and P at joints B and C, respectively, as shown in the figure part (a). The truss bars are made of two L102 × 76 × 6.4 steel angles [see Table F-5(b) available online: cross sectional area of the two angles, $A = 2180$ mm², figure part (b)] having an ultimate stress in tension equal to 390 MPa. The angles are connected to a 12 mm-thick gusset plate at C [figure part (c)] with 16-mm diameter rivets; assume each rivet transfers an equal share of the member force to the gusset plate. The ultimate stresses in shear and bearing for the rivet steel are 190 MPa and 550 MPa, respectively.

Determine the allowable load P_{allow} if a safety factor of 2.5 is desired with respect to the ultimate load that can be carried. (Consider tension in the bars, shear in the rivets, bearing between the rivets and the bars, and also bearing between the rivets and the gusset plate. Disregard friction between the plates and the weight of the truss itself.)

1.7-13 A solid bar of circular cross section (diameter d) has a hole of diameter $d/5$ drilled laterally through the center of the bar (see figure). The allowable average tensile stress on the net cross section of the bar is σ_{allow}.

(a) Obtain a formula for the allowable load P_{allow} that the bar can carry in tension.

(b) Calculate the value of P_{allow} if the bar is made of brass with diameter $d = 1.75$ in. and $\sigma_{allow} = 12$ ksi.

(*Hint*: Use the formulas of Case 15 Appendix E available online.)

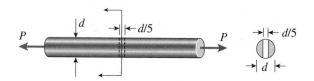

PROB. 1.7-13

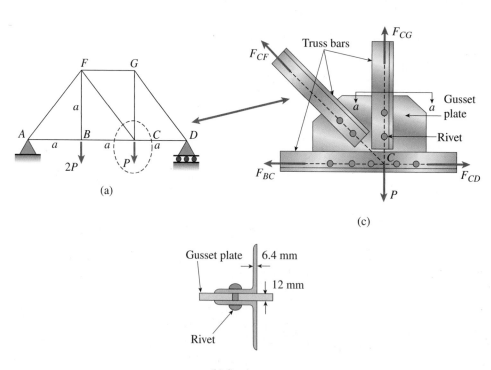

(a)

(c)

(b) Section a–a

PROB. 1.7-12

1.7-14 A solid steel bar of diameter $d_1 = 60$ mm has a hole of diameter $d_2 = 32$ mm drilled through it (see figure). A steel pin of diameter d_2 passes through the hole and is attached to supports.

Determine the maximum permissible tensile load P_{allow} in the bar if the yield stress for shear in the pin is $\tau_Y = 120$ MPa, the yield stress for tension in the bar is $\sigma_Y = 250$ MPa and a factor of safety of 2.0 with respect to yielding is required. (*Hint*: Use the formulas of Case 15, Appendix E available online.)

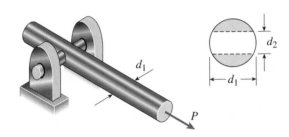

PROB. 1.7-14

1.7-15 A sign of weight W is supported at its base by four bolts anchored in a concrete footing. Wind pressure p acts normal to the surface of the sign; the resultant of the uniform wind pressure is force F at the center of pressure. The wind force is assumed to create equal shear forces $F/4$ in the y-direction at each bolt [see figure parts (a) and (c)]. The overturning effect of the wind force also causes an uplift force R at bolts A and C and a downward force $(-R)$ at bolts B and D [see figure part (b)]. The resulting effects of the wind, and the associated ultimate stresses for each stress condition, are: normal stress in each bolt ($\sigma_u = 60$ ksi); shear through the base plate ($\tau_u = 17$ ksi); horizontal shear and bearing on each bolt ($\tau_{hu} = 25$ ksi and $\sigma_{bu} = 75$ ksi); and bearing on the bottom washer at B (or D) ($\sigma_{bw} = 50$ ksi).

Find the maximum wind pressure p_{max} (psf) that can be carried by the bolted support system for the sign if a safety factor of 2.5 is desired with respect to the ultimate wind load that can be carried.

Use the following numerical data: bolt $d_b = \frac{3}{4}$ in.; washer $d_w = 1.5$ in.; base plate $t_{bp} = 1$ in.; base plate dimensions $h = 14$ in. and $b = 12$ in.; $W = 500$ lb; $H = 17$ ft; sign dimensions ($L_v = 10$ ft. $\times L_h = 12$ ft.); pipe column diameter $d = 6$ in., and pipe column thickness $t = 3/8$ in.

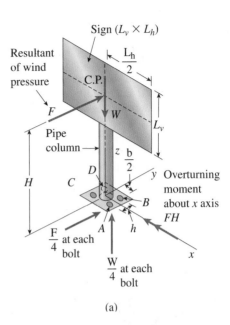

(a)

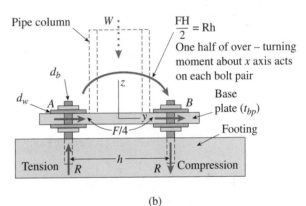

(b)

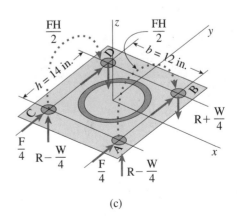

(c)

PROB. 1.7-15

1.7-16 The piston in an engine is attached to a connecting rod AB, which in turn is connected to a crank arm BC (see figure). The piston slides without friction in a cylinder and is subjected to a force P (assumed to be constant) while moving to the right in the figure. The connecting rod, which has diameter d and length L, is attached at both ends by pins. The crank arm rotates about the axle at C with the pin at B moving in a circle of radius R. The axle at C, which is supported by bearings, exerts a resisting moment M against the crank arm.

(a) Obtain a formula for the maximum permissible force P_{allow} based upon an allowable compressive stress σ_c in the connecting rod.

(b) Calculate the force P_{allow} for the following data: $\sigma_c = 160$ MPa, $d = 9.00$ mm, and $R = 0.28L$.

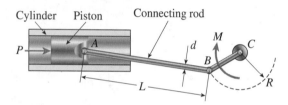

PROB. 1.7-16

Design for Axial Loads and Direct Shear

1.8-1 An aluminum tube is required to transmit an axial tensile force $P = 33$ k [see figure part (a)]. The thickness of the wall of the tube is to be 0.25 in.

(a) What is the minimum required outer diameter d_{min} if the allowable tensile stress is 12,000 psi?

(b) Repeat part (a) if the tube will have a hole of diameter $d/10$ at mid-length [see figure parts (b) and (c)].

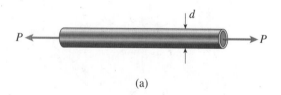

(a)

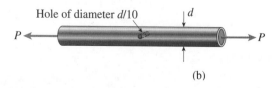

(b)

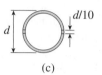

(c)

PROB. 1.8-1

1.8-2 A copper alloy pipe having yield stress $\sigma_Y = 290$ MPa is to carry an axial tensile load $P = 1500$ kN [see figure part (a)]. A factor of safety of 1.8 against yielding is to be used.

(a) If the thickness t of the pipe is to be one-eighth of its outer diameter, what is the minimum required outer diameter d_{min}?

(b) Repeat part (a) if the tube has a hole of diameter $d/10$ drilled through the entire tube as shown in the figure [part (b)].

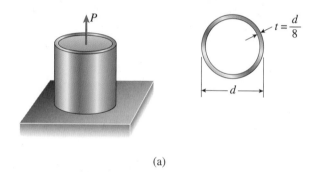

(a)

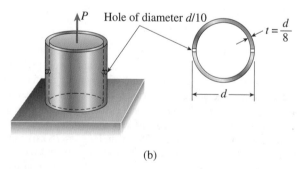

(b)

PROB. 1.8-2

1.8-3 A horizontal beam AB with cross-sectional dimensions ($b = 0.75$ in.) $\times$ ($h = 8.0$ in.) is supported by an inclined strut CD and carries a load $P = 2700$ lb at joint B [see figure part (a)]. The strut, which consists of two bars each of thickness $5b/8$, is connected to the beam by a bolt passing through the three bars meeting at joint C [see figure part (b)].

(a) If the allowable shear stress in the bolt is 13,000 psi, what is the minimum required diameter d_{min} of the bolt at C?

(b) If the allowable bearing stress in the bolt is 19,000 psi, what is the minimum required diameter d_{min} of the bolt at C?

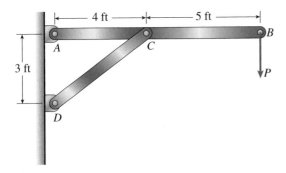

(a)

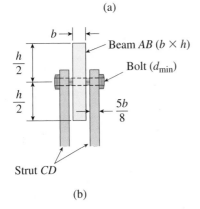

(b)

PROB. 1.8-3

1.8-4 Lateral bracing for an elevated pedestrian walkway is shown in the figure part (a). The thickness of the clevis plate $t_c = 16$ mm and the thickness of the gusset plate $t_g = 20$ mm [see figure part (b)]. The maximum force in the diagonal bracing is expected to be $F = 190$ kN.

If the allowable shear stress in the pin is 90 MPa and the allowable bearing stress between the pin and both the clevis and gusset plates is 150 MPa, what is the minimum required diameter d_{min} of the pin?

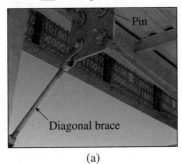

(a)

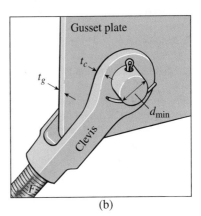

(b)

PROB. 1.8-4

1.8-5 Forces $P_1 = 1500$ lb and $P_2 = 2500$ lb are applied at joint C of plane truss ABC shown in the figure part (a). Member AC has thickness $t_{AC} = 5/16$ in. and member AB is composed of two bars each having thickness $t_{AB}/2 = 3/16$ in. [see figure part (b)]. Ignore the effect of the two plates which make up the pin support at A.

If the allowable shear stress in the pin is 12,000 psi and the allowable bearing stress in the pin is 20,000 psi, what is the minimum required diameter d_{min} of the pin?

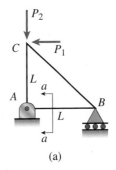

(a)

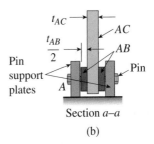

(b)

PROB. 1.8-5

1.8-6 A suspender on a suspension bridge consists of a cable that passes over the main cable (see figure) and supports the bridge deck, which is far below. The suspender is held in position by a metal tie that is prevented from sliding downward by clamps around the suspender cable.

Let P represent the load in each part of the suspender cable, and let θ represent the angle of the suspender cable just above the tie. Finally, let σ_{allow} represent the allowable tensile stress in the metal tie.

(a) Obtain a formula for the minimum required cross-sectional area of the tie.

(b) Calculate the minimum area if $P = 130$ kN, $\theta = 75°$, and $\sigma_{\text{allow}} = 80$ MPa.

points A and B. The cross section is a hollow square with inner dimension $b_1 = 8.5$ in. and outer dimension $b_2 = 10.0$ in. The allowable shear stress in the pin is 8,700 psi, and the allowable bearing stress between the pin and the tube is 13,000 psi.

Determine the minimum diameter of the pin in order to support the weight of the tube. (*Note:* Disregard the rounded corners of the tube when calculating its weight.)

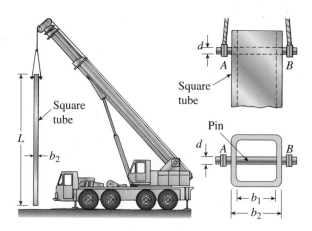

PROB. 1.8-7

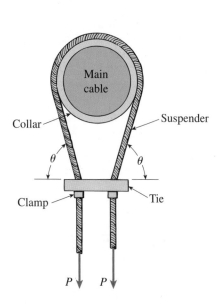

PROB. 1.8-6

1.8-7 A square steel tube of length $L = 20$ ft and width $b_2 = 10.0$ in. is hoisted by a crane (see figure). The tube hangs from a pin of diameter d that is held by the cables at

1.8-8 A cable and pulley system at D is used to bring a 230-kg pole (ACB) to a vertical position as shown in the figure part (a). The cable has tensile force T and is attached at C. The length L of the pole is 6.0 m, the outer diameter is $d = 140$ mm, and the wall thickness $t = 12$ mm. The pole pivots about a pin at A in figure part (b). The allowable shear stress in the pin is 60 MPa and the allowable bearing stress is 90 MPa.

Find the minimum diameter of the pin at A in order to support the weight of the pole in the position shown in the figure part (a).

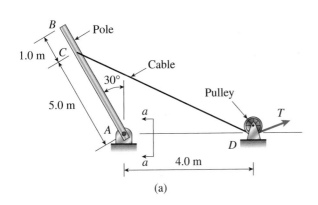

(a)

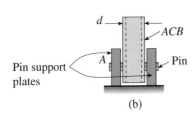

(b)

PROB. 1.8-8

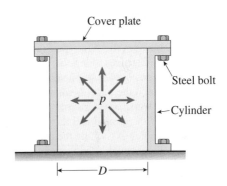

PROB. 1.8-9

1.8-10 A tubular post of outer diameter d_2 is guyed by two cables fitted with turnbuckles (see figure). The cables are tightened by rotating the turnbuckles, thus producing tension in the cables and compression in the post. Both cables are tightened to a tensile force of 110 kN. Also, the angle between the cables and the ground is 60°, and the allowable compressive stress in the post is $\sigma_c = 35$ MPa.

If the wall thickness of the post is 15 mm, what is the minimum permissible value of the outer diameter d_2?

1.8-9 A pressurized circular cylinder has a sealed cover plate fastened with steel bolts (see figure). The pressure p of the gas in the cylinder is 290 psi, the inside diameter D of the cylinder is 10.0 in., and the diameter d_B of the bolts is 0.50 in.

If the allowable tensile stress in the bolts is 10,000 psi, find the number n of bolts needed to fasten the cover.

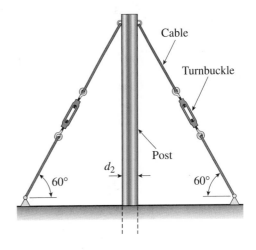

PROB. 1.8-10

1.8-11 A large precast concrete panel for a warehouse is being raised to a vertical position using two sets of cables at two lift lines as shown in the figure part (a). Cable 1 has length $L_1 = 22$ ft and distances along the panel (see figure part (b)) are $a = L_1/2$ and $b = L_1/4$. The cables are attached at lift points B and D and the panel is rotated about its base at A. However, as a worst case, assume that the panel is momentarily lifted off the ground and its total weight must be supported by the cables. Assuming the cable lift forces F at each lift line are about equal, use the simplified model of one half of the panel in figure part (b) to perform your analysis for the lift position shown. The total weight of the panel is $W = 85$ kips. The orientation of the panel is defined by the following angles: $\gamma = 20°$ and $\theta = 10°$.

Find the required cross-sectional area A_C of the cable if its breaking stress is 91 ksi and a factor of safety of 4 with respect to failure is desired.

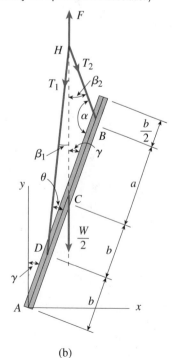

(a)
(Courtesy Tilt-Up Concrete Association)

(b)

PROB. 1.8-11

1.8-12 A steel column of hollow circular cross section is supported on a circular steel base plate and a concrete pedestal (see figure). The column has outside diameter $d = 250$ mm and supports a load $P = 750$ kN.

(a) If the allowable stress in the column is 55 MPa, what is the minimum required thickness t? Based upon your result, select a thickness for the column. (Select a thickness that is an even integer, such as 10, 12, 14, . . . , in units of millimeters.)

(b) If the allowable bearing stress on the concrete pedestal is 11.5 MPa, what is the minimum required diameter D of the base plate if it is designed for the allowable load P_{allow} that the column with the selected thickness can support?

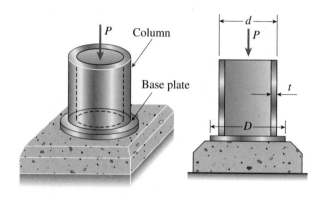

PROB. 1.8-12

1.8-13 An elevated jogging track is supported at intervals by a wood beam AB ($L = 7.5$ ft) which is pinned at A and supported by steel rod BC and a steel washer at B. Both the rod ($d_{BC} = 3/16$ in.) and the washer ($d_B = 1.0$ in.) were designed using a rod tension force of $T_{BC} = 425$ lb. The rod was sized using a factor of safety of 3 against reaching the ultimate stress $\sigma_u = 60$ ksi. An allowable bearing stress $\sigma_{ba} = 565$ psi was used to size the washer at B.

Now, a small platform HF is to be suspended below a section of the elevated track to support some mechanical and electrical equipment. The equipment load is uniform load $q = 50$ lb/ft and concentrated load $W_E = 175$ lb at mid-span of beam HF. The plan is to drill a hole through beam AB at D and install the same rod (d_{BC}) and washer (d_B) at both D and F to support beam HF.

(a) Use σ_u and σ_{ba} to check the proposed design for rod DF and washer d_F; are they acceptable?

(b) Also re-check the normal tensile stress in rod BC and bearing stress at B; if either is inadequate under the additional load from platform HF, redesign them to meet the original design criteria.

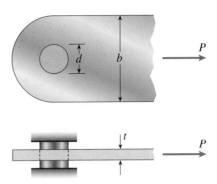

1.8-15 Two bars AB and BC of the same material support a vertical load P (see figure). The length L of the horizontal bar is fixed, but the angle θ can be varied by moving support A vertically and changing the length of bar AC to correspond with the new position of support A. The allowable stresses in the bars are the same in tension and compression.

We observe that when the angle θ is reduced, bar AC becomes shorter but the cross-sectional areas of both bars increase (because the axial forces are larger). The opposite effects occur if the angle θ is increased. Thus, we see that the weight of the structure (which is proportional to the volume) depends upon the angle θ.

Determine the angle θ so that the structure has minimum weight without exceeding the allowable stresses in the bars. (*Note*: The weights of the bars are very small compared to the force P and may be disregarded.)

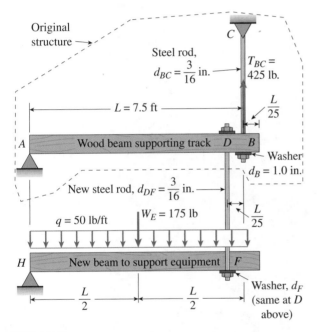

1.8-14 A flat bar of width $b = 60$ mm and thickness $t = 10$ mm is loaded in tension by a force P (see figure). The bar is attached to a support by a pin of diameter d that passes through a hole of the same size in the bar. The allowable tensile stress on the net cross section of the bar is $\sigma_T = 140$ MPa, the allowable shear stress in the pin is $\tau_S = 80$ MPa, and the allowable bearing stress between the pin and the bar is $\sigma_B = 200$ MPa.

(a) Determine the pin diameter d_m for which the load P will be a maximum.

(b) Determine the corresponding value $P_{\max}$ of the load.

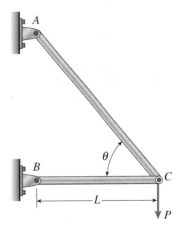

An oil drilling rig is comprised of axially loaded members that must be designed for a variety of loading conditions, including self weight, impact, and temperature effects. (Joe Raedle/Getty Images)

Axially Loaded Members

CHAPTER OVERVIEW

In Chapter 2, we consider several other aspects of axially loaded members, beginning with the determination of changes in lengths caused by loads (Sections 2.2 and 2.3). The calculation of **changes in lengths** is an essential ingredient in the analysis of statically indeterminate structures, a topic we introduce in Section 2.4. If the member is statically indeterminate, we must augment the equations of statical equilibrium with compatibility equations (which rely on **force-displacement relations**) to solve for any unknowns of interest, such as support reactions or internal axial forces in members. Changes in lengths also must be calculated whenever it is necessary to control the displacements of a structure, whether for aesthetic or functional reasons. In Section 2.5, we discuss the **effects of temperature** on the length of a bar, and we introduce the concepts of **thermal stress** and **thermal strain**. Also included in this section is a discussion of the effects of **misfits and prestrains**. Finally, a generalized view of the stresses in axially loaded bars is presented in Section 2.6, where we discuss the **stresses on inclined sections** (as distinct from **cross sections**) of bars. Although only normal stresses act on cross sections of axially loaded bars, both normal and shear stresses act on inclined sections. Stresses on inclined sections of axially loaded members are investigated as a first step toward a more complete consideration of **plane stress states** in later chapters.

Chapter 2 is organized as follows:

2.1 INTRODUCTION

Structural components subjected only to tension or compression are known as **axially loaded members**. Solid bars with straight longitudinal axes are the most common type, although cables and coil springs also carry axial loads. Examples of axially loaded bars are truss members, connecting rods in engines, spokes in bicycle wheels, columns in buildings, and struts in aircraft engine mounts. The stress-strain behavior of such members was discussed in Chapter 1, where we also obtained equations for the stresses acting on cross sections ($\sigma = P/A$) and the strains in longitudinal directions ($\epsilon = \delta/L$).

2.2 CHANGES IN LENGTHS OF AXIALLY LOADED MEMBERS

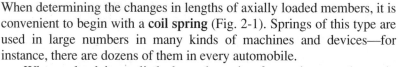

FIG. 2-1 Spring subjected to an axial load P

When determining the changes in lengths of axially loaded members, it is convenient to begin with a **coil spring** (Fig. 2-1). Springs of this type are used in large numbers in many kinds of machines and devices—for instance, there are dozens of them in every automobile.

When a load is applied along the axis of a spring, as shown in Fig. 2-1, the spring gets longer or shorter depending upon the direction of the load. If the load acts away from the spring, the spring elongates and we say that the spring is loaded in *tension*. If the load acts toward the spring, the spring shortens and we say it is in *compression*. However, it should not be inferred from this terminology that the individual coils of a spring are subjected to direct tensile or compressive stresses; rather, the coils act primarily in direct shear and torsion (or twisting). Nevertheless, the overall stretching or shortening of a spring is analogous to the behavior of a bar in tension or compression, and so the same terminology is used.

Springs

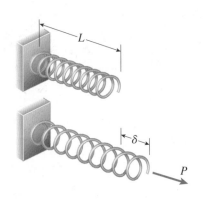

FIG. 2-2 Elongation of an axially loaded spring

The elongation of a spring is pictured in Fig. 2-2, where the upper part of the figure shows a spring in its **natural length** L (also called its *unstressed length*, *relaxed length*, or *free length*), and the lower part of the figure shows the effects of applying a tensile load. Under the action of the force P, the spring lengthens by an amount δ and its final length becomes $L + \delta$. If the material of the spring is **linearly elastic**, the load and elongation will be proportional:

$$P = k\delta \qquad \delta = fP \qquad \text{(2-1a,b)}$$

in which k and f are constants of proportionality.

The constant k is called the **stiffness** of the spring and is defined as the force required to produce a unit elongation, that is, $k = P/\delta$. Similarly, the constant f is known as the **flexibility** and is defined as the elongation produced by a load of unit value, that is, $f = \delta/P$. Although

FIG. 2-3 Prismatic bar of circular cross section

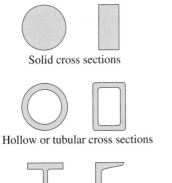

Solid cross sections

Hollow or tubular cross sections

Thin-walled open cross sections

FIG. 2-4 Typical cross sections of structural members

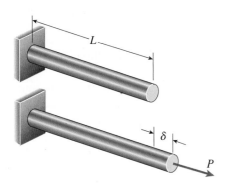

FIG. 2-5 Elongation of a prismatic bar in tension

we used a spring in tension for this discussion, it should be obvious that Eqs. (2-1a) and (2-1b) also apply to springs in compression.

From the preceding discussion it is apparent that the stiffness and flexibility of a spring are the reciprocal of each other:

$$k = \frac{1}{f} \qquad f = \frac{1}{k} \qquad (2\text{-}2\text{a,b})$$

The flexibility of a spring can easily be determined by measuring the elongation produced by a known load, and then the stiffness can be calculated from Eq. (2-2a). Other terms for the stiffness and flexibility of a spring are the **spring constant** and **compliance**, respectively.

The spring properties given by Eqs. (2-1) and (2-2) can be used in the analysis and design of various mechanical devices involving springs, as illustrated later in Example 2-1.

Prismatic Bars

Axially loaded bars elongate under tensile loads and shorten under compressive loads, just as springs do. To analyze this behavior, let us consider the prismatic bar shown in Fig. 2-3. A **prismatic bar** is a structural member having a straight longitudinal axis and constant cross section throughout its length. Although we often use circular bars in our illustrations, we should bear in mind that structural members may have a variety of cross-sectional shapes, such as those shown in Fig. 2-4.

The **elongation** δ of a prismatic bar subjected to a tensile load P is shown in Fig. 2-5. If the load acts through the centroid of the end cross section, the uniform normal stress at cross sections away from the ends is given by the formula $\sigma = P/A$, where A is the cross-sectional area. Furthermore, if the bar is made of a homogeneous material, the axial strain is $\epsilon = \delta/L$, where δ is the elongation and L is the length of the bar.

Let us also assume that the material is **linearly elastic**, which means that it follows Hooke's law. Then the longitudinal stress and strain are related by the equation $\sigma = E\epsilon$, where E is the modulus of elasticity. Combining these basic relationships, we get the following equation for the elongation of the bar:

$$\delta = \frac{PL}{EA} \qquad (2\text{-}3)$$

This equation shows that the elongation is directly proportional to the load P and the length L and inversely proportional to the modulus of elasticity E and the cross-sectional area A. The product EA is known as the **axial rigidity** of the bar.

Although Eq. (2-3) was derived for a member in tension, it applies equally well to a member in compression, in which case δ represents the shortening of the bar. Usually we know by inspection whether a member gets longer or shorter; however, there are occasions when a **sign convention** is needed (for instance, when analyzing a statically indeterminate bar). When that happens, elongation is usually taken as positive and shortening as negative.

The change in length of a bar is normally very small in comparison to its length, especially when the material is a structural metal, such as steel or aluminum. As an example, consider an aluminum strut that is 75.0 in. long and subjected to a moderate compressive stress of 7000 psi. If the modulus of elasticity is 10,500 ksi, the shortening of the strut (from Eq. 2-3 with P/A replaced by σ) is $\delta = 0.050$ in. Consequently, the ratio of the change in length to the original length is 0.05/75, or 1/1500, and the final length is 0.999 times the original length. Under ordinary conditions similar to these, we can use the original length of a bar (instead of the final length) in calculations.

The stiffness and flexibility of a prismatic bar are defined in the same way as for a spring. The stiffness is the force required to produce a unit elongation, or P/δ, and the flexibility is the elongation due to a unit load, or δ/P. Thus, from Eq. (2-3) we see that the **stiffness** and **flexibility** of a prismatic bar are, respectively,

$$k = \frac{EA}{L} \qquad f = \frac{L}{EA} \qquad\qquad (2\text{-}4a,b)$$

Stiffnesses and flexibilities of structural members, including those given by Eqs. (2-4a) and (2-4b), have a special role in the analysis of large structures by computer-oriented methods.

Cables

Cables are used to transmit large tensile forces, for example, when lifting and pulling heavy objects, raising elevators, guying towers, and supporting suspension bridges. Unlike springs and prismatic bars, cables cannot resist compression. Furthermore, they have little resistance to bending and therefore may be curved as well as straight. Nevertheless, a cable is considered to be an axially loaded member because it is subjected only to tensile forces. Because the tensile forces in a cable are directed along the axis, the forces may vary in both direction and magnitude, depending upon the configuration of the cable.

Cables are constructed from a large number of wires wound in some particular manner. While many arrangements are available depending upon how the cable will be used, a common type of cable, shown in Fig. 2-6, is formed by six *strands* wound helically around a central strand. Each strand is in turn constructed of many wires, also wound helically. For this reason, cables are often referred to as **wire rope**.

The cross-sectional area of a cable is equal to the total cross-sectional area of the individual wires, called the **effective area** or **metallic area**. This area is less than the area of a circle having the same

Steel cables on a pulley
(© Barsik/Dreamtime.com)

FIG. 2-6 Typical arrangement of strands and wires in a steel cable

diameter as the cable because there are spaces between the individual wires. For example, the actual cross-sectional area (effective area) of a particular 1.0 inch diameter cable is only 0.471 in.^{2}, whereas the area of a 1.0 in. diameter circle is 0.785 in.^{2}

Under the same tensile load, the elongation of a cable is greater than the elongation of a solid bar of the same material and same metallic cross-sectional area, because the wires in a cable "tighten up" in the same manner as the fibers in a rope. Thus, the modulus of elasticity (called the **effective modulus**) of a cable is less than the modulus of the material of which it is made. The effective modulus of steel cables is about 20,000 ksi (140 GPa), whereas the steel itself has a modulus of about 30,000 ksi (210 GPa).

When determining the **elongation** of a cable from Eq. (2-3), the effective modulus should be used for E and the effective area should be used for A.

In practice, the cross-sectional dimensions and other properties of cables are obtained from the manufacturers. However, for use in solving problems in this book (and definitely *not* for use in engineering applications), we list in Table 2-1 the properties of a particular type of cable. Note that the last column contains the *ultimate load*, which is the load that would cause the cable to break. The *allowable load* is obtained from the ultimate load by applying a safety factor that may range from 3 to 10, depending upon how the cable is to be used. The individual wires in a cable are usually made of high-strength steel, and the calculated tensile stress at the breaking load can be as high as 200,000 psi (1400 MPa).

The following examples illustrate techniques for analyzing simple devices containing springs and bars. The solutions require the use of free-body diagrams, equations of equilibrium, and equations for changes in length. The problems at the end of the chapter provide many additional examples.

TABLE 2-1 PROPERTIES OF STEEL CABLES*

Nominal diameter		Approximate weight		Effective area		Ultimate load	
in.	(mm)	lb/ft	(N/m)	in.^{2}	(mm^2)	lb	(kN)
0.50	(12)	0.42	(6.1)	0.119	(76.7)	23,100	(102)
0.75	(20)	0.95	(13.9)	0.268	(173)	51,900	(231)
1.00	(25)	1.67	(24.4)	0.471	(304)	91,300	(406)
1.25	(32)	2.64	(38.5)	0.745	(481)	144,000	(641)
1.50	(38)	3.83	(55.9)	1.08	(697)	209,000	(930)
1.75	(44)	5.24	(76.4)	1.47	(948)	285,000	(1260)
2.00	(50)	6.84	(99.8)	1.92	(1230)	372,000	(1650)

* To be used solely for solving problems in this book.

Example 2-1

A rigid L-shaped frame ABC consisting of a horizontal arm AB (length $b =$ 10.5 in.) and a vertical arm BC (length $c = 6.4$ in.) is pivoted at point B, as shown in Fig. 2-7a. The pivot is attached to the outer frame BCD, which stands on a laboratory bench. The position of the pointer at C is controlled by a spring (stiffness $k = 4.2$ lb/in.) that is attached to a threaded rod. The position of the threaded rod is adjusted by turning the knurled nut.

The *pitch* of the threads (that is, the distance from one thread to the next) is $p = 1/16$ in., which means that one full revolution of the nut will move the rod by that same amount. Initially, when there is no weight on the hanger, the nut is turned until the pointer at the end of arm BC is directly over the reference mark on the outer frame.

If a weight $W = 2$ lb is placed on the hanger at A, how many revolutions of the nut are required to bring the pointer back to the mark? (Deformations of the

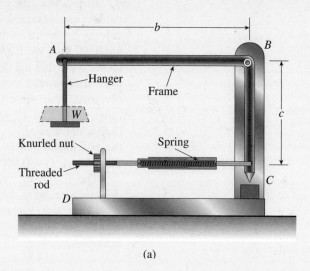

(a)

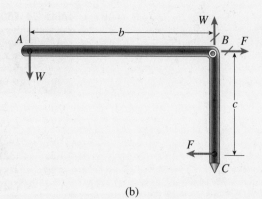

(b)

FIG. 2-7 Example 2-1. (a) Rigid L-shaped frame ABC attached to outer frame BCD by a pivot at B, and (b) free-body diagram of frame ABC

metal parts of the device may be disregarded because they are negligible compared to the change in length of the spring.)

Solution

Inspection of the device (Fig. 2-7a) shows that the weight W acting downward will cause the pointer at C to move to the right. When the pointer moves to the right, the spring stretches by an additional amount—an amount that we can determine from the force in the spring.

To determine the force in the spring, we construct a free-body diagram of frame ABC (Fig. 2-7b). In this diagram, W represents the force applied by the hanger and F represents the force applied by the spring. The reactions at the pivot are indicated with slashes across the arrows (see the discussion of reactions in Section 1.8).

Taking moments about point B gives

$$F = \frac{Wb}{c} \tag{a}$$

The corresponding elongation of the spring (from Eq. 2-1a) is

$$\delta = \frac{F}{k} = \frac{Wb}{ck} \tag{b}$$

To bring the pointer back to the mark, we must turn the nut through enough revolutions to move the threaded rod to the left an amount equal to the elongation of the spring. Since each complete turn of the nut moves the rod a distance equal to the pitch p, the total movement of the rod is equal to np, where n is the number of turns. Therefore,

$$np = \delta = \frac{Wb}{ck} \tag{c}$$

from which we get the following formula for the number of revolutions of the nut:

$$n = \frac{Wb}{ckp} \tag{d} \quad \longleftarrow$$

Numerical results. As the final step in the solution, we substitute the given numerical data into Eq. (d), as follows:

$$n = \frac{Wb}{ckp} = \frac{(2 \text{ lb})(10.5 \text{ in.})}{(6.4 \text{ in.})(4.2 \text{ lb/in.})(1/16 \text{ in.})} = 12.5 \text{ revolutions} \quad \longleftarrow$$

This result shows that if we rotate the nut through 12.5 revolutions, the threaded rod will move to the left an amount equal to the elongation of the spring caused by the 2-lb load, thus returning the pointer to the reference mark.

Example 2-2

The device shown in Fig. 2-8a consists of a horizontal beam *ABC* supported by two vertical bars *BD* and *CE*. Bar *CE* is pinned at both ends but bar *BD* is fixed to the foundation at its lower end. The distance from *A* to *B* is 450 mm and from *B* to *C* is 225 mm. Bars *BD* and *CE* have lengths of 480 mm and 600 mm, respectively, and their cross-sectional areas are 1020 mm^2 and 520 mm^2, respectively. The bars are made of steel having a modulus of elasticity $E = 205$ GPa.

Assuming that beam *ABC* is rigid, find the maximum allowable load P_{max} if the displacement of point *A* is limited to 1.0 mm.

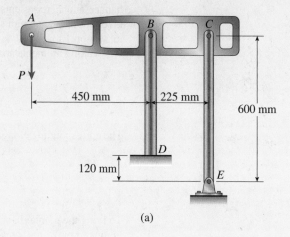

(a)

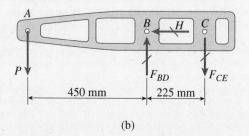

(b)

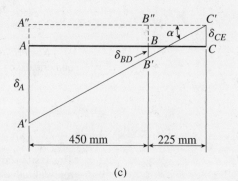

FIG. 2-8 Example 2-2. Horizontal beam *ABC* supported by two vertical bars

(c)

Solution

To find the displacement of point A, we need to know the displacements of points B and C. Therefore, we must find the changes in lengths of bars BD and CE, using the general equation $\delta = PL/EA$ (Eq. 2-3).

We begin by finding the forces in the bars from a free-body diagram of the beam (Fig. 2-8b). Because bar CE is pinned at both ends, it is a "two-force" member and transmits only a vertical force F_{CE} to the beam. However, bar BD can transmit both a vertical force F_{BD} and a horizontal force H. From equilibrium of beam ABC in the horizontal direction, we see that the horizontal force vanishes.

Two additional equations of equilibrium enable us to express the forces F_{BD} and F_{CE} in terms of the load P. Thus, by taking moments about point B and then summing forces in the vertical direction, we find

$$F_{CE} = 2P \qquad F_{BD} = 3P \tag{a}$$

Note that the force F_{CE} acts downward on bar ABC and the force F_{BD} acts upward. Therefore, member CE is in tension and member BD is in compression.

The shortening of member BD is

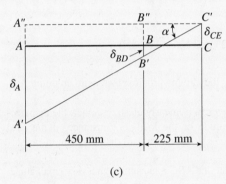

$$\delta_{BD} = \frac{F_{BD} L_{BD}}{E A_{BD}}$$

$$= \frac{(3P)(480 \text{ mm})}{(205 \text{ GPa})(1020 \text{ mm}^2)} = 6.887P \times 10^{-6} \text{ mm} \quad (P = \text{newtons}) \tag{b}$$

FIG. 2-8c (Repeated)

Note that the shortening δ_{BD} is expressed in millimeters provided the load P is expressed in newtons.

Similarly, the lengthening of member CE is

$$\delta_{CE} = \frac{F_{CE} L_{CE}}{E A_{CE}}$$

$$= \frac{(2P)(600 \text{ mm})}{(205 \text{ GPa})(520 \text{ mm}^2)} = 11.26P \times 10^{-6} \text{ mm} \quad (P = \text{newtons}) \tag{c}$$

Again, the displacement is expressed in millimeters provided the load P is expressed in newtons. Knowing the changes in lengths of the two bars, we can now find the displacement of point A.

Displacement diagram. A displacement diagram showing the relative positions of points A, B, and C is sketched in Fig. 2-8c. Line ABC represents the original alignment of the three points. After the load P is applied, member BD shortens by the amount δ_{BD} and point B moves to B'. Also, member CE elongates by the amount δ_{CE} and point C moves to C'. Because the beam ABC is assumed to be rigid, points A', B', and C' lie on a straight line.

For clarity, the displacements are highly exaggerated in the diagram. In reality, line ABC rotates through a very small angle to its new position $A'B'C'$ (see Note 2 at the end of this example).

continued

Using similar triangles, we can now find the relationships between the displacements at points A, B, and C. From triangles $A'A''C'$ and $B'B''C'$ we get

$$\frac{A'A''}{A''C'} = \frac{B'B''}{B''C'} \quad \text{or} \quad \frac{\delta_A + \delta_{CE}}{450 + 225} = \frac{\delta_{BD} + \delta_{CE}}{225} \tag{d}$$

in which all terms are expressed in millimeters.

Substituting for δ_{BD} and δ_{CE} from Eqs. (f) and (g) gives

$$\frac{\delta_A + 11.26P \times 10^{-6}}{450 + 225} = \frac{6.887P \times 10^{-6} + 11.26P \times 10^{-6}}{225}$$

Finally, we substitute for δ_A its limiting value of 1.0 mm and solve the equation for the load P. The result is

$$P = P_{max} = 23{,}200 \text{ N (or 23.2 kN)}$$

When the load reaches this value, the downward displacement at point A is 1.0 mm.

Note 1: Since the structure behaves in a linearly elastic manner, the displacements are proportional to the magnitude of the load. For instance, if the load is one-half of P_{max}, that is, if $P = 11.6$ kN, the downward displacement of point A is 0.5 mm.

Note 2: To verify our premise that line ABC rotates through a very small angle, we can calculate the angle of rotation α from the displacement diagram (Fig. 2-8c), as follows:

$$\tan \alpha = \frac{A'A''}{A''C'} = \frac{\delta_A + \delta_{CE}}{675 \text{ mm}} \tag{e}$$

The displacement δ_A of point A is 1.0 mm, and the elongation δ_{CE} of bar CE is found from Eq. (g) by substituting $P = 23{,}200$ N; the result is $\delta_{CE} = 0.261$ mm. Therefore, from Eq. (i) we get

$$\tan \alpha = \frac{1.0 \text{ mm} + 0.261 \text{ mm}}{675 \text{ mm}} = \frac{1.261 \text{ mm}}{675 \text{ mm}} = 0.001868$$

from which $\alpha = 0.11°$. This angle is so small that if we tried to draw the displacement diagram to scale, we would not be able to distinguish between the original line ABC and the rotated line $A'B'C'$.

Thus, when working with displacement diagrams, we usually can consider the displacements to be very small quantities, thereby simplifying the geometry. In this example we were able to assume that points A, B, and C moved only vertically, whereas if the displacements were large, we would have to consider that they moved along curved paths.

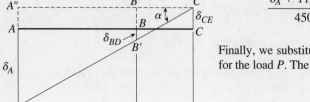

(c)

FIG. 2-8c (Repeated)

2.3 CHANGES IN LENGTHS UNDER NONUNIFORM CONDITIONS

When a prismatic bar of linearly elastic material is loaded only at the ends, we can obtain its change in length from the equation $\delta = PL/EA$, as described in the preceding section. In this section we will see how this same equation can be used in more general situations.

Bars with Intermediate Axial Loads

Suppose, for instance, that a prismatic bar is loaded by one or more axial loads acting at intermediate points along the axis (Fig. 2-9a). We can determine the change in length of this bar by adding algebraically the elongations and shortenings of the individual segments. The procedure is as follows.

1. Identify the segments of the bar (segments *AB*, *BC*, and *CD*) as segments 1, 2, and 3, respectively.
2. Determine the internal axial forces N_1, N_2, and N_3 in segments 1, 2, and 3, respectively, from the free-body diagrams of Figs. 2-9b, c, and d. Note that the internal axial forces are denoted by the letter *N* to distinguish them from the external loads *P*. By summing forces in the vertical direction, we obtain the following expressions for the axial forces:

$$N_1 = -P_B + P_C + P_D \qquad N_2 = P_C + P_D \qquad N_3 = P_D$$

In writing these equations we used the sign convention given in the preceding section (internal axial forces are positive when in tension and negative when in compression).

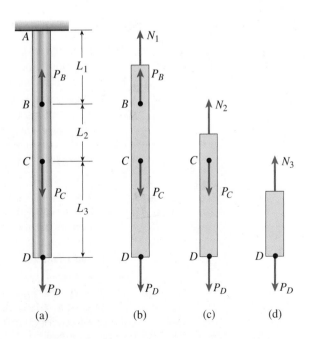

FIG. 2-9 (a) Bar with external loads acting at intermediate points; (b), (c), and (d) free-body diagrams showing the internal axial forces N_1, N_2, and N_3

3.   Determine the changes in the lengths of the segments from Eq. (2-3):

$$\delta_1 = \frac{N_1 L_1}{EA} \qquad \delta_2 = \frac{N_2 L_2}{EA} \qquad \delta_3 = \frac{N_3 L_3}{EA}$$

in which L_1, L_2, and L_3 are the lengths of the segments and EA is the axial rigidity of the bar.

4.   Add δ_1, δ_2, and δ_3 to obtain δ, the change in length of the entire bar:

$$\delta = \sum_{i=1}^{3} \delta_i = \delta_1 + \delta_2 + \delta_3$$

As already explained, the changes in lengths must be added algebraically, with elongations being positive and shortenings negative.

Bars Consisting of Prismatic Segments

This same general approach can be used when the bar consists of several prismatic segments, each having different axial forces, different dimensions, and different materials (Fig. 2-10). The change in length may be obtained from the equation

$$\delta = \sum_{i=1}^{n} \frac{N_i L_i}{E_i A_i} \tag{2-5}$$

in which the subscript i is a numbering index for the various segments of the bar and n is the total number of segments. Note especially that N_i is not an external load but is the internal axial force in segment i.

Bars with Continuously Varying Loads or Dimensions

Sometimes the axial force N and the cross-sectional area A vary continuously along the axis of a bar, as illustrated by the tapered bar of Fig. 2-11a. This bar not only has a continuously varying cross-sectional area but also a continuously varying axial force. In this illustration, the load consists of two parts, a single force P_B acting at end B of the bar and distributed forces $p(x)$ acting along the axis. (A distributed force has units of force per unit distance, such as pounds per inch or newtons per meter.) A distributed axial load may be produced by such factors as centrifugal forces, friction forces, or the weight of a bar hanging in a vertical position.

Under these conditions we can no longer use Eq. (2-5) to obtain the change in length. Instead, we must determine the change in length of a differential element of the bar and then integrate over the length of the bar.

We select a differential element at distance x from the left-hand end of the bar (Fig. 2-11a). The internal axial force $N(x)$ acting at this cross section (Fig. 2-11b) may be determined from equilibrium using either segment AC or segment CB as a free body. In general, this force is a function of x. Also, knowing the dimensions of the bar, we can express the cross-sectional area $A(x)$ as a function of x.

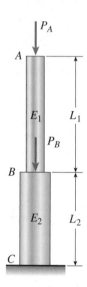

FIG. 2-10 Bar consisting of prismatic segments having different axial forces, different dimensions, and different materials

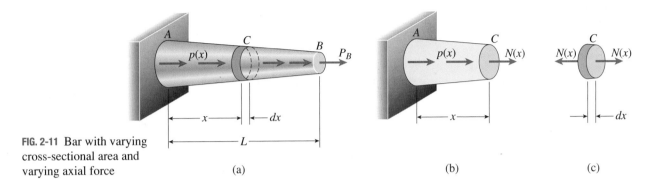

FIG. 2-11 Bar with varying cross-sectional area and varying axial force

(a)   (b)   (c)

The elongation $d\delta$ of the differential element (Fig. 2-11c) may be obtained from the equation $\delta = PL/EA$ by substituting $N(x)$ for P, dx for L, and $A(x)$ for A, as follows:

$$d\delta = \frac{N(x)\,dx}{EA(x)} \tag{2-6}$$

The elongation of the entire bar is obtained by integrating over the length:

$$\delta = \int_0^L d\delta = \int_0^L \frac{N(x)dx}{EA(x)} \tag{2-7}$$

If the expressions for $N(x)$ and $A(x)$ are not too complicated, the integral can be evaluated analytically and a formula for δ can be obtained, as illustrated later in Example 2-4. However, if formal integration is either difficult or impossible, a numerical method for evaluating the integral should be used.

Limitations

Equations (2-5) and (2-7) apply only to bars made of linearly elastic materials, as shown by the presence of the modulus of elasticity E in the formulas. Also, the formula $\delta = PL/EA$ was derived using the assumption that the stress distribution is uniform over every cross section (because it is based on the formula $\sigma = P/A$). This assumption is valid for prismatic bars but not for tapered bars, and therefore Eq. (2-7) gives satisfactory results for a tapered bar only if the angle between the sides of the bar is small.

As an illustration, if the angle between the sides of a bar is 20°, the stress calculated from the expression $\sigma = P/A$ (at an arbitrarily selected cross section) is 3% less than the exact stress for that same cross section (calculated by more advanced methods). For smaller angles, the error is even less. Consequently, we can say that Eq. (2-7) is satisfactory if the angle of taper is small. If the taper is large, more accurate methods of analysis are needed (see Ref. 2-1; a list of references is available online).

The following examples illustrate the determination of changes in lengths of nonuniform bars.

Example 2-3

A vertical steel bar ABC is pin-supported at its upper end and loaded by a force P_1 at its lower end (Fig. 2-12a). A horizontal beam BDE is pinned to the vertical bar at joint B and supported at point D. The beam carries a load P_2 at end E.

The upper part of the vertical bar (segment AB) has length $L_1 = 20.0$ in. and cross-sectional area $A_1 = 0.25$ in.^{2}; the lower part (segment BC) has length $L_2 = 34.8$ in. and area $A_2 = 0.15$ in.^{2} The modulus of elasticity E of the steel is 29.0×10^6 psi. The left- and right-hand parts of beam BDE have lengths $a = 28$ in. and $b = 25$ in., respectively.

Calculate the vertical displacement δ_C at point C if the load $P_1 = 2100$ lb and the load $P_2 = 5600$ lb. (Disregard the weights of the bar and the beam.)

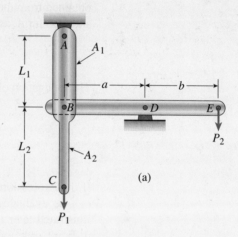

(a)

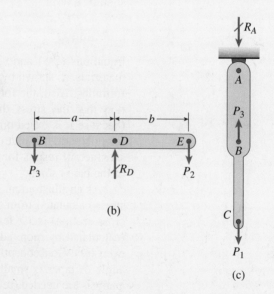

(b)

(c)

FIG. 2-12 Example 2-3. Change in length of a nonuniform bar (bar ABC)

Solution

Axial forces in bar ABC. From Fig. 2-12a, we see that the vertical displacement of point C is equal to the change in length of bar *ABC*. Therefore, we must find the axial forces in both segments of this bar.

The axial force N_2 in the lower segment is equal to the load P_1. The axial force N_1 in the upper segment can be found if we know either the vertical reaction at A or the force applied to the bar by the beam. The latter force can be obtained from a free-body diagram of the beam (Fig. 2-12b), in which the force acting on the beam (from the vertical bar) is denoted P_3 and the vertical reaction at support D is denoted R_D. No horizontal force acts between the bar and the beam, as can be seen from a free-body diagram of the vertical bar itself (Fig. 2-12c). Therefore, there is no horizontal reaction at support D of the beam.

Taking moments about point D for the free-body diagram of the beam (Fig. 2-12b) gives

$$P_3 = \frac{P_2 b}{a} = \frac{(5600 \text{ lb})(25.0 \text{ in.})}{28.0 \text{ in.}} = 5000 \text{ lb} \tag{a}$$

This force acts downward on the beam (Fig. 2-12b) and upward on the vertical bar (Fig. 2-12c).

Now we can determine the downward reaction at support A (Fig. 2-12c):

$$R_A = P_3 - P_1 = 5000 \text{ lb} - 2100 \text{ lb} = 2900 \text{ lb} \tag{b}$$

The upper part of the vertical bar (segment *AB*) is subjected to an axial compressive force N_1 equal to R_A, or 2900 lb. The lower part (segment *BC*) carries an axial tensile force N_2 equal to P_1, or 2100 lb.

Note: As an alternative to the preceding calculations, we can obtain the reaction R_A from a free-body diagram of the entire structure (instead of from the free-body diagram of beam *BDE*).

Changes in length. With tension considered positive, Eq. (2-5) yields

$$\delta = \sum_{i=1}^{n} \frac{N_i L_i}{E_i A_i} = \frac{N_1 L_1}{EA_1} + \frac{N_2 L_2}{EA_2} \tag{c}$$

$$= \frac{(-2900 \text{ lb})(20.0 \text{ in})}{(29.0 \times 10^6 \text{ psi})(0.25 \text{ in.}^2)} + \frac{(2100 \text{ lb})(34.8 \text{ in.})}{(29.0 \times 10^6 \text{ psi})(0.15 \text{ in.}^2)}$$

$$= -0.0080 \text{ in.} + 0.0168 \text{ in.} = 0.0088 \text{ in.}$$

in which δ is the change in length of bar *ABC*. Since δ is positive, the bar elongates. The displacement of point C is equal to the change in length of the bar:

$$\delta_C = 0.0088 \text{ in.}$$

This displacement is downward.

Example 2-4

A tapered bar AB of solid circular cross section and length L (Fig. 2-13a) is supported at end B and subjected to a tensile load P at the free end A. The diameters of the bar at ends A and B are d_A and d_B, respectively.

Determine the elongation of the bar due to the load P, assuming that the angle of taper is small.

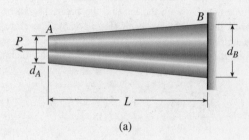

FIG. 2-13 Example 2-4. Change in length of a tapered bar of solid circular cross section

Solution

The bar being analyzed in this example has a constant axial force (equal to the load P) throughout its length. However, the cross-sectional area varies continuously from one end to the other. Therefore, we must use integration (see Eq. 2-7) to determine the change in length.

Cross-sectional area. The first step in the solution is to obtain an expression for the cross-sectional area $A(x)$ at any cross section of the bar. For this purpose, we must establish an origin for the coordinate x. One possibility is to place the origin of coordinates at the free end A of the bar. However, the integrations to be performed will be slightly simplified if we locate the origin of coordinates by extending the sides of the tapered bar until they meet at point O, as shown in Fig. 2-13b.

The distances L_A and L_B from the origin O to ends A and B, respectively, are in the ratio

$$\frac{L_A}{L_B} = \frac{d_A}{d_B} \tag{a}$$

as obtained from similar triangles in Fig. 2-13b. From similar triangles we also get the ratio of the diameter $d(x)$ at distance x from the origin to the diameter d_A at the small end of the bar:

$$\frac{d(x)}{d_A} = \frac{x}{L_A} \quad \text{or} \quad d(x) = \frac{d_A x}{L_A} \tag{b}$$

Therefore, the cross-sectional area at distance x from the origin is

$$A(x) = \frac{\pi [d(x)]^2}{4} = \frac{\pi d_A^2 x^2}{4 L_A^2} \tag{c}$$

Change in length. We now substitute the expression for $A(x)$ into Eq. (2-7) and obtain the elongation δ:

$$\delta = \int \frac{N(x)dx}{EA(x)} = \int_{L_A}^{L_B} \frac{Pdx(4L_A^2)}{E(\pi d_A^2 x^2)} = \frac{4PL_A^2}{\pi E d_A^2} \int_{L_A}^{L_B} \frac{dx}{x^2} \qquad \text{(d)}$$

By performing the integration (see Appendix D for integration formulas available online) and substituting the limits, we get

$$\delta = \frac{4PL_A^2}{\pi E d_A^2} \left[-\frac{1}{x} \right]_{L_A}^{L_B} = \frac{4PL_A^2}{\pi E d_A^2} \left(\frac{1}{L_A} - \frac{1}{L_B} \right) \qquad \text{(e)}$$

This expression for δ can be simplified by noting that

$$\frac{1}{L_A} - \frac{1}{L_B} = \frac{L_B - L_A}{L_A L_B} = \frac{L}{L_A L_B} \qquad \text{(f)}$$

Thus, the equation for δ becomes

$$\delta = \frac{4PL}{\pi E d_A^2} \left(\frac{L_A}{L_B} \right) \qquad \text{(g)}$$

Finally, we substitute $L_A/L_B = d_A/d_B$ (see Eq. a) and obtain

$$\delta = \frac{4PL}{\pi E d_A d_B} \qquad \text{(2-8)} \quad \Longleftarrow$$

This formula gives the elongation of a tapered bar of solid circular cross section. By substituting numerical values, we can determine the change in length for any particular bar.

Note 1: A common mistake is to assume that the elongation of a tapered bar can be determined by calculating the elongation of a prismatic bar that has the same cross-sectional area as the midsection of the tapered bar. Examination of Eq. (2-8) shows that this idea is not valid.

Note 2: The preceding formula for a tapered bar (Eq. 2-8) can be reduced to the special case of a prismatic bar by substituting $d_A = d_B = d$. The result is

$$\delta = \frac{4PL}{\pi E d^2} = \frac{PL}{EA}$$

which we know to be correct.

A general formula such as Eq. (2-8) should be checked whenever possible by verifying that it reduces to known results for *special cases*. If the reduction does not produce a correct result, the original formula is in error. If a correct result is obtained, the original formula may still be incorrect but our confidence in it increases. In other words, this type of check is a necessary but not sufficient condition for the correctness of the original formula.

2.4 STATICALLY INDETERMINATE STRUCTURES

The springs, bars, and cables that we discussed in the preceding sections have one important feature in common—their reactions and internal forces can be determined solely from free-body diagrams and equations of equilibrium. Structures of this type are classified as **statically determinate**. We should note especially that the forces in a statically determinate structure can be found without knowing the properties of the materials. Consider, for instance, the bar AB shown in Fig. 2-14. The calculations for the internal axial forces in both parts of the bar, as well as for the reaction R at the base, are independent of the material of which the bar is made.

Most structures are more complex than the bar of Fig. 2-14, and their reactions and internal forces cannot be found by statics alone. This situation is illustrated in Fig. 2-15, which shows a bar AB fixed at *both* ends. There are now two vertical reactions (R_A and R_B) but only one useful equation of equilibrium—the equation for summing forces in the vertical direction. Since this equation contains two unknowns, it is not sufficient for finding the reactions. Structures of this kind are classified as **statically indeterminate**. To analyze such structures we must supplement the equilibrium equations with additional equations pertaining to the displacements of the structure.

To see how a statically indeterminate structure is analyzed, consider the example of Fig. 2-16a. The prismatic bar AB is attached to rigid supports at both ends and is axially loaded by a force P at an intermediate point C. As already discussed, the reactions R_A and R_B cannot be found by statics alone, because only one **equation of equilibrium** is available:

$$\sum F_{\text{vert}} = 0 \qquad R_A - P + R_B = 0 \qquad \text{(a)}$$

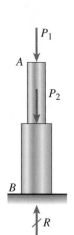

FIG. 2-14 Statically determinate bar

An additional equation is needed in order to solve for the two unknown reactions.

The additional equation is based upon the observation that a bar with both ends fixed does not change in length. If we separate the bar from its supports (Fig. 2-16b), we obtain a bar that is free at both ends and loaded by the three forces, R_A, R_B, and P. These forces cause the bar to change in length by an amount δ_{AB}, which must be equal to zero:

$$\delta_{AB} = 0 \qquad \text{(b)}$$

This equation, called an **equation of compatibility**, expresses the fact that the change in length of the bar must be compatible with the conditions at the supports.

In order to solve Eqs. (a) and (b), we must now express the compatibility equation in terms of the unknown forces R_A and R_B. The relationships between the forces acting on a bar and its changes in length are known as **force-displacement relations**. These relations have various

FIG. 2-15 Statically indeterminate bar

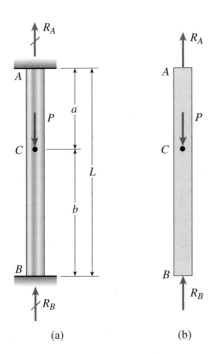

FIG. 2-16 Analysis of a statically indeterminate bar

(a)   (b)

forms depending upon the properties of the material. If the material is linearly elastic, the equation $\delta = PL/EA$ can be used to obtain the force-displacement relations.

Let us assume that the bar of Fig. 2-16 has cross-sectional area A and is made of a material with modulus E. Then the changes in lengths of the upper and lower segments of the bar are, respectively,

$$\delta_{AC} = \frac{R_A a}{EA} \qquad \delta_{CB} = -\frac{R_B b}{EA} \qquad \text{(c,d)}$$

where the minus sign indicates a shortening of the bar. Equations (c) and (d) are the force-displacement relations.

We are now ready to solve simultaneously the three sets of equations (the equation of equilibrium, the equation of compatibility, and the force-displacement relations). In this illustration, we begin by combining the force-displacement relations with the equation of compatibility:

$$\delta_{AB} = \delta_{AC} + \delta_{CB} = \frac{R_A a}{EA} - \frac{R_B b}{EA} = 0 \qquad \text{(e)}$$

Note that this equation contains the two reactions as unknowns.

The next step is to solve simultaneously the equation of equilibrium (Eq. a) and the preceding equation (Eq. e). The results are

$$R_A = \frac{Pb}{L} \qquad R_B = \frac{Pa}{L} \qquad \text{(2-9a,b)}$$

With the reactions known, all other force and displacement quantities can be determined. Suppose, for instance, that we wish to find the downward displacement δ_C of point C. This displacement is equal to the elongation of segment AC:

$$\delta_C = \delta_{AC} = \frac{R_A a}{EA} = \frac{Pab}{LEA} \qquad \text{(2-10)}$$

Also, we can find the stresses in the two segments of the bar directly from the internal axial forces (e.g., $\sigma_{AC} = R_A/A = Pb/AL$).

General Comments

From the preceding discussion we see that the analysis of a statically indeterminate structure involves setting up and solving equations of equilibrium, equations of compatibility, and force-displacement relations. The equilibrium equations relate the loads acting on the structure to the unknown forces (which may be reactions or internal forces), and the compatibility equations express conditions on the displacements of the structure. The force-displacement relations are expressions that use the dimensions and properties of the structural members to relate the forces and displacements of those members. In the case of axially loaded bars that behave in a linearly elastic manner, the relations are based upon the equation $\delta = PL/EA$. Finally, all three sets of equations may be solved simultaneously for the unknown forces and displacements.

In the engineering literature, various terms are used for the conditions expressed by the equilibrium, compatibility, and force-displacement equations. The equilibrium equations are also known as *static* or *kinetic* equations; the compatibility equations are sometimes called *geometric* equations, *kinematic* equations, or equations of *consistent deformations*; and the force-displacement relations are often referred to as *constitutive relations* (because they deal with the *constitution*, or physical properties, of the materials).

For the relatively simple structures discussed in this chapter, the preceding method of analysis is adequate. However, more formalized approaches are needed for complicated structures. Two commonly used methods, the *flexibility method* (also called the *force method*) and the *stiffness method* (also called the *displacement method*), are described in detail in textbooks on structural analysis. Even though these methods are normally used for large and complex structures requiring the solution of hundreds and sometimes thousands of simultaneous equations, they still are based upon the concepts described previously, that is, equilibrium equations, compatibility equations, and force-displacement relations.[*]

The following two examples illustrate the methodology for analyzing statically indeterminate structures consisting of axially loaded members.

---

[*]From a historical viewpoint, it appears that Euler in 1774 was the first to analyze a statically indeterminate system; he considered the problem of a rigid table with four legs supported on an elastic foundation (Refs. 2-2 and 2-3 available online). The next work was done by the French mathematician and engineer L. M. H. Navier, who in 1825 pointed out that statically indeterminate reactions could be found only by taking into account the elasticity of the structure (Ref. 2-4 available online). Navier solved statically indeterminate trusses and beams.

Example 2-5

A solid circular steel cylinder S is encased in a hollow circular copper tube C (Figs. 2-17a and b). The cylinder and tube are compressed between the rigid plates of a testing machine by compressive forces P. The steel cylinder has cross-sectional area A_s and modulus of elasticity E_s, the copper tube has area A_c and modulus E_c, and both parts have length L.

Determine the following quantities: (a) the compressive forces P_s in the steel cylinder and P_c in the copper tube; (b) the corresponding compressive stresses σ_s and σ_c; and (c) the shortening δ of the assembly.

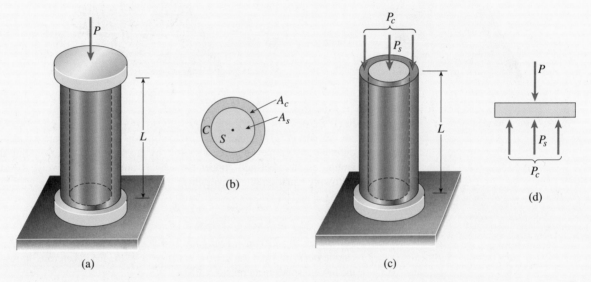

FIG. 2-17 Example 2-5. Analysis of a statically indeterminate structure

Solution

(a) *Compressive forces in the steel cylinder and copper tube.* We begin by removing the upper plate of the assembly in order to expose the compressive forces P_s and P_c acting on the steel cylinder and copper tube, respectively (Fig. 2-17c). The force P_s is the resultant of the uniformly distributed stresses acting over the cross section of the steel cylinder, and the force P_c is the resultant of the stresses acting over the cross section of the copper tube.

Equation of equilibrium. A free-body diagram of the upper plate is shown in Fig. 2-17d. This plate is subjected to the force P and to the unknown compressive forces P_s and P_c ; thus, the equation of equilibrium is

$$\Sigma F_{\text{vert}} = 0 \qquad P_s + P_c - P = 0 \tag{f}$$

This equation, which is the only nontrivial equilibrium equation available, contains two unknowns. Therefore, we conclude that the structure is statically indeterminate.

continued

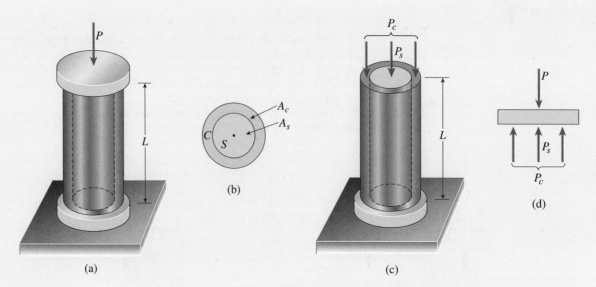

FIG. 2-17 (Repeated)

Equation of compatibility. Because the end plates are rigid, the steel cylinder and copper tube must shorten by the same amount. Denoting the shortenings of the steel and copper parts by δ_s and δ_c, respectively, we obtain the following equation of compatibility:

$$\delta_s = \delta_c \tag{g}$$

Force-displacement relations. The changes in lengths of the cylinder and tube can be obtained from the general equation $\delta = PL/EA$. Therefore, in this example the force-displacement relations are

$$\delta_s = \frac{P_s L}{E_s A_s} \qquad \delta_c = \frac{P_c L}{E_c A_c} \tag{h,i}$$

Solution of equations. We now solve simultaneously the three sets of equations. First, we substitute the force-displacement relations in the equation of compatibility, which gives

$$\frac{P_s L}{E_s A_s} = \frac{P_c L}{E_c A_c} \tag{j}$$

This equation expresses the compatibility condition in terms of the unknown forces.

Next, we solve simultaneously the equation of equilibrium (Eq. f) and the preceding equation of compatibility (Eq. j) and obtain the axial forces in the steel cylinder and copper tube:

$$P_s = P\left(\frac{E_s A_s}{E_s A_s + E_c A_c}\right) \qquad P_c = P\left(\frac{E_c A_c}{E_s A_s + E_c A_c}\right) \tag{2-11a,b}$$

These equations show that the compressive forces in the steel and copper parts are directly proportional to their respective axial rigidities and inversely proportional to the sum of their rigidities.

(b) *Compressive stresses in the steel cylinder and copper tube.* Knowing the axial forces, we can now obtain the compressive stresses in the two materials:

$$\sigma_s = \frac{P_s}{A_s} = \frac{PE_s}{E_sA_s + E_cA_c} \qquad \sigma_c = \frac{P_c}{A_c} = \frac{PE_c}{E_sA_s + E_cA_c} \qquad \text{(2-12a,b)} \quad \Longleftarrow$$

Note that the ratio σ_s/σ_c of the stresses is equal to the ratio E_s/E_c of the moduli of elasticity, showing that in general the "stiffer" material always has the larger stress.

(c) *Shortening of the assembly.* The shortening δ of the entire assembly can be obtained from either Eq. (h) or Eq. (i). Thus, upon substituting the forces (from Eqs. 2-11a and b), we get

$$\delta = \frac{P_s L}{E_s A_s} = \frac{P_c L}{E_c A_c} = \frac{PL}{E_s A_s + E_c A_c} \qquad \text{(2-13)} \quad \Longleftarrow$$

This result shows that the shortening of the assembly is equal to the total load divided by the sum of the stiffnesses of the two parts (recall from Eq. 2-4a that the stiffness of an axially loaded bar is $k = EA/L$).

Alternative solution of the equations. Instead of substituting the force-displacement relations (Eqs. h and i) into the equation of compatibility, we could rewrite those relations in the form

$$P_s = \frac{E_s A_s}{L}\delta_s \qquad P_c = \frac{E_c A_c}{L}\delta_c \qquad \text{(k, l)}$$

and substitute them into the equation of equilibrium (Eq. f):

$$\frac{E_s A_s}{L}\delta_s + \frac{E_c A_c}{L}\delta_c = P \qquad \text{(m)}$$

This equation expresses the equilibrium condition in terms of the unknown displacements. Then we solve simultaneously the equation of compatibility (Eq. g) and the preceding equation, thus obtaining the displacements:

$$\delta_s = \delta_c = \frac{PL}{E_s A_s + E_c A_c} \qquad \text{(n)}$$

which agrees with Eq. (2-13). Finally, we substitute expression (n) into Eqs. (k) and (l) and obtain the compressive forces P_s and P_c (see Eqs. 2-11a and b).

Note: The alternative method of solving the equations is a simplified version of the stiffness (or displacement) method of analysis, and the first method of solving the equations is a simplified version of the flexibility (or force) method. The names of these two methods arise from the fact that Eq. (m) has displacements as unknowns and stiffnesses as coefficients (see Eq. 2-4a), whereas Eq. (j) has forces as unknowns and flexibilities as coefficients (see Eq. 2-4b).

Example 2-6

A horizontal rigid bar AB is pinned at end A and supported by two wires (CD and EF) at points D and F (Fig. 2-18a). A vertical load P acts at end B of the bar. The bar has length $3b$ and wires CD and EF have lengths L_1 and L_2, respectively. Also, wire CD has diameter d_1 and modulus of elasticity E_1; wire EF has diameter d_2 and modulus E_2.

(a) Obtain formulas for the allowable load P if the allowable stresses in wires CD and EF, respectively, are σ_1 and σ_2. (Disregard the weight of the bar itself.)

(b) Calculate the allowable load P for the following conditions: Wire CD is made of aluminum with modulus $E_1 = 72$ GPa, diameter $d_1 = 4.0$ mm, and length $L_1 = 0.40$ m. Wire EF is made of magnesium with modulus $E_2 = 45$ GPa, diameter $d_2 = 3.0$ mm, and length $L_2 = 0.30$ m. The allowable stresses in the aluminum and magnesium wires are $\sigma_1 = 200$ MPa and $\sigma_2 = 175$ MPa, respectively.

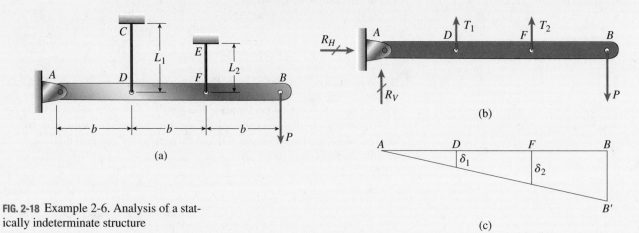

FIG. 2-18 Example 2-6. Analysis of a statically indeterminate structure

Solution

Equation of equilibrium. We begin the analysis by drawing a free-body diagram of bar AB (Fig. 2-18b). In this diagram T_1 and T_2 are the unknown tensile forces in the wires and R_H and R_V are the horizontal and vertical components of the reaction at the support. We see immediately that the structure is statically indeterminate because there are four unknown forces (T_1, T_2, R_H, and R_V) but only three independent equations of equilibrium.

Taking moments about point A (with counterclockwise moments being positive) yields

$$\Sigma M_A = 0 \qquad T_1 b + T_2(2b) - P(3b) = 0 \quad \text{or} \quad T_1 + 2T_2 = 3P \qquad (o)$$

The other two equations, obtained by summing forces in the horizontal direction and summing forces in the vertical direction, are of no benefit in finding T_1 and T_2.

Equation of compatibility. To obtain an equation pertaining to the displacements, we observe that the load P causes bar AB to rotate about the pin support at A, thereby stretching the wires. The resulting displacements are shown in the displacement diagram of Fig. 2-18c, where line AB represents the original position of the rigid bar and line AB' represents the rotated position. The displacements δ_1 and δ_2 are the elongations of the wires. Because these displacements are very small, the bar rotates through a very small angle (shown highly exaggerated in the figure) and we can make calculations on the assumption that points D, F, and B move vertically downward (instead of moving along the arcs of circles).

Because the horizontal distances AD and DF are equal, we obtain the following geometric relationship between the elongations:

$$\delta_2 = 2\delta_1 \tag{p}$$

Equation (p) is the equation of compatibility.

Force-displacement relations. Since the wires behave in a linearly elastic manner, their elongations can be expressed in terms of the unknown forces T_1 and T_2 by means of the following expressions:

$$\delta_1 = \frac{T_1 L_1}{E_1 A_1} \qquad \delta_2 = \frac{T_2 L_2}{E_2 A_2}$$

in which A_1 and A_2 are the cross-sectional areas of wires CD and EF, respectively; that is,

$$A_1 = \frac{\pi d_1^2}{4} \qquad A_2 = \frac{\pi d_2^2}{4}$$

For convenience in writing equations, let us introduce the following notation for the flexibilities of the wires (see Eq. 2-4b):

$$f_1 = \frac{L_1}{E_1 A_1} \qquad f_2 = \frac{L_2}{E_2 A_2} \tag{q,r}$$

Then the force-displacement relations become

$$\delta_1 = f_1 T_1 \qquad \delta_2 = f_2 T_2 \tag{s,t}$$

Solution of equations. We now solve simultaneously the three sets of equations (equilibrium, compatibility, and force-displacement equations). Substituting the expressions from Eqs. (s) and (t) into the equation of compatibility (Eq. p) gives

$$f_2 T_2 = 2 f_1 T_1 \tag{u}$$

The equation of equilibrium (Eq. o) and the preceding equation (Eq. u) each contain the forces T_1 and T_2 as unknown quantities. Solving those two equations simultaneously yields

$$T_1 = \frac{3 f_2 P}{4 f_1 + f_2} \qquad T_2 = \frac{6 f_1 P}{4 f_1 + f_2} \tag{v,w}$$

Knowing the forces T_1 and T_2, we can easily find the elongations of the wires from the force-displacement relations.

continued

(a) *Allowable load P.* Now that the statically indeterminate analysis is completed and the forces in the wires are known, we can determine the permissible value of the load P. The stress σ_1 in wire CD and the stress σ_2 in wire EF are readily obtained from the forces (Eqs. v and w):

$$\sigma_1 = \frac{T_1}{A_1} = \frac{3P}{A_1}\left(\frac{f_2}{4f_1 + f_2}\right) \qquad \sigma_2 = \frac{T_2}{A_2} = \frac{6P}{A_2}\left(\frac{f_1}{4f_1 + f_2}\right)$$

From the first of these equations we solve for the permissible force P_1 based upon the allowable stress σ_1 in wire CD:

$$P_1 = \frac{\sigma_1 A_1(4f_1 + f_2)}{3f_2} \qquad \qquad \text{(2-14a)} \quad \Longleftarrow$$

Similarly, from the second equation we get the permissible force P_2 based upon the allowable stress σ_2 in wire EF:

$$P_2 = \frac{\sigma_2 A_2(4f_1 + f_2)}{6f_1} \qquad \qquad \text{(2-14b)} \quad \Longleftarrow$$

The smaller of these two loads is the maximum allowable load P_{allow}.

(b) *Numerical calculations for the allowable load.* Using the given data and the preceding equations, we obtain the following numerical values:

$$A_1 = \frac{\pi d_1^2}{4} = \frac{\pi (4.0 \text{ mm})^2}{4} = 12.57 \text{ mm}^2$$

$$A_2 = \frac{\pi d_2^2}{4} = \frac{\pi (3.0 \text{ mm})^2}{4} = 7.069 \text{ mm}^2$$

$$f_1 = \frac{L_1}{E_1 A_1} = \frac{0.40 \text{ m}}{(72 \text{ GPa})(12.57 \text{ mm}^2)} = 0.4420 \times 10^{-6} \text{ m/N}$$

$$f_2 = \frac{L_2}{E_2 A_2} = \frac{0.30 \text{ m}}{(45 \text{ GPa})(7.069 \text{ mm}^2)} = 0.9431 \times 10^{-6} \text{ m/N}$$

Also, the allowable stresses are

$$\sigma_1 = 200 \text{ MPa} \qquad \sigma_2 = 175 \text{ MPa}$$

Therefore, substituting into Eqs. (2-14a and b) gives

$$P_1 = 2.41 \text{ kN} \quad P_2 = 1.26 \text{ kN}$$

The first result is based upon the allowable stress σ_1 in the aluminum wire and the second is based upon the allowable stress σ_2 in the magnesium wire. The allowable load is the smaller of the two values:

$$P_{\text{allow}} = 1.26 \text{ kN} \qquad \qquad \Longleftarrow$$

At this load the stress in the magnesium is 175 MPa (the allowable stress) and the stress in the aluminum is $(1.26/2.41)(200 \text{ MPa}) = 105 \text{ MPa}$. As expected, this stress is less than the allowable stress of 200 MPa.

2.5 THERMAL EFFECTS, MISFITS, AND PRESTRAINS

External loads are not the only sources of stresses and strains in a structure. Other sources include *thermal effects* arising from temperature changes, *misfits* resulting from imperfections in construction, and *prestrains* that are produced by initial deformations. Still other causes are settlements (or movements) of supports, inertial loads resulting from accelerating motion, and natural phenomenon such as earthquakes.

Thermal effects, misfits, and prestrains are commonly found in both mechanical and structural systems and are described in this section. As a general rule, they are much more important in the design of statically indeterminate structures than in statically determinate ones.

Thermal Effects

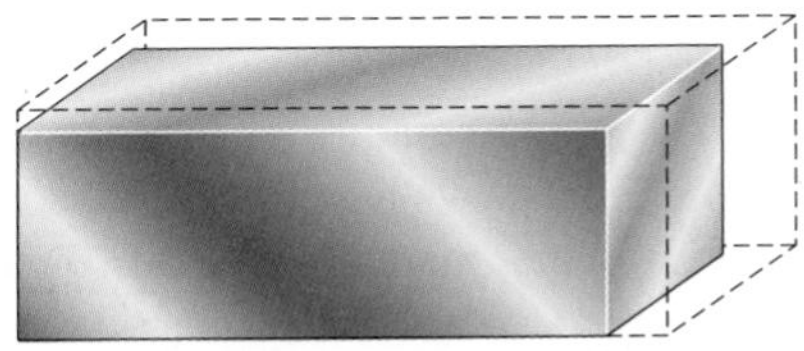

FIG. 2-19 Block of material subjected to an increase in temperature

Changes in temperature produce expansion or contraction of structural materials, resulting in **thermal strains** and **thermal stresses**. A simple illustration of thermal expansion is shown in Fig. 2-19, where the block of material is unrestrained and therefore free to expand. When the block is heated, every element of the material undergoes thermal strains in all directions, and consequently the dimensions of the block increase. If we take corner A as a fixed reference point and let side AB maintain its original alignment, the block will have the shape shown by the dashed lines.

For most structural materials, thermal strain ϵ_T is proportional to the temperature change ΔT; that is,

$$\epsilon_T = \alpha(\Delta T) \tag{2-15}$$

in which α is a property of the material called the **coefficient of thermal expansion**. Since strain is a dimensionless quantity, the coefficient of thermal expansion has units equal to the reciprocal of temperature change. In SI units the dimensions of α can be expressed as either 1/K (the reciprocal of kelvins) or 1/°C (the reciprocal of degrees Celsius). The value of α is the same in both cases because a *change* in temperature is numerically the same in both kelvins and degrees Celsius. In USCS units, the dimensions of α are 1/°F (the reciprocal of degrees Fahrenheit).[*] Typical values of α are listed in Table I-4 of Appendix I (available online).

When a **sign convention** is needed for thermal strains, we usually assume that expansion is positive and contraction is negative.

To demonstrate the relative importance of thermal strains, we will compare thermal strains with load-induced strains in the following manner. Suppose we have an axially loaded bar with longitudinal strains given by the equation $\epsilon = \sigma/E$, where σ is the stress and E is the

---

[*]For a discussion of temperature units and scales, see Section B.4 of Appendix B available online.

modulus of elasticity. Then suppose we have an identical bar subjected to a temperature change ΔT, which means that the bar has thermal strains given by Eq. (2-15). Equating the two strains gives the equation

$$\sigma = E\alpha(\Delta T)$$

From this equation we can calculate the axial stress σ that produces the same strain as does the temperature change ΔT. For instance, consider a stainless steel bar with $E = 30 \times 10^6$ psi and $\alpha = 9.6 \times 10^{-6}/°F$. A quick calculation from the preceding equation for σ shows that a change in temperature of 100°F produces the same strain as a stress of 29,000 psi. This stress is in the range of typical allowable stresses for stainless steel. Thus, a relatively modest change in temperature produces strains of the same magnitude as the strains caused by ordinary loads, which shows that temperature effects can be important in engineering design.

Ordinary structural materials expand when heated and contract when cooled, and therefore an increase in temperature produces a positive thermal strain. Thermal strains usually are reversible, in the sense that the member returns to its original shape when its temperature returns to the original value. However, a few special metallic alloys have recently been developed that do not behave in the customary manner. Instead, over certain temperature ranges their dimensions decrease when heated and increase when cooled.

Water is also an unusual material from a thermal standpoint—it expands when heated at temperatures above 4°C and also expands when cooled below 4°C. Thus, water has its maximum density at 4°C.

Now let us return to the block of material shown in Fig. 2-19. We assume that the material is homogeneous and isotropic and that the temperature increase ΔT is uniform throughout the block. We can calculate the increase in *any* dimension of the block by multiplying the original dimension by the thermal strain. For instance, if one of the dimensions is L, then that dimension will increase by the amount

$$\delta_T = \epsilon_T L = \alpha(\Delta T)L \tag{2-16}$$

Equation (2-16) is a **temperature-displacement relation**, analogous to the force-displacement relations described in the preceding section. It can be used to calculate changes in lengths of structural members subjected to uniform temperature changes, such as the elongation δ_T of the prismatic bar shown in Fig. 2-20. (The transverse dimensions of the bar also change, but these changes are not shown in the figure since they usually have no effect on the axial forces being transmitted by the bar.)

In the preceding discussions of thermal strains, we assumed that the structure had no restraints and was able to expand or contract freely. These conditions exist when an object rests on a frictionless surface or hangs in open space. In such cases no stresses are produced by a uniform temperature change throughout the object, although nonuniform

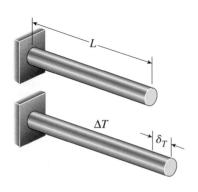

FIG. 2-20 Increase in length of a prismatic bar due to a uniform increase in temperature (Eq. 2-16)

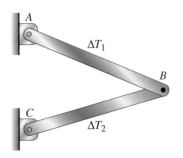

FIG. 2-21 Statically determinate truss with a uniform temperature change in each member

temperature changes may produce internal stresses. However, many structures have supports that prevent free expansion and contraction, in which case **thermal stresses** will develop even when the temperature change is uniform throughout the structure.

To illustrate some of these ideas about thermal effects, consider the two-bar truss *ABC* of Fig. 2-21 and assume that the temperature of bar *AB* is changed by ΔT_1 and the temperature of bar *BC* is changed by ΔT_2. Because the truss is statically determinate, both bars are free to lengthen or shorten, resulting in a displacement of joint *B*. However, there are no stresses in either bar and no reactions at the supports. This conclusion applies generally to **statically determinate structures**; that is, uniform temperature changes in the members produce thermal strains (and the corresponding changes in lengths) without producing any corresponding stresses.

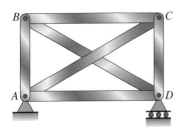

FIG. 2-22 Statically indeterminate truss subjected to temperature changes

Forces can develop in statically indeterminate trusses due to temperature and prestrain
(Barros & Barros/Getty Images)

A **statically indeterminate structure** may or may not develop temperature stresses, depending upon the character of the structure and the nature of the temperature changes. To illustrate some of the possibilities, consider the statically indeterminate truss shown in Fig. 2-22. Because the supports of this structure permit joint *D* to move horizontally, no stresses are developed when the *entire* truss is heated uniformly. All members increase in length in proportion to their original lengths, and the truss becomes slightly larger in size.

However, if some bars are heated and others are not, thermal stresses will develop because the statically indeterminate arrangement of the bars prevents free expansion. To visualize this condition, imagine that just one bar is heated. As this bar becomes longer, it meets resistance from the other bars, and therefore stresses develop in all members.

The analysis of a statically indeterminate structure with temperature changes is based upon the concepts discussed in the preceding section, namely equilibrium equations, compatibility equations, and displacement relations. The principal difference is that we now use temperature-displacement relations (Eq. 2-16) in addition to force-displacement relations (such as $\delta = PL/EA$) when performing the analysis. The following two examples illustrate the procedures in detail.

Example 2-7

A prismatic bar AB of length L is held between immovable supports (Fig. 2-23a). If the temperature of the bar is raised uniformly by an amount ΔT, what thermal stress σ_T is developed in the bar? (Assume that the bar is made of linearly elastic material.)

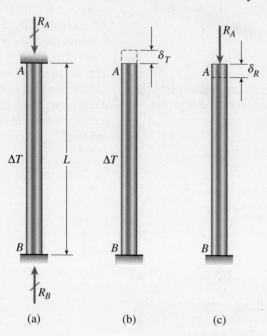

FIG. 2-23 Example 2-7. Statically indeterminate bar with uniform temperature increase ΔT

(a)   (b)   (c)

Solution

Because the temperature increases, the bar tends to elongate but is restrained by the rigid supports at A and B. Therefore, reactions R_A and R_B are developed at the supports, and the bar is subjected to uniform compressive stresses.

Equation of equilibrium. The only forces acting on the bar are the reactions shown in Fig. 2-23a. Therefore, equilibrium of forces in the vertical direction gives

$$\Sigma F_{\text{vert}} = 0 \qquad R_B - R_A = 0 \tag{a}$$

Since this is the only nontrivial equation of equilibrium, and since it contains two unknowns, we see that the structure is statically indeterminate and an additional equation is needed.

Equation of compatibility. The equation of compatibility expresses the fact that the change in length of the bar is zero (because the supports do not move):

$$\delta_{AB} = 0 \tag{b}$$

To determine this change in length, we remove the upper support of the bar and obtain a bar that is fixed at the base and free to displace at the upper end (Figs. 2-23b and c). When only the temperature change is acting (Fig. 2-23b), the bar elongates by an amount δ_T, and when only the reaction R_A is acting, the bar shortens by an amount δ_R (Fig. 2-23c). Thus, the net change in length is $\delta_{AB} = \delta_T - \delta_R$, and the equation of compatibility becomes

$$\delta_{AB} = \delta_T - \delta_R = 0 \qquad\qquad \text{(c)}$$

Displacement relations. The increase in length of the bar due to the temperature change is given by the temperature-displacement relation (Eq. 2-16):

$$\delta_T = \alpha(\Delta T)L \qquad\qquad \text{(d)}$$

in which α is the coefficient of thermal expansion. The decrease in length due to the force R_A is given by the force-displacement relation:

$$\delta_R = \frac{R_A L}{EA} \qquad\qquad \text{(e)}$$

in which E is the modulus of elasticity and A is the cross-sectional area.

Solution of equations. Substituting the displacement relations (d) and (e) into the equation of compatibility (Eq. c) gives the following equation:

$$\delta_T - \delta_R = \alpha(\Delta T)L - \frac{R_A L}{EA} = 0 \qquad\qquad \text{(f)}$$

We now solve simultaneously the preceding equation and the equation of equilibrium (Eq. a) for the reactions R_A and R_B:

$$R_A = R_B = EA\alpha(\Delta T) \qquad\qquad \text{(2-17)}$$

From these results we obtain the thermal stress σ_T in the bar:

$$\sigma_T = \frac{R_A}{A} = \frac{R_B}{A} = E\alpha(\Delta T) \qquad\qquad \text{(2-18)} \quad\Longleftarrow$$

This stress is compressive when the temperature of the bar increases.

Note 1: In this example the reactions are independent of the length of the bar and the stress is independent of both the length and the cross-sectional area (see Eqs. 2-17 and 2-18). Thus, once again we see the usefulness of a symbolic solution, because these important features of the bar's behavior might not be noticed in a purely numerical solution.

Note 2: When determining the thermal elongation of the bar (Eq. d), we assumed that the material was homogeneous and that the increase in temperature was uniform throughout the volume of the bar. Also, when determining the decrease in length due to the reactive force (Eq. e), we assumed linearly elastic behavior of the material. These limitations should always be kept in mind when writing equations such as Eqs. (d) and (e).

Note 3: The bar in this example has zero longitudinal displacements, not only at the fixed ends but also at every cross section. Thus, there are no axial strains in this bar, and we have the special situation of *longitudinal stresses without longitudinal strains*. Of course, there are transverse strains in the bar, from both the temperature change and the axial compression.

Example 2-8

A sleeve in the form of a circular tube of length L is placed around a bolt and fitted between washers at each end (Fig. 2-24a). The nut is then turned until it is just snug. The sleeve and bolt are made of different materials and have different cross-sectional areas. (Assume that the coefficient of thermal expansion α_S of the sleeve is greater than the coefficient α_B of the bolt.)

(a) If the temperature of the entire assembly is raised by an amount ΔT, what stresses σ_S and σ_B are developed in the sleeve and bolt, respectively?

(b) What is the increase δ in the length L of the sleeve and bolt?

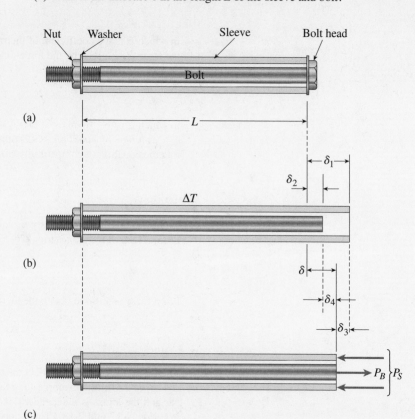

FIG. 2-24 Example 2-8. Sleeve and bolt assembly with uniform temperature increase ΔT

Solution

Because the sleeve and bolt are of different materials, they will elongate by different amounts when heated and allowed to expand freely. However, when they are held together by the assembly, free expansion cannot occur and thermal stresses are developed in both materials. To find these stresses, we use the same concepts as in any statically indeterminate analysis—equilibrium equations, compatibility equations, and displacement relations. However, we cannot formulate these equations until we disassemble the structure.

A simple way to cut the structure is to remove the head of the bolt, thereby allowing the sleeve and bolt to expand freely under the temperature change ΔT

(Fig. 2-24b). The resulting elongations of the sleeve and bolt are denoted δ_1 and δ_2, respectively, and the corresponding *temperature-displacement relations* are

$$\delta_1 = \alpha_S(\Delta T)L \quad \delta_2 = \alpha_B(\Delta T)L \qquad \text{(g,h)}$$

Since α_S is greater than α_B, the elongation δ_1 is greater than δ_2, as shown in Fig. 2-24b.

The axial forces in the sleeve and bolt must be such that they shorten the sleeve and stretch the bolt until the final lengths of the sleeve and bolt are the same. These forces are shown in Fig. 2-24c, where P_S denotes the compressive force in the sleeve and P_B denotes the tensile force in the bolt. The corresponding shortening δ_3 of the sleeve and elongation δ_4 of the bolt are

$$\delta_3 = \frac{P_S L}{E_S A_S} \qquad \delta_4 = \frac{P_B L}{E_B A_B} \qquad \text{(i,j)}$$

in which $E_S A_S$ and $E_B A_B$ are the respective axial rigidities. Equations (i) and (j) are the *load-displacement relations*.

Now we can write an *equation of compatibility* expressing the fact that the final elongation δ is the same for both the sleeve and bolt. The elongation of the sleeve is $\delta_1 - \delta_3$ and of the bolt is $\delta_2 + \delta_4$; therefore,

$$\delta = \delta_1 - \delta_3 = \delta_2 + \delta_4 \qquad \text{(k)}$$

Substituting the temperature-displacement and load-displacement relations (Eqs. g to j) into this equation gives

$$\delta = \alpha_S(\Delta T)L - \frac{P_S L}{E_S A_S} = \alpha_B(\Delta T)L + \frac{P_B L}{E_B A_B} \qquad \text{(l)}$$

from which we get

$$\frac{P_S L}{E_S A_S} + \frac{P_B L}{E_B A_B} = \alpha_S(\Delta T)L - \alpha_B(\Delta T)L \qquad \text{(m)}$$

which is a modified form of the compatibility equation. Note that it contains the forces P_S and P_B as unknowns.

An *equation of equilibrium* is obtained from Fig. 2-24c, which is a free-body diagram of the part of the assembly remaining after the head of the bolt is removed. Summing forces in the horizontal direction gives

$$P_S = P_B \qquad \text{(n)}$$

which expresses the obvious fact that the compressive force in the sleeve is equal to the tensile force in the bolt.

We now solve simultaneously Eqs. (m) and (n) and obtain the axial forces in the sleeve and bolt:

$$P_S = P_B = \frac{(\alpha_S - \alpha_B)(\Delta T)E_S A_S E_B A_B}{E_S A_S + E_B A_B} \qquad \text{(2-19)}$$

When deriving this equation, we assumed that the temperature increased and that the coefficient α_S was greater than the coefficient α_B. Under these conditions, P_S is the compressive force in the sleeve and P_B is the tensile force in the bolt.

continued

The results will be quite different if the temperature increases but the coefficient α_S is less than the coefficient α_B. Under these conditions, a gap will open between the bolt head and the sleeve and there will be no stresses in either part of the assembly.

(a) *Stresses in the sleeve and bolt.* Expressions for the stresses σ_S and σ_B in the sleeve and bolt, respectively, are obtained by dividing the corresponding forces by the appropriate areas:

$$\sigma_S = \frac{P_S}{A_S} = \frac{(\alpha_S - \alpha_B)(\Delta T)E_S E_B A_B}{E_S A_S + E_B A_B} \qquad (2\text{-}20a) \quad \Longleftarrow$$

$$\sigma_B = \frac{P_B}{A_B} = \frac{(\alpha_S - \alpha_B)(\Delta T)E_S A_S E_B}{E_S A_S + E_B A_B} \qquad (2\text{-}20b) \quad \Longleftarrow$$

Under the assumed conditions, the stress σ_S in the sleeve is compressive and the stress σ_B in the bolt is tensile. It is interesting to note that these stresses are independent of the length of the assembly and their magnitudes are inversely proportional to their respective areas (that is, $\sigma_S/\sigma_B = A_B/A_S$).

(b) *Increase in length of the sleeve and bolt.* The elongation δ of the assembly can be found by substituting either P_S or P_B from Eq. (2-19) into Eq. (l), yielding

$$\delta = \frac{(\alpha_S E_S A_S + \alpha_B E_B A_B)(\Delta T)L}{E_S A_S + E_B A_B} \qquad (2\text{-}21) \quad \Longleftarrow$$

With the preceding formulas available, we can readily calculate the forces, stresses, and displacements of the assembly for any given set of numerical data.

Note: As a partial check on the results, we can see if Eqs. (2-19), (2-20), and (2-21) reduce to known values in simplified cases. For instance, suppose that the bolt is rigid and therefore unaffected by temperature changes. We can represent this situation by setting $\alpha_B = 0$ and letting E_B become infinitely large, thereby creating an assembly in which the sleeve is held between rigid supports. Substituting these values into Eqs. (2-19), (2-20), and (2-21), we find

$$P_S = E_S A_S \alpha_S(\Delta T) \qquad \sigma_S = E_S \alpha_S(\Delta T) \qquad \delta = 0$$

These results agree with those of Example 2-7 for a bar held between rigid supports (compare with Eqs. 2-17 and 2-18, and with Eq. b).

As a second special case, suppose that the sleeve and bolt are made of the same material. Then both parts will expand freely and will lengthen the same amount when the temperature changes. No forces or stresses will be developed. To see if the derived equations predict this behavior, we substitute $\alpha_S = \alpha_B = \alpha$ into Eqs. (2-19), (2-20), and (2-21) and obtain

$$P_S = P_B = 0 \qquad \sigma_S = \sigma_B = 0 \qquad \delta = \alpha(\Delta T)L$$

which are the expected results.

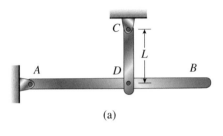

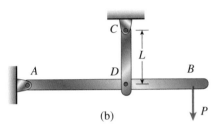

FIG. 2-25 Statically determinate structure with a small misfit

Misfits and Prestrains

Suppose that a member of a structure is manufactured with its length slightly different from its prescribed length. Then the member will not fit into the structure in its intended manner, and the geometry of the structure will be different from what was planned. We refer to situations of this kind as **misfits**. Sometimes misfits are intentionally created in order to introduce strains into the structure at the time it is built. Because these strains exist before any loads are applied to the structure, they are called **prestrains**. Accompanying the prestrains are prestresses, and the structure is said to be **prestressed**. Common examples of prestressing are spokes in bicycle wheels (which would collapse if not prestressed), the pretensioned faces of tennis racquets, shrink-fitted machine parts, and prestressed concrete beams.

If a structure is **statically determinate**, small misfits in one or more members will not produce strains or stresses, although there will be departures from the theoretical configuration of the structure. To illustrate this statement, consider a simple structure consisting of a horizontal beam AB supported by a vertical bar CD (Fig. 2-25a). If bar CD has exactly the correct length L, the beam will be horizontal at the time the structure is built. However, if the bar is slightly longer than intended, the beam will make a small angle with the horizontal. Nevertheless, there will be no strains or stresses in either the bar or the beam attributable to the incorrect length of the bar. Furthermore, if a load P acts at the end of the beam (Fig. 2-25b), the stresses in the structure due to that load will be unaffected by the incorrect length of bar CD.

In general, if a structure is statically determinate, the presence of small misfits will produce small changes in geometry but no strains or stresses. Thus, the effects of a misfit are similar to those of a temperature change.

The situation is quite different if the structure is **statically indeterminate**, because then the structure is not free to adjust to misfits (just as it is not free to adjust to certain kinds of temperature changes). To show this, consider a beam supported by two vertical bars (Fig. 2-26a). If both bars have exactly the correct length L, the structure can be assembled with no strains or stresses and the beam will be horizontal.

Suppose, however, that bar CD is slightly longer than the prescribed length. Then, in order to assemble the structure, bar CD must be compressed by external forces (or bar EF stretched by external forces), the bars must be fitted into place, and then the external forces must be released. As a result, the beam will deform and rotate, bar CD will be in compression, and bar EF will be in tension. In other words, prestrains will exist in all members and the structure will be prestressed, even though no external loads are acting. If a load P is now added (Fig. 2-26b), additional strains and stresses will be produced.

The analysis of a statically indeterminate structure with misfits and prestrains proceeds in the same general manner as described previously for loads and temperature changes. The basic ingredients of the analysis

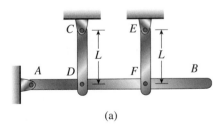

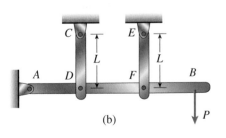

FIG. 2-26 Statically indeterminate structure with a small misfit

are equations of equilibrium, equations of compatibility, force-displacement relations, and (if appropriate) temperature-displacement relations. The methodology is illustrated in Example 2-9.

Bolts and Turnbuckles

Prestressing a structure requires that one or more parts of the structure be stretched or compressed from their theoretical lengths. A simple way to produce a change in length is to tighten a bolt or a turnbuckle. In the case of a **bolt** (Fig. 2-27) each turn of the nut will cause the nut to travel along the bolt a distance equal to the spacing p of the threads (called the *pitch* of the threads). Thus, the distance δ traveled by the nut is

$$\delta = np \tag{2-22}$$

in which n is the number of revolutions of the nut (not necessarily an integer). Depending upon how the structure is arranged, turning the nut can stretch or compress a member.

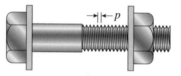

FIG. 2-27 The *pitch* of the threads is the distance from one thread to the next

In the case of a **double-acting turnbuckle** (Fig. 2-28), there are two end screws. Because a right-hand thread is used at one end and a left-hand thread at the other, the device either lengthens or shortens when the buckle is rotated. Each full turn of the buckle causes it to travel a distance p along each screw, where again p is the pitch of the threads. Therefore, if the turnbuckle is tightened by one turn, the screws are drawn closer together by a distance $2p$ and the effect is to shorten the device by $2p$. For n turns, we have

$$\delta = 2np \tag{2-23}$$

Turnbuckles are often inserted in cables and then tightened, thus creating initial tension in the cables, as illustrated in the following example.

FIG. 2-28 Double-acting turnbuckle. (Each full turn of the turnbuckle shortens or lengthens the cable by $2p$, where p is the pitch of the screw threads.)

Example 2-9

The mechanical assembly shown in Fig. 2-29a consists of a copper tube, a rigid end plate, and two steel cables with turnbuckles. The slack is removed from the cables by rotating the turnbuckles until the assembly is snug but with no initial stresses. (Further tightening of the turnbuckles will produce a prestressed condition in which the cables are in tension and the tube is in compression.)

(a) Determine the forces in the tube and cables (Fig. 2-29a) when the turnbuckles are tightened by n turns.

(b) Determine the shortening of the tube.

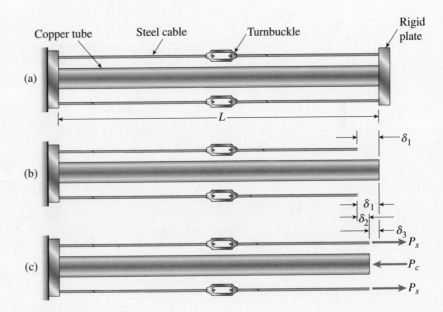

FIG. 2-29 Example 2-9. Statically indeterminate assembly with a copper tube in compression and two steel cables in tension

Solution

We begin the analysis by removing the plate at the right-hand end of the assembly so that the tube and cables are free to change in length (Fig. 2-29b). Rotating the turnbuckles through n turns will shorten the cables by a distance

$$\delta_1 = 2np \tag{o}$$

as shown in Fig. 2-29b.

The tensile forces in the cables and the compressive force in the tube must be such that they elongate the cables and shorten the tube until their final lengths are the same. These forces are shown in Fig. 2-29c, where P_s denotes the tensile force in one of the steel cables and P_c denotes the compressive force in the copper tube. The elongation of a cable due to the force P_s is

$$\delta_2 = \frac{P_s L}{E_s A_s} \tag{p}$$

continued

in which $E_s A_s$ is the axial rigidity and L is the length of a cable. Also, the compressive force P_c in the copper tube causes it to shorten by

$$\delta_3 = \frac{P_c L}{E_c A_c} \tag{q}$$

in which $E_c A_c$ is the axial rigidity of the tube. Equations (p) and (q) are the *load-displacement relations*.

The final shortening of one of the cables is equal to the shortening δ_1 caused by rotating the turnbuckle minus the elongation δ_2 caused by the force P_s. This final shortening of the cable must equal the shortening δ_3 of the tube:

$$\delta_1 - \delta_2 = \delta_3 \tag{r}$$

which is the *equation of compatibility*.

Substituting the turnbuckle relation (Eq. o) and the load-displacement relations (Eqs. p and q) into the preceding equation yields

$$2np - \frac{P_s L}{E_s A_s} = \frac{P_c L}{E_c A_c} \tag{s}$$

or

$$\frac{P_s L}{E_s A_s} + \frac{P_c L}{E_c A_c} = 2np \tag{t}$$

which is a modified form of the compatibility equation. Note that it contains P_s and P_c as unknowns.

From Fig. 2-29c, which is a free-body diagram of the assembly with the end plate removed, we obtain the following equation of equilibrium:

$$2P_s = P_c \tag{u}$$

(a) *Forces in the cables and tube.* Now we solve simultaneously Eqs. (t) and (u) and obtain the axial forces in the steel cables and copper tube, respectively:

$$P_s = \frac{2np E_c A_c E_s A_s}{L(E_c A_c + 2E_s A_s)} \qquad P_c = \frac{4np E_c A_c E_s A_s}{L(E_c A_c + 2E_s A_s)} \qquad \text{(2-24a,b)} \quad \Longleftarrow$$

Recall that the forces P_s are tensile forces and the force P_c is compressive. If desired, the stresses σ_s and σ_c in the steel and copper can now be obtained by dividing the forces P_s and P_c by the cross-sectional areas A_s and A_c, respectively.

(b) *Shortening of the tube.* The decrease in length of the tube is the quantity δ_3 (see Fig. 2-29 and Eq. q):

$$\delta_3 = \frac{P_c L}{E_c A_c} = \frac{4np E_s A_s}{E_c A_c + 2E_s A_s} \qquad \text{(2-25)} \quad \Longleftarrow$$

With the preceding formulas available, we can readily calculate the forces, stresses, and displacements of the assembly for any given set of numerical data.

2.6 STRESSES ON INCLINED SECTIONS

In our previous discussions of tension and compression in axially loaded members, the only stresses we considered were the normal stresses acting on cross sections. These stresses are pictured in Fig. 2-30, where we consider a bar AB subjected to axial loads P.

When the bar is cut at an intermediate cross section by a plane mn (perpendicular to the x axis), we obtain the free-body diagram shown in Fig. 2-30b. The normal stresses acting over the cut section may be calculated from the formula $\sigma_x = P/A$ provided that the stress distribution is uniform over the entire cross-sectional area A. As explained in Chapter 1, this condition exists if the bar is prismatic, the material is homogeneous, the axial force P acts at the centroid of the cross-sectional area, and the cross section is away from any localized stress concentrations. Of course, there are no shear stresses acting on the cut section, because it is perpendicular to the longitudinal axis of the bar.

For convenience, we usually show the stresses in a two-dimensional view of the bar (Fig. 2-30c) rather than the more complex three-dimensional view (Fig. 2-30b). However, when working with two-dimensional figures we must not forget that the bar has a thickness

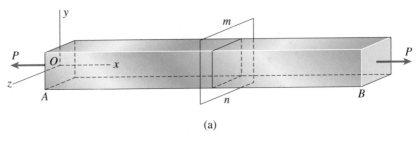

(a)

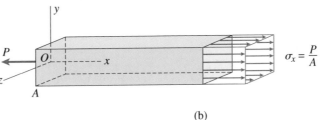

(b)

FIG. 2-30 Prismatic bar in tension showing the stresses acting on cross section mn: (a) bar with axial forces P, (b) three-dimensional view of the cut bar showing the normal stresses, and (c) two-dimensional view

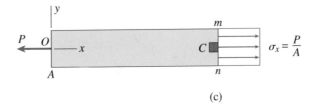

(c)

perpendicular to the plane of the figure. This third dimension must be considered when making derivations and calculations.

Stress Elements

The most useful way of representing the stresses in the bar of Fig. 2-30 is to isolate a small element of material, such as the element labeled C in Fig. 2-30c, and then show the stresses acting on all faces of this element. An element of this kind is called a **stress element**. The stress element at point C is a small rectangular block (it doesn't matter whether it is a cube or a rectangular parallelepiped) with its right-hand face lying in cross section mn.

The dimensions of a stress element are assumed to be infinitesimally small, but for clarity we draw the element to a large scale, as in Fig. 2-31a. In this case, the edges of the element are parallel to the x, y, and z axes, and the only stresses are the normal stresses σ_x acting on the x faces (recall that the x faces have their normals parallel to the x axis). Because it is more convenient, we usually draw a two-dimensional view of the element (Fig. 2-31b) instead of a three-dimensional view.

Stresses on Inclined Sections

The stress element of Fig. 2-31 provides only a limited view of the stresses in an axially loaded bar. To obtain a more complete picture, we need to investigate the stresses acting on **inclined sections**, such as the section cut by the inclined plane pq in Fig. 2-32a. Because the stresses are the same throughout the entire bar, the stresses acting over the inclined section must be uniformly distributed, as pictured in the free-body diagrams of Fig. 2-32b (three-dimensional view) and Fig. 2-32c (two-dimensional view). From the equilibrium of the free body we know that the resultant of the stresses must be a horizontal force P. (The resultant is drawn with a dashed line in Figs. 2-32b and 2-32c.)

FIG. 2-31 Stress element at point C of the axially loaded bar shown in Fig. 2-30c: (a) three-dimensional view of the element, and (b) two-dimensional view of the element

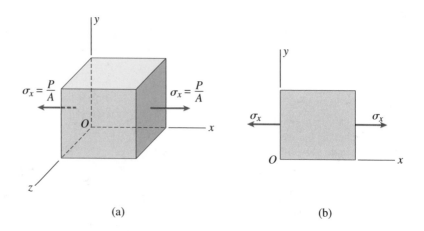

(a)   (b)

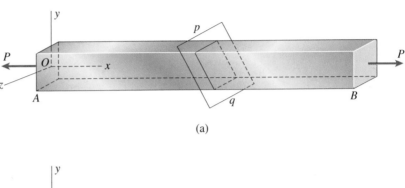

(a)

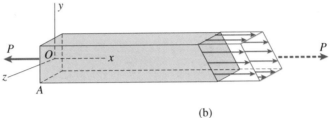

(b)

FIG. 2-32 Prismatic bar in tension
showing the stresses acting on an
inclined section *pq*: (a) bar with axial
forces *P*, (b) three-dimensional view of
the cut bar showing the stresses, and
(c) two-dimensional view

(c)

As a preliminary matter, we need a scheme for specifying the **orien-tation** of the inclined section *pq*. A standard method is to specify the angle θ between the *x* axis and the normal *n* to the section (see Fig. 2-33a on the next page). Thus, the angle θ for the inclined section shown in the figure is approximately 30°. By contrast, cross section *mn* (Fig. 2-30a) has an angle θ equal to zero (because the normal to the section is the *x* axis). For additional examples, consider the stress element of Fig. 2-31. The angle θ for the right-hand face is 0, for the top face is 90° (a longitudinal section of the bar), for the left-hand face is 180°, and for the bottom face is 270° (or −90°).

Let us now return to the task of finding the stresses acting on section *pq* (Fig. 2-33b). As already mentioned, the resultant of these stresses is a force *P* acting in the *x* direction. This resultant may be resolved into two components, a normal force *N* that is perpendicular to the inclined plane *pq* and a shear force *V* that is tangential to it. These force components are

$$N = P \cos \theta \qquad V = P \sin \theta \qquad (2\text{-}26\text{a,b})$$

Associated with the forces *N* and *V* are normal and shear stresses that are uniformly distributed over the inclined section (Figs. 2-33c and d). The

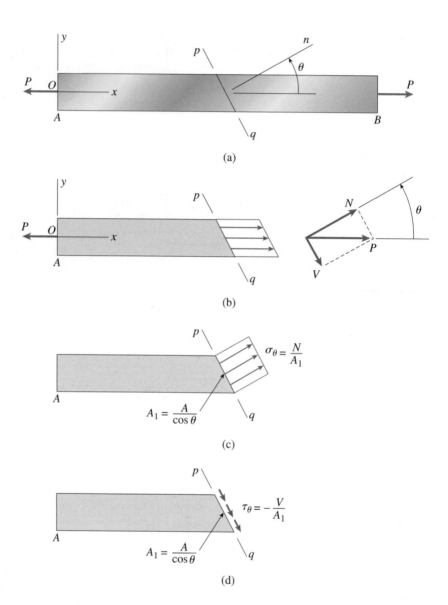

FIG. 2-33 Prismatic bar in tension showing the stresses acting on an inclined section *pq*

normal stress is equal to the normal force N divided by the area of the section, and the shear stress is equal to the shear force V divided by the area of the section. Thus, the stresses are

$$\sigma = \frac{N}{A_1} \qquad \tau = \frac{V}{A_1} \qquad (2\text{-}27a,b)$$

in which A_1 is the area of the inclined section, as follows:

$$A_1 = \frac{A}{\cos \theta} \qquad (2\text{-}28)$$

As usual, A represents the cross-sectional area of the bar. The stresses σ and τ act in the directions shown in Figs. 2-33c and d, that is, in the same directions as the normal force N and shear force V, respectively.

At this point we need to establish a standardized **notation and sign convention** for stresses acting on inclined sections. We will use a subscript θ to indicate that the stresses act on a section inclined at an angle θ (Fig. 2-34), just as we use a subscript x to indicate that the stresses act on a section perpendicular to the x axis (see Fig. 2-30). Normal stresses σ_θ are positive in tension and shear stresses τ_θ are positive when they tend to produce counterclockwise rotation of the material, as shown in Fig. 2-34.

FIG. 2-34 Sign convention for stresses acting on an inclined section. (Normal stresses are positive when in tension and shear stresses are positive when they tend to produce counterclockwise rotation.)

For a bar in tension, the normal force N produces positive normal stresses σ_θ (see Fig. 2-33c) and the shear force V produces negative shear stresses τ_θ (see Fig. 2-33d). These stresses are given by the following equations (see Eqs. 2-26, 2-27, and 2-28):

$$\sigma_\theta = \frac{N}{A_1} = \frac{P}{A}\cos^2\theta \qquad \tau_\theta = -\frac{V}{A_1} = -\frac{P}{A}\sin\theta\cos\theta$$

Introducing the notation $\sigma_x = P/A$, in which σ_x is the normal stress on a cross section, and also using the trigonometric relations

$$\cos^2\theta = \frac{1}{2}(1 + \cos 2\theta) \qquad \sin\theta\cos\theta = \frac{1}{2}(\sin 2\theta)$$

we get the following expressions for the **normal and shear stresses:**

$$\sigma_\theta = \sigma_x\cos^2\theta = \frac{\sigma_x}{2}(1 + \cos 2\theta) \qquad (2\text{-}29a)$$

$$\tau_\theta = -\sigma_x\sin\theta\cos\theta = -\frac{\sigma_x}{2}(\sin 2\theta) \qquad (2\text{-}29b)$$

These equations give the stresses acting on an inclined section oriented at an angle θ to the x axis (Fig. 2-34).

It is important to recognize that Eqs. (2-29a) and (2-29b) were derived only from statics, and therefore they are independent of the material. Thus, these equations are valid for any material, whether it behaves linearly or nonlinearly, elastically or inelastically.

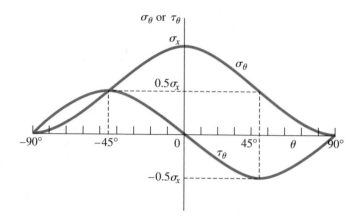

FIG. 2-35 Graph of normal stress σ_θ and shear stress τ_θ versus angle θ of the inclined section (see Fig. 2-34 and Eqs. 2-29a and b)

Maximum Normal and Shear Stresses

The manner in which the stresses vary as the inclined section is cut at various angles is shown in Fig. 2-35. The horizontal axis gives the angle θ as it varies from $-90°$ to $+90°$, and the vertical axis gives the stresses σ_θ and τ_θ. Note that a positive angle θ is measured counterclockwise from the x axis (Fig. 2-34) and a negative angle is measured clockwise.

As shown on the graph, the normal stress σ_θ equals σ_x when $\theta = 0$. Then, as θ increases or decreases, the normal stress diminishes until at $\theta = \pm90°$ it becomes zero, because there are no normal stresses on sections cut parallel to the longitudinal axis. The **maximum normal stress** occurs at $\theta = 0$ and is

$$\sigma_{max} = \sigma_x \tag{2-30}$$

Also, we note that when $\theta = \pm45°$, the normal stress is one-half the maximum value.

The shear stress τ_θ is zero on cross sections of the bar ($\theta = 0$) as well as on longitudinal sections ($\theta = \pm90°$). Between these extremes, the stress varies as shown on the graph, reaching the largest positive value when $\theta = -45°$ and the largest negative value when $\theta = +45°$. These **maximum shear stresses** have the same magnitude:

$$\tau_{max} = \frac{\sigma_x}{2} \tag{2-31}$$

but they tend to rotate the element in opposite directions.

The maximum stresses in a **bar in tension** are shown in Fig. 2-36. Two stress elements are selected—element A is oriented at $\theta = 0°$ and element B is oriented at $\theta = 45°$. Element A has the maximum normal stresses (Eq. 2-30) and element B has the maximum shear stresses (Eq. 2-31). In the case of element A (Fig. 2-36b), the only stresses are the maximum normal stresses (no shear stresses exist on any of the faces).

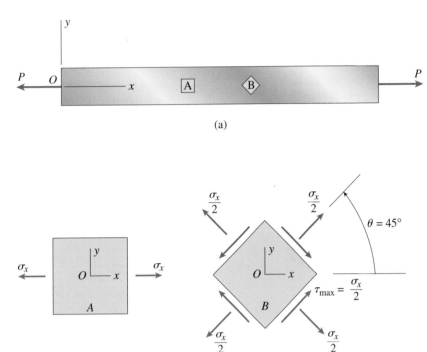

FIG. 2-36 Normal and shear stresses acting on stress elements oriented at $\theta = 0°$ and $\theta = 45°$ for a bar in tension

(a)

(b)

(c)

In the case of element B (Fig. 2-36c), both normal and shear stresses act on all faces (except, of course, the front and rear faces of the element). Consider, for instance, the face at $45°$ (the upper right-hand face). On this face the normal and shear stresses (from Eqs. 2-29a and b) are $\sigma_x/2$ and $-\sigma_x/2$, respectively. Hence, the normal stress is tension (positive) and the shear stress acts clockwise (negative) against the element. The stresses on the remaining faces are obtained in a similar manner by substituting $\theta = 135°$, $-45°$, and $-135°$ into Eqs. (2-29a and b).

Thus, in this special case of an element oriented at $\theta = 45°$, the normal stresses on all four faces are the same (equal to $\sigma_x/2$) and all four shear stresses have the maximum magnitude (equal to $\sigma_x/2$). Also, note that the shear stresses acting on perpendicular planes are equal in magnitude and have directions either toward, or away from, the line of intersection of the planes, as discussed in detail in Section 1.6.

If a bar is loaded in compression instead of tension, the stress σ_x will be compression and will have a negative value. Consequently, all stresses acting on stress elements will have directions opposite to those for a bar in tension. Of course, Eqs. (2-29a and b) can still be used for the calculations simply by substituting σ_x as a negative quantity.

Load

Load

FIG. 2-37 Shear failure along a 45° plane of a wood block loaded in compression

Even though the maximum shear stress in an axially loaded bar is only one-half the maximum normal stress, the shear stress may cause failure if the material is much weaker in shear than in tension. An example of a shear failure is pictured in Fig. 2-37, which shows a block of wood that was loaded in compression and failed by shearing along a 45° plane.

A similar type of behavior occurs in mild steel loaded in tension. During a tensile test of a flat bar of low-carbon steel with polished surfaces, visible *slip bands* appear on the sides of the bar at approximately 45° to the axis (Fig. 2-38). These bands indicate that the material is failing in shear along the planes on which the shear stress is maximum. Such bands were first observed by G. Piobert in 1842 and W. Lüders in 1860 (see Refs. 2-5 and 2-6 available online), and today they are called either *Lüders' bands* or *Piobert's bands*. They begin to appear when the yield stress is reached in the bar (point B in Fig. 1-10 of Section 1.3).

Uniaxial Stress

The state of stress described throughout this section is called **uniaxial stress**, for the obvious reason that the bar is subjected to simple tension or compression in just one direction. The most important orientations of stress elements for uniaxial stress are $\theta = 0$ and $\theta = 45°$ (Fig. 2-36b and c); the former has the maximum normal stress and the latter has the maximum shear stress. If sections are cut through the bar at other angles, the stresses acting on the faces of the corresponding stress elements can be determined from Eqs. (2-29a and b), as illustrated in Examples 2-10 and 2-11 that follow.

Uniaxial stress is a special case of a more general stress state known as *plane stress*, which is described in detail in Chapter 6.

Load

FIG. 2-38 Slip bands (or Lüders' bands) in a polished steel specimen loaded in tension

Load

Example 2-10

A prismatic bar having cross-sectional area $A = 1200 \text{ mm}^2$ is compressed by an axial load $P = 90$ kN (Fig. 2-39a).

(a) Determine the stresses acting on an inclined section pq cut through the bar at an angle $\theta = 25°$.

(b) Determine the complete state of stress for $\theta = 25°$ and show the stresses on a properly oriented stress element.

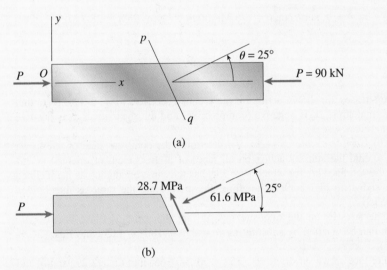

(a)

(b)

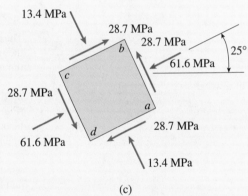

(c)

FIG. 2-39 Example 2-10. Stresses on an inclined section

Solution

(a) *Stresses on the inclined section.* To find the stresses acting on a section at $\theta = 25°$, we first calculate the normal stress σ_x acting on a cross section:

$$\sigma_x = -\frac{P}{A} = -\frac{90\,\text{kN}}{1200\,\text{mm}^2} = -75 \text{ MPa}$$

continued

where the minus sign indicates that the stress is compressive. Next, we calculate the normal and shear stresses from Eqs. (2-29a and b) with $\theta = 25°$, as follows:

$$\sigma_\theta = \sigma_x \cos^2 \theta = (-75 \text{ MPa})(\cos 25°)^2 = -61.6 \text{ MPa} \qquad \Longleftarrow$$

$$\tau_\theta = -\sigma_x \sin \theta \cos \theta = (75 \text{ MPa})(\sin 25°)(\cos 25°) = 28.7 \text{ MPa} \qquad \Longleftarrow$$

These stresses are shown acting on the inclined section in Fig. 2-39b. Note that the normal stress σ_θ is negative (compressive) and the shear stress τ_θ is positive (counterclockwise).

(b) *Complete state of stress.* To determine the complete state of stress, we need to find the stresses acting on all faces of a stress element oriented at 25° (Fig. 2-39c). Face *ab*, for which $\theta = 25°$, has the same orientation as the inclined plane shown in Fig. 2-39b. Therefore, the stresses are the same as those given previously.

The stresses on the opposite face *cd* are the same as those on face *ab*, which can be verified by substituting $\theta = 25° + 180° = 205°$ into Eqs. (2-29a and b).

For face *ad* we substitute $\theta = 25° - 90° = -65°$ into Eqs. (2-29a and b) and obtain

$$\sigma_\theta = -13.4 \text{ MPa} \qquad \tau_\theta = -28.7 \text{ MPa}$$

These same stresses apply to the opposite face *bc*, as can be verified by substituting $\theta = 25° + 90° = 115°$ into Eqs. (2-29a and b). Note that the normal stress is compressive and the shear stress acts clockwise.

The complete state of stress is shown by the stress element of Fig. 2-39c. A sketch of this kind is an excellent way to show the directions of the stresses and the orientations of the planes on which they act.

Example 2-11

A compression bar having a square cross section of width b must support a load $P = 8000$ lb (Fig. 2-40a). The bar is constructed from two pieces of material that are connected by a glued joint (known as a *scarf joint*) along plane pq, which is at an angle $\alpha = 40°$ to the vertical. The material is a structural plastic for which the allowable stresses in compression and shear are 1100 psi and 600 psi, respectively. Also, the allowable stresses in the glued joint are 750 psi in compression and 500 psi in shear.

Determine the minimum width b of the bar.

Solution

For convenience, let us rotate a segment of the bar to a horizontal position (Fig. 2-40b) that matches the figures used in deriving the equations for the stresses on an inclined section (see Figs. 2-33 and 2-34). With the bar in this position, we see that the normal n to the plane of the glued joint (plane pq) makes an angle $\beta = 90° - \alpha$, or $50°$, with the axis of the bar. Since the angle θ is defined as positive when counterclockwise (Fig. 2-34), we conclude that $\theta = -50°$ for the glued joint.

The cross-sectional area of the bar is related to the load P and the stress σ_x acting on the cross sections by the equation

$$A = \frac{P}{\sigma_x} \tag{a}$$

Therefore, to find the required area, we must determine the value of σ_x corresponding to each of the four allowable stresses. Then the smallest value of σ_x will determine the required area. The values of σ_x are obtained by rearranging Eqs. (2-29a and b) as follows:

$$\sigma_x = \frac{\sigma_\theta}{\cos^2\theta} \qquad \sigma_x = -\frac{\tau_\theta}{\sin\theta\cos\theta} \tag{2-32a,b}$$

We will now apply these equations to the glued joint and to the plastic.

(a) *Values of σ_x based upon the allowable stresses in the glued joint.* For compression in the glued joint we have $\sigma_\theta = -750$ psi and $\theta = -50°$. Substituting into Eq. (2-32a), we get

$$\sigma_x = \frac{-750 \text{ psi}}{(\cos -50°)^2} = -1815 \text{ psi} \tag{b}$$

For shear in the glued joint we have an allowable stress of 500 psi. However, it is not immediately evident whether τ_θ is $+500$ psi or -500 psi. One

continued

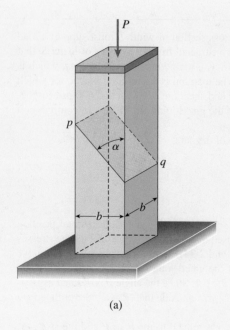

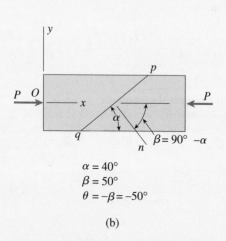

(a)

$\alpha = 40°$

$\beta = 50°$

$\theta = -\beta = -50°$

(b)

FIG. 2-40 Example 2-11. Stresses on an inclined section

approach is to substitute both $+500$ psi and -500 psi into Eq. (2-32b) and then select the value of σ_x that is negative. The other value of σ_x will be positive (tension) and does not apply to this bar. Another approach is to inspect the bar itself (Fig. 2-40b) and observe from the directions of the loads that the shear stress will act clockwise against plane pq, which means that the shear stress is negative. Therefore, we substitute $\tau_\theta = -500$ psi and $\theta = -50°$ into Eq. (2-32b) and obtain

$$\sigma_x = -\frac{-500 \text{ psi}}{(\sin -50°)(\cos -50°)} = -1015 \text{ psi} \qquad \text{(c)}$$

(b) *Values of σ_x based upon the allowable stresses in the plastic.* The maximum compressive stress in the plastic occurs on a cross section. Therefore, since the allowable stress in compression is 1100 psi, we know immediately that

$$\sigma_x = -1100 \text{ psi} \qquad \text{(d)}$$

The maximum shear stress occurs on a plane at $45°$ and is numerically equal to $\sigma_x/2$ (see Eq. 2-31). Since the allowable stress in shear is 600 psi, we obtain

$$\sigma_x = -1200 \text{ psi} \qquad \text{(e)}$$

This same result can be obtained from Eq. (2-32b) by substituting $\tau_\theta = 600$ psi and $\theta = 45°$.

(c) *Minimum width of the bar.* Comparing the four values of σ_x (Eqs. b, c, d, and e), we see that the smallest is $\sigma_x = -1015$ psi. Therefore, this value governs the design. Substituting into Eq. (a), and using only numerical values, we obtain the required area:

$$A = \frac{8000 \text{ lb}}{1015 \text{ psi}} = 7.88 \text{ in.}^2$$

Since the bar has a square cross section ($A = b^2$), the minimum width is

$$b_{\min} = \sqrt{A} = \sqrt{7.88 \text{ in.}^2} = 2.81 \text{ in.}$$

Any width larger than $b_{\min}$ will ensure that the allowable stresses are not exceeded

CHAPTER SUMMARY & REVIEW

In Chapter 2, we investigated the behavior of axially loaded bars acted on by distributed loads, such as self weight, and also temperature changes and prestrains. We developed force-displacement relations for use in computing changes in lengths of bars under both uniform (i.e., constant force over its entire length) and nonuniform conditions (i.e., axial forces, and perhaps also cross-sectional area, vary over the length of the bar). Then, equilibrium and compatibility equations were developed for statically indeterminate structures in a superposition procedure leading to solution for all unknown forces, stresses, etc. We developed equations for normal and shear stresses on inclined sections and, from these equations, found maximum normal and shear stresses along the bar. The major concepts presented in this chapter are as follows:

1. The elongation or shortening (δ) of prismatic bars subjected to tensile or compressive centroidal loads is proportional to both the load (P) and the length (L) of the bar, and inversely proportional to the axial rigidity (EA) of the bar; this relationship is called a **force-displacement relation**.

$$\delta = \frac{PL}{EA}$$

2. Cables are **tension-only elements**, and an effective modulus of elasticity (E_e) and effective cross-sectional area (A_e) should be used to account for the tightening effect that occurs when cables are placed under load.

3. The axial rigidity per unit length of a bar is referred to as its **stiffness** (k), and the inverse relationship is the **flexibility** (f) of the bar.

$$\delta = Pf = \frac{P}{k} \quad f = \frac{L}{EA} = \frac{1}{k}$$

4. The summation of the displacements of the individual segments of a nonprismatic bar equals the elongation or shortening of the entire bar (δ).

$$\delta = \sum_{i=1}^{n} \frac{N_i L_i}{E_i A_i}$$

Free-body diagrams are used to find the axial force (N_i) in each segment i; if axial forces and/or cross-sectional areas vary continuously, an integral expression is required.

$$\delta = \int_0^L d\delta = \int_0^L \frac{N(x)dx}{EA(x)}$$

5. If the bar structure is **statically indeterminate**, additional equations (beyond those available from statics) are required to solve for unknown forces. **Compatibility equations** are used to relate bar displacements to support conditions and thereby generate additional relationships among the unknowns. It is convenient to use a **superposition** of "released" (or statically determinate) structures to represent the actual statically indeterminate bar structure.

continued

6. **Thermal effects** result in displacements proportional to the temperature change (ΔT) and the length (L) of the bar but not stresses in statically determinate structures. The coefficient of thermal expansion (α) of the material also is required to compute axial strains (ϵ_T) and axial displacements (δ_T) due to thermal effects.

$$\epsilon_T = \alpha(\Delta T) \qquad \delta_T = \epsilon_T L = \alpha(\Delta T)L$$

7. **Misfits** and **prestrains** induce axial forces only in statically indeterminate bars.

8. **Maximum normal** (σ_{max}) and **shear stresses** (τ_{max}) can be obtained by considering an inclined stress element for a bar loaded by axial forces. The maximum normal stress occurs along the axis of the bar, but the maximum shear stress occurs at an inclination of 45° to the axis of the bar, and the maximum shear stress is one-half of the maximum normal stress.

$$\sigma_{max} = \sigma_x \qquad \tau_{max} = \frac{\sigma_x}{2}$$

PROBLEMS CHAPTER 2

Changes in Lengths of Axially Loaded Members

2.2-1 The L-shaped arm ABC shown in the figure lies in a vertical plane and pivots about a horizontal pin at A. The arm has constant cross-sectional area and total weight W. A vertical spring of stiffness k supports the arm at point B. Obtain a formula for the elongation of the spring due to the weight of the arm.

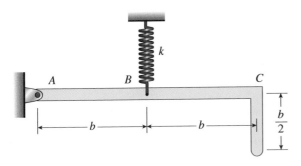

PROB. 2.2-1

2.2-2 A steel cable with nominal diameter 25 mm (see Table 2-1) is used in a construction yard to lift a bridge section weighing 38 kN, as shown in the figure. The cable has an effective modulus of elasticity $E = 140$ GPa.

(a) If the cable is 14 m long, how much will it stretch when the load is picked up?

(b) If the cable is rated for a maximum load of 70 kN, what is the factor of safety with respect to failure of the cable?

PROB. 2.2-2

2.2-3 A steel wire and a copper wire have equal lengths and support equal loads P (see figure). The moduli of elasticity for the steel and copper are $E_s = 30,000$ ksi and $E_c = 18,000$ ksi, respectively.

(a) If the wires have the same diameters, what is the ratio of the elongation of the copper wire to the elongation of the steel wire?

(b) If the wires stretch the same amount, what is the ratio of the diameter of the copper wire to the diameter of the steel wire?

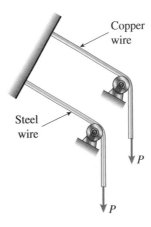

PROB. 2.2-3

2.2-4 By what distance h does the cage shown in the figure move downward when the weight W is placed inside it? (See the figure on the next page.)

Consider only the effects of the stretching of the cable, which has axial rigidity $EA = 10,700$ kN. The pulley at A has diameter $d_A = 300$ mm and the pulley at B has diameter $d_B = 150$ mm. Also, the distance $L_1 = 4.6$ m, the distance $L_2 = 10.5$ m, and the weight $W = 22$ kN. (*Note:* When calculating the length of the cable, include the parts of the cable that go around the pulleys at A and B.)

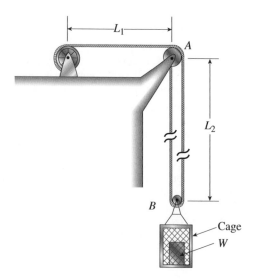

2.2-5 A safety valve on the top of a tank containing steam under pressure p has a discharge hole of diameter d (see figure). The valve is designed to release the steam when the pressure reaches the value p_{max}.

If the natural length of the spring is L and its stiffness is k, what should be the dimension h of the valve? (Express your result as a formula for h.)

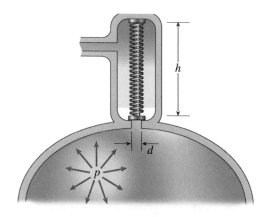

2.2-6 The device shown in the figure consists of a pointer ABC supported by a spring of stiffness $k = 800$ N/m. The spring is positioned at distance $b = 150$ mm from the pinned end A of the pointer. The device is adjusted so that when there is no load P, the pointer reads zero on the angular scale.

If the load $P = 8$ N, at what distance x should the load be placed so that the pointer will read 3° on the scale?

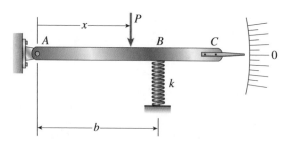

2.2-7 Two rigid bars, AB and CD, rest on a smooth horizontal surface (see figure). Bar AB is pivoted end A, and bar CD is pivoted at end D. The bars are connected to each other by two linearly elastic springs of stiffness k. Before the load P is applied, the lengths of the springs are such that the bars are parallel and the springs are without stress.

Derive a formula for the displacement δ_C at point C when the load P is acting near point B as shown. (Assume that the bars rotate through very small angles under the action of the load P.)

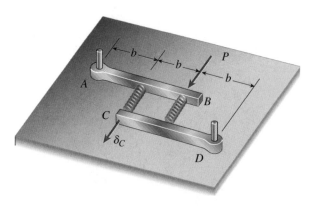

2.2-8 The three-bar truss ABC shown in the figure has a span $L = 3$ m and is constructed of steel pipes having cross-sectional area $A = 3900$ mm^2 and modulus of elasticity $E = 200$ GPa. Identical loads P act both vertically and horizontally at joint C, as shown.

(a) If $P = 650$ kN, what is the horizontal displacement of joint B?

(b) What is the maximum permissible load value P_{max} if the displacement of joint B is limited to 1.5 mm?

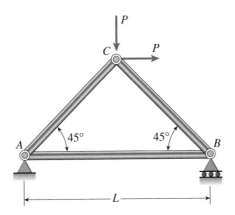

PROB. 2.2-8

2.2-9 An aluminum wire having a diameter $d = 1/10$ in. and length $L = 12$ ft is subjected to a tensile load P (see figure). The aluminum has modulus of elasticity $E = 10,600$ ksi

If the maximum permissible elongation of the wire is 1/8 in. and the allowable stress in tension is 10 ksi, what is the allowable load P_{max}?

PROB. 2.2-9

2.2-10 A uniform bar AB of weight $W = 25$ N is supported by two springs, as shown in the figure. The spring on the left has stiffness $k_1 = 300$ N/m and natural length $L_1 = 250$ mm. The corresponding quantities for the spring on the right are $k_2 = 400$ N/m and $L_2 = 200$ mm. The distance between the springs is $L = 350$ mm, and the spring on the right is suspended from a support that is distance $h = 80$ mm below the point of support for the spring on the left. Neglect the weight of the springs.

(a) At what distance x from the left-hand spring [(figure part (a)] should a load $P = 18$ N be placed in order to bring the bar to a horizontal position?

(b) If P is now removed, what new value of k_1 is required so that the bar [(figure part (a)] will hang in a horizontal position under weight W?

(c) If P is removed and $k_1 = 300$ N/m, what distance b should spring k_1 be moved to the right so that the bar (figure part a) will hang in a horizontal position under weight W?

(d) If the spring on the left is now replaced by two springs in series ($k_1 = 300$N/m, k_3) with overall natural length $L_1 = 250$ mm [(see figure part (b)], what value of k_3 is required so that the bar will hang in a horizontal position under weight W?

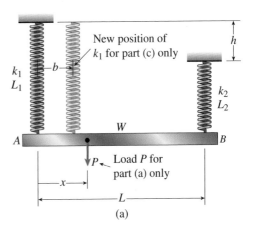

(a)

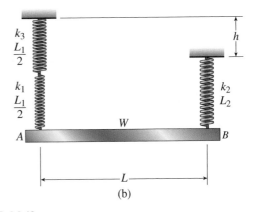

(b)

PROB. 2.2-10

2.2-11 A hollow, circular, cast-iron pipe ($E_c = 12,000$ ksi) supports a brass rod ($E_b = 14,000$ ksi) and weight $W = 2$ kips, as shown in figure on the next page. The outside diameter of the pipe is $d_c = 6$ in.

(a) If the allowable compressive stress in the pipe is 5000 psi and the allowable shortening of the pipe is 0.02 in., what is the minimum required wall thickness $t_{c,min}$? (Include the weights of the rod and steel cap in your calculations.)

(b) What is the elongation of the brass rod δ_r due to both load W and its own weight?

(c) What is the minimum required clearance h?

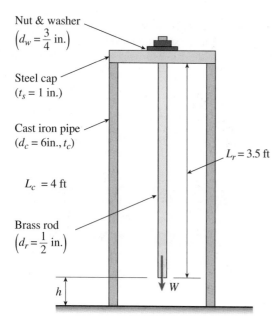

PROB. 2.2-11

2.2-13 A framework ABC consists of two rigid bars AB and BC, each having length b (see the first part of the figure below). The bars have pin connections at A, B, and C and are joined by a spring of stiffness k. The spring is attached at the midpoints of the bars. The framework has a pin support at A and a roller support at C, and the bars are at an angle α to the horizontal.

When a vertical load P is applied at joint B (see the second part of the figure below) the roller support C moves to the right, the spring is stretched, and the angle of the bars decreases from α to the angle θ.

Determine the angle θ and the increase δ in the distance between points A and C. (Use the following data; $b = 8.0$ in., $k = 16$ lb/in., $\alpha = 45°$, and $P = 10$ lb.)

2.2-14 Solve the preceding problem for the following data: $b = 200$ mm, $k = 3.2$ kN/m, $\alpha = 45°$, and $P = 50$ N.

2.2-12 The horizontal rigid beam $ABCD$ is supported by vertical bars BE and CF and is loaded by vertical forces $P_1 = 400$ kN and $P_2 = 360$ kN acting at points A and D, respectively (see figure). Bars BE and CF are made of steel ($E = 200$ GPa) and have cross-sectional areas $A_{BE} = 11,100$ mm^2 and $A_{CF} = 9,280$ mm^2. The distances between various points on the bars are shown in the figure.

Determine the vertical displacements δ_A and δ_D of points A and D, respectively.

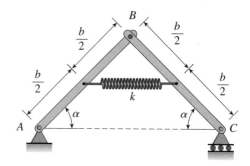

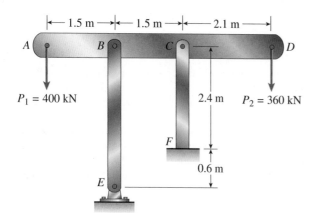

PROB. 2.2-12

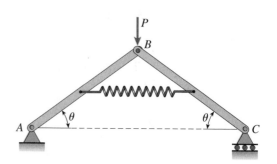

PROBS. 2.2-13 and 2.2-14

Changes in Lengths Under Nonuniform Conditions

2.3-1 Calculate the elongation of a copper bar of solid circular cross section with tapered ends when it is stretched by axial loads of magnitude 3.0 k (see figure).

The length of the end segments is 20 in. and the length of the prismatic middle segment is 50 in. Also, the diameters at cross sections A, B, C, and D are 0.5, 1.0, 1.0, and 0.5 in., respectively, and the modulus of elasticity is 18,000 ksi. (*Hint:* Use the result of Example 2-4.)

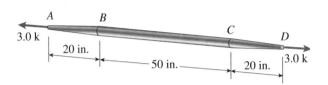

PROB. 2.3-1

2.3-2 A long, rectangular copper bar under a tensile load P hangs from a pin that is supported by two steel posts (see figure). The copper bar has a length of 2.0 m, a cross-sectional area of 4800 mm^2, and a modulus of elasticity $E_c = 120$ GPa. Each steel post has a height of 0.5 m, a cross-sectional area of 4500 mm^2, and a modulus of elasticity $E_s = 200$ GPa.

(a) Determine the downward displacement δ of the lower end of the copper bar due to a load $P = 180$ kN.

(b) What is the maximum permissible load P_{max} if the displacement δ is limited to 1.0 mm?

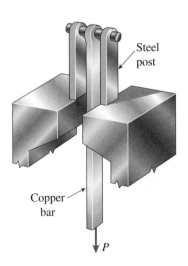

PROB. 2.3-2

2.3-3 A steel bar AD (see figure) has a cross-sectional area of 0.40 in.^{2} and is loaded by forces $P_1 = 2700$ lb, $P_2 = 1800$ lb, and $P_3 = 1300$ lb. The lengths of the segments of the bar are $a = 60$ in., $b = 24$ in., and $c = 36$ in.

(a) Assuming that the modulus of elasticity $E = 30 \times 10^6$ psi, calculate the change in length δ of the bar. Does the bar elongate or shorten?

(b) By what amount P should the load P_3 be increased so that the bar does not change in length when the three loads are applied?

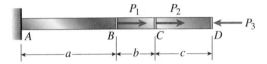

PROB. 2.3-3

2.3-4 A rectangular bar of length L has a slot in the middle half of its length (see figure). The bar has width b, thickness t, and modulus of elasticity E. The slot has width $b/4$.

(a) Obtain a formula for the elongation δ of the bar due to the axial loads P.

(b) Calculate the elongation of the bar if the material is high-strength steel, the axial stress in the middle region is 160 MPa, the length is 750 mm, and the modulus of elasticity is 210 GPa.

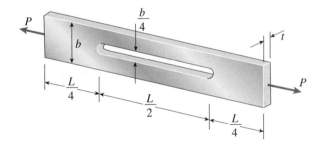

PROBS. 2.3-4 and 2.3-5

2.3-5 Solve the preceding problem if the axial stress in the middle region is 24,000 psi, the length is 30 in., and the modulus of elasticity is 30×10^6 psi.

2.3-6 A two-story building has steel columns AB in the first floor and BC in the second floor, as shown in the figure. The roof load P_1 equals 400 kN and the second-floor load P_2 equals 720 kN. Each column has length $L = 3.75$ m. The cross-sectional areas of the first- and second-floor columns are 11,000 mm^2 and 3,900 mm^2, respectively.

(a) Assuming that $E = 206$ GPa, determine the total shortening δ_{AC} of the two columns due to the combined action of the loads P_1 and P_2.

(b) How much additional load P_0 can be placed at the top of the column (point C) if the total shortening δ_{AC} is not to exceed 4.0 mm?

(c) Finally, if loads P are applied at the ends and $d_{max} = d_2/2$, what is the permissible length x of the hole if shortening is to be limited to 8.0 mm? [(See figure part (c.)]

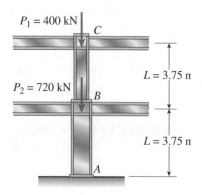

$P_1 = 400$ kN

C

$L = 3.75$ m

$P_2 = 720$ kN

B

$L = 3.75$ m

A

PROB. 2.3-6

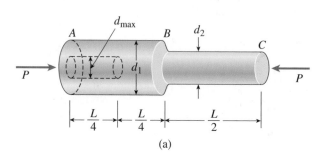

(a)

2.3-7 A steel bar 8.0 ft long has a circular cross section of diameter $d_1 = 0.75$ in. over one-half of its length and diameter $d_2 = 0.5$ in. over the other half (see figure). The modulus of elasticity $E = 30 \times 10^6$ psi.

(a) How much will the bar elongate under a tensile load $P = 5000$ lb?

(b) If the same volume of material is made into a bar of constant diameter d and length 8.0 ft, what will be the elongation under the same load P?

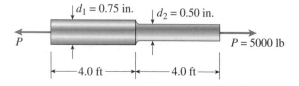

$d_1 = 0.75$ in. $d_2 = 0.50$ in.

P $P = 5000$ lb

—4.0 ft—   —4.0 ft—

PROB. 2.3-7

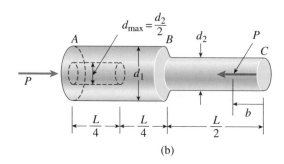

(b)

2.3-8 A bar ABC of length L consists of two parts of equal lengths but different diameters. Segment AB has diameter $d_1 = 100$ mm, and segment BC has diameter $d_2 = 60$ mm. Both segments have length $L/2 = 0.6$ m. A longitudinal hole of diameter d is drilled through segment AB for one-half of its length (distance $L/4 = 0.3$ m). The bar is made of plastic having modulus of elasticity $E = 4.0$ GPa. Compressive loads $P = 110$ kN act at the ends of the bar.

(a) If the shortening of the bar is limited to 8.0 mm, what is the maximum allowable diameter d_{max} of the hole? (See figure part a.)

(b) Now, if d_{max} is instead set at $d_2/2$, at what distance b from end C should load P be applied to limit the bar shortening to 8.0 mm? [(See figure part (b.)]

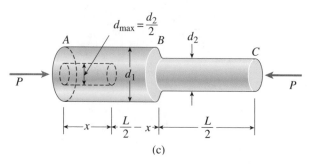

(c)

PROB. 2.3-8

2.3-9 A wood pile, driven into the earth, supports a load P entirely by friction along its sides (see figure). The friction force f per unit length of pile is assumed to be uniformly distributed over the surface of the pile. The pile has length L, cross-sectional area A, and modulus of elasticity E.

(a) Derive a formula for the shortening δ of the pile in terms of P, L, E, and A.

(b) Draw a diagram showing how the compressive stress σ_c varies throughout the length of the pile.

2.3-10 Consider the copper tubes joined below using a "sweated" joint. Use the properties and dimensions given.

(a) Find the total elongation of segment 2-3-4 ($\delta_{2\text{-}4}$) for an applied tensile force of $P = 5$ kN. Use $E_c = 120$ GPa.

(b) If the yield strength in shear of the tin-lead solder is $\tau_y = 30$ MPa and the tensile yield strength of the copper is $\sigma_y = 200$ MPa, what is the maximum load P_{max} that can be applied to the joint if the desired factor of safety in shear is $\text{FS}_\tau = 2$ and in tension is $\text{FS}_\sigma = 1.7$?

(c) Find the value of L_2 at which tube and solder capacities are equal.

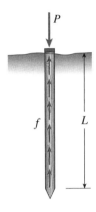

PROB. 2.3-9

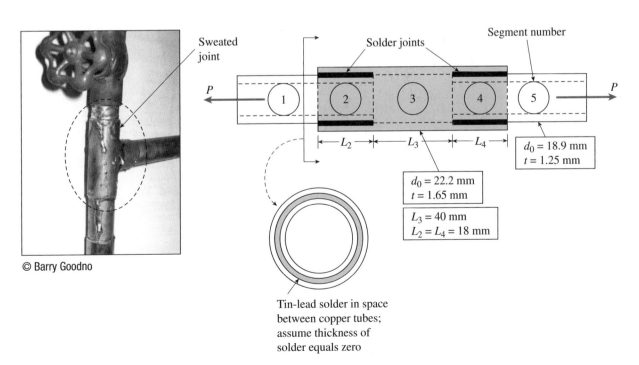

© Barry Goodno

Tin-lead solder in space between copper tubes; assume thickness of solder equals zero

PROB. 2.3-10

2.3-11 The nonprismatic cantilever circular bar shown has an internal cylindrical hole of diameter $d/2$ from 0 to x, so the net area of the cross section for Segment 1 is $(3/4)A$. Load P is applied at x, and load $P/2$ is applied at $x = L$. Assume that E is constant.

(a) Find reaction force R_1 at support.

(b) Find internal axial forces N_i in segments 1 and 2.

(c) Find x required to obtain axial displacement at joint 3 of $\delta_3 = PL/EA$.

(d) In (c), what is the displacement at joint 2, δ_2?

(e) If P acts at $x = 2L/3$ and $P/2$ at joint 3 is replaced by βP, find β so that $\delta_3 = PL/EA$.

(f) Draw the *axial force* (AFD: $N(x)$, $0 \leq x \leq L$) and *axial displacement* (ADD: $\delta(x)$, $0 \leq x \leq L$) *diagrams* using results from (b) through (d) above.

2.3-12 A prismatic bar AB of length L, cross-sectional area A, modulus of elasticity E, and weight W hangs vertically under its own weight (see figure).

(a) Derive a formula for the downward displacement δ_C of point C, located at distance h from the lower end of the bar.

(b) What is the elongation δ_B of the entire bar?

(c) What is the ratio β of the elongation of the upper half of the bar to the elongation of the lower half of the bar?

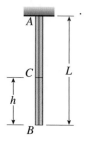

PROB. 2.3-12

2.3-13 A flat bar of rectangular cross section, length L, and constant thickness t is subjected to tension by forces P (see figure). The width of the bar varies linearly from b_1 at the smaller end to b_2 at the larger end. Assume that the angle of taper is small.

(a) Derive the following formula for the elongation of the bar:

$$\delta = \frac{PL}{Et(b_2 - b_1)} \ln \frac{b_2}{b_1}$$

(b) Calculate the elongation, assuming $L = 5$ ft, $t = 1.0$ in., $P = 25$ k, $b_1 = 4.0$ in., $b_2 = 6.0$ in., and $E = 30 \times 10^6$ psi.

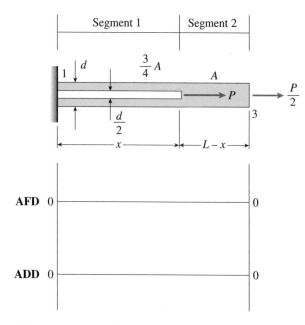

PROB. 2.3-11

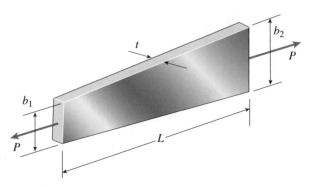

PROB. 2.3-13

2.3-14 A post *AB* supporting equipment in a laboratory is tapered uniformly throughout its height *H* (see figure). The cross sections of the post are square, with dimensions $b \times b$ at the top and $1.5b \times 1.5b$ at the base.

Derive a formula for the shortening δ of the post due to the compressive load *P* acting at the top. (Assume that the angle of taper is small and disregard the weight of the post itself.)

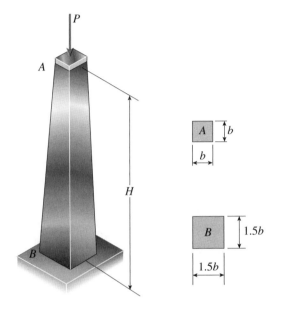

PROB. 2.3-14

2.3-15 A long, slender bar in the shape of a right circular cone with length *L* and base diameter *d* hangs vertically under the action of its own weight (see figure). The weight of the cone is *W* and the modulus of elasticity of the material is *E*.

Derive a formula for the increase δ in the length of the bar due to its own weight. (Assume that the angle of taper of the cone is small.)

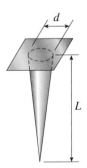

PROB. 2.3-15

2.3-16 A uniformly tapered plastic tube *AB* of circular cross section and length *L* is shown in the figure. The average diameters at the ends are d_A and $d_B = 2d_A$. Assume *E* is constant. Find the elongation δ of the tube when it is subjected to loads *P* acting at the ends. Use the following numerical data: $d_A = 35$ mm, $L = 300$ mm, $E = 2.1$ GPa, $P = 25$ kN. Consider the following cases:

(a) A hole of *constant* diameter d_A is drilled from *B* toward *A* to form a hollow section of length $x = L/2$;

(b) A hole of *variable* diameter $d(x)$ is drilled from *B* toward *A* to form a hollow section of length $x = L/2$ and constant thickness $t = d_A/20$.

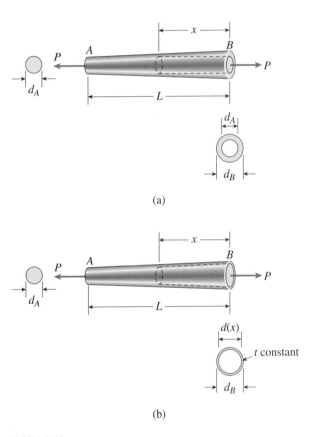

(a)

(b)

PROB. 2.3-16

2.3-17 The main cables of a suspension bridge [see part (a) of the figure on the next page] follow a curve that is nearly parabolic because the primary load on the cables is the weight of the bridge deck, which is uniform in intensity along the horizontal. Therefore, let us represent the central region *AOB* of one of the main cables [see part (b) of the figure] as a parabolic cable supported at points *A* and *B* and carrying a uniform load of intensity q along the horizontal. The span of the cable is L, the sag is h, the axial rigidity is EA, and the origin of coordinates is at midspan.

(a) Derive the following formula for the elongation of cable AOB shown in part (b) of the figure:

$$\delta = \frac{qL^3}{8hEA}\left(1 + \frac{16h^2}{3L^2}\right)$$

(b) Calculate the elongation δ of the central span of one of the main cables of the Golden Gate Bridge, for which the dimensions and properties are $L = 4200$ ft, $h = 470$ ft, $q = 12,700$ lb/ft, and $E = 28,800,000$ psi. The cable consists of 27,572 parallel wires of diameter 0.196 in.

Hint: Determine the tensile force T at any point in the cable from a free-body diagram of part of the cable; then determine the elongation of an element of the cable of length ds; finally, integrate along the curve of the cable to obtain an equation for the elongation δ.

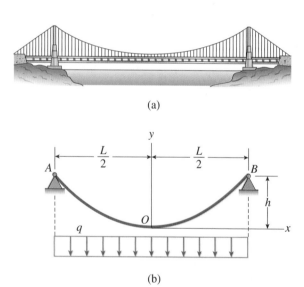

(a)

(b)

PROB. 2.3-17

2.3-18 A bar ABC revolves in a horizontal plane about a vertical axis at the midpoint C (see figure). The bar, which has length $2L$ and cross-sectional area A, revolves at constant angular speed ω. Each half of the bar (AC and BC) has weight W_1 and supports a weight W_2 at its end.

Derive the following formula for the elongation of one-half of the bar (that is, the elongation of either AC or BC):

$$\delta = \frac{L^2\omega^2}{3gEA}(W_1 + 3W_2)$$

in which E is the modulus of elasticity of the material of the bar and g is the acceleration of gravity.

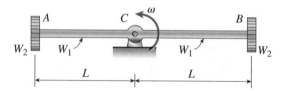

PROB. 2.3-18

Statically Indeterminate Structures

2.4-1 The assembly shown in the figure consists of a brass core (diameter $d_1 = 0.25$ in.) surrounded by a steel shell (inner diameter $d_2 = 0.28$ in., outer diameter $d_3 = 0.35$ in.). A load P compresses the core and shell, which have length $L = 4.0$ in. The moduli of elasticity of the brass and steel are $E_b = 15 \times 10^6$ psi and $E_s = 30 \times 10^6$ psi, respectively.

(a) What load P will compress the assembly by 0.003 in.?

(b) If the allowable stress in the steel is 22 ksi and the allowable stress in the brass is 16 ksi, what is the allowable compressive load P_{allow}? (*Suggestion:* Use the equations derived in Example 2-5.)

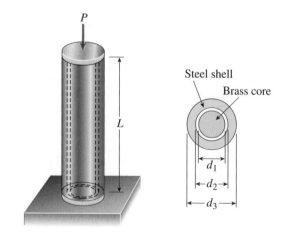

PROB. 2.4-1

2.4-2 A cylindrical assembly consisting of a brass core and an aluminum collar is compressed by a load P (see figure). The length of the aluminum collar and brass core is 350 mm, the diameter of the core is 25 mm, and the outside diameter of the collar is 40 mm. Also, the moduli of elasticity of the aluminum and brass are 72 GPa and 100 GPa, respectively.

(a) If the length of the assembly decreases by 0.1% when the load P is applied, what is the magnitude of the load?

(b) What is the maximum permissible load P_{max} if the allowable stresses in the aluminum and brass are 80 MPa and 120 MPa, respectively? (*Suggestion:* Use the equations derived in Example 2-5.)

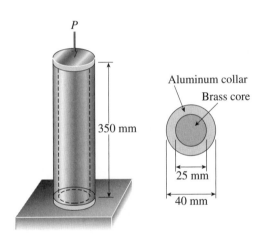

PROB. 2.4-2

2.4-3 Three prismatic bars, two of material A and one of material B, transmit a tensile load P (see figure). The two outer bars (material A) are identical. The cross-sectional area of the middle bar (material B) is 50% larger than the cross-sectional area of one of the outer bars. Also, the modulus of elasticity of material A is twice that of material B.

(a) What fraction of the load P is transmitted by the middle bar?

(b) What is the ratio of the stress in the middle bar to the stress in the outer bars?

(c) What is the ratio of the strain in the middle bar to the strain in the outer bars?

PROB. 2.4-3

2.4-4 A circular bar ACB of diameter d having a cylindrical hole of length x and diameter $d/2$ from A to C is held between rigid supports at A and B. A load P acts at $L/2$ from ends A and B. Assume E is constant.

(a) Obtain formulas for the reactions R_A and R_B at supports A and B, respectively, due to the load P (see figure part a).

(b) Obtain a formula for the displacement δ at the point of load application (see figure part a).

(c) For what value of x is $R_B = (6/5) R_A$? (See figure part a.)

(d) Repeat (a) if the bar is now tapered linearly from A to B as shown in figure part b and $x = L/2$.

(e) Repeat (a) if the bar is now rotated to a vertical position, load P is removed, and the bar is hanging under its own weight (assume mass density $= \rho$). [(See figure part (c.)] Assume that $x = L/2$.

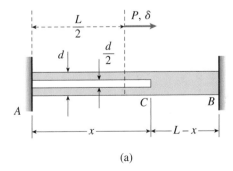

(a)

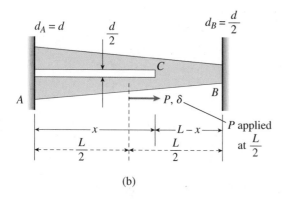

(b)

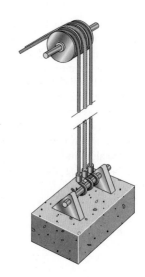

PROB. 2.4-5

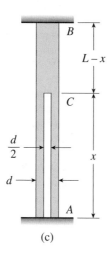

(c)

PROB. 2.4-4

2.4-6 A plastic rod AB of length $L = 0.5$ m has a diameter $d_1 = 30$ mm (see figure). A plastic sleeve CD of length $c = 0.3$ m and outer diameter $d_2 = 45$ mm is securely bonded to the rod so that no slippage can occur between the rod and the sleeve. The rod is made of an acrylic with modulus of elasticity $E_1 = 3.1$ GPa and the sleeve is made of a polyamide with $E_2 = 2.5$ GPa.

(a) Calculate the elongation δ of the rod when it is pulled by axial forces $P = 12$ kN.

(b) If the sleeve is extended for the full length of the rod, what is the elongation?

(c) If the sleeve is removed, what is the elongation?

2.4-5 Three steel cables jointly support a load of 12 k (see figure). The diameter of the middle cable is 3/4 in. and the diameter of each outer cable is 1/2 in. The tensions in the cables are adjusted so that each cable carries one-third of the load (i.e., 4 k). Later, the load is increased by 9 k to a total load of 21 k.

(a) What percent of the total load is now carried by the middle cable?

(b) What are the stresses σ_M and σ_O in the middle and outer cables, respectively? (*Note:* See Table 2-1 in Section 2.2 for properties of cables.)

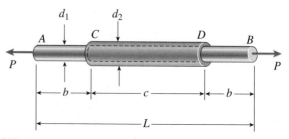

PROB. 2.4-6

2.4-7 The axially loaded bar *ABCD* shown in the figure is held between rigid supports. The bar has cross-sectional area A_1 from *A* to *C* and $2A_1$ from *C* to *D*.

(a) Derive formulas for the reactions R_A and R_D at the ends of the bar.

(b) Determine the displacements δ_B and δ_C at points *B* and *C*, respectively.

(c) Draw an axial-displacement diagram (ADD) in which the abscissa is the distance from the left-hand support to any point in the bar and the ordinate is the horizontal displacement δ at that point.

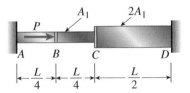

PROB. 2.4-7

2.4-8 The fixed-end bar *ABCD* consists of three prismatic segments, as shown in the figure. The end segments have cross-sectional area $A_1 = 840 \text{ mm}^2$ and length $L_1 = 200$ mm. The middle segment has cross-sectional area $A_2 = 1260 \text{ mm}^2$ and length $L_2 = 250$ mm. Loads P_B and P_C are equal to 25.5 kN and 17.0 kN, respectively.

(a) Determine the reactions R_A and R_D at the fixed supports.

(b) Determine the compressive axial force F_{BC} in the middle segment of the bar.

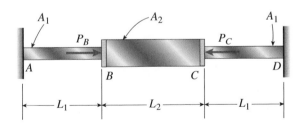

PROB. 2.4-8

2.4-9 The aluminum and steel pipes shown in the figure are fastened to rigid supports at ends *A* and *B* and to a rigid plate *C* at their junction. The aluminum pipe is twice as long as the steel pipe. Two equal and symmetrically placed loads *P* act on the plate at *C*.

(a) Obtain formulas for the axial stresses σ_a and σ_s in the aluminum and steel pipes, respectively.

(b) Calculate the stresses for the following data: $P = 12$ k, cross-sectional area of aluminum pipe $A_a = 8.92 \text{ in.}^2$,

cross-sectional area of steel pipe $A_s = 1.03 \text{ in.}^2$, modulus of elasticity of aluminum $E_a = 10 \times 10^6$ psi, and modulus of elasticity of steel $E_s = 29 \times 10^6$ psi.

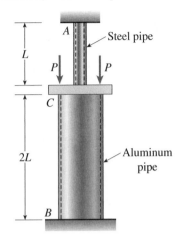

PROB. 2.4-9

2.4-10 A nonprismatic bar *ABC* is composed of two segments: *AB* of length L_1 and cross-sectional area A_1; and *BC* of length L_2 and cross-sectional area A_2. The modulus of elasticity *E*, mass density ρ, and acceleration of gravity *g* are constants. Initially, bar *ABC* is horizontal and then is restrained at *A* and *C* and rotated to a vertical position. The bar then hangs vertically under its own weight (see figure). Let $A_1 = 2A_2 = A$ and $L_1 = \frac{3}{5} L, L_2 = \frac{2}{5} L$.

(a) Obtain formulas for the reactions R_A and R_C at supports *A* and *C*, respectively, due to gravity.

(b) Derive a formula for the downward displacement δ_B of point *B*.

(c) Find expressions for the axial stresses a small distance above points *B* and *C*, respectively.

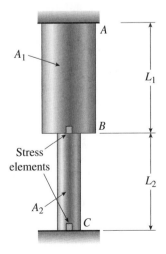

PROB. 2.4-10

2.4-11 A *bimetallic* bar (or composite bar) of square cross section with dimensions $2b \times 2b$ is constructed of two different metals having moduli of elasticity E_1 and E_2 (see figure). The two parts of the bar have the same cross-sectional dimensions. The bar is compressed by forces P acting through rigid end plates. The line of action of the loads has an eccentricity e of such magnitude that each part of the bar is stressed uniformly in compression.

(a) Determine the axial forces P_1 and P_2 in the two parts of the bar.

(b) Determine the eccentricity e of the loads.

(c) Determine the ratio σ_1/σ_2 of the stresses in the two parts of the bar.

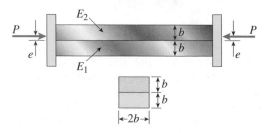

PROB. 2.4-11

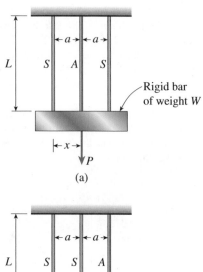

(a)

(b)

PROB. 2.4-12

2.4-12 A rigid bar of weight $W = 800$ N hangs from three equally spaced vertical wires (length $L = 150$ mm, spacing $a = 50$ mm): two of steel and one of aluminum. The wires also support a load P acting on the bar. The diameter of the steel wires is $d_s = 2$ mm, and the diameter of the aluminum wire is $d_a = 4$ mm. Assume $E_s = 210$ GPa and $E_a = 70$ GPa.

(a) What load P_{allow} can be supported *at the midpoint of the bar* ($x = a$) if the allowable stress in the steel wires is 220 MPa and in the aluminum wire is 80 MPa? [See figure part (a).]

(b) What is P_{allow} if the load is positioned at $x = a/2$? [See figure part (a).]

(c) Repeat (b) above if the second and third wires are *switched* as shown in figure part (b).

2.4-13 A horizontal rigid bar of weight $W = 7200$ lb is supported by three slender circular rods that are equally spaced (see figure on the next page). The two outer rods are made of aluminum ($E_1 = 10 \times 10^6$ psi) with diameter $d_1 = 0.4$ in. and length $L_1 = 40$ in. The inner rod is magnesium ($E_2 = 6.5 \times 10^6$ psi) with diameter d_2 and length L_2. The allowable stresses in the aluminum and magnesium are 24,000 psi and 13,000 psi, respectively.

If it is desired to have all three rods loaded to their maximum allowable values, what should be the diameter d_2 and length L_2 of the middle rod?

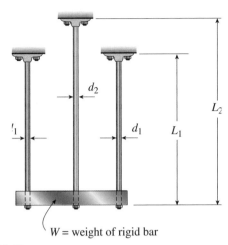

W = weight of rigid bar

PROB. 2.4-13

2.4-14 A circular steel bar *ABC* (*E* = 200 GPa) has cross-sectional area A_1 from *A* to *B* and cross-sectional area A_2 from *B* to *C* (see figure). The bar is supported rigidly at end *A* and is subjected to a load *P* equal to 40 kN at end *C*. A circular steel collar *BD* having cross-sectional area A_3 supports the bar at *B*. The collar fits snugly at *B* and *D* when there is no load.

Determine the elongation δ_{AC} of the bar due to the load *P*. (Assume $L_1 = 2L_3 = 250$ mm, $L_2 = 225$ mm, $A_1 = 2A_3 = 960$ mm^2, and $A_2 = 300$ mm^2.)

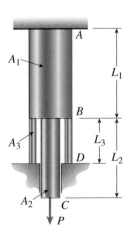

PROB. 2.4-14

2.4-15 A rigid bar *AB* of length *L* = 66 in. is hinged to a support at *A* and supported by two vertical wires attached at points *C* and *D* (see figure). Both wires have the same cross-sectional area (*A* = 0.0272 in.^{2}) and are made of the same material (modulus *E* = 30 × 10^6 psi). The wire at *C* has length *h* = 18 in. and the wire at *D* has length twice that amount. The horizontal distances are *c* = 20 in. and *d* = 50 in.

(a) Determine the tensile stresses σ_C and σ_D in the wires due to the load *P* = 340 lb acting at end *B* of the bar.

(b) Find the downward displacement δ_B at end *B* of the bar.

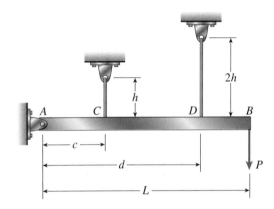

PROB. 2.4-15

2.4-16 A rigid bar *ABCD* is pinned at point *B* and supported by springs at *A* and *D* (see figure). The springs at *A* and *D* have stiffnesses k_1 = 10 kN/m and k_2 = 25 kN/m, respectively, and the dimensions *a*, *b*, and *c* are 250 mm, 500 mm, and 200 mm, respectively. A load *P* acts at point *C*.

If the angle of rotation of the bar due to the action of the load *P* is limited to 3°, what is the maximum permissible load P_{max}?

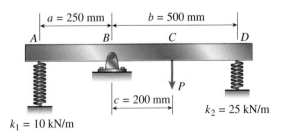

PROB. 2.4-16

2.4-17 A trimetallic bar is uniformly compressed by an axial force $P = 9$ kips applied through a rigid end plate (see figure). The bar consists of a circular steel core surrounded by brass and copper tubes. The steel core has diameter 1.25 in., the brass tube has outer diameter 1.75 in., and the copper tube has outer diameter 2.25 in. The corresponding moduli of elasticity are $E_s = 30,000$ ksi, $E_b = 16,000$ ksi, and $E_c = 18,000$ ksi.

Calculate the compressive stresses σ_s, σ_b, and σ_c in the steel, brass, and copper, respectively, due to the force P.

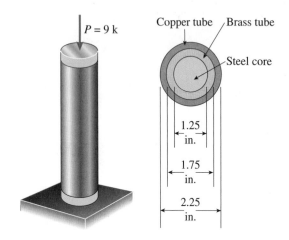

PROB. 2.4-17

Thermal Effects

2.5-1 The rails of a railroad track are welded together at their ends (to form continuous rails and thus eliminate the clacking sound of the wheels) when the temperature is 60°F.

What compressive stress σ is produced in the rails when they are heated by the sun to 120°F if the coefficient of thermal expansion $\alpha = 6.5 \times 10^{-6}/°F$ and the modulus of elasticity $E = 30 \times 10^6$ psi?

2.5-2 An aluminum pipe has a length of 60 m at a temperature of 10°C. An adjacent steel pipe at the same temperature is 5 mm longer than the aluminum pipe.

At what temperature (degrees Celsius) will the aluminum pipe be 15 mm longer than the steel pipe? (Assume that the coefficients of thermal expansion of aluminum and steel are $\alpha_a = 23 \times 10^{-6}/°C$ and $\alpha_s = 12 \times 10^{-6}/°C$, respectively.)

2.5-3 A rigid bar of weight $W = 750$ lb hangs from three equally spaced wires, two of steel and one of aluminum (see figure). The diameter of the wires is 1/8 in. Before they were loaded, all three wires had the same length.

What temperature increase ΔT in all three wires will result in the entire load being carried by the steel wires? (Assume $E_s = 30 \times 10^6$ psi, $\alpha_s = 6.5 \times 10^{-6}/°F$, and $\alpha_a = 12 \times 10^{-6}/°F$.)

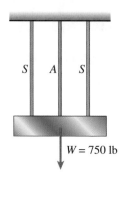

PROB. 2.5-3

2.5-4 A steel rod of 15-mm diameter is held snugly (but without any initial stresses) between rigid walls by the arrangement shown in the figure. (For the steel rod, use $\alpha = 12 \times 10^{-6}/°C$ and $E = 200$ GPa.)

(a) Calculate the temperature drop ΔT (degrees Celsius) at which the average shear stress in the 12-mm diameter bolt becomes 45 MPa.

(b) What are the average bearing stresses in the bolt and clevis at A and the washer ($d_w = 20$ mm) and wall ($t = 18$ mm) at B?

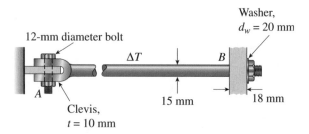

PROB. 2.5-4

2.5-5 A bar AB of length L is held between rigid supports and heated nonuniformly in such a manner that the temperature increase ΔT at distance x from end A is given by the expression $\Delta T = \Delta T_B x^3 / L^3$, where ΔT_B is the increase in temperature at end B of the bar [see figure part (a)].

(a) Derive a formula for the compressive stress σ_c in the bar. (Assume that the material has modulus of elasticity E and coefficient of thermal expansion α).

(b) Now modify the formula in (a) if the rigid support at A is replaced by an elastic support at A having a spring constant k (see figure part b). Assume that only bar AB is subject to the temperature increase.

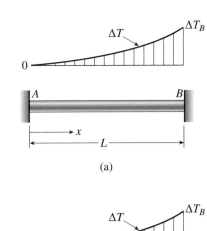

(a)

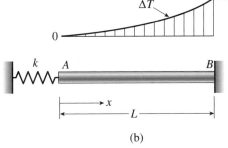

(b)

PROB. 2.5-5

2.5-6 A plastic bar ACB having two different solid circular cross sections is held between rigid supports as shown in the figure. The diameters in the left- and right-hand parts are 50 mm and 75 mm, respectively. The corresponding lengths are 225 mm and 300 mm. Also, the modulus of elasticity E is 6.0 GPa, and the coefficient of thermal expansion α is $100 \times 10^{-6}/°C$. The bar is subjected to a uniform temperature increase of 30°C.

(a) Calculate the following quantities: (1) the compressive force N in the bar; (2) the maximum compressive stress σ_c; and (3) the displacement δ_C of point C.

(b) Repeat (a) if the rigid support at A is replaced by an elastic support having spring constant $k = 50$ MN/m (see figure part b; assume that only the bar ACB is subject to the temperature increase).

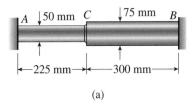

(a)

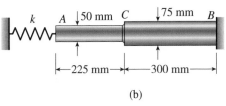

(b)

PROB. 2.5-6

2.5-7 A circular steel rod AB (diameter $d_1 = 1.0$ in., length $L_1 = 3.0$ ft) has a bronze sleeve (outer diameter $d_2 = 1.25$ in., length $L_2 = 1.0$ ft) shrunk onto it so that the two parts are securely bonded (see figure).

Calculate the total elongation δ of the steel bar due to a temperature rise $\Delta T = 500°F$. (Material properties are as follows: for steel, $E_s = 30 \times 10^6$ psi and $\alpha_s = 6.5 \times 10^{-6}/°F$; for bronze, $E_b = 15 \times 10^6$ psi and $\alpha_b = 11 \times 10^{-6}/°F$.)

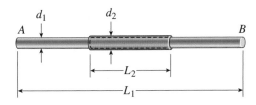

PROB. 2.5-7

2.5-8 A brass sleeve S is fitted over a steel bolt B (see figure), and the nut is tightened until it is just snug. The bolt has a diameter $d_B = 25$ mm, and the sleeve has inside and outside diameters $d_1 = 26$ mm and $d_2 = 36$ mm, respectively.

Calculate the temperature rise ΔT that is required to produce a compressive stress of 25 MPa in the sleeve.

(Use material properties as follows: for the sleeve, $\alpha_S = 21 \times 10^{-6}/°C$ and $E_S = 100$ GPa; for the bolt, $\alpha_B = 10 \times 10^{-6}/°C$ and $E_B = 200$ GPa.) (*Suggestion:* Use the results of Example 2-8.)

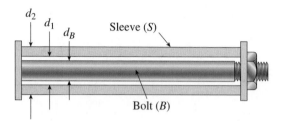

PROB. 2.5-8

2.5-9 Rectangular bars of copper and aluminum are held by pins at their ends, as shown in the figure. Thin spacers provide a separation between the bars. The copper bars have cross-sectional dimensions 0.5 in. × 2.0 in., and the aluminum bar has dimensions 1.0 in. × 2.0 in.

Determine the shear stress in the 7/16 in. diameter pins if the temperature is raised by 100°F. (For copper, $E_c = 18{,}000$ ksi and $\alpha_c = 9.5 \times 10^{-6}/°F$; for aluminum, $E_a = 10{,}000$ ksi and $\alpha_a = 13 \times 10^{-6}/°F$.) *Suggestion:* Use the results of Example 2-8.

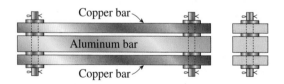

PROB. 2.5-9F

2.5-10 A rigid bar *ABCD* is pinned at end *A* and supported by two cables at points *B* and *C* (see figure). The cable at *B* has nominal diameter $d_B = 12$ mm and the cable at *C* has nominal diameter $d_C = 20$ mm. A load *P* acts at end *D* of the bar.

What is the allowable load *P* if the temperature rises by 60°C and each cable is required to have a factor of safety of at least 5 against its ultimate load?

(*Note:* The cables have effective modulus of elasticity $E = 140$ GPa and coefficient of thermal expansion $\alpha = 12 \times 10^{-6}/°C$. Other properties of the cables can be found in Table 2-1, Section 2.2.)

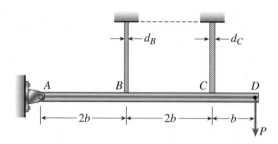

PROB. 2.5-10

2.5-11 A rigid triangular frame is pivoted at *C* and held by two identical horizontal wires at points *A* and *B* (see figure). Each wire has axial rigidity $EA = 120$ k and coefficient of thermal expansion $\alpha = 12.5 \times 10^{-6}/°F$.

(a) If a vertical load $P = 500$ lb acts at point *D*, what are the tensile forces T_A and T_B in the wires at *A* and *B*, respectively?

(b) If, while the load *P* is acting, both wires have their temperatures raised by 180°F, what are the forces T_A and T_B?

(c) What further increase in temperature will cause the wire at *B* to become slack?

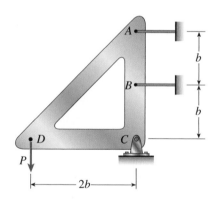

PROB. 2.5-11

Misfits and Prestrains

2.5-12 A steel wire *AB* is stretched between rigid supports (see figure). The initial prestress in the wire is 42 MPa when the temperature is 20°C.

(a) What is the stress σ in the wire when the temperature drops to 0°C?

(b) At what temperature *T* will the stress in the wire become zero? (Assume $\alpha = 14 \times 10^{-6}/°C$ and $E = 200$ GPa.)

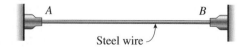

PROB. 2.5-12

2.5-13 A copper bar *AB* of length 25 in. and diameter 2 in. is placed in position at room temperature with a gap of 0.008 in. between end *A* and a rigid restraint (see figure). The bar is supported at end *B* by an elastic spring with spring constant $k = 1.2 \times 10^6$ lb/in.

(a) Calculate the axial compressive stress σ_c in the bar if the temperature *of the bar only* rises 50°F. (For copper, use $\alpha = 9.6 \times 10^{-6}/°F$ and $E = 16 \times 10^6$ psi.)

(b) What is the force in the spring? (Neglect gravity effects.)

(c) Repeat (a) if $k \to \infty$.

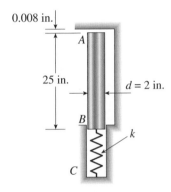

PROB. 2.5-13

2.5-14 A bar *AB* having length *L* and axial rigidity *EA* is fixed at end *A* (see figure). At the other end a small gap of dimension *s* exists between the end of the bar and a rigid surface. A load *P* acts on the bar at point *C*, which is two-thirds of the length from the fixed end.

If the support reactions produced by the load *P* are to be equal in magnitude, what should be the size *s* of the gap?

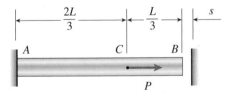

PROB. 2.5-14

2.5-15 Pipe 2 has been inserted snugly into Pipe 1, but the holes for a connecting pin do not line up: there is a gap *s*. The user decides to apply *either* force P_1 to Pipe 1 *or* force P_2 to Pipe 2, whichever is smaller. Determine the following using the numerical properties in the box.

(a) If only P_1 is applied, find P_1 (kips) required to close gap *s*; if a pin is then inserted and P_1 removed, what are reaction forces R_A and R_B for this load case?

(b) If only P_2 is applied, find P_2 (kips) required to close gap *s*; if a pin is inserted and P_2 removed, what are reaction forces R_A and R_B for this load case?

(c) What is the maximum *shear* stress in the pipes, for the loads in (a) and (b)?

(d) If a temperature increase ΔT is to be applied to the entire structure to close gap *s* (*instead of applying forces P_1 and P_2*), find the ΔT required to close the gap. If a pin is inserted after the gap has closed, what are reaction forces R_A and R_B for this case?

(e) Finally, if the structure (with pin inserted) then cools to the *original* ambient temperature, what are reaction forces R_A and R_B?

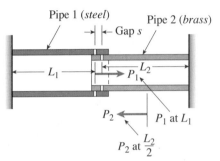

Numerical properties
$E_1 = 30{,}000$ ksi, $E_2 = 14{,}000$ ksi
$\alpha_1 = 6.5 \times 10^{-6}/°F$, $\alpha_2 = 11 \times 10^{-6}/°F$
Gap $s = 0.05$ in.
$L_1 = 56$ in., $d_1 = 6$ in., $t_1 = 0.5$ in., $A_1 = 8.64$ in.^{2}
$L_2 = 36$ in., $d_2 = 5$ in., $t_2 = 0.25$ in., $A_2 = 3.73$ in.^{2}

PROB. 2.5-15

2.5-16 A nonprismatic bar *ABC* made up of segments *AB* (length L_1, cross-sectional area A_1) and *BC* (length L_2, cross-sectional area A_2) is fixed at end *A* and free at end *C* (see figure). The modulus of elasticity of the bar is *E*. A small gap of dimension *s* exists between the end of the bar and an elastic spring of length L_3 and spring constant k_3. If bar *ABC* only (*not the spring*) is subjected to temperature increase ΔT determine the following.

(a) Write an expression for reaction forces R_A and R_D if the elongation of *ABC* exceeds gap length *s*.

(b) Find expressions for the displacements of points *B* and *C* if the elongation of *ABC* exceeds gap length *s*.

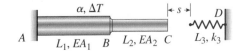

PROB. 2.5-16

2.5-17 Wires B and C are attached to a support at the left-hand end and to a pin-supported rigid bar at the right-hand end (see figure). Each wire has cross-sectional area $A = 0.03$ in.^{2} and modulus of elasticity $E = 30 \times 10^6$ psi. When the bar is in a vertical position, the length of each wire is $L = 80$ in. However, before being attached to the bar, the length of wire B was 79.98 in. and of wire C was 79.95 in.

Find the tensile forces T_B and T_C in the wires under the action of a force $P = 700$ lb acting at the upper end of the bar.

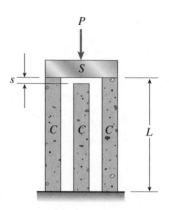

PROB. 2.5-18

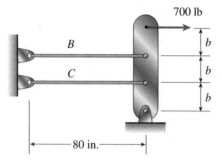

PROB. 2.5-17

2.5-19 A capped cast-iron pipe is compressed by a brass rod, as shown. The nut is turned until it is just snug, then add an additional quarter turn to pre-compress the CI pipe. The pitch of the threads of the bolt is $p = 52$ mils (a mil is one-thousandth of an inch). Use the numerical properties provided.

(a) What stresses σ_p and σ_r will be produced in the cast-iron pipe and brass rod, respectively, by the additional quarter turn of the nut?

(b) Find the bearing stress σ_b beneath the washer and the shear stress τ_c in the steel cap.

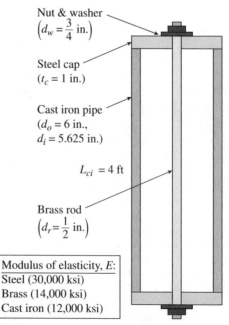

2.5-18 A rigid steel plate is supported by three posts of high-strength concrete each having an effective cross-sectional area $A = 40,000$ mm^2 and length $L = 2$ m (see figure). Before the load P is applied, the middle post is shorter than the others by an amount $s = 1.0$ mm.

Determine the maximum allowable load P_{allow} if the allowable compressive stress in the concrete is $\sigma_{\text{allow}} = 20$ MPa. (Use $E = 30$ GPa for concrete.)

PROB. 2.5-19

2.5-20 A plastic cylinder is held snugly between a rigid plate and a foundation by two steel bolts (see figure).

Determine the compressive stress σ_p in the plastic when the nuts on the steel bolts are tightened by one complete turn.

Data for the assembly are as follows: length $L = 200$ mm, pitch of the bolt threads $p = 1.0$ mm, modulus of elasticity for steel $E_s = 200$ GPa, modulus of elasticity for the plastic $E_p = 7.5$ GPa, cross-sectional area of one bolt $A_s = 36.0$ mm^2, and cross-sectional area of the plastic cylinder $A_p = 960$ mm^2.

PROBS. 2.5-20 and 2.5-21

2.5-21 Solve the preceding problem if the data for the assembly are as follows: length $L = 10$ in., pitch of the bolt threads $p = 0.058$ in., modulus of elasticity for steel $E_s = 30 \times 10^6$ psi, modulus of elasticity for the plastic $E_p = 500$ ksi, cross-sectional area of one bolt $A_s = 0.06$ in.^{2}, and cross-sectional area of the plastic cylinder $A_p = 1.5$ in.^{2}

2.5-22 Consider the sleeve made from two copper tubes joined by tin-lead solder over distance s. The sleeve has brass caps at both ends, which are held in place by a steel bolt and washer with the nut turned just snug at the outset. Then, two "loadings" are applied: $n = 1/2$ turn applied to the nut; at the same time the internal temperature is raised by $\Delta T = 30°C$.

(a) Find the forces in the sleeve and bolt, P_s and P_B, due to both the prestress in the bolt and the temperature increase. For copper, use $E_c = 120$ GPa and $\alpha_c = 17 \times 10^{-6}/°C$; for steel, use $E_s = 200$ GPa and $\alpha_s = 12 \times 10^{-6}/°C$. The pitch of the bolt threads is $p = 1.0$ mm. Assume $s = 26$ mm and bolt diameter $d_b = 5$ mm.

(b) Find the required length of the solder joint, s, if shear stress in the sweated joint cannot exceed the allowable shear stress $\tau_{aj} = 18.5$ MPa.

(c) What is the final elongation of the entire assemblage due to both temperature change ΔT and the initial prestress in the bolt?

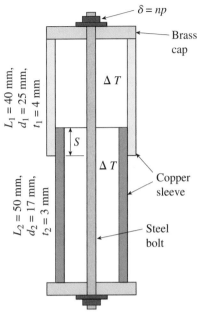

$\delta = np$

Brass cap

$L_1 = 40$ mm,
$d_1 = 25$ mm,
$t_1 = 4$ mm

ΔT

S

$L_2 = 50$ mm,
$d_2 = 17$ mm,
$t_2 = 3$ mm

ΔT

Copper sleeve

Steel bolt

PROB. 2.5-22

2.5-23 A polyethylene tube (length L) has a cap which when installed compresses a spring (with undeformed length $L_1 > L$) by amount $\delta = (L_1 - L)$. Ignore deformations of the cap and base. Use the force at the base of the spring as the redundant. Use numerical properties in the boxes given.

(a) What is the resulting force in the spring, F_k?

(b) What is the resulting force in the tube, F_t?

(c) What is the final length of the tube, L_f?

(d) What temperature change ΔT inside the tube will result in zero force in the spring?

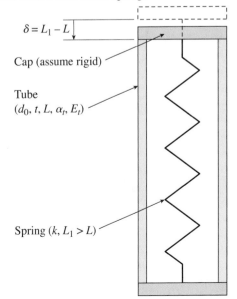

$\delta = L_1 - L$

Cap (assume rigid)

Tube $(d_0, t, L, \alpha_t, E_t)$

Spring $(k, L_1 > L)$

Modulus of elasticity
Polyethylene tube ($E_t = 100$ ksi)

Coefficients of thermal expansion
$\alpha_t = 80 \times 10^{-6}/°F$, $\alpha_k = 6.5 \times 10^{-6}/°F$

Properties and dimensions
$d_0 = 6$ in. $t = \dfrac{1}{8}$ in.
$L_1 = 12.125$ in. $> L = 12$ in. $k = 1.5 \dfrac{\text{kip}}{\text{in.}}$

PROB. 2.5-23

2.5-24 Prestressed concrete beams are sometimes manufactured in the following manner. High-strength steel wires are stretched by a jacking mechanism that applies a force Q, as represented schematically in part (a) of the figure. Concrete is then poured around the wires to form a beam, as shown in part (b).

After the concrete sets properly, the jacks are released and the force Q is removed [see part (c) of the figure]. Thus, the beam is left in a prestressed condition, with the wires in tension and the concrete in compression.

Let us assume that the prestressing force Q produces in the steel wires an initial stress $\sigma_0 = 620$ MPa. If the moduli of elasticity of the steel and concrete are in the ratio 12:1 and the cross-sectional areas are in the ratio 1:50, what are the final stresses σ_s and σ_c in the two materials?

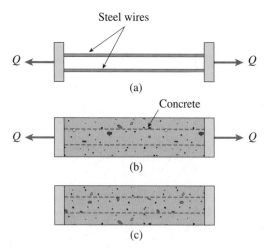

PROB. 2.5-24

2.5-25 A polyethylene tube (length L) has a cap which is held in place by a spring (with undeformed length $L_1 < L$). After installing the cap, the spring is post-tensioned by turning an adjustment screw by amount δ. Ignore deformations of the cap and base. Use the force at the base of the spring as the redundant. Use numerical properties in the boxes below.

(a) What is the resulting force in the spring, F_k?
(b) What is the resulting force in the tube, F_t?
(c) What is the final length of the tube, L_f?
(d) What temperature change ΔT inside the tube will result in zero force in the spring?

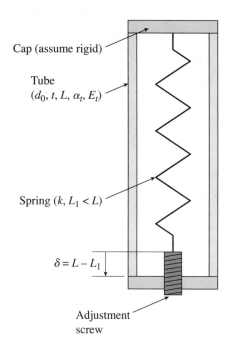

Modulus of elasticity
Polyethylene tube ($E_t = 100$ ksi)

Coefficients of thermal expansion
$\alpha_t = 80 \times 10^{-6}/°F$, $\alpha_k = 6.5 \times 10^{-6}/°F$

Properties and dimensions
$d_0 = 6$ in. $t = \dfrac{1}{8}$ in.
$L_1 = 12$ in. $> L = 11.875$ in. $k = 1.5 \dfrac{\text{kip}}{\text{in.}}$

Stresses on Inclined Sections

2.6-1 A steel bar of rectangular cross section (1.5 in. × 2.0 in.) carries a tensile load P (see figure). The allowable stresses in tension and shear are 14,500 psi and 7,100 psi, respectively. Determine the maximum permissible load P_{max}.

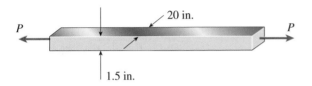

PROB. 2.6-1

2.6-2 A circular steel rod of diameter d is subjected to a tensile force $P = 3.5$ kN (see figure). The allowable stresses in tension and shear are 118 MPa and 48 MPa, respectively. What is the minimum permissible diameter d_{min} of the rod?

PROB. 2.6-2

2.6-3 A standard brick (dimensions 8 in. × 4 in. × 2.5 in.) is compressed lengthwise by a force P, as shown in the figure. If the ultimate shear stress for brick is 1200 psi and the ultimate compressive stress is 3600 psi, what force P_{max} is required to break the brick?

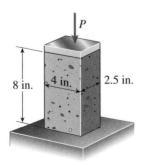

PROB. 2.6-3

2.6-4 A brass wire of diameter $d = 2.42$ mm is stretched tightly between rigid supports so that the tensile force is $T = 98$ N (see figure). The coefficient of thermal expansion for the wire is $19.5 \times 10^{-6}/°C$ and the modulus of elasticity is $E = 110$ GPa.

(a) What is the maximum permissible temperature drop ΔT if the allowable shear stress in the wire is 60 MPa?

(b) At what temperature change does the wire go slack?

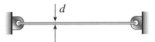

PROBS. 2.6-4 and 2.6-5

2.6-5 A brass wire of diameter $d = 1/16$ in. is stretched between rigid supports with an initial tension T of 37 lb (see figure). Assume that the coefficient of thermal expansion is $10.6 \times 10^{-6}/°F$ and the modulus of elasticity is 15×10^6 psi.)

(a) If the temperature is lowered by 60°F, what is the maximum shear stress τ_{max} in the wire?

(b) If the allowable shear stress is 10,000 psi, what is the maximum permissible temperature drop?

(c) At what temperature change ΔT does the wire go slack?

2.6-6 A steel bar with diameter $d = 12$ mm is subjected to a tensile load $P = 9.5$ kN (see figure).

(a) What is the maximum normal stress σ_{max} in the bar?

(b) What is the maximum shear stress τ_{max}?

(c) Draw a stress element oriented at 45° to the axis of the bar and show all stresses acting on the faces of this element.

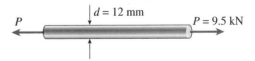

PROB. 2.6-6

2.6-7 During a tension test of a mild-steel specimen (see figure), the extensometer shows an elongation of 0.00120 in. with a gage length of 2 in. Assume that the steel is stressed below the proportional limit and that the modulus of elasticity $E = 30 \times 10^6$ psi.

(a) What is the maximum normal stress σ_{max} in the specimen?

(b) What is the maximum shear stress τ_{max}?

(c) Draw a stress element oriented at an angle of 45° to the axis of the bar and show all stresses acting on the faces of this element.

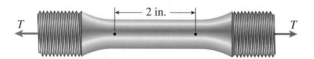

PROB. 2.6-7

2.6-8 A copper bar with a rectangular cross section is held without stress between rigid supports (see figure). Subsequently, the temperature of the bar is raised 50°C.

Determine the stresses on all faces of the elements A and B, and show these stresses on sketches of the elements. (Assume $\alpha = 17.5 \times 10^{-6}$/°C and $E = 120$ GPa.)

PROB. 2.6-8

2.6-9 The bottom chord AB in a small truss ABC (see figure) is fabricated from a W8 $\times$ 28 wide-flange steel section. The cross-sectional area $A = 8.25$ in.^{2} [(Appendix F, Table F-1 (a) available online)] and each of the three applied loads $P = 45$ k. First, find member force N_{AB}; then, determine the normal and shear stresses acting on all faces of stress elements located in the web of member AB and oriented at (a) an angle $\theta = 0°$, (b) an angle $\theta = 30°$, and (c) an angle $\theta = 45°$. In each case, show the stresses on a sketch of a properly oriented element.

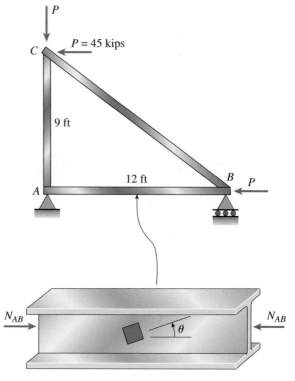

PROB. 2.6-9

2.6-10 A plastic bar of diameter $d = 32$ mm is compressed in a testing device by a force $P = 190$ N applied as shown in the figure.

(a) Determine the normal and shear stresses acting on all faces of stress elements oriented at (1) an angle $\theta = 0°$, (2) an angle $\theta = 22.5°$, and (3) an angle $\theta = 45°$. In each case, show the stresses on a sketch of a properly oriented element. What are σ_{max} and τ_{max}?

(b) Find σ_{max} and τ_{max} in the plastic bar if a re-centering spring of stiffness k is inserted into the testing device, as shown in the figure. The spring stiffness is 1/6 of the axial stiffness of the plastic bar.

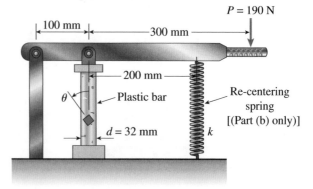

PROB. 2.6-10

2.6-11 A plastic bar of rectangular cross section ($b = 1.5$ in. and $h = 3$ in.) fits snugly between rigid supports at room temperature (68°F) but with no initial stress (see figure). When the temperature of the bar is raised to 160°F, the compressive stress on an inclined plane pq at midspan becomes 1700 psi.

(a) What is the shear stress on plane pq? (Assume $\alpha = 60 \times 10^{-6}/°F$ and $E = 450 \times 10^3$ psi.)

(b) Draw a stress element oriented to plane pq and show the stresses acting on all faces of this element.

(c) If the allowable normal stress is 3400 psi and the allowable shear stress is 1650 psi, what is the maximum load P (in +x *direction*) which can be added at the quarter point (in addition to thermal effects above) without exceeding allowable stress values in the bar?

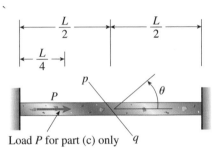

PROBS. 2.6-11

2.6-12 A copper bar of rectangular cross section ($b = 18$ mm and $h = 40$ mm) is held snugly (but without any initial stress) between rigid supports (see figure). The allowable stresses on the inclined plane pq at midspan, for which $\theta = 55°$, are specified as 60 MPa in compression and 30 MPa in shear.

(a) What is the maximum permissible temperature rise ΔT if the allowable stresses on plane pq are not to be exceeded? (Assume $\alpha = 17 \times 10^{-6}/°C$ and $E = 120$ GPa.)

(b) If the temperature increases by the maximum permissible amount, what are the stresses on plane pq?

(c) If the temperature rise $\Delta T = 28°C$, how far to the right of end A (distance βL, expressed as a fraction of length L) can load $P = 15$ kN be applied without exceeding allowable stress values in the bar? Assume that $\sigma_a = 75$ MPa and $\tau_a = 35$ MPa.

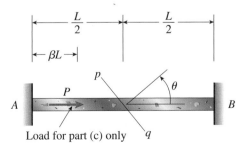

PROBS. 2.6-12

2.6-13 A circular brass bar of diameter d is member AC in truss ABC which has load $P = 5000$ lb applied at joint C. Bar AC is composed of two segments brazed together on a plane pq making an angle $\alpha = 36°$ with the axis of the bar (see figure). The allowable stresses in the brass are 13,500 psi in tension and 6500 psi in shear. On the brazed joint, the allowable stresses are 6000 psi in tension and 3000 psi in shear. What is the tensile force N_{AC} in bar AC? What is the minimum required diameter d_{min} of bar AC?

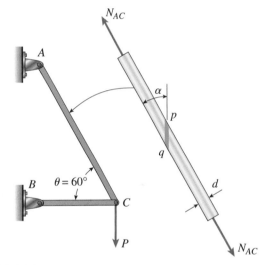

PROB. 2.6-13

2.6-14 Two boards are joined by gluing along a scarf joint, as shown in the figure. For purposes of cutting and gluing, the angle α between the plane of the joint and the faces of the boards must be between 10° and 40°. Under a tensile load P, the normal stress in the boards is 4.9 MPa.

(a) What are the normal and shear stresses acting on the glued joint if $\alpha = 20°$?

(b) If the allowable shear stress on the joint is 2.25 MPa, what is the largest permissible value of the angle α?

(c) For what angle α will the shear stress on the glued joint be numerically equal to twice the normal stress on the joint?

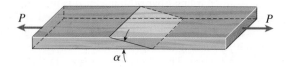

PROB. 2.6-14

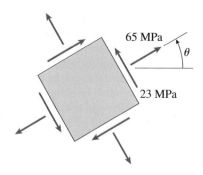

65 MPa

θ

23 MPa

PROB. 2.6-16

2.6-15 Acting on the sides of a stress element cut from a bar in uniaxial stress are tensile stresses of 10,000 psi and 5000 psi, as shown in the figure.

(a) Determine the angle θ and the shear stress τ_θ and show all stresses on a sketch of the element.

(b) Determine the maximum normal stress σ_{max} and the maximum shear stress τ_{max} in the material.

2.6-17 The normal stress on plane pq of a prismatic bar in tension (see figure) is found to be 7500 psi. On plane rs, which makes an angle $\beta = 30°$ with plane pq, the stress is found to be 2500 psi.

Determine the maximum normal stress σ_{max} and maximum shear stress τ_{max} in the bar.

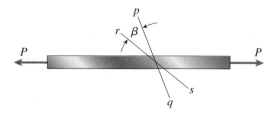

PROB. 2.6-17

2.6-18 A tension member is to be constructed of two pieces of plastic glued along plane pq (see figure). For purposes of cutting and gluing, the angle θ must be between 25° and 45°. The allowable stresses on the glued joint in tension and shear are 5.0 MPa and 3.0 MPa, respectively.

(a) Determine the angle θ so that the bar will carry the largest load P. (Assume that the strength of the glued joint controls the design.)

(b) Determine the maximum allowable load P_{max} if the cross-sectional area of the bar is 225 mm^2.

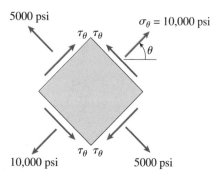

5000 psi

$\sigma_\theta = 10,000$ psi

τ_θ τ_θ

θ

10,000 psi

τ_θ τ_θ

5000 psi

PROB. 2.6-15

2.6-16 A prismatic bar is subjected to an axial force that produces a tensile stress $\sigma_\theta = 65$ MPa and a shear stress $\tau_\theta = 23$ MPa on a certain inclined plane (see figure). Determine the stresses acting on all faces of a stress element oriented at $\theta = 30°$ and show the stresses on a sketch of the element.

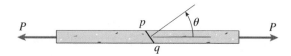

PROB. 2.6-18

2.6-19 A nonprismatic bar 1–2–3 of rectangular cross section (cross-sectional area A) and two materials is held snugly (but without any initial stress) between rigid supports (see figure). The allowable stresses in compression and in shear are specified as σ_a and τ_a, respectively. Use the following numerical data: (*Data*: $b_1 = 4b_2/3 = b$; $A_1 = 2A_2 = A$; $E_1 = 3E_2/4 = E$; $\alpha_1 = 5\alpha_2/4 = \alpha$; $\sigma_{a1} = 4\sigma_{a2}/3 = \sigma_a$, $\tau_{a1} = 2\sigma_{a1}/5$, $\tau_{a2} = 3\sigma_{a2}/5$; let $\sigma_a = 11$ ksi, $P = 12$ kips, $A = 6$ in.^{2}, $b = 8$ in., $E = 30{,}000$ ksi, $\alpha = 6.5 \times 10^{-6}/°F$; $\gamma_1 = 5\gamma_2/3 = \gamma = 490$ lb/ft^3).

(a) If load P is applied at joint 2 as shown, find an expression for the maximum permissible temperature rise ΔT_{max} so that the allowable stresses are not to be exceeded at either location A or B.

(b) If load P is removed and the bar is now rotated to a vertical position where it hangs under its own weight (load intensity $= w_1$ in segment 1–2 and w_2 in segment 2–3), find an expression for the maximum permissible temperature rise ΔT_{max} so that the allowable stresses are not exceeded at either location 1 or 3. Locations 1 and 3 are each a short distance from the supports at 1 and 3, respectively.

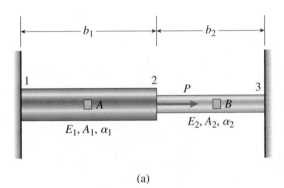

(a)

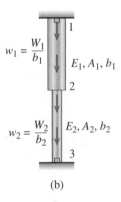

(b)

PROB. 2.6-19

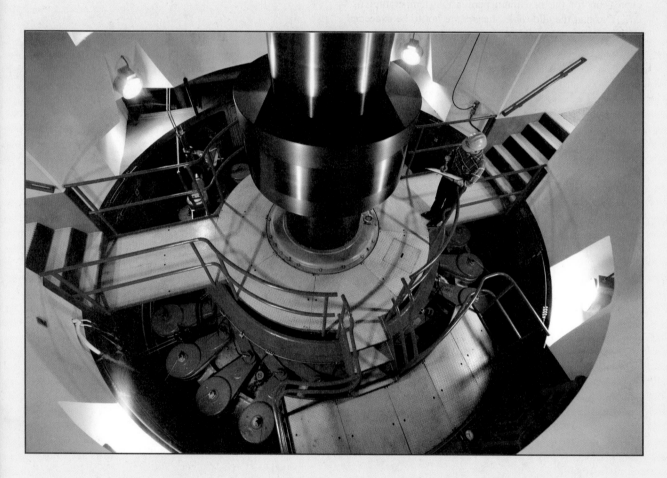

Circular shafts are essential components in machines and devices for power generation and transmission. (Harold Sund/Getty Images)

Torsion

CHAPTER OVERVIEW

Chapter 3 is concerned with the twisting of circular bars and hollow shafts acted upon by torsional moments. First, we consider **uniform torsion** which refers to the case in which torque is constant over the length of a prismatic shaft, while **nonuniform torsion** describes cases in which the torsional moment and/or the torsional rigidity of the cross section varies over the length. As for the case of axial deformations, we must relate stress and strain and also applied loading and deformation. For torsion, recall that Hooke's Law for shear states that shearing stresses, τ, are proportional to shearing strains, γ, with the constant of proportionality being G, the shearing modulus of elasticity. Both shearing stresses and shearing strains vary linearly with increasing radial distance in the cross section, as described by the **torsion formula**. The angle of twist, φ, is proportional to the internal torsional moment and the torsional flexibility of the circular bar. Most of the discussion in this chapter is devoted to linear elastic behavior and small rotations of statically determinate members. However, if the bar is **statically indeterminate**, we must augment the equations of statical equilibrium with compatibility equations (which rely on **torque-displacement relations**) to solve for any unknowns of interest, such as support moments or internal torsional moments in members. Stresses on inclined sections also are investigated as a first step toward a more complete consideration of plane stress states in later chapters.

The topics in Chapter 3 are organized as follows:

3.1 INTRODUCTION

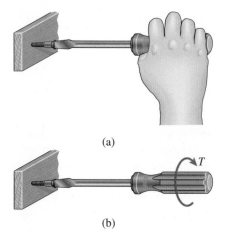

FIG. 3-1 Torsion of a screwdriver due to a torque T applied to the handle

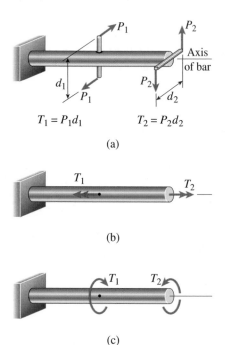

$$T_1 = P_1d_1 \qquad\qquad T_2 = P_2d_2$$

(a)

(b)

(c)

FIG. 3-2 Circular bar subjected to torsion by torques T_1 and T_2

In Chapters 1 and 2, we discussed the behavior of the simplest type of structural member—namely, a straight bar subjected to axial loads. Now we consider a slightly more complex type of behavior known as **torsion**. Torsion refers to the twisting of a straight bar when it is loaded by moments (or torques) that tend to produce rotation about the longitudinal axis of the bar. For instance, when you turn a screwdriver (Fig. 3-1a), your hand applies a torque T to the handle (Fig. 3-1b) and twists the shank of the screwdriver. Other examples of bars in torsion are drive shafts in automobiles, axles, propeller shafts, steering rods, and drill bits.

An idealized case of torsional loading is pictured in Fig. 3-2a, which shows a straight bar supported at one end and loaded by two pairs of equal and opposite forces. The first pair consists of the forces P_1 acting near the midpoint of the bar and the second pair consists of the forces P_2 acting at the end. Each pair of forces forms a **couple** that tends to twist the bar about its longitudinal axis. As we know from statics, the **moment of a couple** is equal to the product of one of the forces and the perpendicular distance between the lines of action of the forces; thus, the first couple has a moment $T_1 = P_1d_1$ and the second has a moment $T_2 = P_2d_2$.

Typical USCS **units** for moment are the pound-foot (lb-ft) and the pound-inch (lb-in.). The SI unit for moment is the newton meter (N·m).

The moment of a couple may be represented by a **vector** in the form of a double-headed arrow (Fig. 3-2b). The arrow is perpendicular to the plane containing the couple, and therefore in this case both arrows are parallel to the axis of the bar. The direction (or *sense*) of the moment is indicated by the *right-hand rule* for moment vectors—namely, using your right hand, let your fingers curl in the direction of the moment, and then your thumb will point in the direction of the vector.

An alternative representation of a moment is a curved arrow acting in the direction of rotation (Fig. 3-2c). Both the curved arrow and vector representations are in common use, and both are used in this book. The choice depends upon convenience and personal preference.

Moments that produce twisting of a bar, such as the moments T_1 and T_2 in Fig. 3-2, are called **torques** or **twisting moments**. Cylindrical members that are subjected to torques and transmit power through rotation are called **shafts**; for instance, the drive shaft of an automobile or the propeller shaft of a ship. Most shafts have circular cross sections, either solid or tubular.

In this chapter we begin by developing formulas for the deformations and stresses in circular bars subjected to torsion. We then analyze the state of stress known as *pure shear* and obtain the relationship between the moduli of elasticity E and G in tension and shear, respectively. Next, we analyze rotating shafts and determine the power they transmit. Finally, we cover several additional topics related to torsion, namely, statically indeterminate members, strain energy, thin-walled tubes of noncircular cross section, and stress concentrations.

3.2 TORSIONAL DEFORMATIONS OF A CIRCULAR BAR

We begin our discussion of torsion by considering a prismatic bar of circular cross section twisted by torques T acting at the ends (Fig. 3-3a). Since every cross section of the bar is identical, and since every cross section is subjected to the same internal torque T, we say that the bar is in **pure torsion**. From considerations of symmetry, it can be proved that cross sections of the bar do not change in shape as they rotate about the longitudinal axis. In other words, all cross sections remain plane and circular and all radii remain straight. Furthermore, if the angle of rotation between one end of the bar and the other is small, neither the length of the bar nor its radius will change.

To aid in visualizing the deformation of the bar, imagine that the left-hand end of the bar (Fig. 3-3a) is fixed in position. Then, under the action of the torque T, the right-hand end will rotate (with respect to the left-hand end) through a small angle ϕ, known as the **angle of twist** (or *angle of rotation*). Because of this rotation, a straight longitudinal line pq on the surface of the bar will become a helical curve pq', where q' is the position of point q after the end cross section has rotated through the angle ϕ (Fig. 3-3b).

The angle of twist changes along the axis of the bar, and at intermediate cross sections it will have a value $\phi(x)$ that is between zero at the left-hand end and ϕ at the right-hand end. If every cross section of the bar has the same radius and is subjected to the same torque (pure torsion), the angle $\phi(x)$ will vary linearly between the ends.

Shear Strains at the Outer Surface

Now consider an element of the bar between two cross sections distance dx apart (see Fig. 3-4a on the next page). This element is shown enlarged in Fig. 3-4b. On its outer surface we identify a small element $abcd$, with sides ab and cd that initially are parallel to the longitudinal axis. During twisting of the bar, the right-hand cross section rotates with respect to the left-hand cross section through a small angle of twist $d\phi$, so that points b and c move to b' and c', respectively. The lengths of the sides of the element, which is now element $ab'c'd$, do not change during this small rotation.

However, the angles at the corners of the element (Fig. 3-4b) are no longer equal to 90°. The element is therefore in a state of **pure shear**,

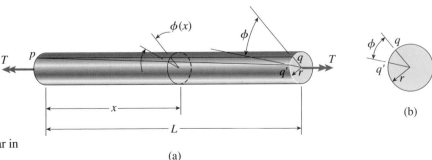

FIG. 3-3 Deformations of a circular bar in pure torsion

(a)

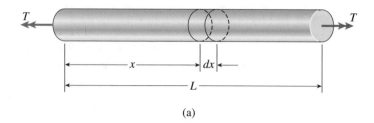

(a)

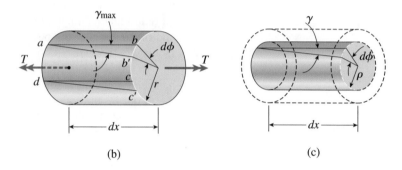

FIG. 3-4 Deformation of an element of length dx cut from a bar in torsion

(b)

(c)

which means that the element is subjected to shear strains but no normal strains (see Fig. 1-28 of Section 1.6). The magnitude of the shear strain at the outer surface of the bar, denoted γ_{max}, is equal to the decrease in the angle at point a, that is, the decrease in angle bad. From Fig. 3-4b we see that the decrease in this angle is

$$\gamma_{max} = \frac{bb'}{ab} \qquad \text{(a)}$$

where γ_{max} is measured in radians, bb' is the distance through which point b moves, and ab is the length of the element (equal to dx). With r denoting the radius of the bar, we can express the distance bb' as $rd\phi$, where $d\phi$ also is measured in radians. Thus, the preceding equation becomes

$$\gamma_{max} = \frac{rd\phi}{dx} \qquad \text{(b)}$$

This equation relates the shear strain at the outer surface of the bar to the angle of twist.

The quantity $d\phi/dx$ is the rate of change of the angle of twist ϕ with respect to the distance x measured along the axis of the bar. We will denote $d\phi/dx$ by the symbol θ and refer to it as the **rate of twist**, or the **angle of twist per unit length**:

$$\theta = \frac{d\phi}{dx} \qquad \text{(3-1)}$$

With this notation, we can now write the equation for the shear strain at the outer surface (Eq. b) as follows:

(b)

FIG. 3-3b (Repeated)

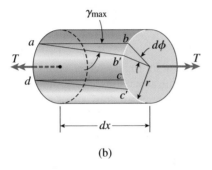

(b)

FIG. 3-4b (Repeated)

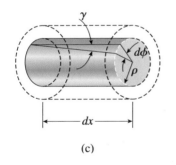

(c)

FIG. 3-4c (Repeated)

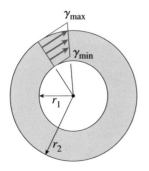

FIG. 3-5 Shear strains in a circular tube

$$\gamma_{max} = \frac{r\,d\phi}{dx} = r\theta \tag{3-2}$$

For convenience, we discussed a bar in pure torsion when deriving Eqs. (3-1) and (3-2). However, both equations are valid in more general cases of torsion, such as when the rate of twist θ is not constant but varies with the distance x along the axis of the bar.

In the special case of pure torsion, the rate of twist is equal to the total angle of twist ϕ divided by the length L, that is, $\theta = \phi/L$. Therefore, *for pure torsion only*, we obtain

$$\gamma_{max} = r\theta = \frac{r\phi}{L} \tag{3-3}$$

This equation can be obtained directly from the geometry of Fig. 3-3a by noting that γ_{max} is the angle between lines pq and pq', that is, γ_{max} is the angle qpq'. Therefore, $\gamma_{max}L$ is equal to the distance qq' at the end of the bar. But since the distance qq' also equals $r\phi$ (Fig. 3-3b), we obtain $r\phi = \gamma_{max}L$, which agrees with Eq. (3-3).

Shear Strains Within the Bar

The shear strains within the interior of the bar can be found by the same method used to find the shear strain γ_{max} at the surface. Because radii in the cross sections of a bar remain straight and undistorted during twisting, we see that the preceding discussion for an element *abcd* at the outer surface (Fig. 3-4b) will also hold for a similar element situated on the surface of an interior cylinder of radius ρ (Fig. 3-4c). Thus, interior elements are also in pure shear with the corresponding shear strains given by the equation (compare with Eq. 3-2):

$$\gamma = \rho\theta = \frac{\rho}{r}\,\gamma_{max} \tag{3-4}$$

This equation shows that the shear strains in a circular bar vary linearly with the radial distance ρ from the center, with the strain being zero at the center and reaching a maximum value γ_{max} at the outer surface.

Circular Tubes

A review of the preceding discussions will show that the equations for the shear strains (Eqs. 3-2 to 3-4) apply to **circular tubes** (Fig. 3-5) as well as to solid circular bars. Figure 3-5 shows the linear variation in shear strain between the maximum strain at the outer surface and the minimum strain at the interior surface. The equations for these strains are as follows:

$$\gamma_{max} = \frac{r_2\phi}{L} \qquad \gamma_{min} = \frac{r_1}{r_2}\,\gamma_{max} = \frac{r_1\phi}{L} \tag{3-5a,b}$$

in which r_1 and r_2 are the inner and outer radii, respectively, of the tube.

All of the preceding equations for the strains in a circular bar are based upon geometric concepts and do not involve the material properties. Therefore, the equations are valid for any material, whether it behaves elastically or inelastically, linearly or nonlinearly. However, the equations are limited to bars having small angles of twist and small strains.

3.3 CIRCULAR BARS OF LINEARLY ELASTIC MATERIALS

Now that we have investigated the shear strains in a circular bar in torsion (see Figs. 3-3 to 3-5), we are ready to determine the directions and magnitudes of the corresponding shear stresses. The directions of the stresses can be determined by inspection, as illustrated in Fig. 3-6a. We observe that the torque T tends to rotate the right-hand end of the bar counterclockwise when viewed from the right. Therefore the shear stresses τ acting on a stress element located on the surface of the bar will have the directions shown in the figure.

For clarity, the stress element shown in Fig. 3-6a is enlarged in Fig. 3-6b, where both the shear strain and the shear stresses are shown. As explained previously in Section 2.6, we customarily draw stress elements in two dimensions, as in Fig. 3-6b, but we must always remember that stress elements are actually three-dimensional objects with a thickness perpendicular to the plane of the figure.

The magnitudes of the shear stresses can be determined from the strains by using the stress-strain relation for the material of the bar. If the material is linearly elastic, we can use **Hooke's law in shear** (Eq. 1-14):

$$\tau = G\gamma \tag{3-6}$$

in which G is the shear modulus of elasticity and γ is the shear strain in radians. Combining this equation with the equations for the shear strains (Eqs. 3-2 and 3-4), we get

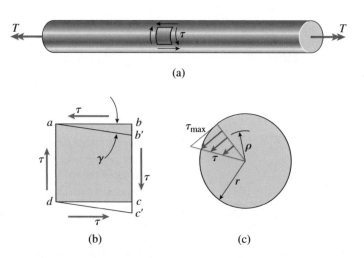

FIG. 3-6 Shear stresses in a circular bar in torsion

$$\tau_{max} = Gr\theta \qquad \tau = G\rho\theta = \frac{\rho}{r}\,\tau_{max} \qquad\qquad (3\text{-}7a,b)$$

in which τ_{max} is the shear stress at the outer surface of the bar (radius r), τ is the shear stress at an interior point (radius ρ), and θ is the rate of twist. (In these equations, θ has units of radians per unit of length.)

Equations (3-7a) and (3-7b) show that the shear stresses vary linearly with the distance from the center of the bar, as illustrated by the triangular stress diagram in Fig. 3-6c. This linear variation of stress is a consequence of Hooke's law. If the stress-strain relation is nonlinear, the stresses will vary nonlinearly and other methods of analysis will be needed.

The shear stresses acting on a cross-sectional plane are accompanied by shear stresses of the same magnitude acting on longitudinal planes (Fig. 3-7). This conclusion follows from the fact that equal shear stresses always exist on mutually perpendicular planes, as explained in Section 1.6. If the material of the bar is weaker in shear on longitudinal planes than on cross-sectional planes, as is typical of wood when the grain runs parallel to the axis of the bar, the first cracks due to torsion will appear on the surface in the longitudinal direction.

The state of pure shear at the surface of a bar (Fig. 3-6b) is equivalent to equal tensile and compressive stresses acting on an element oriented at an angle of 45°, as explained later in Section 3.5. Therefore, a rectangular element with sides at 45° to the axis of the shaft will be subjected to tensile and compressive stresses, as shown in Fig. 3-8. If a torsion bar is made of a material that is weaker in tension than in shear, failure will occur in tension along a helix inclined at 45° to the axis, as you can demonstrate by twisting a piece of classroom chalk.

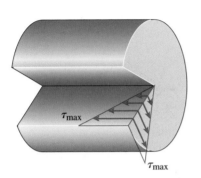

FIG. 3-7 Longitudinal and transverse shear stresses in a circular bar subjected to torsion

FIG. 3-8 Tensile and compressive stresses acting on a stress element oriented at 45° to the longitudinal axis

The Torsion Formula

The next step in our analysis is to determine the relationship between the shear stresses and the torque T. Once this is accomplished, we will be able to calculate the stresses and strains in a bar due to any set of applied torques.

The distribution of the shear stresses acting on a cross section is pictured in Figs. 3-6c and 3-7. Because these stresses act continuously around the cross section, they have a resultant in the form of a moment—a moment equal to the torque T acting on the bar. To determine this resultant, we consider an element of area dA located at radial distance ρ from the axis of the bar (Fig. 3-9). The shear force acting on this element is equal to $\tau\,dA$, where τ is the shear stress at radius ρ. The moment of this force about the axis of the bar is equal to the force times its distance from the center, or $\tau\rho\,dA$. Substituting for the shear stress τ from Eq. (3-7b), we can express this elemental moment as

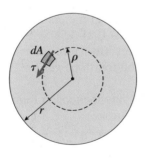

FIG. 3-9 Determination of the resultant of the shear stresses acting on a cross section

$$dM = \tau\rho dA = \frac{\tau_{max}}{r}\,\rho^2 dA$$

The resultant moment (equal to the torque T) is the summation over the entire cross-sectional area of all such elemental moments:

$$T = \int_A dM = \frac{\tau_{max}}{r} \int_A \rho^2 dA = \frac{\tau_{max}}{r} I_P \tag{3-8}$$

in which

$$I_P = \int_A \rho^2 dA \tag{3-9}$$

is the **polar moment of inertia** of the circular cross section.

For a **circle** of radius r and diameter d, the polar moment of inertia is

$$I_P = \frac{\pi r^4}{2} = \frac{\pi d^4}{32} \tag{3-10}$$

as given in Appendix E, Case 9 (available online). Note that moments of inertia have units of length to the fourth power.[*]

An expression for the maximum shear stress can be obtained by rearranging Eq. (3-8), as follows:

$$\tau_{max} = \frac{Tr}{I_P} \tag{3-11}$$

This equation, known as the **torsion formula**, shows that the maximum shear stress is proportional to the applied torque T and inversely proportional to the polar moment of inertia I_P.

Typical **units** used with the torsion formula are as follows. In SI, the torque T is usually expressed in newton meters (N·m), the radius r in meters (m), the polar moment of inertia I_P in meters to the fourth power (m^4), and the shear stress τ in pascals (Pa). If USCS units are used, T is often expressed in pound-feet (lb-ft) or pound-inches (lb-in.), r in inches (in.), I_P in inches to the fourth power (in.^{4}), and τ in pounds per square inch (psi).

Substituting $r = d/2$ and $I_P = \pi d^4/32$ into the torsion formula, we get the following equation for the maximum stress:

$$\tau_{max} = \frac{16T}{\pi d^3} \tag{3-12}$$

This equation applies only to bars of *solid circular cross section*, whereas the torsion formula itself (Eq. 3-11) applies to both solid bars and circular tubes, as explained later. Equation (3-12) shows that the shear stress is inversely proportional to the cube of the diameter. Thus, if the diameter is doubled, the stress is reduced by a factor of eight.

---

[*]Polar moments of inertia are discussed in Section 10.6 of Chapter 10 (available online).

The shear stress at distance ρ from the center of the bar is

$$\tau = \frac{\rho}{r}\, \tau_{\text{max}} = \frac{T\rho}{I_P} \tag{3-13}$$

which is obtained by combining Eq. (3-7b) with the torsion formula (Eq. 3-11). Equation (3-13) is a *generalized torsion formula*, and we see once again that the shear stresses vary linearly with the radial distance from the center of the bar.

Angle of Twist

The angle of twist of a bar of linearly elastic material can now be related to the applied torque T. Combining Eq. (3-7a) with the torsion formula, we get

$$\theta = \frac{T}{GI_P} \tag{3-14}$$

in which θ has units of radians per unit of length. This equation shows that the rate of twist θ is directly proportional to the torque T and inversely proportional to the product GI_P, known as the **torsional rigidity** of the bar.

For a bar in **pure torsion**, the total angle of twist ϕ, equal to the rate of twist times the length of the bar (that is, $\phi = \theta L$), is

$$\phi = \frac{TL}{GI_P} \tag{3-15}$$

in which ϕ is measured in radians. The use of the preceding equations in both analysis and design is illustrated later in Examples 3-1 and 3-2.

The quantity GI_P/L, called the **torsional stiffness** of the bar, is the torque required to produce a unit angle of rotation. The **torsional flexibility** is the reciprocal of the stiffness, or L/GI_P, and is defined as the angle of rotation produced by a unit torque. Thus, we have the following expressions:

$$k_T = \frac{GI_P}{L} \qquad f_T = \frac{L}{GI_P} \tag{a,b}$$

These quantities are analogous to the axial stiffness $k = EA/L$ and axial flexibility $f = L/EA$ of a bar in tension or compression (compare with Eqs. 2-4a and 2-4b). Stiffnesses and flexibilities have important roles in structural analysis.

The equation for the angle of twist (Eq. 3-15) provides a convenient way to determine the shear modulus of elasticity G for a material. By conducting a torsion test on a circular bar, we can measure the angle of twist ϕ produced by a known torque T. Then the value of G can be calculated from Eq. (3-15).

Circular Tubes

Circular tubes are more efficient than solid bars in resisting torsional loads. As we know, the shear stresses in a solid circular bar are maximum at the outer boundary of the cross section and zero at the center. Therefore, most of the material in a solid shaft is stressed significantly below the maximum shear stress. Furthermore, the stresses near the center of the cross section have a smaller moment arm ρ for use in determining the torque (see Fig. 3-9 and Eq. 3-8).

By contrast, in a typical hollow tube most of the material is near the outer boundary of the cross section where both the shear stresses and the moment arms are highest (Fig. 3-10). Thus, if weight reduction and savings of material are important, it is advisable to use a circular tube. For instance, large drive shafts, propeller shafts, and generator shafts usually have hollow circular cross sections.

The analysis of the torsion of a circular tube is almost identical to that for a solid bar. The same basic expressions for the shear stresses may be used (for instance, Eqs. 3-7a and 3-7b). Of course, the radial distance ρ is limited to the range r_1 to r_2, where r_1 is the inner radius and r_2 is the outer radius of the bar (Fig. 3-10).

The relationship between the torque T and the maximum stress is given by Eq. (3-8), but the limits on the integral for the polar moment of inertia (Eq. 3-9) are $\rho = r_1$ and $\rho = r_2$. Therefore, the polar moment of inertia of the cross-sectional area of a tube is

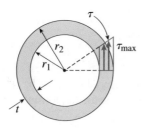

FIG. 3-10 Circular tube in torsion

$$I_P = \frac{\pi}{2}(r_2^4 - r_1^4) = \frac{\pi}{32}(d_2^4 - d_1^4) \qquad (3\text{-}16)$$

The preceding expressions can also be written in the following forms:

$$I_P = \frac{\pi rt}{2}(4r^2 + t^2) = \frac{\pi dt}{4}(d^2 + t^2) \qquad (3\text{-}17)$$

in which r is the *average radius* of the tube, equal to $(r_1 + r_2)/2$; d is the *average diameter*, equal to $(d_1 + d_2)/2$; and t is the *wall thickness* (Fig. 3-10), equal to $r_2 - r_1$. Of course, Eqs. (3-16) and (3-17) give the same results, but sometimes the latter is more convenient.

If the tube is relatively thin so that the wall thickness t is small compared to the average radius r, we may disregard the terms t^2 in Eq. (3-17). With this simplification, we obtain the following *approximate formulas* for the polar moment of inertia:

$$I_P \approx 2\pi r^3 t = \frac{\pi d^3 t}{4} \qquad (3\text{-}18)$$

These expressions are given in Case 22 of Appendix E (available online).

Reminders: In Eqs. 3-17 and 3-18, the quantities r and d are the average radius and diameter, not the maximums. Also, Eqs. 3-16 and 3-17 are exact; Eq. 3-18 is approximate.

The torsion formula (Eq. 3-11) may be used for a circular tube of linearly elastic material provided I_P is evaluated according to Eq. (3-16), Eq. (3-17), or, if appropriate, Eq. (3-18). The same comment applies to the general equation for shear stress (Eq. 3-13), the equations for rate of twist and angle of twist (Eqs. 3-14 and 3-15), and the equations for stiffness and flexibility (Eqs. a and b).

The shear stress distribution in a tube is pictured in Fig. 3-10. From the figure, we see that the average stress in a thin tube is nearly as great as the maximum stress. This means that a hollow bar is more efficient in the use of material than is a solid bar, as explained previously and as demonstrated later in Examples 3-2 and 3-3.

When designing a circular tube to transmit a torque, we must be sure that the thickness t is large enough to prevent wrinkling or buckling of the wall of the tube. For instance, a maximum value of the radius to thickness ratio, such as $(r_2/t)_{max} = 12$, may be specified. Other design considerations include environmental and durability factors, which also may impose requirements for minimum wall thickness. These topics are discussed in courses and textbooks on mechanical design.

Limitations

The equations derived in this section are limited to bars of circular cross section (either solid or hollow) that behave in a linearly elastic manner. In other words, the loads must be such that the stresses do not exceed the proportional limit of the material. Furthermore, the equations for stresses are valid only in parts of the bars away from stress concentrations (such as holes and other abrupt changes in shape) and away from cross sections where loads are applied.

Finally, it is important to emphasize that the equations for the torsion of circular bars and tubes cannot be used for bars of other shapes. Noncircular bars, such as rectangular bars and bars having I-shaped cross sections, behave quite differently than do circular bars. For instance, their cross sections do *not* remain plane and their maximum stresses are *not* located at the farthest distances from the midpoints of the cross sections. Thus, these bars require more advanced methods of analysis, such as those presented in books on theory of elasticity and advanced mechanics of materials.[*]

---

[*]The torsion theory for circular bars originated with the work of the famous French scientist C. A. de Coulomb (1736–1806); further developments were due to Thomas Young and A. Duleau (Ref. 3-1). The general theory of torsion (for bars of any shape) is due to the most famous elastician of all time, Barré de Saint-Venant (1797–1886); see Ref. 2-10. A list of references is available online.

Example 3-1

A solid steel bar of circular cross section (Fig. 3-11) has diameter $d = 1.5$ in., length $L = 54$ in., and shear modulus of elasticity $G = 11.5 \times 10^6$ psi. The bar is subjected to torques T acting at the ends.

(a) If the torques have magnitude $T = 250$ lb-ft, what is the maximum shear stress in the bar? What is the angle of twist between the ends?

(b) If the allowable shear stress is 6000 psi and the allowable angle of twist is 2.5°, what is the maximum permissible torque?

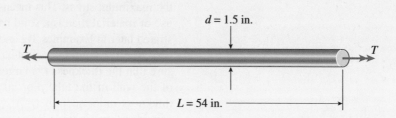

$d = 1.5$ in.

T T

$L = 54$ in.

FIG. 3-11 Example 3-1. Bar in pure torsion

Solution

(a) *Maximum shear stress and angle of twist.* Because the bar has a solid circular cross section, we can find the maximum shear stress from Eq. (3-12), as follows:

$$\tau_{max} = \frac{16T}{\pi d^3} = \frac{16(250\ \text{lb-ft})(12\ \text{in./ft})}{\pi(1.5\ \text{in.})^3} = 4530\ \text{psi} \quad \Longleftarrow$$

In a similar manner, the angle of twist is obtained from Eq. (3-15) with the polar moment of inertia given by Eq. (3-10):

$$I_P = \frac{\pi d^4}{32} = \frac{\pi(1.5\ \text{in.})^4}{32} = 0.4970\ \text{in.}^4$$

$$\phi = \frac{TL}{GI_P} = \frac{(250\ \text{lb-ft})(12\ \text{in./ft})(54\ \text{in.})}{(11.5 \times 10^6\ \text{psi})(0.4970\ \text{in.}^4)} = 0.02834\ \text{rad} = 1.62° \quad \Longleftarrow$$

Thus, the analysis of the bar under the action of the given torque is completed.

(b) *Maximum permissible torque.* The maximum permissible torque is determined either by the allowable shear stress or by the allowable angle of twist. Beginning with the shear stress, we rearrange Eq. (3-12) and calculate as follows:

$$T_1 = \frac{\pi d^3 \tau_{allow}}{16} = \frac{\pi}{16}(1.5\ \text{in.})^3(6000\ \text{psi}) = 3980\ \text{lb-in.} = 331\ \text{lb-ft}$$

Any torque larger than this value will result in a shear stress that exceeds the allowable stress of 6000 psi.

Using a rearranged Eq. (3-15), we now calculate the torque based upon the angle of twist:

Ship drive shaft is a key part of the propulsion system
(Louie Psihoyos/Science Faction)

$$T_2 = \frac{GI_P\phi_{allow}}{L} = \frac{(11.5 \times 10^6 \text{ psi})(0.4970 \text{ in.}^4)(2.5°)(\pi \text{ rad}/180°)}{54 \text{ in.}}$$

$$= 4618 \text{ lb-in.} = 385 \text{ lb-ft}$$

Any torque larger than T_2 will result in the allowable angle of twist being exceeded. The maximum permissible torque is the smaller of T_1 and T_2:

$$T_{max} = 331 \text{ lb-ft}$$

In this example, the allowable shear stress provides the limiting condition.

Example 3-2

Complex crank shaft
(Peter Ginter/Science Faction)

A steel shaft is to be manufactured either as a solid circular bar or as a circular tube (Fig. 3-12). The shaft is required to transmit a torque of 1200 N·m without exceeding an allowable shear stress of 40 MPa nor an allowable rate of twist of 0.75°/m. (The shear modulus of elasticity of the steel is 78 GPa.)

(a) Determine the required diameter d_0 of the solid shaft.

(b) Determine the required outer diameter d_2 of the hollow shaft if the thickness t of the shaft is specified as one-tenth of the outer diameter.

(c) Determine the ratio of diameters (that is, the ratio d_2/d_0) and the ratio of weights of the hollow and solid shafts.

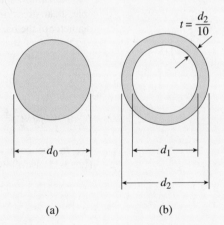

(a)   (b)

FIG. 3-12 Example 3-2. Torsion of a steel shaft

Solution

(a) *Solid shaft.* The required diameter d_0 is determined either from the allowable shear stress or from the allowable rate of twist. In the case of the allowable shear stress we rearrange Eq. (3-12) and obtain

$$d_0^3 = \frac{16T}{\pi\tau_{allow}} = \frac{16(1200 \text{ N·m})}{\pi(40 \text{ MPa})} = 152.8 \times 10^{-6} \text{ m}^3$$

continued

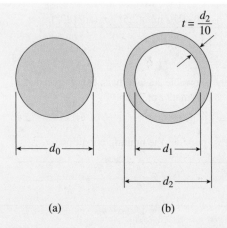

(a)   (b)

FIG. 3-12 (Repeated)

from which we get

$$d_0 = 0.0535 \text{ m} = 53.5 \text{ mm}$$

In the case of the allowable rate of twist, we start by finding the required polar moment of inertia (see Eq. 3-14):

$$I_P = \frac{T}{G\theta_{\text{allow}}} = \frac{1200 \text{ N·m}}{(78 \text{ GPa})(0.75°/\text{m})(\pi \text{ rad}/180°)} = 1175 \times 10^{-9} \text{ m}^4$$

Since the polar moment of inertia is equal to $\pi d^4/32$, the required diameter is

$$d_0^4 = \frac{32 I_P}{\pi} = \frac{32(1175 \times 10^{-9} \text{ m}^4)}{\pi} = 11.97 \times 10^{-6} \text{ m}^4$$

or

$$d_0 = 0.0588 \text{ m} = 58.8 \text{ mm}$$

Comparing the two values of d_0, we see that the rate of twist governs the design and the required diameter of the solid shaft is

$$d_0 = 58.8 \text{ mm}$$

In a practical design, we would select a diameter slightly larger than the calculated value of d_0; for instance, 60 mm.

(b) *Hollow shaft.* Again, the required diameter is based upon either the allowable shear stress or the allowable rate of twist. We begin by noting that the outer diameter of the bar is d_2 and the inner diameter is

$$d_1 = d_2 - 2t = d_2 - 2(0.1 d_2) = 0.8 d_2$$

Thus, the polar moment of inertia (Eq. 3-16) is

$$I_P = \frac{\pi}{32} (d_2^4 - d_1^4) = \frac{\pi}{32} \left[d_2^4 - (0.8 d_2)^4 \right] = \frac{\pi}{32} (0.5904 d_2^4) = 0.05796 d_2^4$$

In the case of the allowable shear stress, we use the torsion formula (Eq. 3-11) as follows:

$$\tau_{\text{allow}} = \frac{Tr}{I_P} = \frac{T(d_2/2)}{0.05796 d_2^4} = \frac{T}{0.1159 d_2^3}$$

Rearranging, we get

$$d_2^3 = \frac{T}{0.1159 \tau_{\text{allow}}} = \frac{1200 \text{ N·m}}{0.1159(40 \text{ MPa})} = 258.8 \times 10^{-6} \text{ m}^3$$

Solving for d_2 gives

$$d_2 = 0.0637 \text{ m} = 63.7 \text{ mm}$$

which is the required outer diameter based upon the shear stress.

In the case of the allowable rate of twist, we use Eq. (3-14) with θ replaced by θ_{allow} and I_P replaced by the previously obtained expression; thus,

$$\theta_{\text{allow}} = \frac{T}{G(0.05796d_2^4)}$$

from which

$$d_2^4 = \frac{T}{0.05796G\theta_{\text{allow}}}$$

$$= \frac{1200 \text{ N·m}}{0.05796(78 \text{ GPa})(0.75°/\text{m})(\pi \text{ rad}/180°)} = 20.28 \times 10^{-6} \text{ m}^4$$

Solving for d_2 gives

$$d_2 = 0.0671 \text{ m} = 67.1 \text{ mm}$$

which is the required diameter based upon the rate of twist.

Comparing the two values of d_2, we see that the rate of twist governs the design and the required outer diameter of the hollow shaft is

$$d_2 = 67.1 \text{ mm} \qquad \Longleftarrow$$

The inner diameter d_1 is equal to $0.8d_2$, or 53.7 mm. (As practical values, we might select $d_2 = 70$ mm and $d_1 = 0.8d_2 = 56$ mm.)

(c) *Ratios of diameters and weights.* The ratio of the outer diameter of the hollow shaft to the diameter of the solid shaft (using the calculated values) is

$$\frac{d_2}{d_0} = \frac{67.1 \text{ mm}}{58.8 \text{ mm}} = 1.14 \qquad \Longleftarrow$$

Since the weights of the shafts are proportional to their cross-sectional areas, we can express the ratio of the weight of the hollow shaft to the weight of the solid shaft as follows:

$$\frac{W_{\text{hollow}}}{W_{\text{solid}}} = \frac{A_{\text{hollow}}}{A_{\text{solid}}} = \frac{\pi(d_2^2 - d_1^2)/4}{\pi d_0^2/4} = \frac{d_2^2 - d_1^2}{d_0^2}$$

$$= \frac{(67.1 \text{ mm})^2 - (53.7 \text{ mm})^2}{(58.8 \text{ mm})^2} = 0.47 \qquad \Longleftarrow$$

These results show that the hollow shaft uses only 47% as much material as does the solid shaft, while its outer diameter is only 14% larger.

Note: This example illustrates how to determine the required sizes of both solid bars and circular tubes when allowable stresses and allowable rates of twist are known. It also illustrates the fact that circular tubes are more efficient in the use of materials than are solid circular bars.

Example 3-3

A hollow shaft and a solid shaft constructed of the same material have the same length and the same outer radius R (Fig. 3-13). The inner radius of the hollow shaft is $0.6R$.

(a) Assuming that both shafts are subjected to the same torque, compare their shear stresses, angles of twist, and weights.

(b) Determine the strength-to-weight ratios for both shafts.

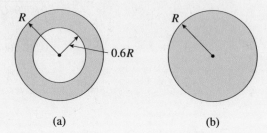

FIG. 3-13 Example 3-3. Comparison of hollow and solid shafts

(a)   (b)

Solution

(a) *Comparison or shear stresses.* The maximum shear stresses, given by the torsion formula (Eq. 3-11), are proportional to $1/I_P$ inasmuch as the torques and radii are the same. For the hollow shaft, we get

$$I_P = \frac{\pi R^4}{2} - \frac{\pi (0.6R)^4}{2} = 0.4352 \pi R^4$$

and for the solid shaft,

$$I_P = \frac{\pi R^4}{2} = 0.5 \pi R^4$$

Therefore, the ratio β_1 of the maximum shear stress in the hollow shaft to that in the solid shaft is

$$\beta_1 = \frac{\tau_H}{\tau_S} = \frac{0.5 \pi R^4}{0.4352 \pi R^4} = 1.15 \qquad \Longleftarrow$$

where the subscripts H and S refer to the hollow shaft and the solid shaft, respectively.

Comparison of angles of twist. The angles of twist (Eq. 3-15) are also proportional to $1/I_P$, because the torques T, lengths L, and moduli of elasticity G are the same for both shafts. Therefore, their ratio is the same as for the shear stresses:

$$\beta_2 = \frac{\phi_H}{\phi_S} = \frac{0.5 \pi R^4}{0.4352 \pi R^4} = 1.15 \qquad \Longleftarrow$$

Comparison of weights. The weights of the shafts are proportional to their cross-sectional areas; consequently, the weight of the solid shaft is proportional to πR^2 and the weight of the hollow shaft is proportional to

$$\pi R^2 - \pi(0.6R)^2 = 0.64\,\pi R^2$$

Therefore, the ratio of the weight of the hollow shaft to the weight of the solid shaft is

$$\beta_3 = \frac{W_H}{W_S} = \frac{0.64\pi R^2}{\pi R^2} = 0.64 \qquad \Longleftarrow$$

From the preceding ratios we again see the inherent advantage of hollow shafts. In this example, the hollow shaft has 15% greater stress and 15% greater angle of rotation than the solid shaft but 36% less weight.

(b) *Strength-to-weight ratios.* The relative efficiency of a structure is sometimes measured by its *strength-to-weight ratio*, which is defined for a bar in torsion as the allowable torque divided by the weight. The allowable torque for the hollow shaft of Fig. 3-13a (from the torsion formula) is

$$T_H = \frac{\tau_{\max}I_P}{R} = \frac{\tau_{\max}(0.4352\,\pi R^4)}{R} = 0.4352\,\pi R^3\tau_{\max}$$

and for the solid shaft is

$$T_S = \frac{\tau_{\max}I_P}{R} = \frac{\tau_{\max}(0.5\,\pi R^4)}{R} = 0.5\,\pi R^3\tau_{\max}$$

The weights of the shafts are equal to the cross-sectional areas times the length L times the weight density γ of the material:

$$W_H = 0.64\,\pi R^2 L\gamma \qquad W_S = \pi R^2 L\gamma$$

Thus, the strength-to-weight ratios S_H and S_S for the hollow and solid bars, respectively, are

$$S_H = \frac{T_H}{W_H} = 0.68\,\frac{\tau_{\max}R}{\gamma L} \qquad S_S = \frac{T_S}{W_S} = 0.5\,\frac{\tau_{\max}R}{\gamma L} \qquad \Longleftarrow$$

In this example, the strength-to-weight ratio of the hollow shaft is 36% greater than the strength-to-weight ratio for the solid shaft, demonstrating once again the relative efficiency of hollow shafts. For a thinner shaft, the percentage will increase; for a thicker shaft, it will decrease.

3.4 NONUNIFORM TORSION

As explained in Section 3.2, *pure torsion* refers to torsion of a prismatic bar subjected to torques acting only at the ends. **Nonuniform torsion** differs from pure torsion in that the bar need not be prismatic and the applied torques may act anywhere along the axis of the bar. Bars in nonuniform torsion can be analyzed by applying the formulas of pure torsion to finite segments of the bar and then adding the results, or by applying the formulas to differential elements of the bar and then integrating.

To illustrate these procedures, we will consider three cases of nonuniform torsion. Other cases can be handled by techniques similar to those described here.

Case 1. *Bar consisting of prismatic segments with constant torque throughout each segment* (Fig. 3-14). The bar shown in part (a) of the figure has two different diameters and is loaded by torques acting at points *A*, *B*, *C*, and *D*. Consequently, we divide the bar into segments in such a way that each segment is prismatic and subjected to a constant torque. In this example, there are three such segments, *AB*, *BC*, and *CD*. Each segment is in pure torsion, and therefore all of the formulas derived in the preceding section may be applied to each part separately.

The first step in the analysis is to determine the magnitude and direction of the internal torque in each segment. Usually the torques can be determined by inspection, but if necessary they can be found by cutting sections through the bar, drawing free-body diagrams, and solving equations of equilibrium. This process is illustrated in parts (b), (c), and (d) of the figure. The first cut is made anywhere in segment *CD*, thereby exposing the internal torque T_{CD}. From the free-body diagram (Fig. 3-14b), we see that T_{CD} is equal to $-T_1 - T_2 + T_3$. From the next diagram we see that T_{BC} equals $-T_1 - T_2$, and from the last we find that T_{AB} equals $-T_1$. Thus,

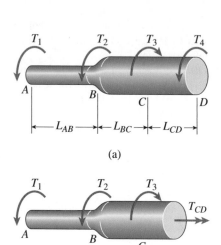

(a)

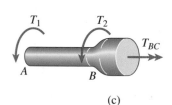

(b)

(c)

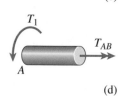

(d)

FIG. 3-14 Bar in nonuniform torsion (Case 1)

$$T_{CD} = -T_1 - T_2 + T_3 \qquad T_{BC} = -T_1 - T_2 \qquad T_{AB} = -T_1 \quad \text{(a,b,c)}$$

Each of these torques is constant throughout the length of its segment.

When finding the shear stresses in each segment, we need only the magnitudes of these internal torques, since the directions of the stresses are not of interest. However, when finding the angle of twist for the entire bar, we need to know the direction of twist in each segment in order to combine the angles of twist correctly. Therefore, we need to establish a *sign convention* for the internal torques. A convenient rule in many cases is the following: *An internal torque is positive when its vector points away from the cut section and negative when its vector points toward the section.* Thus, all of the internal torques shown in Figs. 3-14b, c, and d are pictured in their positive directions. If the calculated torque (from Eq. a, b, or c) turns out to have a positive sign, it means that the torque acts in the assumed direction; if the torque has a negative sign, it acts in the opposite direction.

The maximum shear stress in each segment of the bar is readily obtained from the torsion formula (Eq. 3-11) using the appropriate cross-sectional dimensions and internal torque. For instance, the maximum stress in segment *BC* (Fig. 3-14) is found using the diameter of that segment and the torque T_{BC} calculated from Eq. (b). The maximum stress in the entire bar is the largest stress from among the stresses calculated for each of the three segments.

The angle of twist for each segment is found from Eq. (3-15), again using the appropriate dimensions and torque. The total angle of twist of one end of the bar with respect to the other is then obtained by algebraic summation, as follows:

$$\phi = \phi_1 + \phi_2 + \ldots + \phi_n \tag{3-19}$$

where ϕ_1 is the angle of twist for segment 1, ϕ_2 is the angle for segment 2, and so on, and n is the total number of segments. Since each angle of twist is found from Eq. (3-15), we can write the general formula

$$\phi = \sum_{i=1}^{n} \phi_i = \sum_{i=1}^{n} \frac{T_i L_i}{G_i (I_P)_i} \tag{3-20}$$

in which the subscript i is a numbering index for the various segments. For segment i of the bar, T_i is the internal torque (found from equilibrium, as illustrated in Fig. 3-14), L_i is the length, G_i is the shear modulus, and $(I_P)_i$ is the polar moment of inertia. Some of the torques (and the corresponding angles of twist) may be positive and some may be negative. By summing *algebraically* the angles of twist for all segments, we obtain the total angle of twist ϕ between the ends of the bar. The process is illustrated later in Example 3-4.

Case 2. *Bar with continuously varying cross sections and constant torque* (Fig. 3-15). When the torque is constant, the maximum shear stress in a solid bar always occurs at the cross section having the smallest diameter, as shown by Eq. (3-12). Furthermore, this observation usually holds for tubular bars. If this is the case, we only need to investigate the smallest cross section in order to calculate the maximum shear stress. Otherwise, it may be necessary to evaluate the stresses at more than one location in order to obtain the maximum.

To find the angle of twist, we consider an element of length dx at distance x from one end of the bar (Fig. 3-15). The differential angle of rotation $d\phi$ for this element is

$$d\phi = \frac{T dx}{G I_P(x)} \tag{d}$$

in which $I_P(x)$ is the polar moment of inertia of the cross section at distance x from the end. The angle of twist for the entire bar is the summation of the differential angles of rotation:

$$\phi = \int_0^L d\phi = \int_0^L \frac{T dx}{G I_P(x)} \tag{3-21}$$

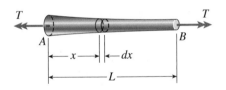

FIG. 3-15 Bar in nonuniform torsion (Case 2)

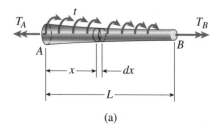

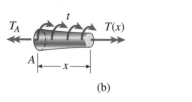

FIG. 3-16 Bar in nonuniform torsion (Case 3)

If the expression for the polar moment of inertia $I_P(x)$ is not too complex, this integral can be evaluated analytically, as in Example 3-5. In other cases, it must be evaluated numerically.

Case 3. *Bar with continuously varying cross sections and continuously varying torque* (Fig. 3-16). The bar shown in part (a) of the figure is subjected to a *distributed torque* of intensity t per unit distance along the axis of the bar. As a result, the internal torque $T(x)$ varies continuously along the axis (Fig. 3-16b). The internal torque can be evaluated with the aid of a free-body diagram and an equation of equilibrium. As in Case 2, the polar moment of inertia $I_P(x)$ can be evaluated from the cross-sectional dimensions of the bar.

Knowing both the torque and polar moment of inertia as functions of x, we can use the torsion formula to determine how the shear stress varies along the axis of the bar. The cross section of maximum shear stress can then be identified, and the maximum shear stress can be determined.

The angle of twist for the bar of Fig. 3-16a can be found in the same manner as described for Case 2. The only difference is that the torque, like the polar moment of inertia, also varies along the axis. Consequently, the equation for the angle of twist becomes

$$\phi = \int_0^L d\phi = \int_0^L \frac{T(x)\,dx}{GI_P(x)} \tag{3-22}$$

This integral can be evaluated analytically in some cases, but usually it must be evaluated numerically.

Limitations

The analyses described in this section are valid for bars made of linearly elastic materials with circular cross sections (either solid or hollow). Also, the stresses determined from the torsion formula are valid in regions of the bar *away* from stress concentrations, which are high localized stresses that occur wherever the diameter changes abruptly and wherever concentrated torques are applied. However, stress concentrations have relatively little effect on the angle of twist, and therefore the equations for ϕ are generally valid.

Finally, we must keep in mind that the torsion formula and the formulas for angles of twist were derived for prismatic bars. We can safely apply them to bars with varying cross sections only when the changes in diameter are small and gradual. As a rule of thumb, the formulas given here are satisfactory as long as the angle of taper (the angle between the sides of the bar) is less than $10°$.

Example 3-4

A solid steel shaft *ABCDE* (Fig. 3-17) having diameter $d = 30$ mm turns freely in bearings at points *A* and *E*. The shaft is driven by a gear at *C*, which applies a torque $T_2 = 450$ N·m in the direction shown in the figure. Gears at *B* and *D* are driven by the shaft and have resisting torques $T_1 = 275$ N·m and $T_3 = 175$ N·m, respectively, acting in the opposite direction to the torque T_2. Segments *BC* and *CD* have lengths $L_{BC} = 500$ mm and $L_{CD} = 400$ mm, respectively, and the shear modulus $G = 80$ GPa.

Determine the maximum shear stress in each part of the shaft and the angle of twist between gears *B* and *D*.

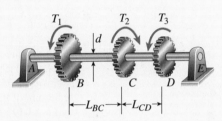

FIG. 3-17 Example 3-4. Steel shaft in torsion

Solution

Each segment of the bar is prismatic and subjected to a constant torque (Case 1). Therefore, the first step in the analysis is to determine the torques acting in the segments, after which we can find the shear stresses and angles of twist.

Torques acting in the segments. The torques in the end segments (*AB* and *DE*) are zero since we are disregarding any friction in the bearings at the supports. Therefore, the end segments have no stresses and no angles of twist.

The torque T_{CD} in segment *CD* is found by cutting a section through the segment and constructing a free-body diagram, as in Fig. 3-18a. The torque is assumed to be positive, and therefore its vector points away from the cut section. From equilibrium of the free body, we obtain

$$T_{CD} = T_2 - T_1 = 450\ \text{N·m} - 275\ \text{N·m} = 175\ \text{N·m}$$

The positive sign in the result means that T_{CD} acts in the assumed positive direction.

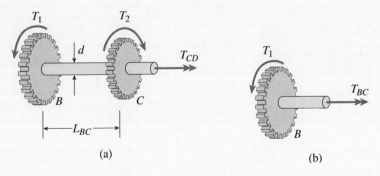

FIG. 3-18 Free-body diagrams for Example 3-4

(a)

(b)

continued

The torque in segment BC is found in a similar manner, using the free-body diagram of Fig. 3-18b:

$$T_{BC} = -T_1 = -275 \text{ N·m}$$

Note that this torque has a negative sign, which means that its direction is opposite to the direction shown in the figure.

Shear stresses. The maximum shear stresses in segments BC and CD are found from the modified form of the torsion formula (Eq. 3-12); thus,

$$\tau_{BC} = \frac{16T_{BC}}{\pi d^3} = \frac{16(275 \text{ N·m})}{\pi(30 \text{ mm})^3} = 51.9 \text{ MPa}$$

$$\tau_{CD} = \frac{16T_{CD}}{\pi d^3} = \frac{16(175 \text{ N·m})}{\pi(30 \text{ mm})^3} = 33.0 \text{ MPa}$$

Since the directions of the shear stresses are not of interest in this example, only absolute values of the torques are used in the preceding calculations.

Angles of twist. The angle of twist ϕ_{BD} between gears B and D is the algebraic sum of the angles of twist for the intervening segments of the bar, as given by Eq. (3-19); thus,

$$\phi_{BD} = \phi_{BC} + \phi_{CD}$$

When calculating the individual angles of twist, we need the moment of inertia of the cross section:

$$I_P = \frac{\pi d^4}{32} = \frac{\pi(30 \text{ mm})^4}{32} = 79{,}520 \text{ mm}^4$$

Now we can determine the angles of twist, as follows:

$$\phi_{BC} = \frac{T_{BC} L_{BC}}{G I_P} = \frac{(-275 \text{ N·m})(500 \text{ mm})}{(80 \text{ GPa})(79{,}520 \text{ mm}^4)} = -0.0216 \text{ rad}$$

$$\phi_{CD} = \frac{T_{CD} L_{CD}}{G I_P} = \frac{(175 \text{ N·m})(400 \text{ mm})}{(80 \text{ GPa})(79{,}520 \text{ mm}^4)} = 0.0110 \text{ rad}$$

Note that in this example the angles of twist have opposite directions. Adding algebraically, we obtain the total angle of twist:

$$\phi_{BD} = \phi_{BC} + \phi_{CD} = -0.0216 + 0.0110 = -0.0106 \text{ rad} = -0.61°$$

The minus sign means that gear D rotates clockwise (when viewed from the right-hand end of the shaft) with respect to gear B. However, for most purposes only the absolute value of the angle of twist is needed, and therefore it is sufficient to say that the angle of twist between gears B and D is $0.61°$. The angle of twist between the two ends of a shaft is sometimes called the *wind-up*.

Notes: The procedures illustrated in this example can be used for shafts having segments of different diameters or of different materials, as long as the dimensions and properties remain constant within each segment.

Only the effects of torsion are considered in this example and in the problems at the end of the chapter. Bending effects are considered later, beginning with Chapter 4.

Example 3-5

A tapered bar AB of solid circular cross section is twisted by torques T applied at the ends (Fig. 3-19). The diameter of the bar varies linearly from d_A at the left-hand end to d_B at the right-hand end, with d_B assumed to be greater than d_A.

(a) Determine the maximum shear stress in the bar.

(b) Derive a formula for the angle of twist of the bar.

Solution

(a) *Shear stresses.* Since the maximum shear stress at any cross section in a solid bar is given by the modified torsion formula (Eq. 3-12), we know immediately that the maximum shear stress occurs at the cross section having the smallest diameter, that is, at end A (see Fig. 3-19):

$$\tau_{max} = \frac{16T}{\pi d_A^3}$$

(b) *Angle of twist.* Because the torque is constant and the polar moment of inertia varies continuously with the distance x from end A (Case 2), we will use Eq. (3-21) to determine the angle of twist. We begin by setting up an expression for the diameter d at distance x from end A:

$$d = d_A + \frac{d_B - d_A}{L}x \qquad (3\text{-}23)$$

in which L is the length of the bar. We can now write an expression for the polar moment of inertia:

$$I_P(x) = \frac{\pi d^4}{32} = \frac{\pi}{32}\left(d_A + \frac{d_B - d_A}{L}x\right)^4 \qquad (3\text{-}24)$$

Substituting this expression into Eq. (3-21), we get a formula for the angle of twist:

$$\phi = \int_0^L \frac{T\,dx}{GI_P(x)} = \frac{32T}{\pi G}\int_0^L \frac{dx}{\left(d_A + \dfrac{d_B - d_A}{L}x\right)^4} \qquad (3\text{-}25)$$

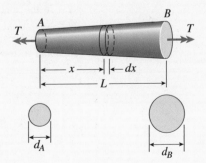

FIG. 3-19 Example 3-5. Tapered bar in torsion

continued

To evaluate the integral in this equation, we note that it is of the form

$$\int \frac{dx}{(a + bx)^4}$$

in which

$$a = d_A \qquad b = \frac{d_B - d_A}{L} \qquad\qquad\qquad \text{(e,f)}$$

With the aid of a table of integrals (see Appendix D available online), we find

$$\int \frac{dx}{(a + bx)^4} = -\frac{1}{3b(a + bx)^3}$$

This integral is evaluated in our case by substituting for x the limits 0 and L and substituting for a and b the expressions in Eqs. (e) and (f). Thus, the integral in Eq. (3-25) equals

$$\frac{L}{3(d_B - d_A)}\left(\frac{1}{d_A^3} - \frac{1}{d_B^3}\right) \qquad\qquad\qquad \text{(g)}$$

Replacing the integral in Eq. (3-25) with this expression, we obtain

$$\phi = \frac{32TL}{3\pi G(d_B - d_A)}\left(\frac{1}{d_A^3} - \frac{1}{d_B^3}\right) \qquad \text{(3-26)} \quad \Longleftarrow$$

which is the desired equation for the angle of twist of the tapered bar.

A convenient form in which to write the preceding equation is

$$\phi = \frac{TL}{G(I_P)_A}\left(\frac{\beta^2 + \beta + 1}{3\beta^3}\right) \qquad\qquad \text{(3-27)}$$

in which

$$\beta = \frac{d_B}{d_A} \qquad (I_P)_A = \frac{\pi d_A^4}{32} \qquad\qquad \text{(3-28)}$$

The quantity β is the ratio of end diameters and $(I_P)_A$ is the polar moment of inertia at end A.

In the special case of a prismatic bar, we have $\beta = 1$ and Eq. (3-27) gives $\phi = TL/G(I_P)_A$, as expected. For values of β greater than 1, the angle of rotation decreases because the larger diameter at end B produces an increase in the torsional stiffness (as compared to a prismatic bar).

3.5 STRESSES AND STRAINS IN PURE SHEAR

When a circular bar, either solid or hollow, is subjected to torsion, shear stresses act over the cross sections and on longitudinal planes, as illustrated previously in Fig. 3-7. We will now examine in more detail the stresses and strains produced during twisting of a bar.

We begin by considering a stress element *abcd* cut between two cross sections of a bar in torsion (Figs. 3-20a and b). This element is in a state of **pure shear**, because the only stresses acting on it are the shear stresses τ on the four side faces (see the discussion of shear stresses in Section 1.6.)

The directions of these shear stresses depend upon the directions of the applied torques *T*. In this discussion, we assume that the torques rotate the right-hand end of the bar clockwise when viewed from the right (Fig. 3-20a); hence the shear stresses acting on the element have the directions shown in the figure. This same state of stress exists for a similar element cut from the interior of the bar, except that the magnitudes of the shear stresses are smaller because the radial distance to the element is smaller.

The directions of the torques shown in Fig. 3-20a are intentionally chosen so that the resulting shear stresses (Fig. 3-20b) are positive according to the sign convention for shear stresses described previously in Section 1.6. This **sign convention** is repeated here:

A shear stress acting on a positive face of an element is positive if it acts in the positive direction of one of the coordinate axes and negative if it acts in the negative direction of an axis. Conversely, a shear stress acting on a negative face of an element is positive if it acts in the negative direction of one of the coordinate axes and negative if it acts in the positive direction of an axis.

Applying this sign convention to the shear stresses acting on the stress element of Fig. 3-20b, we see that all four shear stresses are positive. For instance, the stress on the right-hand face (which is a positive face because the *x* axis is directed to the right) acts in the positive direction of the *y* axis; therefore, it is a positive shear stress. Also, the stress on the left-hand face (which is a negative face) acts in the negative direction of the *y* axis; therefore, it is a positive shear stress. Analogous comments apply to the remaining stresses.

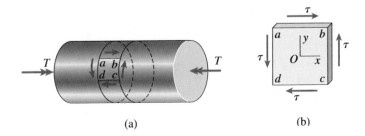

FIG. 3-20 Stresses acting on a stress element cut from a bar in torsion (pure shear)

(a)

(b)

Stresses on Inclined Planes

We are now ready to determine the stresses acting on *inclined planes* cut through the stress element in pure shear. We will follow the same approach as the one we used in Section 2.6 for investigating the stresses in uniaxial stress.

A two-dimensional view of the stress element is shown in Fig. 3-21a. As explained previously in Section 2.6, we usually draw a two-dimensional view for convenience, but we must always be aware that the element has a third dimension (thickness) perpendicular to the plane of the figure.

We now cut from the element a wedge-shaped (or "triangular") stress element having one face oriented at an angle θ to the x axis (Fig. 3-21b). Normal stresses σ_θ and shear stresses τ_θ act on this inclined face and are shown in their positive directions in the figure. The **sign convention** for stresses σ_θ and τ_θ was described previously in Section 2.6 and is repeated here:

Normal stresses σ_θ are positive in tension and shear stresses τ_θ are positive when they tend to produce counterclockwise rotation of the material. (Note that this sign convention for the shear stress τ_θ acting on an inclined plane is different from the sign convention for ordinary shear stresses τ that act on the sides of rectangular elements oriented to a set of xy axes.)

The horizontal and vertical faces of the triangular element (Fig. 3-21b) have positive shear stresses τ acting on them, and the front and rear faces of the element are free of stress. Therefore, all stresses acting on the element are visible in this figure.

The stresses σ_θ and τ_θ may now be determined from the equilibrium of the triangular element. The *forces* acting on its three side faces can be obtained by multiplying the stresses by the areas over which they act. For instance, the force on the left-hand face is equal to τA_0, where A_0 is the area of the vertical face. This force acts in the negative y direction and is shown in the *free-body diagram* of Fig. 3-21c. Because the thickness of the element in the z direction is constant, we see that the area of the bottom face is $A_0 \tan \theta$ and the area of the inclined face is A_0

FIG. 3-21 Analysis of stresses on inclined planes: (a) element in pure shear, (b) stresses acting on a triangular stress element, and (c) forces acting on the triangular stress element (free-body diagram)

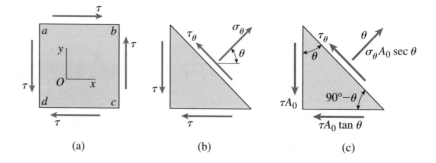

(a)　　　　　　(b)　　　　　　(c)

sec θ. Multiplying the stresses acting on these faces by the correspon-ding areas enables us to obtain the remaining forces and thereby complete the free-body diagram (Fig. 3-21c).

We are now ready to write two equations of equilibrium for the trian-gular element, one in the direction of σ_θ and the other in the direction of τ_θ. When writing these equations, the forces acting on the left-hand and bottom faces must be resolved into components in the directions of σ_θ and τ_θ. Thus, the first equation, obtained by summing forces in the direc-tion of σ_θ, is

$$\sigma_\theta A_0 \sec \theta = \tau A_0 \sin \theta + \tau A_0 \tan \theta \cos \theta$$

or

$$\sigma_\theta = 2\tau \sin \theta \cos \theta \tag{3-29a}$$

The second equation is obtained by summing forces in the direction of τ_θ:

$$\tau_\theta A_0 \sec \theta = \tau A_0 \cos \theta - \tau A_0 \tan \theta \sin \theta$$

or

$$\tau_\theta = \tau(\cos^2\theta - \sin^2\theta) \tag{3-29b}$$

These equations can be expressed in simpler forms by introducing the following trigonometric identities (see Appendix D available online):

$$\sin 2\theta = 2 \sin \theta \cos \theta \qquad \cos 2\theta = \cos^2 \theta - \sin^2 \theta$$

Then the equations for σ_θ and τ_θ become

$$\boxed{\sigma_\theta = \tau \sin 2\theta \qquad \tau_\theta = \tau \cos 2\theta} \tag{3-30a,b}$$

Equations (3-30a and b) give the normal and shear stresses acting on any inclined plane in terms of the shear stresses τ acting on the x and y planes (Fig. 3-21a) and the angle θ defining the orientation of the inclined plane (Fig. 3-21b).

The manner in which the stresses σ_θ and τ_θ vary with the orientation of the inclined plane is shown by the graph in Fig. 3-22, which is a plot of Eqs. (3-30a and b). We see that for $\theta = 0$, which is the right-hand face of the stress element in Fig. 3-21a, the graph gives $\sigma_\theta = 0$ and $\tau_\theta = \tau$. This

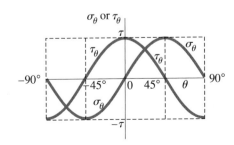

FIG. 3-22 Graph of normal stresses σ_θ and shear stresses τ_θ versus angle θ of the inclined plane

latter result is expected, because the shear stress τ acts counterclockwise against the element and therefore produces a positive shear stress τ_θ.

For the top face of the element ($\theta = 90°$), we obtain $\sigma_\theta = 0$ and $\tau_\theta = -\tau$. The minus sign for τ_θ means that it acts clockwise against the element, that is, to the right on face ab (Fig. 3-21a), which is consistent with the direction of the shear stress τ. Note that the numerically largest shear stresses occur on the planes for which $\theta = 0$ and $90°$, as well as on the opposite faces ($\theta = 180°$ and $270°$).

From the graph we see that the normal stress σ_θ reaches a maximum value at $\theta = 45°$. At that angle, the stress is positive (tension) and equal numerically to the shear stress τ. Similarly, σ_θ has its minimum value (which is compressive) at $\theta = -45°$. At both of these $45°$ angles, the shear stress τ_θ is equal to zero. These conditions are pictured in Fig. 3-23 which shows stress elements oriented at $\theta = 0$ and $\theta = 45°$. The element at $45°$ is acted upon by equal tensile and compressive stresses in perpendicular directions, with no shear stresses.

Note that the normal stresses acting on the $45°$ element (Fig. 3-23b) correspond to an element subjected to shear stresses τ acting in the directions shown in Fig. 3-23a. If the shear stresses acting on the element of Fig. 3-23a are reversed in direction, the normal stresses acting on the $45°$ planes also will change directions.

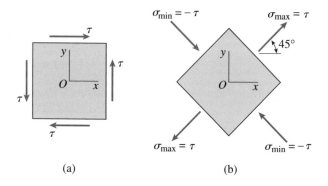

FIG. 3-23 Stress elements oriented at $\theta = 0$ and $\theta = 45°$ for pure shear

(a)   (b)

If a stress element is oriented at an angle other than $45°$, both normal and shear stresses will act on the inclined faces (see Eqs. 3-30a and b and Fig. 3-22). Stress elements subjected to these more general conditions are discussed in detail in Chapter 6.

The equations derived in this section are valid for a stress element in pure shear regardless of whether the element is cut from a bar in torsion or from some other structural element. Also, since Eqs. (3-30) were derived from equilibrium only, they are valid for any material, whether or not it behaves in a linearly elastic manner.

The existence of maximum tensile stresses on planes at $45°$ to the x axis (Fig. 3-23b) explains why bars in torsion that are made of materials that are brittle and weak in tension fail by cracking along

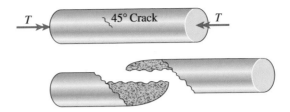

FIG. 3-24 Torsion failure of a brittle material by tension cracking along a 45° helical surface

a 45° helical surface (Fig. 3-24). As mentioned in Section 3.3, this type of failure is readily demonstrated by twisting a piece of classroom chalk.

Strains in Pure Shear

Let us now consider the strains that exist in an element in pure shear. For instance, consider the element in pure shear shown in Fig. 3-23a. The corresponding shear strains are shown in Fig. 3-25a, where the deformations are highly exaggerated. The shear strain γ is the change in angle between two lines that were originally perpendicular to each other, as discussed previously in Section 1.6. Thus, the decrease in the angle at the lower left-hand corner of the element is the shear strain γ (measured in radians). This same change in angle occurs at the upper right-hand corner, where the angle decreases, and at the other two corners, where the angles increase. However, the lengths of the sides of the element, including the thickness perpendicular to the plane of the paper, do not change when these shear deformations occur. Therefore, the element changes its shape from a rectangular parallelepiped (Fig. 3-23a) to an oblique parallelepiped (Fig. 3-25a). This change in shape is called a **shear distortion**.

If the material is linearly elastic, the shear strain for the element oriented at $\theta = 0$ (Fig. 3-25a) is related to the shear stress by Hooke's law in shear:

$$\gamma = \frac{\tau}{G} \tag{3-31}$$

where, as usual, the symbol G represents the shear modulus of elasticity.

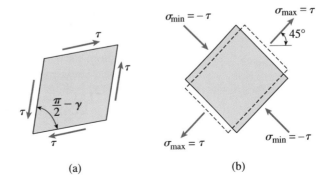

FIG. 3-25 Strains in pure shear: (a) shear distortion of an element oriented at $\theta = 0$, and (b) distortion of an element oriented at $\theta = 45°$

Next, consider the strains that occur in an element oriented at $\theta = 45°$ (Fig. 3-25b). The tensile stresses acting at 45° tend to elongate the element in that direction. Because of the Poisson effect, they also tend to shorten it in the perpendicular direction (the direction where $\theta = 135°$ or $-45°$). Similarly, the compressive stresses acting at 135° tend to shorten the element in that direction and elongate it in the 45° direction. These dimensional changes are shown in Fig. 3-25b, where the dashed lines show the deformed element. Since there are no shear distortions, the element remains a rectangular parallelepiped even though its dimensions have changed.

If the material is linearly elastic and follows Hooke's law, we can obtain an equation relating strain to stress for the element at $\theta = 45°$ (Fig. 3-25b). The tensile stress σ_{max} acting at $\theta = 45°$ produces a positive normal strain in that direction equal to σ_{max}/E. Since $\sigma_{max} = \tau$, we can also express this strain as τ/E. The stress σ_{max} also produces a negative strain in the perpendicular direction equal to $-\nu\tau/E$, where ν is Poisson's ratio. Similarly, the stress $\sigma_{min} = -\tau$ (at $\theta = 135°$) produces a negative strain equal to $-\tau/E$ in that direction and a positive strain in the perpendicular direction (the 45° direction) equal to $\nu\tau/E$. Therefore, the normal strain in the 45° direction is

$$\epsilon_{max} = \frac{\tau}{E} + \frac{\nu\tau}{E} = \frac{\tau}{E}(1 + \nu) \tag{3-32}$$

which is positive, representing elongation. The strain in the perpendicular direction is a negative strain of the same amount. In other words, pure shear produces elongation in the 45° direction and shortening in the 135° direction. These strains are consistent with the shape of the deformed element of Fig. 3-25a, because the 45° diagonal has lengthened and the 135° diagonal has shortened.

In the next section we will use the geometry of the deformed element to relate the shear strain γ (Fig. 3-25a) to the normal strain ϵ_{max} in the 45° direction (Fig. 3-25b). In so doing, we will derive the following relationship:

$$\epsilon_{max} = \frac{\gamma}{2} \tag{3-33}$$

This equation, in conjunction with Eq. (3-31), can be used to calculate the maximum shear strains and maximum normal strains in pure torsion when the shear stress τ is known.

Example 3-6

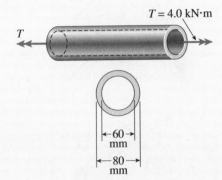

$T = 4.0$ kN·m

←60→
mm

←80→
mm

FIG. 3-26 Example 3-6. Circular tube in torsion

A circular tube with an outside diameter of 80 mm and an inside diameter of 60 mm is subjected to a torque $T = 4.0$ kN·m (Fig. 3-26). The tube is made of aluminum alloy 7075-T6.

(a) Determine the maximum shear, tensile, and compressive stresses in the tube and show these stresses on sketches of properly oriented stress elements.

(b) Determine the corresponding maximum strains in the tube and show these strains on sketches of the deformed elements.

Solution

(a) *Maximum stresses.* The maximum values of all three stresses (shear, tensile, and compressive) are equal numerically, although they act on different planes. Their magnitudes are found from the torsion formula:

$$\tau_{max} = \frac{Tr}{I_P} = \frac{(4000 \text{ N·m})(0.040 \text{ m})}{\frac{\pi}{32}\left[(0.080 \text{ m})^4 - (0.060 \text{ m})^4\right]} = 58.2 \text{ MPa}$$

The maximum shear stresses act on cross-sectional and longitudinal planes, as shown by the stress element in Fig. 3-27a, where the x axis is parallel to the longitudinal axis of the tube.

The maximum tensile and compressive stresses are

$$\sigma_t = 58.2 \text{ MPa} \qquad \sigma_c = -58.2 \text{ MPa}$$

These stresses act on planes at 45° to the axis (Fig. 3-27b).

(b) *Maximum strains.* The maximum shear strain in the tube is obtained from Eq. (3-31). The shear modulus of elasticity is obtained from Table I-2, Appendix I (available online), as $G = 27$ GPa. Therefore, the maximum shear strain is

$$\gamma_{max} = \frac{\tau_{max}}{G} = \frac{58.2 \text{ MPa}}{27 \text{ GPa}} = 0.0022 \text{ rad}$$

The deformed element is shown by the dashed lines in Fig. 3-27c.

The magnitude of the maximum normal strains (from Eq. 3-33) is

$$\epsilon_{max} = \frac{\gamma_{max}}{2} = 0.0011$$

Thus, the maximum tensile and compressive strains are

$$\epsilon_t = 0.0011 \qquad \epsilon_c = -0.0011$$

The deformed element is shown by the dashed lines in Fig. 3-27d for an element with sides of unit length.

continued

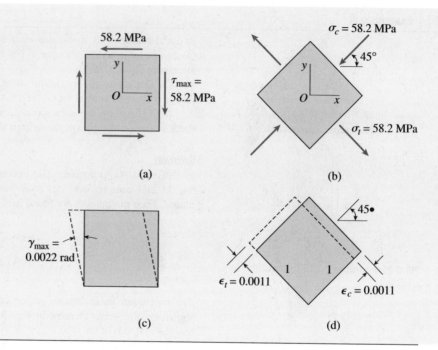

FIG. 3-27 Stress and strain elements for the tube of Example 3-6: (a) maximum shear stresses, (b) maximum tensile and compressive stresses; (c) maximum shear strains, and (d) maximum tensile and compressive strains

3.6 RELATIONSHIP BETWEEN MODULI OF ELASTICITY *E* AND *G*

An important relationship between the moduli of elasticity E and G can be obtained from the equations derived in the preceding section. For this purpose, consider the stress element $abcd$ shown in Fig. 3-28a on the next page. The front face of the element is assumed to be square, with the length of each side denoted as h. When this element is subjected to pure shear by stresses τ, the front face distorts into a rhombus (Fig. 3-28b) with sides of length h and with shear strain $\gamma = \tau/G$. Because of the distortion, diagonal bd is lengthened and diagonal ac is shortened. The length of diagonal bd is equal to its initial length $\sqrt{2}\,h$ times the factor $1 + \epsilon_{max}$, where ϵ_{max} is the normal strain in the 45° direction; thus,

$$L_{bd} = \sqrt{2}\,h(1 + \epsilon_{max}) \qquad\qquad \text{(a)}$$

This length can be related to the shear strain γ by considering the geometry of the deformed element.

To obtain the required geometric relationships, consider triangle abd (Fig. 3-28c) which represents one-half of the rhombus pictured in Fig. 3-28b. Side bd of this triangle has length L_{bd} (Eq. a), and the other sides have length h. Angle adb of the triangle is equal to one-half of angle adc of the rhombus, or $\pi/4 - \gamma/2$. The angle abd in the triangle is the same. Therefore, angle dab of the triangle equals $\pi/2 + \gamma$. Now using the law of cosines (see Appendix D online) for triangle abd, we get

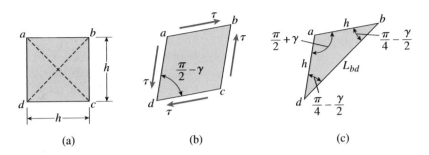

FIG. 3-28 Geometry of deformed element in pure shear

(a)   (b)   (c)

$$L_{bd}^2 = h^2 + h^2 - 2h^2 \cos\left(\frac{\pi}{2} + \gamma\right)$$

Substituting for L_{bd} from Eq. (a) and simplifying, we get

$$(1 + \epsilon_{max})^2 = 1 - \cos\left(\frac{\pi}{2} + \gamma\right)$$

By expanding the term on the left-hand side, and also observing that $\cos(\pi/2 + \gamma) = -\sin\gamma$, we obtain

$$1 + 2\epsilon_{max} + \epsilon_{max}^2 = 1 + \sin\gamma$$

Because ϵ_{max} and γ are very small strains, we can disregard ϵ_{max}^2 in comparison with $2\epsilon_{max}$ and we can replace $\sin\gamma$ by γ. The resulting expression is

$$\epsilon_{max} = \frac{\gamma}{2} \tag{3-34}$$

which establishes the relationship already presented in Section 3.5 as Eq. (3-33).

The shear strain γ appearing in Eq. (3-34) is equal to τ/G by Hooke's law (Eq. 3-31) and the normal strain ϵ_{max} is equal to $\tau(1 + \nu)/E$ by Eq. (3-32). Making both of these substitutions in Eq. (3-34) yields

$$G = \frac{E}{2(1 + \nu)} \tag{3-35}$$

We see that E, G, and ν are not independent properties of a linearly elastic material. Instead, if any two of them are known, the third can be calculated from Eq. (3-35).

Typical values of E, G, and ν are listed in Table I-2, Appendix I (available online).

3.7 TRANSMISSION OF POWER BY CIRCULAR SHAFTS

The most important use of circular shafts is to transmit mechanical power from one device or machine to another, as in the drive shaft of an automobile, the propeller shaft of a ship, or the axle of a bicycle. The power is transmitted through the rotary motion of the shaft, and the amount of power transmitted depends upon the magnitude of the torque and the speed of rotation. A common design problem is to determine the required size of a shaft so that it will transmit a specified amount of power at a specified rotational speed without exceeding the allowable stresses for the material.

Let us suppose that a motor-driven shaft (Fig. 3-29) is rotating at an angular speed ω, measured in radians per second (rad/s). The shaft transmits a torque T to a device (not shown in the figure) that is performing useful work. The torque applied by the shaft to the external device has the same sense as the angular speed ω, that is, its vector points to the left. However, the torque shown in the figure is the torque exerted *on the shaft* by the device, and so its vector points in the opposite direction.

In general, the work W done by a torque of constant magnitude is equal to the product of the torque and the angle through which it rotates; that is,

$$W = T\psi \tag{3-36}$$

where ψ is the angle of rotation in radians.

Power is the *rate* at which work is done, or

$$P = \frac{dW}{dt} = T\frac{d\psi}{dt} \tag{3-37}$$

in which P is the symbol for power and t represents time. The rate of change $d\psi/dt$ of the angular displacement ψ is the angular speed ω, and therefore the preceding equation becomes

$$P = T\omega \qquad (\omega = \text{rad/s}) \tag{3-38}$$

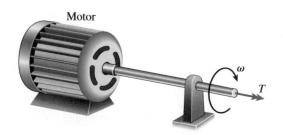

FIG. 3-29 Shaft transmitting a constant torque T at an angular speed ω

This formula, which is familiar from elementary physics, gives the power transmitted by a rotating shaft transmitting a constant torque T.

The **units** to be used in Eq. (3-38) are as follows. If the torque T is expressed in newton meters, then the power is expressed in watts (W). One watt is equal to one newton meter per second (or one joule per second). If T is expressed in pound-feet, then the power is expressed in foot-pounds per second.[*]

Angular speed is often expressed as the frequency f of rotation, which is the number of revolutions per unit of time. The unit of frequency is the hertz (Hz), equal to one revolution per second (s^{-1}). Inasmuch as one revolution equals 2π radians, we obtain

$$\omega = 2\pi f \qquad (\omega = \text{rad/s}, \ f = \text{Hz} = s^{-1}) \qquad (3\text{-}39)$$

The expression for power (Eq. 3-38) then becomes

$$P = 2\pi f T \qquad (f = \text{Hz} = s^{-1}) \qquad (3\text{-}40)$$

Another commonly used unit is the number of revolutions per minute (rpm), denoted by the letter n. Therefore, we also have the following relationships:

$$n = 60 f \qquad (3\text{-}41)$$

and

$$P = \frac{2\pi n T}{60} \qquad (n = \text{rpm}) \qquad (3\text{-}42)$$

In Eqs. (3-40) and (3-42), the quantities P and T have the same units as in Eq. (3-38); that is, P has units of watts if T has units of newton meters, and P has units of foot-pounds per second if T has units of pound-feet.

In U.S. engineering practice, power is sometimes expressed in horse-power (hp), a unit equal to 550 ft-lb/s. Therefore, the horsepower H being transmitted by a rotating shaft is

$$H = \frac{2\pi n T}{60(550)} = \frac{2\pi n T}{33,000} \qquad (n = \text{rpm}, \ T = \text{lb-ft}, \ H = \text{hp}) \qquad (3\text{-}43)$$

One horsepower is approximately 746 watts.

The preceding equations relate the torque acting in a shaft to the power transmitted by the shaft. Once the torque is known, we can determine the shear stresses, shear strains, angles of twist, and other desired quantities by the methods described in Sections 3.2 through 3.5.

The following examples illustrate some of the procedures for analyzing rotating shafts.

---

[*]See Table B-1, Appendix B (available online), for units of work and power.

Example 3-7

A motor driving a solid circular steel shaft transmits 40 hp to a gear at B (Fig. 3-30). The allowable shear stress in the steel is 6000 psi.

(a) What is the required diameter d of the shaft if it is operated at 500 rpm?
(b) What is the required diameter d if it is operated at 3000 rpm?

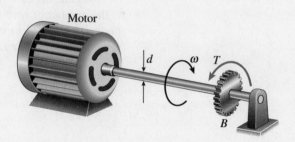

Motor

B

FIG. 3-30 Example 3-7. Steel shaft in torsion

Solution

(a) *Motor operating at 500 rpm.* Knowing the horsepower and the speed of rotation, we can find the torque T acting on the shaft by using Eq. (3-43). Solving that equation for T, we get

$$T = \frac{33,000H}{2\pi n} = \frac{33,000(40 \text{ hp})}{2\pi(500 \text{ rpm})} = 420.2 \text{ lb-ft} = 5042 \text{ lb-in.}$$

This torque is transmitted by the shaft from the motor to the gear.

The maximum shear stress in the shaft can be obtained from the modified torsion formula (Eq. 3-12):

$$\tau_{max} = \frac{16T}{\pi d^3}$$

Solving that equation for the diameter d, and also substituting τ_{allow} for τ_{max}, we get

$$d^3 = \frac{16T}{\pi \tau_{allow}} = \frac{16(5042 \text{ lb-in.})}{\pi(6000 \text{ psi})} = 4.280 \text{ in.}^3$$

from which

$$d = 1.62 \text{ in.}$$

The diameter of the shaft must be at least this large if the allowable shear stress is not to be exceeded.

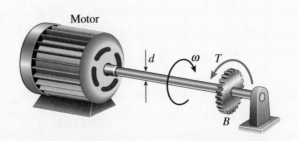

Motor

FIG. 3-30 (Repeated)

(b) *Motor operating at 3000 rpm.* Following the same procedure as in part (a), we obtain

$$T = \frac{33,000H}{2\pi n} = \frac{33,000(40 \text{ hp})}{2\pi(3000 \text{ rpm})} = 70.03 \text{ lb-ft} = 840.3 \text{ lb-in.}$$

$$d^3 = \frac{16T}{\pi\tau_{\text{allow}}} = \frac{16(840.3 \text{ lb-in.})}{\pi(6000 \text{ psi})} = 0.7133 \text{ in.}^3$$

$$d = 0.89 \text{ in.}$$

which is less than the diameter found in part (a).

This example illustrates that the higher the speed of rotation, the smaller the required size of the shaft (for the same power and the same allowable stress).

Example 3-8

A solid steel shaft *ABC* of 50 mm diameter (Fig. 3-31a) is driven at *A* by a motor that transmits 50 kW to the shaft at 10 Hz. The gears at *B* and *C* drive machinery requiring power equal to 35 kW and 15 kW, respectively.

Compute the maximum shear stress τ_{max} in the shaft and the angle of twist ϕ_{AC} between the motor at *A* and the gear at *C*. (Use *G* = 80 GPa.)

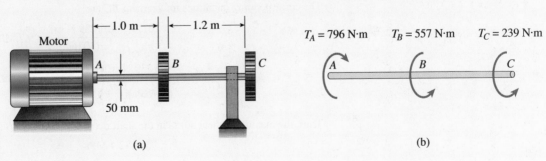

Motor

←1.0 m→ ←1.2 m→

A   B   C

50 mm

(a)

$T_A = 796$ N·m $T_B = 557$ N·m $T_C = 239$ N·m

A   B   C

(b)

FIG. 3-31 Example 3-8. Steel shaft in torsion

continued

Solution

Torques acting on the shaft. We begin the analysis by determining the torques applied to the shaft by the motor and the two gears. Since the motor supplies 50 kW at 10 Hz, it creates a torque T_A at end A of the shaft (Fig. 3-31b) that we can calculate from Eq. (3-40):

$$T_A = \frac{P}{2\pi f} = \frac{50\ kW}{2\pi(10\ Hz)} = 796\ N\cdot m$$

In a similar manner, we can calculate the torques T_B and T_C applied by the gears to the shaft:

$$T_B = \frac{P}{2\pi f} = \frac{35\ kW}{2\pi(10\ Hz)} = 557\ N\cdot m$$

$$T_C = \frac{P}{2\pi f} = \frac{15\ kW}{2\pi(10\ Hz)} = 239\ N\cdot m$$

These torques are shown in the free-body diagram of the shaft (Fig. 3-31b). Note that the torques applied by the gears are opposite in direction to the torque applied by the motor. (If we think of T_A as the "load" applied to the shaft by the motor, then the torques T_B and T_C are the "reactions" of the gears.)

The internal torques in the two segments of the shaft are now found (by inspection) from the free-body diagram of Fig. 3-31b:

$$T_{AB} = 796\ N\cdot m \qquad T_{BC} = 239\ N\cdot m$$

Both internal torques act in the same direction, and therefore the angles of twist in segments AB and BC are additive when finding the total angle of twist. (To be specific, both torques are positive according to the sign convention adopted in Section 3.4.)

Shear stresses and angles of twist. The shear stress and angle of twist in segment AB of the shaft are found in the usual manner from Eqs. (3-12) and (3-15):

$$\tau_{AB} = \frac{16T_{AB}}{\pi d^3} = \frac{16(796\ N\cdot m)}{\pi(50\ mm)^3} = 32.4\ MPa$$

$$\phi_{AB} = \frac{T_{AB}L_{AB}}{GI_P} = \frac{(796\ N\cdot m)(1.0\ m)}{(80\ GPa)\left(\dfrac{\pi}{32}\right)(50\ mm)^4} = 0.0162\ rad$$

The corresponding quantities for segment BC are

$$\tau_{BC} = \frac{16T_{BC}}{\pi d^3} = \frac{16(239\ N\cdot m)}{\pi(50\ mm)^3} = 9.7\ MPa$$

$$\phi_{BC} = \frac{T_{BC}L_{BC}}{GI_P} = \frac{(239\ N\cdot m)(1.2\ m)}{(80\ GPa)\left(\dfrac{\pi}{32}\right)(50\ mm)^4} = 0.0058\ rad$$

Thus, the maximum shear stress in the shaft occurs in segment AB and is

$$\tau_{max} = 32.4\ MPa \qquad \Longleftarrow$$

Also, the total angle of twist between the motor at A and the gear at C is

$$\phi_{AC} = \phi_{AB} + \phi_{BC} = 0.0162\ rad + 0.0058\ rad = 0.0220\ rad = 1.26° \qquad \Longleftarrow$$

As explained previously, both parts of the shaft twist in the same direction, and therefore the angles of twist are added.

3.8 STATICALLY INDETERMINATE TORSIONAL MEMBERS

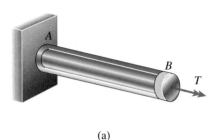

(a)

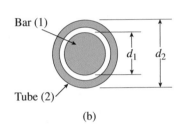

Bar (1)

Tube (2)

(b)

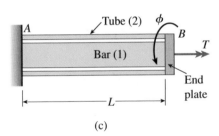

(c)

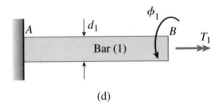

(d)

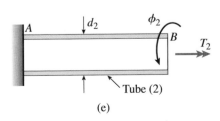

(e)

FIG. 3-32 Statically indeterminate bar in torsion

The bars and shafts described in the preceding sections of this chapter are *statically determinate* because all internal torques and all reactions can be obtained from free-body diagrams and equations of equilibrium. However, if additional restraints, such as fixed supports, are added to the bars, the equations of equilibrium will no longer be adequate for determining the torques. The bars are then classified as **statically indeterminate**. Torsional members of this kind can be analyzed by supplementing the equilibrium equations with compatibility equations pertaining to the rotational displacements. Thus, the general method for analyzing statically indeterminate torsional members is the same as described in Section 2.4 for statically indeterminate bars with axial loads.

The first step in the analysis is to write **equations of equilibrium**, obtained from free-body diagrams of the given physical situation. The unknown quantities in the equilibrium equations are torques, either internal torques or reaction torques.

The second step in the analysis is to formulate **equations of compatibility**, based upon physical conditions pertaining to the angles of twist. As a consequence, the compatibility equations contain angles of twist as unknowns.

The third step is to relate the angles of twist to the torques by **torque-displacement relations**, such as $\phi = TL/GI_P$. After introducing these relations into the compatibility equations, they too become equations containing torques as unknowns. Therefore, the last step is to obtain the unknown torques by solving simultaneously the equations of equilibrium and compatibility.

To illustrate the method of solution, we will analyze the composite bar *AB* shown in Fig. 3-32a. The bar is attached to a fixed support at end *A* and loaded by a torque *T* at end *B*. Furthermore, the bar consists of two parts: a solid bar and a tube (Figs. 3-32b and c), with both the solid bar and the tube joined to a rigid end plate at *B*.

For convenience, we will identify the solid bar and tube (and their properties) by the numerals 1 and 2, respectively. For instance, the diameter of the solid bar is denoted d_1 and the outer diameter of the tube is denoted d_2. A small gap exists between the bar and the tube, and therefore the inner diameter of the tube is slightly larger than the diameter d_1 of the bar.

When the torque *T* is applied to the composite bar, the end plate rotates through a small angle ϕ (Fig. 3-32c) and torques T_1 and T_2 are developed in the solid bar and the tube, respectively (Figs. 3-32d and e). From equilibrium we know that the sum of these torques equals the applied load, and so the *equation of equilibrium* is

$$T_1 + T_2 = T \qquad (a)$$

Because this equation contains two unknowns (T_1 and T_2), we recognize that the composite bar is statically indeterminate.

To obtain a second equation, we must consider the rotational displacements of both the solid bar and the tube. Let us denote the angle of twist of the solid bar (Fig. 3-32d) by ϕ_1 and the angle of twist of the tube by ϕ_2 (Fig. 3-32e). These angles of twist must be equal because the bar and tube are securely joined to the end plate and rotate with it; consequently, the *equation of compatibility* is

$$\phi_1 = \phi_2 \tag{b}$$

The angles ϕ_1 and ϕ_2 are related to the torques T_1 and T_2 by the *torque-displacement relations*, which in the case of linearly elastic materials are obtained from the equation $\phi = TL/GI_P$. Thus,

$$\phi_1 = \frac{T_1 L}{G_1 I_{P1}} \qquad \phi_2 = \frac{T_2 L}{G_2 I_{P2}} \tag{c,d}$$

in which G_1 and G_2 are the shear moduli of elasticity of the materials and I_{P1} and I_{P2} are the polar moments of inertia of the cross sections.

When the preceding expressions for ϕ_1 and ϕ_2 are substituted into Eq. (b), the equation of compatibility becomes

$$\frac{T_1 L}{G_1 I_{P1}} = \frac{T_2 L}{G_2 I_{P2}} \tag{e}$$

We now have two equations (Eqs. a and e) with two unknowns, so we can solve them for the torques T_1 and T_2. The results are

$$T_1 = T\left(\frac{G_1 I_{P1}}{G_1 I_{P1} + G_2 I_{P2}}\right) \qquad T_2 = T\left(\frac{G_2 I_{P2}}{G_1 I_{P1} + G_2 I_{P2}}\right) \tag{3-44a,b}$$

With these torques known, the essential part of the statically indeterminate analysis is completed. All other quantities, such as stresses and angles of twist, can now be found from the torques.

The preceding discussion illustrates the general methodology for analyzing a statically indeterminate system in torsion. In the following example, this same approach is used to analyze a bar that is fixed against rotation at both ends. In the example and in the problems, we assume that the bars are made of linearly elastic materials. However, the general methodology is also applicable to bars of nonlinear materials—the only change is in the torque-displacement relations.

Example 3-9

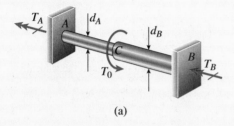

(a)

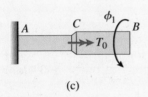

(b)

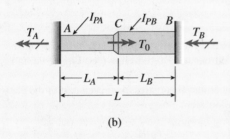

(c)

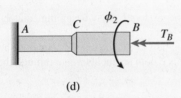

(d)

FIG. 3-33 Example 3-9. Statically indeterminate bar in torsion

The bar *ACB* shown in Figs. 3-33a and b is fixed at both ends and loaded by a torque T_0 at point *C*. Segments *AC* and *CB* of the bar have diameters d_A and d_B, lengths L_A and L_B, and polar moments of inertia I_{PA} and I_{PB}, respectively. The material of the bar is the same throughout both segments.

Obtain formulas for (a) the reactive torques T_A and T_B at the ends, (b) the maximum shear stresses τ_{AC} and τ_{CB} in each segment of the bar, and (c) the angle of rotation ϕ_C at the cross section where the load T_0 is applied.

Solution

Equation of equilibrium. The load T_0 produces reactions T_A and T_B at the fixed ends of the bar, as shown in Figs. 3-33a and b. Thus, from the equilibrium of the bar we obtain

$$T_A + T_B = T_0 \tag{f}$$

Because there are two unknowns in this equation (and no other useful equations of equilibrium), the bar is statically indeterminate.

Equation of compatibility. We now separate the bar from its support at end *B* and obtain a bar that is fixed at end *A* and free at end *B* (Figs. 3-33c and d). When the load T_0 acts alone (Fig. 3-33c), it produces an angle of twist at end *B* that we denote as ϕ_1. Similarly, when the reactive torque T_B acts alone, it produces an angle ϕ_2 (Fig. 3-33d). The angle of twist at end *B* in the original bar, equal to the sum of ϕ_1 and ϕ_2, is zero. Therefore, the equation of compatibility is

$$\phi_1 + \phi_2 = 0 \tag{g}$$

Note that ϕ_1 and ϕ_2 are assumed to be positive in the direction shown in the figure.

Torque-displacement equations. The angles of twist ϕ_1 and ϕ_2 can be expressed in terms of the torques T_0 and T_B by referring to Figs. 3-33c and d and using the equation $\phi = TL/GI_P$. The equations are as follows:

$$\phi_1 = \frac{T_0 L_A}{GI_{PA}} \qquad \phi_2 = -\frac{T_B L_A}{GI_{PA}} - \frac{T_B L_B}{GI_{PB}} \tag{h,i}$$

The minus signs appear in Eq. (i) because T_B produces a rotation that is opposite in direction to the positive direction of ϕ_2 (Fig. 3-33d).

We now substitute the angles of twist (Eqs. h and i) into the compatibility equation (Eq. g) and obtain

$$\frac{T_0 L_A}{GI_{PA}} - \frac{T_B L_A}{GI_{PA}} - \frac{T_B L_B}{GI_{PB}} = 0$$

or

$$\frac{T_B L_A}{I_{PA}} + \frac{T_B L_B}{I_{PB}} = \frac{T_0 L_A}{I_{PA}} \tag{j}$$

Solution of equations. The preceding equation can be solved for the torque T_B, which then can be substituted into the equation of equilibrium (Eq. f) to obtain the torque T_A. The results are

$$T_A = T_0\left(\frac{L_B I_{PA}}{L_B I_{PA} + L_A I_{PB}}\right) \qquad T_B = T_0\left(\frac{L_A I_{PB}}{L_B I_{PA} + L_A I_{PB}}\right) \qquad \text{(3-45a,b)} \impliedby$$

Thus, the reactive torques at the ends of the bar have been found, and the statically indeterminate part of the analysis is completed.

As a special case, note that if the bar is prismatic ($I_{PA} = I_{PB} = I_P$) the preceding results simplify to

$$T_A = \frac{T_0 L_B}{L} \qquad T_B = \frac{T_0 L_A}{L} \qquad \text{(3-46a,b)}$$

where L is the total length of the bar. These equations are analogous to those for the reactions of an axially loaded bar with fixed ends (see Eqs. 2-9a and 2-9b).

Maximum shear stresses. The maximum shear stresses in each part of the bar are obtained directly from the torsion formula:

$$\tau_{AC} = \frac{T_A d_A}{2 I_{PA}} \qquad \tau_{CB} = \frac{T_B d_B}{2 I_{PB}}$$

Substituting from Eqs. (3-45a) and (3-45b) gives

$$\tau_{AC} = \frac{T_0 L_B d_A}{2(L_B I_{PA} + L_A I_{PB})} \qquad \tau_{CB} = \frac{T_0 L_A d_B}{2(L_B I_{PA} + L_A I_{PB})} \qquad \text{(3-47a,b)} \impliedby$$

By comparing the product $L_B d_A$ with the product $L_A d_B$, we can immediately determine which segment of the bar has the larger stress.

Angle of rotation. The angle of rotation ϕ_C at section C is equal to the angle of twist of either segment of the bar, since both segments rotate through the same angle at section C. Therefore, we obtain

$$\phi_C = \frac{T_A L_A}{G I_{PA}} = \frac{T_B L_B}{G I_{PB}} = \frac{T_0 L_A L_B}{G(L_B I_{PA} + L_A I_{PB})} \qquad \text{(3-48)} \impliedby$$

In the special case of a prismatic bar ($I_{PA} = I_{PB} = I_P$), the angle of rotation at the section where the load is applied is

$$\phi_C = \frac{T_0 L_A L_B}{G L I_P} \qquad \text{(3-49)}$$

This example illustrates not only the analysis of a statically indeterminate bar but also the techniques for finding stresses and angles of rotation. In addition, note that the results obtained in this example are valid for a bar consisting of either solid or tubular segments.

In Chapter 3, we investigated the behavior of bars and hollow tubes acted on by concentrated torques or distributed torsional moments as well as prestrain effects. We developed torque-displacement relations for use in computing angles of twist of bars under both uniform (i.e., constant torsional moment over its entire length) and nonuniform conditions (i.e., torques, and perhaps also polar moment of inertia, vary over the length of the bar). Then, equilibrium and compatibility equations were developed for statically indeterminate structures in a superposition procedure leading to solution for all unknown torques, rotational displacements, stresses, etc. Starting with a state of pure shear on stress elements aligned with the axis of the bar, we then developed equations for normal and shear stresses on inclined sections. A number of advanced topics were presented in the last parts of the chapter. The major concepts presented in this chapter are as follows:

1. For circular bars and tubes, the **shearing stress** (τ) and **strain** (γ) vary linearly with radial distance from the center of the cross-section.

$$\tau = (\rho/r)\tau_{max} \quad \gamma = (\rho/r)\gamma_{max}$$

2. The **torsion formula** defines the relation between shear stress and torsional moment. Maximum shear stress τ_{max} occurs on the outer surface of the bar or tube and depends on torsional moment T, radial distance r, and second moment of inertia of the cross section I_p, known as polar moment of inertia for circular cross sections. Thin-walled tubes are seen to be more efficient in torsion, because the available material is more uniformly stressed than solid circular bars.

$$\tau_{max} = \frac{Tr}{I_P}$$

3. The angle of twist ϕ of prismatic circular bars subjected to torsional moment(s) is proportional to both the torque T and the length of the bar L, and inversely proportional to the torsional rigidity (GI_p) of the bar; this relationship is called the *torque-displacement relation*.

$$\phi = \frac{TL}{GI_P}$$

4. The angle of twist per unit length of a bar is referred to as its **torsional flexibility** (f_T), and the inverse relationship is the torsional **stiffness** ($k_T = 1/f_T$) of the bar or shaft.

$$k_T = \frac{GI_P}{L} \qquad f_T = \frac{L}{GI_P}$$

5. The summation of the twisting deformations of the individual segments of a nonprismatic shaft equals the twist of the entire bar (ϕ). Free-body diagrams are used to find the torsional moments (T_i) in each segment i.

$$\phi = \sum_{i=1}^{n} \phi_i = \sum_{i=1}^{n} \frac{T_i L_i}{G_i(I_P)_i}$$

If torsional moments and/or cross sectional properties (I_p) vary continuously, an integral expression is required.

$$\phi = \int_0^L d\phi = \int_0^L \frac{T(x)\,dx}{GI_P(x)}$$

6. If the bar structure is **statically indeterminate**, additional equations are required to solve for unknown moments. **Compatibility equations** are used to relate bar rotations to support conditions and thereby generate additional relationships among the unknowns. It is convenient to use a **superposition** of "released" (or statically determinate) structures to represent the actual statically indeterminate bar structure.

7. **Misfits** and **prestrains** induce torsional moments only in statically indeterminate bars or shafts.

8. A circular shaft is subjected to **pure shear** due to torsional moments. **Maximum normal** and **shear stresses** can be obtained by considering an inclined stress element. The maximum shear stress occurs on an element aligned with the axis of the bar, but the maximum normal stress occurs at an inclination of 45° to the axis of the bar, and the maximum normal stress is equal to the maximum shear stress

$$\sigma_{max} = \tau$$

We can also find a relationship between the maximum shear and normal strains for the case of pure shear:

$$\epsilon_{max} = \gamma_{max}/2$$

9. Circular shafts are commonly used to transmit mechanical power from one device or machine to another. If the torque T is expressed in newton meters and n is the shaft rpm, the power P is expressed in watts as

$$P = \frac{2\pi nT}{60}$$

In US customary units, torque T is given in ft-lb and power may be given in horsepower (hp), H, as

$$H = \frac{2\pi nT}{33,000}$$

PROBLEMS CHAPTER 3

Torsional Deformations

3.2-1 A copper rod of length $L = 18.0$ in. is to be twisted by torques T (see figure) until the angle of rotation between the ends of the rod is $3.0°$.

If the allowable shear strain in the copper is 0.0006 rad, what is the maximum permissible diameter of the rod?

PROBS. 3.2-1 and 3.2-2

3.2-2 A plastic bar of diameter $d = 56$ mm is to be twisted by torques T (see figure) until the angle of rotation between the ends of the bar is $4.0°$.

If the allowable shear strain in the plastic is 0.012 rad, what is the minimum permissible length of the bar?

3.2-3 A circular aluminum tube subjected to pure torsion by torques T (see figure) has an outer radius r_2 equal to 1.5 times the inner radius r_1.

(a) If the maximum shear strain in the tube is measured as 400×10^{-6} rad, what is the shear strain γ_1 at the inner surface?

(b) If the maximum allowable rate of twist is 0.125 degrees per foot and the maximum shear strain is to be kept at 400×10^{-6} rad by adjusting the torque T, what is the minimum required outer radius $(r_2)_{min}$?

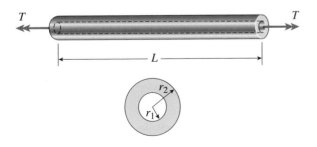

PROBS. 3.2-3, 3.2-4, and 3.2-5

3.2-4 A circular steel tube of length $L = 1.0$ m is loaded in torsion by torques T (see figure).

(a) If the inner radius of the tube is $r_1 = 45$ mm and the measured angle of twist between the ends is $0.5°$, what is the shear strain γ_1 (in radians) at the inner surface?

(b) If the maximum allowable shear strain is 0.0004 rad and the angle of twist is to be kept at $0.45°$ by adjusting the torque T, what is the maximum permissible outer radius $(r_2)_{max}$?

3.2-5 Solve the preceding problem if the length $L = 56$ in., the inner radius $r_1 = 1.25$ in., the angle of twist is $0.5°$, and the allowable shear strain is 0.0004 rad.

Circular Bars and Tubes

3.3-1 A prospector uses a hand-powered winch (see figure) to raise a bucket of ore in his mine shaft. The axle of the winch is a steel rod of diameter $d = 0.625$ in. Also, the distance from the center of the axle to the center of the lifting rope is $b = 4.0$ in.

If the weight of the loaded bucket is $W = 100$ lb, what is the maximum shear stress in the axle due to torsion?

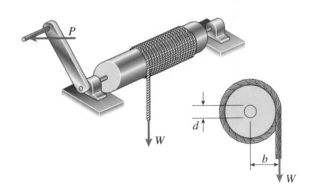

PROB. 3.3-1

3.3-2 When drilling a hole in a table leg, a furniture maker uses a hand-operated drill (see figure) with a bit of diameter $d = 4.0$ mm.

(a) If the resisting torque supplied by the table leg is equal to 0.3 N·m, what is the maximum shear stress in the drill bit?

(b) If the shear modulus of elasticity of the steel is $G = 75$ GPa, what is the rate of twist of the drill bit (degrees per meter)?

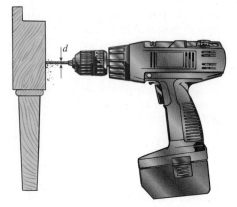

PROB. 3.3-2

3.3-3 While removing a wheel to change a tire, a driver applies forces $P = 25$ lb at the ends of two of the arms of a lug wrench (see figure). The wrench is made of steel with shear modulus of elasticity $G = 11.4 \times 10^6$ psi. Each arm of the wrench is 9.0 in. long and has a solid circular cross section of diameter $d = 0.5$ in.

(a) Determine the maximum shear stress in the arm that is turning the lug nut (arm A).

(b) Determine the angle of twist (in degrees) of this same arm.

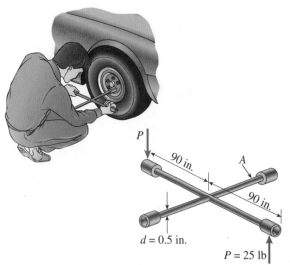

PROB. 3.3-3

3.3-4 An aluminum bar of solid circular cross section is twisted by torques T acting at the ends (see figure). The dimensions and shear modulus of elasticity are as follows: $L = 1.4$ m, $d = 32$ mm, and $G = 28$ GPa.

(a) Determine the torsional stiffness of the bar.

(b) If the angle of twist of the bar is 5°, what is the maximum shear stress? What is the maximum shear strain (in radians)?

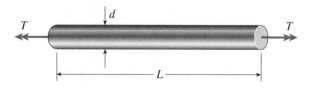

PROB. 3.3-4

3.3-5 A high-strength steel drill rod used for boring a hole in the earth has a diameter of 0.5 in. (see figure).The allowable shear stress in the steel is 40 ksi and the shear modulus of elasticity is 11,600 ksi.

What is the minimum required length of the rod so that one end of the rod can be twisted 30° with respect to the other end without exceeding the allowable stress?

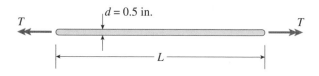

PROB. 3.3-5

3.3-6 The steel shaft of a socket wrench has a diameter of 8.0 mm. and a length of 200 mm (see figure).

If the allowable stress in shear is 60 MPa, what is the maximum permissible torque T_{max} that may be exerted with the wrench?

Through what angle ϕ (in degrees) will the shaft twist under the action of the maximum torque? (Assume $G = 78$ GPa and disregard any bending of the shaft.)

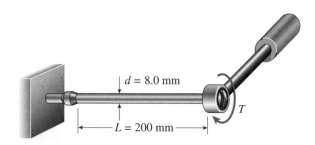

PROB. 3.3-6

3.3-7 A circular tube of aluminum is subjected to torsion by torques T applied at the ends (see figure). The bar is 24 in. long, and the inside and outside diameters are 1.25 in. and 1.75 in., respectively. It is determined by measurement that the angle of twist is 4° when the torque is 6200 lb-in.

Calculate the maximum shear stress τ_{max} in the tube, the shear modulus of elasticity G, and the maximum shear strain γ_{max} (in radians).

3.3-9 Three identical circular disks A, B, and C are welded to the ends of three identical solid circular bars (see figure). The bars lie in a common plane and the disks lie in planes perpendicular to the axes of the bars. The bars are welded at their intersection D to form a rigid connection. Each bar has diameter $d_1 = 0.5$ in. and each disk has diameter $d_2 = 3.0$ in.

Forces P_1, P_2, and P_3 act on disks A, B, and C, respectively, thus subjecting the bars to torsion. If $P_1 = 28$ lb, what is the maximum shear stress τ_{max} in any of the three bars?

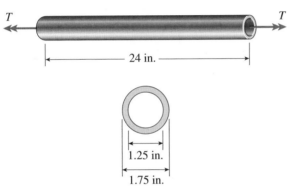

PROB. 3.3-7

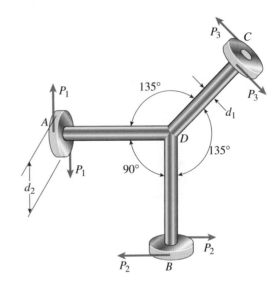

PROB. 3.3-9

3.3-8 A propeller shaft for a small yacht is made of a solid steel bar 104 mm in diameter. The allowable stress in shear is 48 MPa, and the allowable rate of twist is 2.0° in 3.5 meters.

Assuming that the shear modulus of elasticity is $G = 80$ GPa, determine the maximum torque T_{max} that can be applied to the shaft.

3.3-10 The steel axle of a large winch on an ocean liner is subjected to a torque of 1.65 kN·m (see figure). What is the minimum required diameter d_{min} if the allowable shear stress is 48 MPa and the allowable rate of twist is 0.75°/m? (Assume that the shear modulus of elasticity is 80 GPa.)

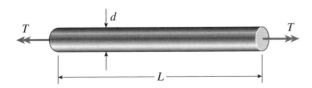

PROB. 3.3-8

PROB. 3.3-10

3.3-11 A hollow steel shaft used in a construction auger has outer diameter $d_2 = 6.0$ in. and inner diameter $d_1 = 4.5$ in. (see figure on the next page). The steel has shear modulus of elasticity $G = 11.0 \times 10^6$ psi.

For an applied torque of 150 k-in., determine the following quantities:

(a) shear stress τ_2 at the outer surface of the shaft,
(b) shear stress τ_1 at the inner surface, and
(c) rate of twist θ (degrees per unit of length).

Also, draw a diagram showing how the shear stresses vary in magnitude along a radial line in the cross section.

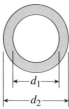

PROBS. 3.3-11 and 3.3-12

3.3-12 Solve the preceding problem if the shaft has outer diameter $d_2 = 150$ mm and inner diameter $d_1 = 100$ mm. Also, the steel has shear modulus of elasticity $G = 75$ GPa and the applied torque is 16 kN·m.

3.3-13 A vertical pole of solid circular cross section is twisted by horizontal forces $P = 1100$ lb acting at the ends of a horizontal arm AB (see figure). The distance from the outside of the pole to the line of action of each force is $c = 5.0$ in.

If the allowable shear stress in the pole is 4500 psi, what is the minimum required diameter d_{min} of the pole?

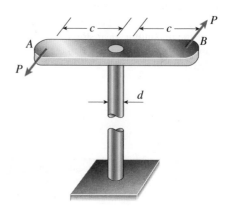

PROBS. 3.3-13 and 3.3-14

3.3-14 Solve the preceding problem if the horizontal forces have magnitude $P = 5.0$ kN, the distance $c = 125$ mm, and the allowable shear stress is 30 MPa.

3.3-15 A solid brass bar of diameter $d = 1.25$ in. is subjected to torques T_1, as shown in part (a) of the figure. The allowable shear stress in the brass is 12 ksi.

(a) What is the maximum permissible value of the torques T_1?

(b) If a hole of diameter 0.625 in. is drilled longitudinally through the bar, as shown in part (b) of the figure, what is the maximum permissible value of the torques T_2?

(c) What is the percent decrease in torque and the percent decrease in weight due to the hole?

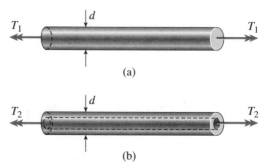

PROB. 3.3-15

3.3-16 A hollow aluminum tube used in a roof structure has an outside diameter $d_2 = 104$ mm and an inside diameter $d_1 = 82$ mm (see figure). The tube is 2.75 m long, and the aluminum has shear modulus $G = 28$ GPa.

(a) If the tube is twisted in pure torsion by torques acting at the ends, what is the angle of twist (in degrees) when the maximum shear stress is 48 MPa?

(b) What diameter d is required for a solid shaft (see figure) to resist the same torque with the same maximum stress?

(c) What is the ratio of the weight of the hollow tube to the weight of the solid shaft?

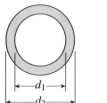

PROB. 3.3-16

3.3-17 A circular tube of inner radius r_1 and outer radius r_2 is subjected to a torque produced by forces $P = 900$ lb (see figure). The forces have their lines of action at a distance $b = 5.5$ in. from the outside of the tube.

If the allowable shear stress in the tube is 6300 psi and the inner radius $r_1 = 1.2$ in., what is the minimum permissible outer radius r_2?

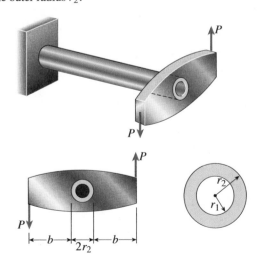

PROB. 3.3-17

Nonuniform Torsion

3.4-1 A stepped shaft *ABC* consisting of two solid circular segments is subjected to torques T_1 and T_2 acting in opposite directions, as shown in the figure. The larger segment of the shaft has diameter $d_1 = 2.25$ in. and length $L_1 = 30$ in.; the smaller segment has diameter $d_2 = 1.75$ in. and length $L_2 = 20$ in. The material is steel with shear modulus $G = 11 \times 10^6$ psi, and the torques are $T_1 = 20{,}000$ lb-in. and $T_2 = 8000$ lb-in.

Calculate the following quantities: (a) the maximum shear stress τ_{max} in the shaft, and (b) the angle of twist ϕ_C (in degrees) at end *C*.

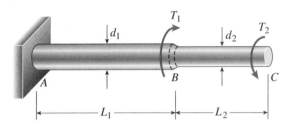

PROB. 3.4-1

3.4-2 A circular tube of outer diameter $d_3 = 70$ mm and inner diameter $d_2 = 60$ mm is welded at the right-hand end to a fixed plate and at the left-hand end to a rigid end plate (see figure). A solid circular bar of diameter $d_1 = 40$ mm is inside of, and concentric with, the tube. The bar passes through a hole in the fixed plate and is welded to the rigid end plate.

The bar is 1.0 m long and the tube is half as long as the bar. A torque $T = 1000$ N·m acts at end *A* of the bar. Also, both the bar and tube are made of an aluminum alloy with shear modulus of elasticity $G = 27$ GPa.

(a) Determine the maximum shear stresses in both the bar and tube.

(b) Determine the angle of twist (in degrees) at end *A* of the bar.

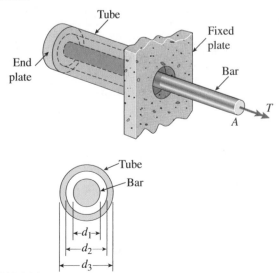

PROB. 3.4-2

3.4-3 A stepped shaft *ABCD* consisting of solid circular segments is subjected to three torques, as shown in the figure. The torques have magnitudes 12.5 k-in., 9.8 k-in., and 9.2 k-in. The length of each segment is 25 in. and the diameters of the segments are 3.5 in., 2.75 in., and 2.5 in. The material is steel with shear modulus of elasticity $G = 11.6 \times 10^3$ ksi.
  (a) Calculate the maximum shear stress τ_{max} in the shaft.
  (b) Calculate the angle of twist φ_D (in degrees) at end *D*.

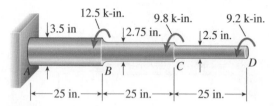

PROB. 3.4-3

3.4-4 A solid circular bar *ABC* consists of two segments, as shown in the figure. One segment has diameter $d_1 = 56$ mm and length $L_1 = 1.45$ m; the other segment has diameter $d_2 = 48$ mm and length $L_2 = 1.2$ m.
  What is the allowable torque T_{allow} if the shear stress is not to exceed 30 MPa and the angle of twist between the ends of the bar is not to exceed 1.25°? (Assume $G = 80$ GPa.)

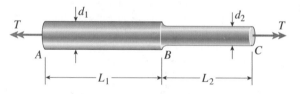

PROB. 3.4-4

3.4-5 A hollow tube *ABCDE* constructed of monel metal is subjected to five torques acting in the directions shown in the figure. The magnitudes of the torques are $T_1 = 1000$ lb-in., $T_2 = T_4 = 500$ lb-in., and $T_3 = T_5 = 800$ lb-in. The tube has an outside diameter $d_2 = 1.0$ in. The allowable shear stress is 12,000 psi and the allowable rate of twist is 2.0°/ft.
  Determine the maximum permissible inside diameter d_1 of the tube.

| T_1 | T_2 | T_3 | T_4 | T_5 |
| 1000 lb-in. | 500 lb-in. | 800 lb-in. | 500 lb-in. | 800 lb-in. |

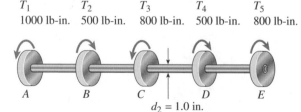

$d_2 = 1.0$ in.

PROB. 3.4-5

3.4-6 A shaft of solid circular cross section consisting of two segments is shown in the first part of the figure. The left-hand segment has diameter 80 mm and length 1.2 m; the right-hand segment has diameter 60 mm and length 0.9 m.
  Shown in the second part of the figure is a hollow shaft made of the same material and having the same length. The thickness *t* of the hollow shaft is *d*/10, where *d* is the outer diameter. Both shafts are subjected to the same torque.
  If the hollow shaft is to have the same torsional stiffness as the solid shaft, what should be its outer diameter *d*?

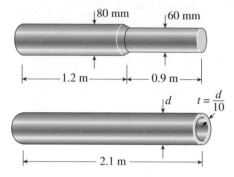

PROB. 3.4-6

3.4-7 Four gears are attached to a circular shaft and transmit the torques shown in the figure. The allowable shear stress in the shaft is 10,000 psi.
  (a) What is the required diameter *d* of the shaft if it has a solid cross section?
  (b) What is the required outside diameter *d* if the shaft is hollow with an inside diameter of 1.0 in.?

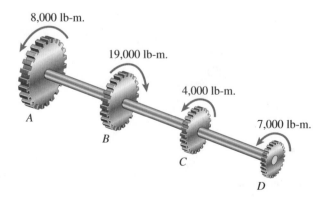

PROB. 3.4-7

3.4-8 A tapered bar AB of solid circular cross section is twisted by torques T (see figure). The diameter of the bar varies linearly from d_A at the left-hand end to d_B at the right-hand end.

For what ratio d_B/d_A will the angle of twist of the tapered bar be one-half the angle of twist of a prismatic bar of diameter d_A? (The prismatic bar is made of the same material, has the same length, and is subjected to the same torque as the tapered bar.) *Hint*: Use the results of Example 3-5.

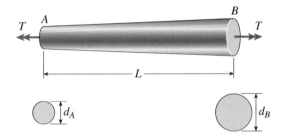

PROBS. 3.4-8, 3.4-9, and 3.4-10

3.4-9 A tapered bar AB of solid circular cross section is twisted by torques $T = 36,000$ lb-in. (see figure). The diameter of the bar varies linearly from d_A at the left-hand end to d_B at the right-hand end. The bar has length $L = 4.0$ ft and is made of an aluminum alloy having shear modulus of elasticity $G = 3.9 \times 10^6$ psi. The allowable shear stress in the bar is 15,000 psi and the allowable angle of twist is 3.0°.

If the diameter at end B is 1.5 times the diameter at end A, what is the minimum required diameter d_A at end A? (*Hint:* Use the results of Example 3-5).

3.4-10 The bar shown in the figure is tapered linearly from end A to end B and has a solid circular cross section. The diameter at the smaller end of the bar is $d_A = 25$ mm and the length is $L = 300$ mm. The bar is made of steel with shear modulus of elasticity $G = 82$ GPa.

If the torque $T = 180$ N·m and the allowable angle of twist is 0.3°, what is the minimum allowable diameter d_B at the larger end of the bar? (*Hint:* Use the results of Example 3-5.)

3.4-11 The nonprismatic cantilever circular bar shown has an internal cylindrical hole from 0 to x, so the net polar moment of inertia of the cross section for segment 1 is $(7/8)I_p$. Torque T is applied at x and torque $T/2$ is applied at $x = L$. Assume that G is constant.

(a) Find reaction moment R_1.

(b) Find internal torsional moments T_i in segments 1 & 2.

(c) Find x required to obtain twist at joint 3 of $\varphi_3 = TL/GI_p$

(d) What is the rotation at joint 2, φ_2?

(e) Draw the torsional moment (TMD: $T(x), 0 \le x \le L$) and displacement (TDD: $\varphi(x), 0 \le x \le L$) diagrams.

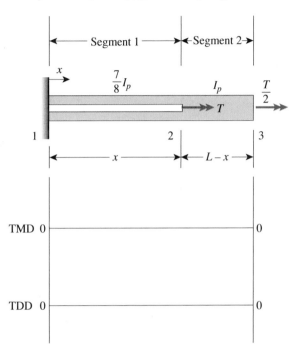

PROB. 3.4-11

3.4-12 A uniformly tapered tube AB of hollow circular cross section is shown in the figure. The tube has constant wall thickness t and length L. The average diameters at the ends are d_A and $d_B = 2d_A$. The polar moment of inertia may be represented by the approximate formula $I_P \approx \pi d^3 t/4$ (see Eq. 3-18).

Derive a formula for the angle of twist ϕ of the tube when it is subjected to torques T acting at the ends.

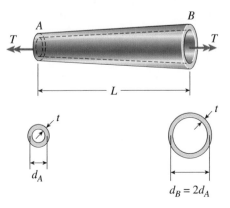

PROB. 3.4-12

3.4-13 A uniformly tapered aluminum-alloy tube AB of circular cross section and length L is shown in the figure. The outside diameters at the ends are d_A and $d_B = 2d_A$. A hollow section of length $L/2$ and constant thickness $t = d_A/10$ is cast into the tube and extends from B halfway toward A.

(a) Find the angle of twist φ of the tube when it is subjected to torques T acting at the ends. Use numerical values as follows: $d_A = 2.5$ in., $L = 48$ in., $G = 3.9 \times 10^6$ psi, and $T = 40,000$ in-lb.

(b) Repeat (a) if the hollow section has constant diameter d_A. (See figure part b.)

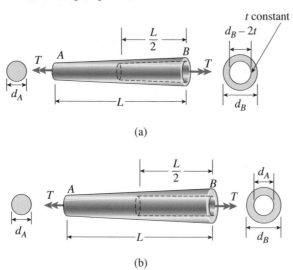

(a)

(b)

PROB. 3.4-13

3.4-14 For the *thin* nonprismatic steel pipe of constant thickness t and variable diameter d shown with applied torques at joints 2 and 3, determine the following.

(a) Find reaction moment R_1.

(b) Find an expression for twist rotation φ_3 at joint 3. Assume that G is constant.

(c) Draw the torsional moment diagram (TMD: $T(x)$, $0 \leq x \leq L$).

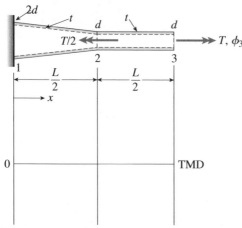

PROB. 3.4-14

3.4-15 A mountain-bike rider going uphill applies torque $T = Fd$ ($F = 15$ lb, d $= 4$ in.) to the end of the handlebars $ABCD$ (by pulling on the handlebar extenders DE). Consider the right half of the handlebar assembly only (assume the bars are fixed at the fork at A). Segments AB and CD are prismatic with lengths $L_1 = 2$ in. and $L_3 = 8.5$ in., and with outer diameters and thicknesses $d_{01} = 1.25$ in., $t_{01} = 0.125$ in., and $d_{03} = 0.87$ in., $t_{03} = 0.115$ in., respectively as shown. Segment BC of length $L_2 = 1.2$ in., however, is tapered, and outer diameter and thickness vary linearly between dimensions at B and C.

Consider torsion effects only. Assume $G = 4000$ ksi is constant.

Derive an integral expression for the angle of twist φ_D of half of the handlebar tube when it is subjected to torque $T = Fd$ acting at the end. Evaluate φ_D for the given numerical values.

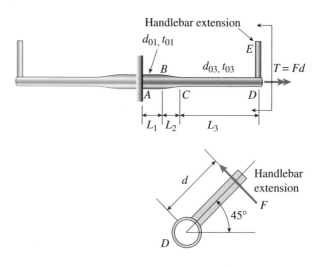

(Bontrager Race XXX Lite Flat Handlebar, used Courtesy of Bontrager)

(© Barry Goodno)

PROB. 3.4-15

3.4-16 A prismatic bar AB of length L and solid circular cross section (diameter d) is loaded by a distributed torque of constant intensity t per unit distance (see figure).

(a) Determine the maximum shear stress τ_{max} in the bar.

(b) Determine the angle of twist ϕ between the ends of the bar.

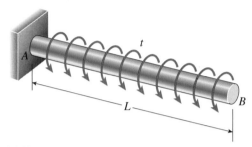

PROB. 3.4-16

3.4-17 A prismatic bar AB of solid circular cross section (diameter d) is loaded by a distributed torque (see figure). The intensity of the torque, that is, the torque per unit distance, is denoted $t(x)$ and varies linearly from a maximum value t_A at end A to zero at end B. Also, the length of the bar is L and the shear modulus of elasticity of the material is G.

(a) Determine the maximum shear stress τ_{max} in the bar.

(b) Determine the angle of twist ϕ between the ends of the bar.

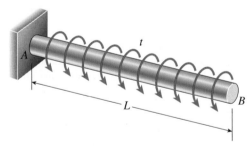

PROB. 3.4-17

3.4-18 A nonprismatic bar ABC of solid circular cross section is loaded by distributed torques (see figure). The intensity of the torques, that is, the torque per unit distance, is denoted $t(x)$ and varies linearly from zero at A to a maximum value T_0/L at B. Segment BC has linearly distributed torque of intensity $t(x) = T_0/3L$ of opposite sign to that applied along AB. Also, the polar moment of inertia of AB is twice that of BC, and the shear modulus of elasticity of the material is G.

(a) Find reaction torque R_A.

(b) Find internal torsional moments $T(x)$ in segments AB and BC.

(c) Find rotation ϕ_C.

(d) Find the maximum shear stress τ_{max} and its location along the bar.

(e) Draw the torsional moment diagram (TMD: $T(x)$, $0 \leq x \leq L$).

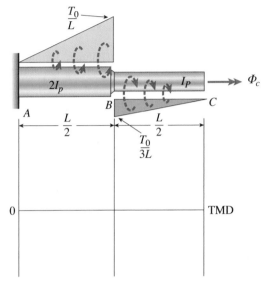

PROB. 3.4-18

3.4-19 A magnesium-alloy wire of diameter $d = 4$ mm and length L rotates inside a flexible tube in order to open or close a switch from a remote location (see figure). A torque T is applied manually (either clockwise or counterclockwise) at end B, thus twisting the wire inside the tube. At the other end A, the rotation of the wire operates a handle that opens or closes the switch.

A torque $T_0 = 0.2$ N·m is required to operate the switch. The torsional stiffness of the tube, combined with friction between the tube and the wire, induces a distributed torque of constant intensity $t = 0.04$ N·m/m (torque per unit distance) acting along the entire length of the wire.

(a) If the allowable shear stress in the wire is $\tau_{allow} = 30$ MPa, what is the longest permissible length L_{max} of the wire?

(b) If the wire has length $L = 4.0$ m and the shear modulus of elasticity for the wire is $G = 15$ GPa, what is the angle of twist ϕ (in degrees) between the ends of the wire?

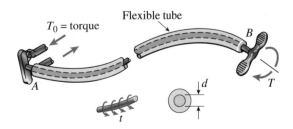

PROB. 3.4-19

3.4-20 Two hollow tubes are connected by a pin at B which is inserted into a hole drilled through both tubes at B (see cross-section view at B). Tube BC fits snugly into tube AB but neglect any friction on the interface. Tube inner and outer diameters d_i ($i = 1, 2, 3$) and pin diameter d_p are labeled in the figure. Torque T_0 is applied at joint C. The shear modulus of elasticity of the material is G.

Find expressions for the maximum torque $T_{0,max}$ which can be applied at C for each of the following conditions.

(a) The shear in the connecting pin is less than some allowable value ($\tau_{pin} < \tau_{p,allow}$).

(b) The shear in tube AB or BC is less than some allowable value ($\tau_{tube} < \tau_{t,allow}$).

(c) What is the maximum rotation ϕ_C for each of cases (a) and (b) above?

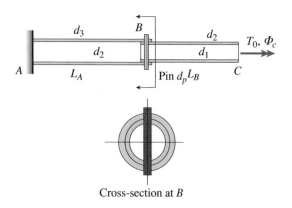

Cross-section at B

PROB. 3.4-20

Pure Shear

3.5-1 A hollow aluminum shaft (see figure) has outside diameter $d_2 = 4.0$ in. and inside diameter $d_1 = 2.0$ in. When twisted by torques T, the shaft has an angle of twist per unit distance equal to $0.54°/ft$. The shear modulus of elasticity of the aluminum is $G = 4.0 \times 10^6$ psi.

(a) Determine the maximum tensile stress σ_{max} in the shaft.

(b) Determine the magnitude of the applied torques T.

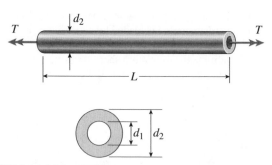

PROBS. 3.5-1, 3.5-2, and 3.5-3

3.5-2 A hollow steel bar ($G = 80$ GPa) is twisted by torques T (see figure). The twisting of the bar produces a maximum shear strain $\gamma_{max} = 640 \times 10^{-6}$ rad. The bar has outside and inside diameters of 150 mm and 120 mm, respectively.

(a) Determine the maximum tensile strain in the bar.

(b) Determine the maximum tensile stress in the bar.

(c) What is the magnitude of the applied torques T?

3.5-3 A tubular bar with outside diameter $d_2 = 4.0$ in. is twisted by torques $T = 70.0$ k-in. (see figure). Under the action of these torques, the maximum tensile stress in the bar is found to be 6400 psi.

(a) Determine the inside diameter d_1 of the bar.

(b) If the bar has length $L = 48.0$ in. and is made of aluminum with shear modulus $G = 4.0 \times 10^6$ psi, what is the angle of twist ϕ (in degrees) between the ends of the bar?

(c) Determine the maximum shear strain γ_{max} (in radians)?

3.5-4 A solid circular bar of diameter $d = 50$ mm (see figure) is twisted in a testing machine until the applied torque reaches the value $T = 500$ N·m. At this value of torque, a strain gage oriented at $45°$ to the axis of the bar gives a reading $\epsilon = 339 \times 10^{-6}$.

What is the shear modulus G of the material?

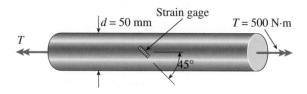

PROB. 3.5-4

3.5-5 A steel tube ($G = 11.5 \times 10^6$ psi) has an outer diameter $d_2 = 2.0$ in. and an inner diameter $d_1 = 1.5$ in. When twisted by a torque T, the tube develops a maximum normal strain of 170×10^{-6}.

What is the magnitude of the applied torque T?

3.5-6 A solid circular bar of steel ($G = 78$ GPa) transmits a torque $T = 360$ N·m. The allowable stresses in tension, compression, and shear are 90 MPa, 70 MPa, and 40 MPa, respectively. Also, the allowable tensile strain is 220×10^{-6}. Determine the minimum required diameter d of the bar.

3.5-7 The normal strain in the 45° direction on the surface of a circular tube (see figure) is 880×10^{-6} when the torque $T = 750$ lb-in. The tube is made of copper alloy with $G = 6.2 \times 10^{6}$ psi.

If the outside diameter d_2 of the tube is 0.8 in., what is the inside diameter d_1?

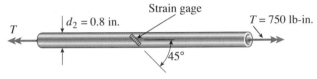

PROB. 3.5-7

3.5-8 An aluminum tube has inside diameter $d_1 = 50$ mm, shear modulus of elasticity $G = 27$ GPa, and torque $T = 4.0$ kN·m. The allowable shear stress in the aluminum is 50 MPa and the allowable normal strain is 900×10^{-6}.

Determine the required outside diameter d_2.

3.5-9 A solid steel bar ($G = 11.8 \times 10^{6}$ psi) of diameter $d = 2.0$ in. is subjected to torques $T = 8.0$ k-in. acting in the directions shown in the figure.

(a) Determine the maximum shear, tensile, and compressive stresses in the bar and show these stresses on sketches of properly oriented stress elements.

(b) Determine the corresponding maximum strains (shear, tensile, and compressive) in the bar and show these strains on sketches of the deformed elements.

PROB. 3.5-9

3.5-10 A solid aluminum bar ($G = 27$ GPa) of diameter $d = 40$ mm is subjected to torques $T = 300$ N·m acting in the directions shown in the figure.

(a) Determine the maximum shear, tensile, and compressive stresses in the bar and show these stresses on sketches of properly oriented stress elements.

(b) Determine the corresponding maximum strains (shear, tensile, and compressive) in the bar and show these strains on sketches of the deformed elements.

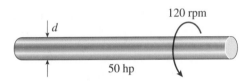

PROB. 3.5-10

Transmission of Power

3.7-1 A generator shaft in a small hydroelectric plant turns at 120 rpm and delivers 50 hp (see figure).

(a) If the diameter of the shaft is $d = 3.0$ in., what is the maximum shear stress τ_{max} in the shaft?

(b) If the shear stress is limited to 4000 psi, what is the minimum permissible diameter d_{min} of the shaft?

PROB. 3.7-1

3.7-2 A motor drives a shaft at 12 Hz and delivers 20 kW of power (see figure).

(a) If the shaft has a diameter of 30 mm, what is the maximum shear stress τ_{max} in the shaft?

(b) If the maximum allowable shear stress is 40 MPa, what is the minimum permissible diameter d_{min} of the shaft?

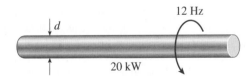

PROB. 3.7-2

3.7-3 The propeller shaft of a large ship has outside diameter 18 in. and inside diameter 12 in., as shown in the figure. The shaft is rated for a maximum shear stress of 4500 psi.

(a) If the shaft is turning at 100 rpm, what is the maximum horsepower that can be transmitted without exceeding the allowable stress?

(b) If the rotational speed of the shaft is doubled but the power requirements remain unchanged, what happens to the shear stress in the shaft?

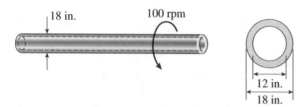

PROB. 3.7-3

3.7-4 The drive shaft for a truck (outer diameter 60 mm and inner diameter 40 mm) is running at 2500 rpm (see figure).

(a) If the shaft transmits 150 kW, what is the maximum shear stress in the shaft?

(b) If the allowable shear stress is 30 MPa, what is the maximum power that can be transmitted?

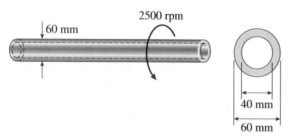

PROB. 3.7-4

3.7-5 A hollow circular shaft for use in a pumping station is being designed with an inside diameter equal to 0.75 times the outside diameter. The shaft must transmit 400 hp at 400 rpm without exceeding the allowable shear stress of 6000 psi.

Determine the minimum required outside diameter d.

3.7-6 A tubular shaft being designed for use on a construction site must transmit 120 kW at 1.75 Hz. The inside diameter of the shaft is to be one-half of the outside diameter.

If the allowable shear stress in the shaft is 45 MPa, what is the minimum required outside diameter d?

3.7-7 A propeller shaft of solid circular cross section and diameter d is spliced by a collar of the same material (see figure). The collar is securely bonded to both parts of the shaft.

What should be the minimum outer diameter d_1 of the collar in order that the splice can transmit the same power as the solid shaft?

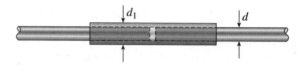

PROB. 3.7-7

3.7-8 What is the maximum power that can be delivered by a hollow propeller shaft (outside diameter 50 mm, inside diameter 40 mm, and shear modulus of elasticity 80 GPa) turning at 600 rpm if the allowable shear stress is 100 MPa and the allowable rate of twist is 3.0°/m?

3.7-9 A motor delivers 275 hp at 1000 rpm to the end of a shaft (see figure). The gears at B and C take out 125 and 150 hp, respectively.

Determine the required diameter d of the shaft if the allowable shear stress is 7500 psi and the angle of twist between the motor and gear C is limited to 1.5°. (Assume $G = 11.5 \times 10^6$ psi, $L_1 = 6$ ft, and $L_2 = 4$ ft.)

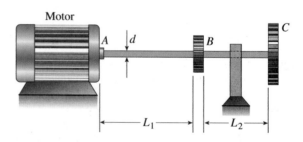

PROBS. 3.7-9 and 3.7-10

3.7-10 The shaft ABC shown in the figure is driven by a motor that delivers 300 kW at a rotational speed of 32 Hz. The gears at B and C take out 120 and 180 kW, respectively. The lengths of the two parts of the shaft are $L_1 = 1.5$ m and $L_2 = 0.9$ m.

Determine the required diameter d of the shaft if the allowable shear stress is 50 MPa, the allowable angle of twist between points A and C is 4.0°, and $G = 75$ GPa.

Statically Indeterminate Torsional Members

3.8-1 A solid circular bar *ABCD* with fixed supports is acted upon by torques T_0 and $2T_0$ at the locations shown in the figure.

Obtain a formula for the maximum angle of twist ϕ_{max} of the bar. (*Hint:* Use Eqs. 3-46a and b of Example 3-9 to obtain the reactive torques.)

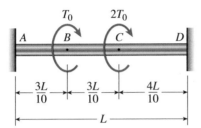

PROB. 3.8-1

3.8-2 A solid circular bar *ABCD* with fixed supports at ends *A* and *D* is acted upon by two equal and oppositely directed torques T_0, as shown in the figure. The torques are applied at points *B* and *C*, each of which is located at distance *x* from one end of the bar. (The distance *x* may vary from zero to $L/2$.)

(a) For what distance *x* will the angle of twist at points *B* and *C* be a maximum?

(b) What is the corresponding angle of twist ϕ_{max}? (*Hint:* Use Eqs. 3-46a and b of Example 3-9 to obtain the reactive torques.)

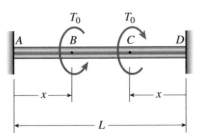

PROB. 3.8-2

3.8-3 A solid circular shaft *AB* of diameter *d* is fixed against rotation at both ends (see figure). A circular disk is attached to the shaft at the location shown.

What is the largest permissible angle of rotation ϕ_{max} of the disk if the allowable shear stress in the shaft is τ_{allow}? (Assume that $a > b$. Also, use Eqs. 3-46a and b of Example 3-9 to obtain the reactive torques.)

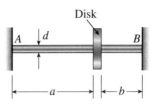

PROB. 3.8-3

3.8-4 A hollow steel shaft *ACB* of outside diameter 50 mm and inside diameter 40 mm is held against rotation at ends *A* and *B* (see figure). Horizontal forces *P* are applied at the ends of a vertical arm that is welded to the shaft at point *C*.

Determine the allowable value of the forces *P* if the maximum permissible shear stress in the shaft is 45 MPa. (*Hint:* Use Eqs. 3-46a and b of Example 3-9 to obtain the reactive torques.)

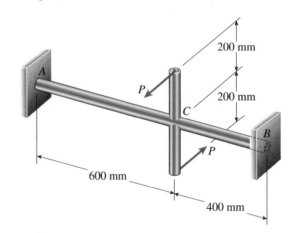

PROB. 3.8-4

3.8-5 A stepped shaft *ACB* having solid circular cross sections with two different diameters is held against rotation at the ends (see figure).

If the allowable shear stress in the shaft is 6000 psi, what is the maximum torque $(T_0)_{max}$ that may be applied at section *C*? (*Hint:* Use Eqs. 3-45a and b of Example 3-9 to obtain the reactive torques.)

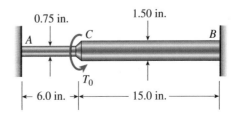

PROB. 3.8-5

3.8-6 A stepped shaft *ACB* having solid circular cross sections with two different diameters is held against rotation at the ends (see figure).

If the allowable shear stress in the shaft is 43 MPa, what is the maximum torque $(T_0)_{max}$ that may be applied at section *C*? (*Hint:* Use Eqs. 3-45a and b of Example 3-9 to obtain the reactive torques.)

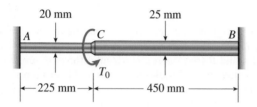

PROB. 3.8-6

3.8-7 A stepped shaft *ACB* is held against rotation at ends *A* and *B* and subjected to a torque T_0 acting at section *C* (see figure). The two segments of the shaft (*AC* and *CB*) have diameters d_A and d_B, respectively, and polar moments of inertia I_{PA} and I_{PB}, respectively. The shaft has length *L* and segment *AC* has length *a*.

(a) For what ratio *a/L* will the maximum shear stresses be the same in both segments of the shaft?

(b) For what ratio *a/L* will the internal torques be the same in both segments of the shaft? (*Hint:* Use Eqs. 3-45a and b of Example 3-9 to obtain the reactive torques.)

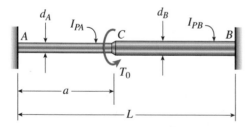

PROB. 3.8-7

3.8-8 A circular bar *AB* of length *L* is fixed against rotation at the ends and loaded by a distributed torque *t(x)* that varies linearly in intensity from zero at end *A* to t_0 at end *B* (see figure).

Obtain formulas for the fixed-end torques T_A and T_B.

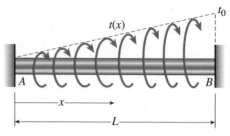

PROB. 3.8-8

3.8-9 A circular bar *AB* with ends fixed against rotation has a hole extending for half of its length (see figure). The outer diameter of the bar is $d_2 = 3.0$ in. and the diameter of the hole is $d_1 = 2.4$ in. The total length of the bar is $L = 50$ in.

At what distance *x* from the left-hand end of the bar should a torque T_0 be applied so that the reactive torques at the supports will be equal?

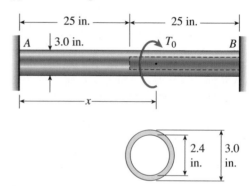

PROB. 3.8-9

3.8-10 A solid steel bar of diameter $d_1 = 25.0$ mm is enclosed by a steel tube of outer diameter $d_3 = 37.5$ mm and inner diameter $d_2 = 30.0$ mm (see figure). Both bar and tube are held rigidly by a support at end *A* and joined securely to a rigid plate at end *B*. The composite bar, which has a length $L = 550$ mm, is twisted by a torque $T = 400$ N·m acting on the end plate.

(a) Determine the maximum shear stresses τ_1 and τ_2 in the bar and tube, respectively.

(b) Determine the angle of rotation ϕ (in degrees) of the end plate, assuming that the shear modulus of the steel is $G = 80$ GPa.

(c) Determine the torsional stiffness k_T of the composite bar. (*Hint:* Use Eqs. 3-44a and b to find the torques in the bar and tube.)

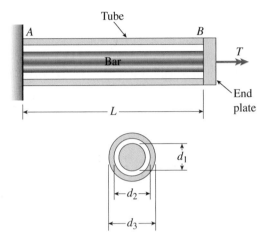

PROBS. 3.8-10 and 3.8-11

3.8-11 A solid steel bar of diameter $d_1 = 1.50$ in. is enclosed by a steel tube of outer diameter $d_3 = 2.25$ in. and inner diameter $d_2 = 1.75$ in. (see figure). Both bar and tube are held rigidly by a support at end A and joined securely to a rigid plate at end B. The composite bar, which has length $L = 30.0$ in., is twisted by a torque $T = 5000$ lb-in. acting on the end plate.

(a) Determine the maximum shear stresses τ_1 and τ_2 in the bar and tube, respectively.

(b) Determine the angle of rotation ϕ (in degrees) of the end plate, assuming that the shear modulus of the steel is $G = 11.6 \times 10^6$ psi.

(c) Determine the torsional stiffness k_T of the composite bar. (*Hint:* UseEqs. 3-44a and b to find the torques in the bar and tube.)

3.8-12 The composite shaft shown in the figure is manufactured by shrink-fitting a steel sleeve over a brass core so that the two parts act as a single solid bar in torsion. The outer diameters of the two parts are $d_1 = 40$ mm for the brass core

and $d_2 = 50$ mm for the steel sleeve. The shear moduli of elasticity are $G_b = 36$ GPa for the brass and $G_s = 80$ GPa for the steel.

Assuming that the allowable shear stresses in the brass and steel are $\tau_b = 48$ MPa and $\tau_s = 80$ MPa, respectively, determine the maximum permissible torque T_{max} that may be applied to the shaft. (*Hint:* Use Eqs. 3-44a and b to find the torques.)

PROBS. 3.8-12 and 3.8-13

3.8-13 The composite shaft shown in the figure is manufactured by shrink-fitting a steel sleeve over a brass core so that the two parts act as a single solid bar in torsion. The outer diameters of the two parts are $d_1 = 1.6$ in. for the brass core and $d_2 = 2.0$ in. for the steel sleeve. The shear moduli of elasticity are $G_b = 5400$ ksi for the brass and $G_s = 12,000$ ksi for the steel.

Assuming that the allowable shear stresses in the brass and steel are $\tau_b = 4500$ psi and $\tau_s = 7500$ psi, respectively, determine the maximum permissible torque T_{max} that may be applied to the shaft. (*Hint:* Use Eqs. 3-44a and b to find the torques.)

3.8-14 A steel shaft (G_s = 80 GPa) of total length L = 3.0 m is encased for one-third of its length by a brass sleeve (G_b = 40 GPa) that is securely bonded to the steel (see figure). The outer diameters of the shaft and sleeve are d_1 = 70 mm and d_2 = 90 mm, respectively.

(a) Determine the allowable torque T_1 that may be applied to the ends of the shaft if the angle of twist between the ends is limited to 8.0°.

(b) Determine the allowable torque T_2 if the shear stress in the brass is limited to τ_b = 70 MPa.

(c) Determine the allowable torque T_3 if the shear stress in the steel is limited to τ_s = 110 MPa.

(d) What is the maximum allowable torque T_{max} if all three of the preceding conditions must be satisfied?

3.8-15 A uniformly tapered aluminum-alloy tube AB of circular cross section and length L is fixed against rotation at A and B, as shown in the figure. The outside diameters at the ends are d_A and d_B = $2d_A$. A hollow section of length $L/2$ and constant thickness $t = d_A/10$ is cast into the tube and extends from B halfway toward A. Torque T_0 is applied at $L/2$.

(a) Find the reactive torques at the supports, T_A and T_B. Use numerical values as follows: d_A = 2.5 in., L = 48 in., G = 3.9 × 10⁶ psi, T_0 = 40,000 in-lb.

(b) Repeat (a) if the hollow section has constant diameter d_A.

3.8-16 A hollow circular tube A (outer diameter d_A, wall thickness t_A) fits over the end of a circular tube B (d_B, t_B), as shown in the figure. The far ends of both tubes are fixed. Initially, a hole through tube B makes an angle β with a line through two holes in tube A. Then tube B is twisted until the holes are aligned, and a pin (diameter d_p) is placed through the holes. When tube B is released, the system returns to equilibrium. Assume that G is constant.

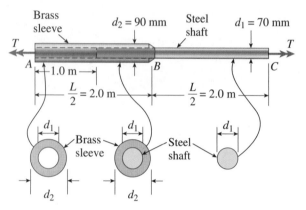

PROB. 3.8-14

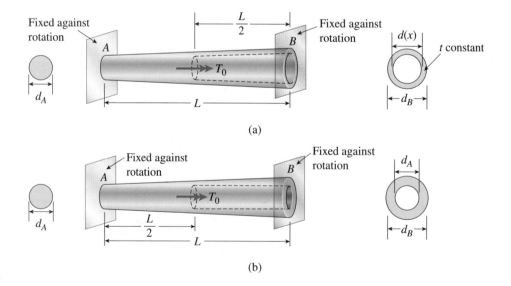

PROB. 3.8-15

(a) Use superposition to find the reactive torques T_A and T_B at the supports.

(b) Find an expression for the maximum value of β if the shear stress in the pin, τ_p, cannot exceed $\tau_{p,\text{allow}}$.

(c) Find an expression for the maximum value of β if the shear stress in the tubes, τ_t, cannot exceed $\tau_{t,\text{allow}}$.

(d) Find an expression for the maximum value of β if the bearing stress in the pin at C cannot exceed $\sigma_{b,\text{allow}}$.

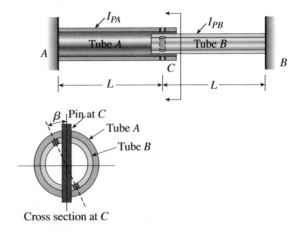

Cross section at C

PROB. 3.8-16

Shear forces and bending moments govern the design of beams in a variety of structures such as building frames and bridges.
(© Jupiter Images, 2007)

Shear Forces and Bending Moments

CHAPTER OVERVIEW

Chapter 4 begins with a review of two-dimensional beam and frame analysis which you learned in your first course in mechanics, **Statics**. First, various types of beams, loadings, and support conditions are defined for typical structures, such as cantilever and simple beams. **Applied loads** may be concentrated (either a force or moment) or distributed. **Support conditions** include clamped, roller, pinned, and sliding supports. The number and arrangement of supports must produce a stable structure model that is either statically determinate or statically indeterminate. We will study only statically determinate beam structures in this chapter.

The focus in this chapter are the **internal stress resultants** (axial N, shear V, and moment M) at any point in the structure. In some structures, internal **"releases"** are introduced into the structure at specified points to control the magnitude of N, V, or M in certain members, and must be included in the analytical model. At these release points, N, V, or M may be considered to have a value of zero. Graphical displays or **diagrams** showing the variation of N, V, and M over the entire structure are very useful in beam and frame design (as we will see in Chapter 5), because these diagrams quickly identify locations and values of maximum axial force, shear, and moment needed for design.

The above topics on beams and frames are discussed in Chapter 4 as follows:

4.1 INTRODUCTION

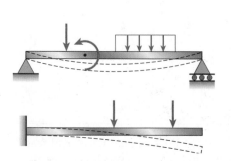

Structural members are usually classified according to the types of loads that they support. For instance, an *axially loaded bar* supports forces having their vectors directed along the axis of the bar, and a *bar in torsion* supports torques (or couples) having their moment vectors directed along the axis. In this chapter, we begin our study of **beams** (Fig. 4-1), which are structural members subjected to lateral loads, that is, forces or moments having their vectors perpendicular to the axis of the bar.

The beams shown in Fig. 4-1 are classified as *planar structures* because they lie in a single plane. If all loads act in that same plane, and if all deflections (shown by the dashed lines) occur in that plane, then we refer to that plane as the **plane of bending**.

FIG. 4-1 Examples of beams subjected to lateral loads

In this chapter we discuss shear forces and bending moments in beams, and we will show how these quantities are related to each other and to the loads. Finding the shear forces and bending moments is an essential step in the design of any beam. We usually need to know not only the maximum values of these quantities, but also the manner in which they vary along the axis. Once the shear forces and bending moments are known, we can find the stresses, strains, and deflections, as discussed later in Chapters 5, 6 and 8.

4.2 TYPES OF BEAMS, LOADS, AND REACTIONS

Beams are usually described by the manner in which they are supported. For instance, a beam with a pin support at one end and a roller support at the other (Fig. 4-2a) is called a **simply supported beam** or a **simple beam**. The essential feature of a **pin support** is that it prevents

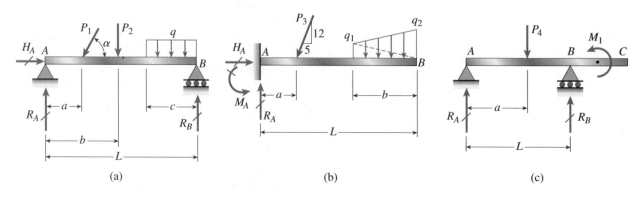

(a)   (b)   (c)

FIG. 4-2 Types of beams: (a) simple beam, (b) cantilever beam, and (c) beam with an overhang

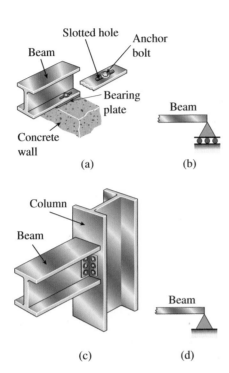

FIG. 4-3 Beam supported on a wall:
(a) actual construction, and
(b) representation as a roller support.
Beam-to-column connection: (c) actual
construction, and (d) representation as a
pin support.

Beam-to-column connection with one beam
attached to column flange and other attached
to column web
(Joe Gough/Shutterstock)

translation at the end of a beam but does not prevent rotation. Thus, end
A of the beam of Fig. 4-2a cannot move horizontally or vertically but the
axis of the beam can rotate in the plane of the figure. Consequently, a
pin support is capable of developing a force reaction with both horizontal and vertical components (H_A and R_A), but it cannot develop a
moment reaction.

At end B of the beam (Fig. 4-2a) the **roller support** prevents translation in the vertical direction but not in the horizontal direction; hence
this support can resist a vertical force (R_B) but not a horizontal force. Of
course, the axis of the beam is free to rotate at B just as it is at A. The vertical reactions at roller supports and pin supports may act *either* upward
or downward, and the horizontal reaction at a pin support may act either
to the left or to the right. In the figures, reactions are indicated by slashes
across the arrows in order to distinguish them from loads, as explained
previously in Section 1.8.

The beam shown in Fig. 4-2b, which is fixed at one end and free at
the other, is called a **cantilever beam**. At the **fixed support** (or *clamped
support*) the beam can neither translate nor rotate, whereas at the free end
it may do both. Consequently, both force and moment reactions may exist
at the fixed support.

The third example in the figure is a **beam with an overhang** (Fig. 4-2c).
This beam is simply supported at points A and B (that is, it has a pin support
at A and a roller support at B) but it also projects beyond the support at B.
The overhanging segment BC is similar to a cantilever beam except that the
beam axis may rotate at point B.

When drawing sketches of beams, we identify the supports by
conventional symbols, such as those shown in Fig. 4-2. These symbols
indicate the manner in which the beam is restrained, and therefore they
also indicate the nature of the reactive forces and moments. However,
the symbols do not represent the actual physical construction. For instance,
consider the examples shown in Fig. 4-3. Part (a) of the figure shows a
wide-flange beam supported on a concrete wall and held down by
anchor bolts that pass through slotted holes in the lower flange of the
beam. This connection restrains the beam against vertical movement
(either upward or downward) but does not prevent horizontal movement. Also, any restraint against rotation of the longitudinal axis of the
beam is small and ordinarily may be disregarded. Consequently, this
type of support is usually represented by a roller, as shown in part (b)
of the figure.

The second example (Fig. 4-3c) is a beam-to-column connection in
which the beam is attached to the column flange by bolted angles. (See
photo.) This type of support is usually assumed to restrain the beam against
horizontal and vertical movement but not against rotation (restraint against
rotation is slight because both the angles and the column can bend).
Thus, this connection is usually represented as a pin support for the beam
(Fig. 4-3d).

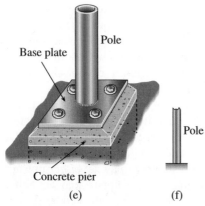

Base plate

Pole

Pole

Concrete pier

(e)   (f)

FIG. 4-3 Pole anchored to a concrete pier: (e) actual construction, and (f) representation as a fixed support

The last example (Fig. 4-3e) is a metal pole welded to a base plate that is anchored to a concrete pier embedded deep in the ground. Since the base of the pole is fully restrained against both translation and rotation, it is represented as a fixed support (Fig. 4-3f).

The task of representing a real structure by an **idealized model**, as illustrated by the beams shown in Fig. 4-2, is an important aspect of engineering work. The model should be simple enough to facilitate mathematical analysis and yet complex enough to represent the actual behavior of the structure with reasonable accuracy. Of course, every model is an approximation to nature. For instance, the actual supports of a beam are never perfectly rigid, and so there will always be a small amount of translation at a pin support and a small amount of rotation at a fixed support. Also, supports are never entirely free of friction, and so there will always be a small amount of restraint against translation at a roller support. In most circumstances, especially for statically determinate beams, these deviations from the idealized conditions have little effect on the action of the beam and can safely be disregarded.

Types of Loads

Several types of loads that act on beams are illustrated in Fig. 4-2. When a load is applied over a very small area it may be idealized as a **concentrated load**, which is a single force. Examples are the loads P_1, P_2, P_3, and P_4 in the figure. When a load is spread along the axis of a beam, it is represented as a **distributed load**, such as the load q in part (a) of the figure. Distributed loads are measured by their **intensity**, which is expressed in units of force per unit distance (for example, newtons per meter or pounds per foot). A **uniformly distributed load**, or **uniform load**, has constant intensity q per unit distance (Fig. 4-2a). A varying load has an intensity that changes with distance along the axis; for instance, the **linearly varying load** of Fig. 4-2b has an intensity that varies linearly from q_1 to q_2. Another kind of load is a **couple**, illustrated by the couple of moment M_1 acting on the overhanging beam (Fig. 4-2c).

As mentioned in Section 4.1, we assume in this discussion that the loads act in the plane of the figure, which means that all forces must have their vectors in the plane of the figure and all couples must have their moment vectors perpendicular to the plane of the figure. Furthermore, the beam itself must be symmetric about that plane, which means that every cross section of the beam must have a vertical axis of symmetry. Under these conditions, the beam will deflect only in the *plane of bending* (the plane of the figure).

Reactions

Finding the reactions is usually the first step in the analysis of a beam. Once the reactions are known, the shear forces and bending moments can be found, as described later in this chapter. If a beam is supported in a statically determinate manner, all reactions can be found from free-body diagrams and equations of equilibrium.

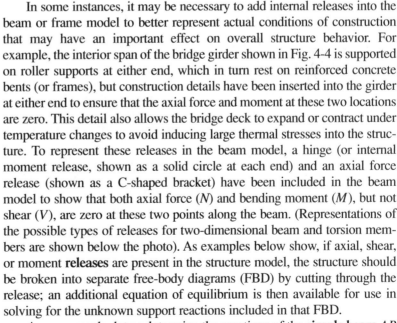

Internal releases and end supports in model of bridge beam
(Courtesy of the National Information Service for Earthquake
Engineering EERC, University of California, Berkeley.)

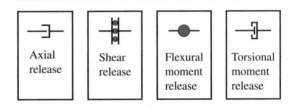

FIG. 4-4 Types of internal member
releases for two-dimensional beam and
frame members

In some instances, it may be necessary to add internal releases into the
beam or frame model to better represent actual conditions of construction
that may have an important effect on overall structure behavior. For
example, the interior span of the bridge girder shown in Fig. 4-4 is supported
on roller supports at either end, which in turn rest on reinforced concrete
bents (or frames), but construction details have been inserted into the girder
at either end to ensure that the axial force and moment at these two locations
are zero. This detail also allows the bridge deck to expand or contract under
temperature changes to avoid inducing large thermal stresses into the structure. To represent these releases in the beam model, a hinge (or internal
moment release, shown as a solid circle at each end) and an axial force
release (shown as a C-shaped bracket) have been included in the beam
model to show that both axial force (N) and bending moment (M), but not
shear (V), are zero at these two points along the beam. (Representations of
the possible types of releases for two-dimensional beam and torsion members are shown below the photo). As examples below show, if axial, shear,
or moment **releases** are present in the structure model, the structure should
be broken into separate free-body diagrams (FBD) by cutting through the
release; an additional equation of equilibrium is then available for use in
solving for the unknown support reactions included in that FBD.

As an example, let us determine the reactions of the **simple beam** AB
of Fig. 4-2a. This beam is loaded by an inclined force P_1, a vertical
force P_2, and a uniformly distributed load of intensity q. We begin by
noting that the beam has three unknown reactions: a horizontal force H_A
at the pin support, a vertical force R_A at the pin support, and a vertical

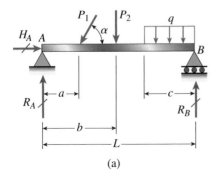

(a)

FIG. 4-2a Simple beam (Repeated)

force R_B at the roller support. For a planar structure, such as this beam, we know from statics that we can write three independent equations of equilibrium. Thus, since there are three unknown reactions and three equations, the beam is statically determinate.

The equation of horizontal equilibrium is

$$\Sigma F_{\text{horiz}} = 0 \quad H_A - P_1 \cos \alpha = 0$$

from which we get

$$H_A = P_1 \cos \alpha$$

This result is so obvious from an inspection of the beam that ordinarily we would not bother to write the equation of equilibrium.

To find the vertical reactions R_A and R_B we write equations of moment equilibrium about points B and A, respectively, with counterclockwise moments being positive:

$$\Sigma M_B = 0 \quad -R_A L + (P_1 \sin \alpha)(L - a) + P_2(L - b) + qc^2/2 = 0$$

$$\Sigma M_A = 0 \quad R_B L - (P_1 \sin \alpha)(a) - P_2 b - qc(L - c/2) = 0$$

Solving for R_A and R_B, we get

$$R_A = \frac{(P_1 \sin \alpha)(L - a)}{L} + \frac{P_2(L - b)}{L} + \frac{qc^2}{2L}$$

$$R_B = \frac{(P_1 \sin \alpha)(a)}{L} + \frac{P_2 b}{L} + \frac{qc(L - c/2)}{L}$$

As a check on these results we can write an equation of equilibrium in the vertical direction and verify that it reduces to an identity.

If the beam structure in Fig. 4-2a is modified to replace the roller support at B with a pin support, it is now one degree statically indeterminate. However, if an axial force release is inserted into the model, as shown in Fig. 4-5 just to the left of the point of application of load P_1, the beam still can be analyzed using the laws of statics alone because the release provides one additional equilibrium equation. The beam must be cut at the release to expose the internal stress resultants N, V, and M; but now $N = 0$ at the release, so reactions $H_A = 0$ and $H_B = P_1 \cos \alpha$.

As a second example, consider the **cantilever beam** of Fig. 4-2b. The loads consist of an inclined force P_3 and a linearly varying distributed load. The latter is represented by a trapezoidal diagram of load intensity that varies from q_1 to q_2. The reactions at the fixed support are a horizontal force H_A, a vertical force R_A, and a couple M_A. Equilibrium of forces in the horizontal direction gives

$$H_A = \frac{5P_3}{13}$$

and equilibrium in the vertical direction gives

$$R_A = \frac{12P_3}{13} + \left(\frac{q_1 + q_2}{2}\right)b$$

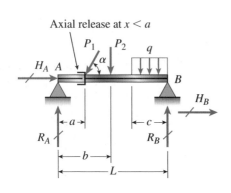

FIG. 4-5 Simple beam with axial release

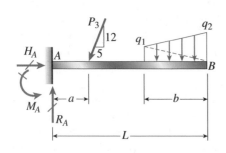

(b)

FIG. 4-2b Cantilever beam (Repeated)

In finding this reaction we used the fact that the resultant of the distributed load is equal to the area of the trapezoidal loading diagram.

The moment reaction M_A at the fixed support is found from an equation of equilibrium of moments. In this example we will sum moments about point A in order to eliminate both H_A and R_A from the moment equation. Also, for the purpose of finding the moment of the distributed load, we will divide the trapezoid into two triangles, as shown by the dashed line in Fig. 4-2b. Each load triangle can be replaced by its resultant, which is a force having its magnitude equal to the area of the triangle and having its line of action through the centroid of the triangle. Thus, the moment about point A of the lower triangular part of the load is

$$\left(\frac{q_1 b}{2}\right)\left(L - \frac{2b}{3}\right)$$

in which $q_1 b/2$ is the resultant force (equal to the area of the triangular load diagram) and $L - 2b/3$ is the moment arm (about point A) of the resultant.

The moment of the upper triangular portion of the load is obtained by a similar procedure, and the final equation of moment equilibrium (counterclockwise is positive) is

$$\sum M_A = 0 \qquad M_A - \left(\frac{12P_3}{13}\right)a - \frac{q_1 b}{2}\left(L - \frac{2b}{3}\right) - \frac{q_2 b}{2}\left(L - \frac{b}{3}\right) = 0$$

from which

$$M_A = \frac{12P_3 a}{13} + \frac{q_1 b}{2}\left(L - \frac{2b}{3}\right) + \frac{q_2 b}{2}\left(L - \frac{b}{3}\right)$$

Since this equation gives a positive result, the reactive moment M_A acts in the assumed direction, that is, counterclockwise. (The expressions for R_A and M_A can be checked by taking moments about end B of the beam and verifying that the resulting equation of equilibrium reduces to an identity.)

If the cantilever beam structure in Fig. 4-2b is modified to add a roller support at B, it is now referred to as a one degree statically indeterminate "propped" cantilever beam. However, if a moment release is inserted into the model as shown in Fig. 4-6, just to the right of the point of application of load P_3, the beam can still be analyzed using the laws of statics alone because the release provides one additional equilibrium equation. The beam must be cut at the release to expose the internal stress resultants N, V, and M; now $M = 0$ at the release so reaction R_B can be computed by summing moments in the right-hand free-body diagram. Once R_B is known, reaction R_A can

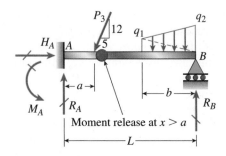

FIG. 4-6 Propped cantilever beam with moment release

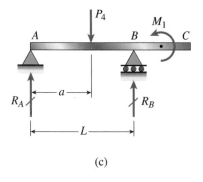

(c)

FIG. 4-2c Beam with an overhang (Repeated)

once again be computed by summing vertical forces, and reaction moment M_A can be obtained by summing moments about point A. Results are summarized in Fig. 4-6. Note that reaction H_A is unchanged from that reported above for the original cantilever beam structure in Fig. 4-2b.

$$R_B = \frac{\frac{1}{2}q_1 b\left(L - a - \frac{2}{3}b\right) + \frac{1}{2}q_2 b\left(L - a - \frac{b}{3}\right)}{L - a}$$

$$R_A = \frac{12}{13}P_3 + \left(\frac{q_1 + q_2}{2}\right)(b) - R_B$$

$$R_A = \frac{1}{78}\frac{-72P_3 L + 72P_3 a - 26 q_1 b^2 - 13 q_2 b^2}{-L + a}$$

$$M_A = \frac{12}{13}P_3 a + q_1\frac{b}{2}\left(L - \frac{2}{3}b\right) + q_2\frac{b}{2}\left(L - \frac{b}{3}\right) - R_B L$$

$$M_A = \frac{1}{78} a\frac{-72P_3 L + 72P_3 a - 26 q_1 b^2 - 13 q_2 b^2}{-L + a}$$

The **beam with an overhang** (Fig. 4-2c) supports a vertical force P_4 and a couple of moment M_1. Since there are no horizontal forces acting on the beam, the horizontal reaction at the pin support is nonexistent and we do not need to show it on the free-body diagram. In arriving at this conclusion, we made use of the equation of equilibrium for forces in the horizontal direction. Consequently, only two independent equations of equilibrium remain—either two moment equations or one moment equation plus the equation for vertical equilibrium.

Let us arbitrarily decide to write two moment equations, the first for moments about point B and the second for moments about point A, as follows (counterclockwise moments are positive):

$$\sum M_B = 0 \qquad -R_A L + P_4(L - a) + M_1 = 0$$
$$\sum M_A = 0 \qquad -P_4 a + R_B L + M_1 = 0$$

Therefore, the reactions are

$$R_A = \frac{P_4(L - a)}{L} + \frac{M_1}{L} \qquad R_B = \frac{P_4 a}{L} - \frac{M_1}{L}$$

Again, summation of forces in the vertical direction provides a check on these results.

If the beam structure with an overhang in Fig. 4-2c is modified to add a roller support at C, it is now a one degree statically indeterminate two-span beam. However, if a shear release is inserted into the model as shown in Fig. 4-7, just to the left of support B, the beam can be analyzed using the

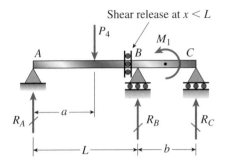

Shear release at $x < L$

FIG. 4-7 Modified beam with overhang—add shear release

laws of statics alone because the release provides one additional equilibrium equation. The beam must be cut at the release to expose the internal stress resultants N, V, and M; now $V = 0$ at the release so reaction R_A can be computed by summing forces in the left-hand free-body diagram. R_A is readily seen to be equal to P_4. Once R_A is known, reaction R_C can be computed by summing moments about joint B, and reaction R_B can be obtained by summing all vertical forces. Results are summarized below.

$$R_A = P_4$$

$$R_C = \frac{P_4 a - M_1}{b}$$

$$R_B = P_4 - R_A - R_C$$

$$R_B = \frac{M_1 - P_4 a}{b}$$

The preceding discussion illustrates how the reactions of statically determinate beams are calculated from equilibrium equations. We have intentionally used symbolic examples rather than numerical examples in order to show how the individual steps are carried out.

4.3 SHEAR FORCES AND BENDING MOMENTS

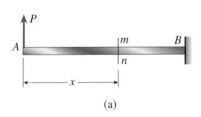

(a)

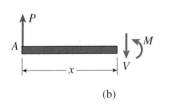

(b)

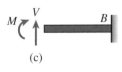

(c)

FIG. 4-8 Shear force V and bending moment M in a beam

When a beam is loaded by forces or couples, stresses and strains are created throughout the interior of the beam. To determine these stresses and strains, we first must find the internal forces and internal couples that act on cross sections of the beam.

As an illustration of how these internal quantities are found, consider a cantilever beam AB loaded by a force P at its free end (Fig. 4-8a). We cut through the beam at a cross section mn located at distance x from the free end and isolate the left-hand part of the beam as a free body (Fig. 4-8b). The free body is held in equilibrium by the force P and by the stresses that act over the cut cross section. These stresses represent the action of the right-hand part of the beam on the left-hand part. At this stage of our discussion we do not know the distribution of the stresses acting over the cross section; all we know is that the resultant of these stresses must be such as to maintain equilibrium of the free body.

From statics, we know that the resultant of the stresses acting on the cross section can be reduced to a **shear force** V and a **bending moment** M (Fig. 4-8b). Because the load P is transverse to the axis of the beam, no axial force exists at the cross section. Both the shear force and the bending moment act in the plane of the beam, that is, the vector for the shear force lies in the plane of the figure and the vector for the moment is perpendicular to the plane of the figure.

Shear forces and bending moments, like axial forces in bars and internal torques in shafts, are the resultants of stresses distributed over the cross section. Therefore, these quantities are known collectively as **stress resultants**.

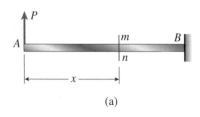

(a)

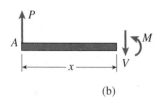

(b)

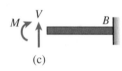

(c)

FIG. 4-8 (Repeated)

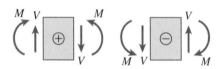

FIG. 4-9 Sign conventions for shear force V and bending moment M

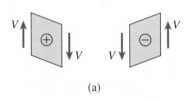

(a)

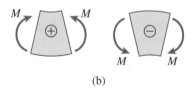

(b)

FIG. 4-10 Deformations (highly exaggerated) of a beam element caused by (a) shear forces, and (b) bending moments

The stress resultants in statically determinate beams can be calculated from equations of equilibrium. In the case of the cantilever beam of Fig. 4-8a, we use the free-body diagram of Fig. 4-8b. Summing forces in the vertical direction and also taking moments about the cut section, we get

$$\Sigma F_{\text{vert}} = 0 \qquad P - V = 0 \ \text{or} \ V = P$$

$$\Sigma M = 0 \qquad M - Px = 0 \ \text{or} \ M = Px$$

where x is the distance from the free end of the beam to the cross section where V and M are being determined. Thus, through the use of a free-body diagram and two equations of equilibrium, we can calculate the shear force and bending moment without difficulty.

Sign Conventions

Let us now consider the sign conventions for shear forces and bending moments. It is customary to assume that shear forces and bending moments are positive when they act in the directions shown in Fig. 4-8b. Note that the shear force tends to rotate the material clockwise and the bending moment tends to compress the upper part of the beam and elongate the lower part. Also, in this instance, the shear force acts downward and the bending moment acts counterclockwise.

The action of these *same* stress resultants against the right-hand part of the beam is shown in Fig. 4-8c. The directions of both quantities are now reversed—the shear force acts upward and the bending moment acts clockwise. However, the shear force still tends to rotate the material clockwise and the bending moment still tends to compress the upper part of the beam and elongate the lower part.

Therefore, we must recognize that the algebraic sign of a stress resultant is determined by how it deforms the material on which it acts, rather than by its direction in space. In the case of a beam, *a positive shear force acts clockwise against the material* (Figs. 4-8b and c) *and a negative shear force acts counterclockwise against the material. Also, a positive bending moment compresses the upper part of the beam* (Figs. 4-8b and c) *and a negative bending moment compresses the lower part.*

To make these conventions clear, both positive and negative shear forces and bending moments are shown in Fig. 4-9. The forces and moments are shown acting on an element of a beam cut out between two cross sections that are a small distance apart.

The *deformations* of an element caused by both positive and negative shear forces and bending moments are sketched in Fig. 4-10. We see that a positive shear force tends to deform the element by causing the right-hand face to move downward with respect to the left-hand face, and, as already mentioned, a positive bending moment compresses the upper part of a beam and elongates the lower part.

Sign conventions for stress resultants are called **deformation sign conventions** because they are based upon how the material is deformed. For instance, we previously used a deformation sign convention in dealing with axial forces in a bar. We stated that an axial force producing elongation

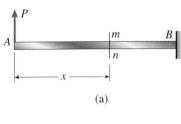

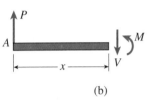

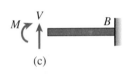

FIG. 4-8 (Repeated)

(or tension) in a bar is positive and an axial force producing shortening (or compression) is negative. Thus, the sign of an axial force depends upon how it deforms the material, not upon its direction in space.

By contrast, when writing equations of equilibrium we use **static sign conventions**, in which forces are positive or negative according to their directions along the coordinate axes. For instance, if we are summing forces in the y direction, forces acting in the positive direction of the y axis are taken as positive and forces acting in the negative direction are taken as negative.

As an example, consider Fig. 4-8b, which is a free-body diagram of part of the cantilever beam. Suppose that we are summing forces in the vertical direction and that the y axis is positive upward. Then the load P is given a positive sign in the equation of equilibrium because it acts upward. However, the shear force V (which is a *positive* shear force) is given a negative sign because it acts downward (that is, in the negative direction of the y axis). This example shows the distinction between the deformation sign convention used for the shear force and the static sign convention used in the equation of equilibrium.

The following examples illustrate the techniques for handling sign conventions and determining shear forces and bending moments in beams. The general procedure consists of constructing free-body diagrams and solving equations of equilibrium.

Example 4-1

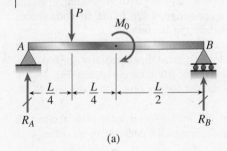

(a)

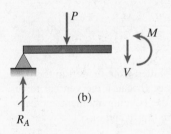

(b)

FIG. 4-11 Example 4-1. Shear forces and bending moment in a simple beam

A simple beam AB supports two loads, a force P and a couple M_0, acting as shown in Fig. 4-11a.

Find the shear force V and bending moment M in the beam at cross sections located as follows: (a) a small distance to the left of the midpoint of the beam, and (b) a small distance to the right of the midpoint of the beam.

Solution

Reactions. The first step in the analysis of this beam is to find the reactions R_A and R_B at the supports. Taking moments about ends B and A gives two equations of equilibrium, from which we find, respectively,

$$R_A = \frac{3P}{4} - \frac{M_0}{L} \qquad R_B = \frac{P}{4} + \frac{M_0}{L} \qquad \text{(a)}$$

(a) *Shear force and bending moment to the left of the midpoint.* We cut the beam at a cross section just to the left of the midpoint and draw a free-body diagram of either half of the beam. In this example, we choose the left-hand half of the beam as the free body (Fig. 4-11b). This free body is held in equilibrium by the load P, the reaction R_A, and the two unknown stress resultants—the shear force V and the bending moment M, both of which are shown in their positive directions (see Fig. 4-9). The couple M_0 does not act on the free body because the beam is cut to the left of its point of application.

continued

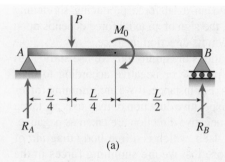

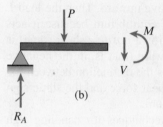

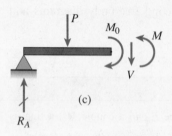

FIG. 4-11 Example 4-1. Shear forces and bending moment in a simple beam (parts (a) and (b) repeated)

Summing forces in the vertical direction (upward is positive) gives

$$\Sigma F_{\text{vert}} = 0 \qquad R_A - P - V = 0$$

from which we get the shear force:

$$V = R_A - P = -\frac{P}{4} - \frac{M_0}{L} \qquad \text{(b)}$$

This result shows that when P and M_0 act in the directions shown in Fig. 4-11a, the shear force (at the selected location) is negative and acts in the opposite direction to the positive direction assumed in Fig. 4-11b.

Taking moments about an axis through the cross section where the beam is cut (see Fig. 4-11b) gives

$$\Sigma M = 0 \qquad -R_A\left(\frac{L}{2}\right) + P\left(\frac{L}{4}\right) + M = 0$$

in which counterclockwise moments are taken as positive. Solving for the bending moment M, we get

$$M = R_A\left(\frac{L}{2}\right) - P\left(\frac{L}{4}\right) = \frac{PL}{8} - \frac{M_0}{2} \qquad \text{(c)}$$

The bending moment M may be either positive or negative, depending upon the magnitudes of the loads P and M_0. If it is positive, it acts in the direction shown in the figure; if it is negative, it acts in the opposite direction.

(b) *Shear force and bending moment to the right of the midpoint*. In this case we cut the beam at a cross section just to the right of the midpoint and again draw a free-body diagram of the part of the beam to the left of the cut section (Fig. 4-11c). The difference between this diagram and the former one is that the couple M_0 now acts on the free body.

From two equations of equilibrium, the first for forces in the vertical direction and the second for moments about an axis through the cut section, we obtain

$$V = -\frac{P}{4} - \frac{M_0}{L} \qquad M = \frac{PL}{8} + \frac{M_0}{2} \qquad \text{(d,e)}$$

These results show that when the cut section is shifted from the left to the right of the couple M_0, the shear force does not change (because the vertical forces acting on the free body do not change) but the bending moment increases algebraically by an amount equal to M_0 (compare Eqs. c and e).

Example 4-2

A cantilever beam that is free at end A and fixed at end B is subjected to a distributed load of linearly varying intensity q (Fig. 4-12a). The maximum intensity of the load occurs at the fixed support and is equal to q_0.

Find the shear force V and bending moment M at distance x from the free end of the beam.

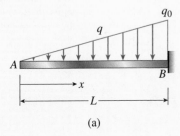

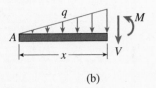

FIG. 4-12 Example 4-2. Shear force and bending moment in a cantilever beam

Solution

Shear force. We cut through the beam at distance x from the left-hand end and isolate part of the beam as a free body (Fig. 4-12b). Acting on the free body are the distributed load q, the shear force V, and the bending moment M. Both unknown quantities (V and M) are assumed to be positive.

The intensity of the distributed load at distance x from the end is

$$q = \frac{q_0 x}{L} \tag{4-1}$$

Therefore, the total downward load on the free body, equal to the area of the triangular loading diagram (Fig. 4-12b), is

$$\frac{1}{2}\left(\frac{q_0 x}{L}\right)(x) = \frac{q_0 x^2}{2L}$$

From an equation of equilibrium in the vertical direction we find

$$V = -\frac{q_0 x^2}{2L} \tag{4-2a}$$

continued

At the free end A $(x = 0)$ the shear force is zero, and at the fixed end B $(x = L)$ the shear force has its maximum value:

$$V_{\max} = -\frac{q_0 L}{2} \tag{4-2b}$$

which is numerically equal to the total downward load on the beam. The minus signs in Eqs. (4-2a) and (4-2b) show that the shear forces act in the opposite direction to that pictured in Fig. 4-12b.

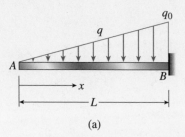

(a)

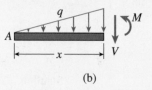

FIG. 4-12 (Repeated)

(b)

Bending moment. To find the bending moment M in the beam (Fig. 4-12b), we write an equation of moment equilibrium about an axis through the cut section. Recalling that the moment of a triangular load is equal to the area of the loading diagram times the distance from its centroid to the axis of moments, we obtain the following equation of equilibrium (counterclockwise moments are positive):

$$\Sigma M = 0 \qquad M + \frac{1}{2}\left(\frac{q_0 x}{L}\right)(x)\left(\frac{x}{3}\right) = 0$$

from which we get

$$M = -\frac{q_0 x^3}{6L} \tag{4-3a}$$

At the free end of the beam $(x = 0)$, the bending moment is zero, and at the fixed end $(x = L)$ the moment has its numerically largest value:

$$M_{\max} = -\frac{q_0 L^2}{6} \tag{4-3b}$$

The minus signs in Eqs. (4-3a) and (4-3b) show that the bending moments act in the opposite direction to that shown in Fig. 4-12b.

Example 4-3

A simple beam with an overhang is supported at points A and B (Fig. 4-13a). A uniform load of intensity $q = 200$ lb/ft acts throughout the length of the beam and a concentrated load $P = 14$ k acts at a point 9 ft from the left-hand support. The span length is 24 ft and the length of the overhang is 6 ft.

Calculate the shear force V and bending moment M at cross section D located 15 ft from the left-hand support.

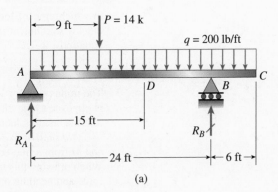

(a)

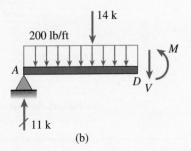

(b)

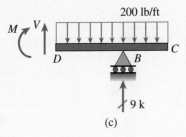

(c)

FIG. 4-13 Example 4-3. Shear force and bending moment in a beam with an overhang

Solution

Reactions. We begin by calculating the reactions R_A and R_B from equations of equilibrium for the entire beam considered as a free body. Thus, taking moments about the supports at B and A, respectively, we find

$$R_A = 11 \text{ k} \quad R_B = 9 \text{ k}$$

continued

Shear force and bending moment at section D. Now we make a cut at section D and construct a free-body diagram of the left-hand part of the beam (Fig. 4-13b). When drawing this diagram, we assume that the unknown stress resultants V and M are positive.

The equations of equilibrium for the free body are as follows:

$$\Sigma F_{vert} = 0 \quad 11 \text{ k} - 14 \text{ k} - (0.200 \text{ k/ft})(15 \text{ ft}) - V = 0$$

$$\Sigma M_D = 0 \quad -(11 \text{ k})(15 \text{ ft}) + (14 \text{ k})(6 \text{ ft}) + (0.200 \text{ k/ft})(15 \text{ ft})(7.5 \text{ ft}) + M = 0$$

in which upward forces are taken as positive in the first equation and counter-clockwise moments are taken as positive in the second equation. Solving these equations, we get

$$V = -6 \text{ k} \qquad M = 58.5 \text{ k-ft}$$

The minus sign for V means that the shear force is negative, that is, its direction is opposite to the direction shown in Fig. 4-13b. The positive sign for M means that the bending moment acts in the direction shown in the figure.

Alternative free-body diagram. Another method of solution is to obtain V and M from a free-body diagram of the right-hand part of the beam (Fig. 4-13c). When drawing this free-body diagram, we again assume that the unknown shear force and bending moment are positive. The two equations of equilibrium are

$$\Sigma F_{vert} = 0 \quad V + 9 \text{ k} - (0.200 \text{ k/ft})(15 \text{ ft}) = 0$$

$$\Sigma M_D = 0 \qquad -M + (9 \text{ k})(9 \text{ ft}) - (0.200 \text{ k/ft})(15 \text{ ft})(7.5 \text{ ft}) = 0$$

from which

$$V = -6 \text{ k} \quad M = 58.5 \text{ k-ft}$$

as before. As often happens, the choice between free-body diagrams is a matter of convenience and personal preference.

4.4 RELATIONSHIPS BETWEEN LOADS, SHEAR FORCES, AND BENDING MOMENTS

We will now obtain some important relationships between loads, shear forces, and bending moments in beams. These relationships are quite useful when investigating the shear forces and bending moments throughout the entire length of a beam, and they are especially helpful when constructing shear-force and bending-moment diagrams (Section 4.5).

As a means of obtaining the relationships, let us consider an element of a beam cut out between two cross sections that are distance dx apart (Fig. 4-14). The load acting on the top surface of the element may be a distributed load, a concentrated load, or a couple, as shown in Figs. 4-14a, b, and c, respectively. The **sign conventions** for these loads are as follows: *Distributed loads and concentrated loads are positive when they act downward on the beam and negative when they act upward. A couple acting as a load on a beam is positive when it is counterclockwise and negative when it is clockwise.* If other sign conventions

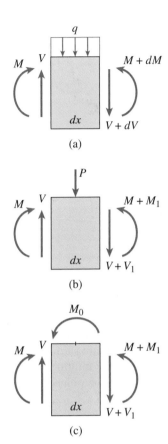

FIG. 4-14 Element of a beam used in deriving the relationships between loads, shear forces, and bending moments. (All loads and stress resultants are shown in their positive directions.)

are used, changes may occur in the signs of the terms appearing in the equations derived in this section.

The shear forces and bending moments acting on the sides of the element are shown in their positive directions in Fig. 4-10. In general, the shear forces and bending moments vary along the axis of the beam. Therefore, their values on the right-hand face of the element may be different from their values on the left-hand face.

In the case of a distributed load (Fig. 4-14a) the increments in V and M are infinitesimal, and so we denote them by dV and dM, respectively. The corresponding stress resultants on the right-hand face are $V + dV$ and $M + dM$.

In the case of a concentrated load (Fig. 4-14b) or a couple (Fig. 4-14c) the increments may be finite, and so they are denoted V_1 and M_1. The corresponding stress resultants on the right-hand face are $V + V_1$ and $M + M_1$.

For each type of loading we can write two equations of equilibrium for the element—one equation for equilibrium of forces in the vertical direction and one for equilibrium of moments. The first of these equations gives the relationship between the load and the shear force, and the second gives the relationship between the shear force and the bending moment.

Distributed Loads (Fig. 4-14a)

The first type of loading is a distributed load of intensity q, as shown in Fig. 4-14a. We will consider first its relationship to the shear force and second its relationship to the bending moment.

Shear Force. Equilibrium of forces in the vertical direction (upward forces are positive) gives

$$\sum F_{\text{vert}} = 0 \quad V - q\,dx - (V + dV) = 0$$

or

$$\frac{dV}{dx} = -q \tag{4-4}$$

From this equation we see that the rate of change of the shear force at any point on the axis of the beam is equal to the negative of the intensity of the distributed load at that same point. (*Note:* If the sign convention for the distributed load is reversed, so that q is positive upward instead of downward, then the minus sign is omitted in the preceding equation.)

Some useful relations are immediately obvious from Eq. (4-4). For instance, if there is no distributed load on a segment of the beam (that is, if $q = 0$), then $dV/dx = 0$ and the shear force is constant in that part of the beam. Also, if the distributed load is uniform along part of the beam ($q = $ constant), then dV/dx is also constant and the shear force varies linearly in that part of the beam.

As a demonstration of Eq. (4-4), consider the cantilever beam with a linearly varying load that we discussed in Example 4-2 of the preceding section (see Fig. 4-12). The load on the beam (from Eq. 4-1) is

$$q = \frac{q_0 x}{L}$$

which is positive because it acts downward. Also, the shear force (Eq. 4-2a) is

$$V = -\frac{q_0 x^2}{2L}$$

Taking the derivative dV/dx gives

$$\frac{dV}{dx} = \frac{d}{dx}\left(-\frac{q_0 x^2}{2L}\right) = -\frac{q_0 x}{L} = -q$$

which agrees with Eq. (4-4).

A useful relationship pertaining to the shear forces at two different cross sections of a beam can be obtained by integrating Eq. (4-4) along the axis of the beam. To obtain this relationship, we multiply both sides of Eq. (4-4) by dx and then integrate between any two points A and B on the axis of the beam; thus,

$$\int_A^B dV = -\int_A^B q\, dx \tag{a}$$

where we are assuming that x increases as we move from point A to point B. The left-hand side of this equation equals the difference $(V_B - V_A)$ of the shear forces at B and A. The integral on the right-hand side represents the area of the loading diagram between A and B, which in turn is equal to the magnitude of the resultant of the distributed load acting between points A and B. Thus, from Eq. (a) we get

$$V_B - V_A = -\int_A^B q\, dx$$

$$= -(\text{area of the loading diagram between } A \text{ and } B) \tag{4-5}$$

In other words, the change in shear force between two points along the axis of the beam is equal to the negative of the total downward load between those points. The area of the loading diagram may be positive (if q acts downward) or negative (if q acts upward).

Because Eq. (4-4) was derived for an element of the beam subjected *only* to a distributed load (or to no load), we cannot use Eq. (4-4) at a point where a concentrated load is applied (because the *intensity* of

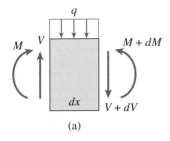

(a)

FIG. 4-14a (Repeated)

load is not defined for a concentrated load). For the same reason, we cannot use Eq. (4-5) if a concentrated load P acts on the beam between points A and B.

Bending Moment. Let us now consider the moment equilibrium of the beam element shown in Fig. 4-14a. Summing moments about an axis at the left-hand side of the element (the axis is perpendicular to the plane of the figure), and taking counterclockwise moments as positive, we obtain

$$\sum M = 0 \qquad -M - q\,dx\left(\frac{dx}{2}\right) - (V + dV)dx + M + dM = 0$$

Discarding products of differentials (because they are negligible compared to the other terms), we obtain the following relationship:

$$\frac{dM}{dx} = V \tag{4-6}$$

This equation shows that the rate of change of the bending moment at any point on the axis of a beam is equal to the shear force at that same point. For instance, if the shear force is zero in a region of the beam, then the bending moment is constant in that same region.

Equation (4-6) applies only in regions where distributed loads (or no loads) act on the beam. At a point where a concentrated load acts, a sudden change (or discontinuity) in the shear force occurs and the derivative dM/dx is undefined at that point.

Again using the cantilever beam of Fig. 4-12 as an example, we recall that the bending moment (Eq. 4-3a) is

$$M = -\frac{q_0 x^3}{6L}$$

Therefore, the derivative dM/dx is

$$\frac{dM}{dx} = \frac{d}{dx}\left(-\frac{q_0 x^3}{6L}\right) = -\frac{q_0 x^2}{2L}$$

which is equal to the shear force in the beam (see Eq. 4-2a).

Integrating Eq. (4-6) between two points A and B on the beam axis gives

$$\int_A^B dM = \int_A^B V\,dx \tag{b}$$

The integral on the left-hand side of this equation is equal to the difference $(M_B - M_A)$ of the bending moments at points B and A. To interpret the integral on the right-hand side, we need to consider V as a function of x and visualize a shear-force diagram showing the variation of V with x. Then we see that the integral on the right-hand side represents the area below the shear-force diagram between A and B. Therefore, we can express Eq. (b) in the following manner:

$$M_B - M_A = \int_A^B V \, dx$$

$$= \text{(area of the shear-force diagram between } A \text{ and } B) \quad (4\text{-}7)$$

This equation is valid even when concentrated loads act on the beam between points A and B. However, it is not valid if a couple acts between A and B. A couple produces a sudden change in the bending moment, and the left-hand side of Eq. (b) cannot be integrated across such a discontinuity.

Concentrated Loads (Fig. 4-14b)

Now let us consider a concentrated load P acting on the beam element (Fig. 4-14b). From equilibrium of forces in the vertical direction, we get

$$V - P - (V + V_1) = 0 \quad \text{or} \quad V_1 = -P \quad (4\text{-}8)$$

This result means that an abrupt change in the shear force occurs at any point where a concentrated load acts. As we pass from left to right through the point of load application, the shear force decreases by an amount equal to the magnitude of the downward load P.

From equilibrium of moments about the left-hand face of the element (Fig. 4-14b), we get

$$-M - P\left(\frac{dx}{2}\right) - (V + V_1)dx + M + M_1 = 0$$

or

$$M_1 = P\left(\frac{dx}{2}\right) + V \, dx + V_1 \, dx \quad (c)$$

Since the length dx of the element is infinitesimally small, we see from this equation that the increment M_1 in the bending moment is also infinitesimally small. *Thus, the bending moment does not change as we pass through the point of application of a concentrated load.*

Even though the bending moment M does not change at a concentrated load, its rate of change dM/dx undergoes an abrupt change. At the left-hand side of the element (Fig. 4-14b), the rate of change of the bending moment (see Eq. 4-6) is $dM/dx = V$. At the right-hand side, the rate of change is $dM/dx = V + V_1 = V - P$. *Therefore, at the point of application of a concentrated load P, the rate of change dM/dx of the bending moment decreases abruptly by an amount equal to P.*

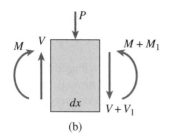

(b)

FIG. 4-14b (Repeated)

Loads in the Form of Couples (Fig. 4-14c)

The last case to be considered is a load in the form of a couple M_0 (Fig. 4-14c). From equilibrium of the element in the vertical direction we obtain $V_1 = 0$, which shows that *the shear force does not change at the point of application of a couple.*

Equilibrium of moments about the left-hand side of the element gives

$$-M + M_0 - (V + V_1)dx + M + M_1 = 0$$

Disregarding terms that contain differentials (because they are negligible compared to the finite terms), we obtain

$$M_1 = -M_0 \qquad (4\text{-}9)$$

This equation shows that the bending moment decreases by M_0 as we move from left to right through the point of load application. *Thus, the bending moment changes abruptly at the point of application of a couple.*

Equations (4-4) through (4-9) are useful when making a complete investigation of the shear forces and bending moments in a beam, as discussed in the next section.

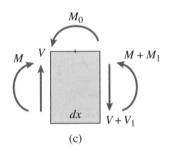

(c)

FIG. 4-14c (Repeated)

4.5 SHEAR-FORCE AND BENDING-MOMENT DIAGRAMS

When designing a beam, we usually need to know how the shear forces and bending moments vary throughout the length of the beam. Of special importance are the maximum and minimum values of these quantities. Information of this kind is usually provided by graphs in which the shear force and bending moment are plotted as ordinates and the distance x along the axis of the beam is plotted as the abscissa. Such graphs are called **shear-force and bending-moment diagrams**.

To provide a clear understanding of these diagrams, we will explain in detail how they are constructed and interpreted for three basic loading conditions—a single concentrated load, a uniform load, and several concentrated loads. In addition, Examples 4-4 to 4-7 at the end of the section provide detailed illustration of the techniques for handling various kinds of loads, including the case of a couple acting as a load on a beam.

Concentrated Load

Let us begin with a simple beam AB supporting a concentrated load P (Fig. 4-15a). The load P acts at distance a from the left-hand support and distance b from the right-hand support. Considering the entire beam as a free body, we can readily determine the reactions of the beam from equilibrium; the results are

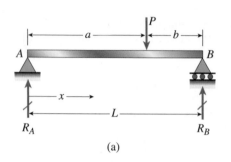

(a)

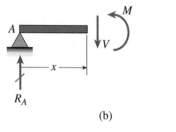

(b)

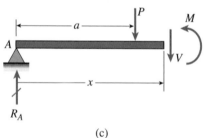

(c)

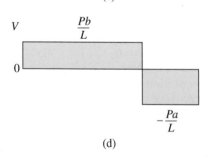

(d)

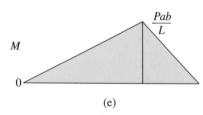

(e)

FIG. 4-15 Shear-force and bending-moment diagrams for a simple beam with a concentrated load

$$R_A = \frac{Pb}{L} \qquad R_B = \frac{Pa}{L} \qquad \text{(4-10a,b)}$$

We now cut through the beam at a cross section to the left of the load P and at distance x from the support at A. Then we draw a free-body diagram of the left-hand part of the beam (Fig. 4-15b). From the equations of equilibrium for this free body, we obtain the shear force V and bending moment M at distance x from the support:

$$V = R_A = \frac{Pb}{L} \qquad M = R_A x = \frac{Pbx}{L} \qquad (0 < x < a) \qquad \text{(4-11a,b)}$$

These expressions are valid only for the part of the beam to the left of the load P.

Next, we cut through the beam to the right of the load P (that is, in the region $a < x < L$) and again draw a free-body diagram of the left-hand part of the beam (Fig. 4-15c). From the equations of equilibrium for this free body, we obtain the following expressions for the shear force and bending moment:

$$V = R_A - P = \frac{Pb}{L} - P = -\frac{Pa}{L} \qquad (a < x < L) \qquad \text{(4-12a)}$$

$$M = R_A x - P(x - a) = \frac{Pbx}{L} - P(x - a)$$

$$= \frac{Pa}{L}(L - x) \qquad (a < x < L) \qquad \text{(4-12b)}$$

Note that these equations are valid only for the right-hand part of the beam.

The equations for the shear forces and bending moments (Eqs. 4-11 and 4-12) are plotted below the sketches of the beam. Figure 4-15d is the *shear-force diagram* and Fig. 4-15e is the *bending-moment diagram*.

From the first diagram we see that the shear force at end A of the beam ($x = 0$) is equal to the reaction R_A. Then it remains constant to the point of application of the load P. At that point, the shear force decreases abruptly by an amount equal to the load P. In the right-hand part of the beam, the shear force is again constant but equal numerically to the reaction at B.

As shown in the second diagram, the bending moment in the left-hand part of the beam increases linearly from zero at the support to Pab/L at the concentrated load ($x = a$). In the right-hand part, the bending moment is again a linear function of x, varying from Pab/L at $x = a$ to zero at the support ($x = L$). Thus, the maximum bending moment is

$$M_{max} = \frac{Pab}{L} \qquad \text{(4-13)}$$

and occurs under the concentrated load.

When deriving the expressions for the shear force and bending moment to the right of the load P (Eqs. 4-12a and b), we considered the equilibrium of the left-hand part of the beam (Fig. 4-15c). This free body is acted upon by the forces R_A and P in addition to V and M. It is slightly simpler in this particular example to consider the right-hand portion of the beam as a free body, because then only one force (R_B) appears in the equilibrium equations (in addition to V and M). Of course, the final results are unchanged.

Certain characteristics of the shear-force and bending moment diagrams (Figs. 4-15d and e) may now be seen. We note first that the slope dV/dx of the shear-force diagram is zero in the regions $0 < x < a$ and $a < x < L$, which is in accord with the equation $dV/dx = -q$ (Eq. 4-4). Also, in these same regions the slope dM/dx of the bending moment diagram is equal to V (Eq. 4-6). To the left of the load P, the slope of the moment diagram is positive and equal to Pb/L; to the right, it is negative and equal to $-Pa/L$. Thus, at the point of application of the load P there is an abrupt change in the shear-force diagram (equal to the magnitude of the load P) and a corresponding change in the slope of the bending-moment diagram.

Now consider the *area* of the shear-force diagram. As we move from $x = 0$ to $x = a$, the area of the shear-force diagram is $(Pb/L)a$, or Pab/L. This quantity represents the increase in bending moment between these same two points (see Eq. 4-7). From $x = a$ to $x = L$, the area of the shear-force diagram is $-Pab/L$, which means that in this region the bending moment decreases by that amount. Consequently, the bending moment is zero at end B of the beam, as expected.

If the bending moments at both ends of a beam are zero, as is usually the case with a simple beam, then the area of the shear-force diagram between the ends of the beam must be zero provided no couples act on the beam (see the discussion in Section 4.4 following Eq. 4-7).

As mentioned previously, the maximum and minimum values of the shear forces and bending moments are needed when designing beams. For a simple beam with a single concentrated load, the maximum shear force occurs at the end of the beam nearest to the concentrated load and the maximum bending moment occurs under the load itself.

Uniform Load

A simple beam with a uniformly distributed load of constant intensity q is shown in Fig. 4-16a on the next page. Because the beam and its loading are symmetric, we see immediately that each of the reactions (R_A and R_B) is equal to $qL/2$. Therefore, the shear force and bending moment at distance x from the left-hand end are

$$V = R_A - qx = \frac{qL}{2} - qx \tag{4-14a}$$

$$M = R_A x - qx\left(\frac{x}{2}\right) = \frac{qLx}{2} - \frac{qx^2}{2} \tag{4-14b}$$

These equations, which are valid throughout the length of the beam, are plotted as shear-force and bending moment diagrams in Figs. 4-16b and c, respectively.

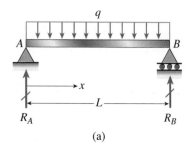

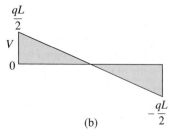

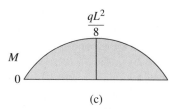

FIG. 4-16 Shear-force and bending-moment diagrams for a simple beam with a uniform load

The shear-force diagram consists of an inclined straight line having ordinates at $x = 0$ and $x = L$ equal numerically to the reactions. The slope of the line is $-q$, as expected from Eq. (4-4). The bending-moment diagram is a parabolic curve that is symmetric about the midpoint of the beam. At each cross section the slope of the bending-moment diagram is equal to the shear force (see Eq. 4-6):

$$\frac{dM}{dx} = \frac{d}{dx}\left(\frac{qLx}{2} - \frac{qx^2}{2}\right) = \frac{qL}{2} - qx = V$$

The maximum value of the bending moment occurs at the midpoint of the beam where both dM/dx and the shear force V are equal to zero. Therefore, we substitute $x = L/2$ into the expression for M and obtain

$$M_{max} = \frac{qL^2}{8} \tag{4-15}$$

as shown on the bending-moment diagram.

The diagram of load intensity (Fig. 4-16a) has area qL, and according to Eq. (4-5) the shear force V must decrease by this amount as we move along the beam from A to B. We can see that this is indeed the case, because the shear force decreases from $qL/2$ to $-qL/2$.

The area of the shear-force diagram between $x = 0$ and $x = L/2$ is $qL^2/8$, and we see that this area represents the increase in the bending moment between those same two points (Eq. 4-7). In a similar manner, the bending moment decreases by $qL^2/8$ in the region from $x = L/2$ to $x = L$.

Several Concentrated Loads

If several concentrated loads act on a simple beam (Fig. 4-17a), expressions for the shear forces and bending moments may be determined for each segment of the beam between the points of load application. Again using free-body diagrams of the left-hand part of the beam and measuring the distance x from end A, we obtain the following equations for the first segment of the beam:

$$V = R_A \qquad M = R_A x \qquad (0 < x < a_1) \tag{4-16a,b}$$

For the second segment, we get

$$V = R_A - P_1 \qquad M = R_A x - P_1(x - a_1) \qquad (a_1 < x < a_2) \tag{4-17a,b}$$

For the third segment of the beam, it is advantageous to consider the right-hand part of the beam rather than the left, because fewer loads act on the corresponding free body. Hence, we obtain

$$V = -R_B + P_3 \tag{4-18a}$$

$$M = R_B(L - x) - P_3(L - b_3 - x) \qquad (a_2 < x < a_3) \tag{4-18b}$$

Finally, for the fourth segment of the beam, we obtain

$$V = -R_B \qquad M = R_B(L - x) \qquad (a_3 < x < L) \tag{4-19a,b}$$

Equations (4-16) through (4-19) can be used to construct the shear-force and bending-moment diagrams (Figs. 4-17b and c).

From the shear-force diagram we note that the shear force is constant in each segment of the beam and changes abruptly at every load point, with the amount of each change being equal to the load. Also, the bending moment in each segment is a linear function of x, and therefore the corresponding part of the bending-moment diagram is an inclined straight line. To assist in drawing these lines, we obtain the bending moments under the concentrated loads by substituting $x = a_1$, $x = a_2$, and $x = a_3$ into Eqs. (4-16b), (4-17b), and (4-18b), respectively. In this manner we obtain the following bending moments:

$$M_1 = R_A a_1 \quad M_2 = R_A a_2 - P_1(a_2 - a_1) \quad M_3 = R_B b_3 \quad \text{(4-20a,b,c)}$$

Knowing these values, we can readily construct the bending-moment diagram by connecting the points with straight lines.

At each discontinuity in the shear force, there is a corresponding change in the slope dM/dx of the bending-moment diagram. Also, the change in bending moment between two load points equals the area of the shear-force diagram between those same two points (see Eq. 4-7). For example, the change in bending moment between loads P_1 and P_2 is $M_2 - M_1$. Substituting from Eqs. (4-20a and b), we get

$$M_2 - M_1 = (R_A - P_1)(a_2 - a_1)$$

which is the area of the rectangular shear-force diagram between $x = a_1$ and $x = a_2$.

The maximum bending moment in a beam having only concentrated loads *must* occur under one of the loads or at a reaction. To show this, recall that the slope of the bending-moment diagram is equal to the shear force. Therefore, whenever the bending moment has a maximum or minimum value, the derivative dM/dx (and hence the shear force) must change sign. However, in a beam with only concentrated loads, the shear force can change sign only under a load.

FIG. 4-17 Shear-force and bending-moment diagrams for a simple beam with several concentrated loads

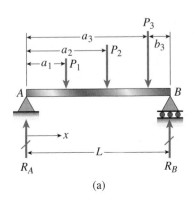

(a)

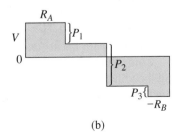

(b)

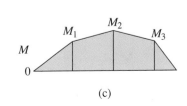

(c)

If, as we proceed along the x axis, the shear force changes from positive to negative (as in Fig. 4-17b), then the slope in the bending moment diagram also changes from positive to negative. Therefore, we must have a maximum bending moment at this cross section. Conversely, a change in shear force from a negative to a positive value indicates a minimum bending moment. Theoretically, the shear-force diagram can intersect the horizontal axis at several points, although this is quite unlikely. Corresponding to each such intersection point, there is a local maximum or minimum in the bending-moment diagram. The values of all local maximums and minimums must be determined in order to find the maximum positive and negative bending moments in a beam.

General Comments

In our discussions we frequently use the terms "maximum" and "minimum" with their common meanings of "largest" and "smallest." Consequently, we refer to "the maximum bending moment in a beam" regardless of whether the bending-moment diagram is described by a smooth, continuous function (as in Fig. 4-16c) or by a series of lines (as in Fig. 4-17c).

Furthermore, we often need to distinguish between positive and negative quantities. Therefore, we use expressions such as "maximum positive moment" and "maximum negative moment." In both of these cases, the expression refers to the numerically largest quantity; that is, the term "maximum negative moment" really means "numerically largest negative moment." Analogous comments apply to other beam quantities, such as shear forces and deflections.

The maximum positive and negative bending moments in a beam may occur at the following places: (1) a cross section where a concentrated load is applied and the shear force changes sign (see Figs. 4-15 and 4-17), (2) a cross section where the shear force equals zero (see Fig. 4-16), (3) a point of support where a vertical reaction is present, and (4) a cross section where a couple is applied. The preceding discussions and the following examples illustrate all of these possibilities.

When several loads act on a beam, the shear-force and bending-moment diagrams can be obtained by superposition (or summation) of the diagrams obtained for each of the loads acting separately. For instance, the shear-force diagram of Fig. 4-17b is actually the sum of three separate diagrams, each of the type shown in Fig. 4-15d for a single concentrated load. We can make an analogous comment for the bending-moment diagram of Fig. 4-17c. Superposition of shear-force and bending-moment diagrams is permissible because shear forces and bending moments in statically determinate beams are linear functions of the applied loads.

Computer programs are readily available for drawing shear-force and bending-moment diagrams. After you have developed an understanding of the nature of the diagrams by constructing them manually, you should feel secure in using computer programs to plot the diagrams and obtain numerical results.

Example 4-4

Draw the shear-force and bending-moment diagrams for a simple beam with a uniform load of intensity q acting over part of the span (Fig. 4-18a).

Solution

Reactions. We begin the analysis by determining the reactions of the beam from a free-body diagram of the entire beam (Fig. 4-18a). The results are

$$R_A = \frac{qb(b + 2c)}{2L} \qquad R_B = \frac{qb(b + 2a)}{2L} \qquad (4\text{-}21a,b)$$

Shear forces and bending moments. To obtain the shear forces and bending moments for the entire beam, we must consider the three segments of the beam individually. For each segment we cut through the beam to expose the shear force V and bending moment M. Then we draw a free-body diagram containing V and M as unknown quantities. Lastly, we sum forces in the vertical direction to obtain the shear force and take moments about the cut section to obtain the bending moment. The results for all three segments are as follows:

$$V = R_A \qquad M = R_A x \qquad (0 < x < a) \qquad (4\text{-}22a,b)$$

$$V = R_A - q(x - a) \qquad M = R_A x - \frac{q(x - a)^2}{2} \qquad (a < x < a + b) \qquad (4\text{-}23a,b)$$

$$V = -R_B \qquad M = R_B(L - x) \qquad (a + b < x < L) \qquad (4\text{-}24a,b)$$

These equations give the shear force and bending moment at every cross section of the beam. As a partial check on these results, we can apply Eq. (4-4) to the shear forces and Eq. (4-6) to the bending moments and verify that the equations are satisfied.

We now construct the shear-force and bending-moment diagrams (Figs. 4-18b and c) from Eqs. (4-22) through (4-24). The shear-force diagram consists of horizontal straight lines in the unloaded regions of the beam and an

FIG. 4-18 Example 4-4. Simple beam with a uniform load over part of the span

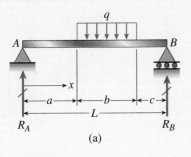
(a)

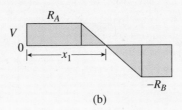
(b)

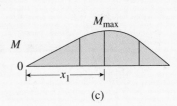
(c)

continued

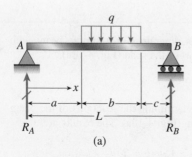

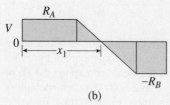

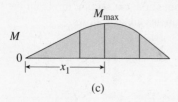

FIG. 4-18 Example 4-4. Simple beam with a uniform load over part of the span (Repeated)

inclined straight line with negative slope in the loaded region, as expected from the equation $dV/dx = -q$.

The bending-moment diagram consists of two inclined straight lines in the unloaded portions of the beam and a parabolic curve in the loaded portion. The inclined lines have slopes equal to R_A and $-R_B$, respectively, as expected from the equation $dM/dx = V$. Also, each of these inclined lines is tangent to the parabolic curve at the point where it meets the curve. This conclusion follows from the fact that there are no abrupt changes in the magnitude of the shear force at these points. Hence, from the equation $dM/dx = V$, we see that the slope of the bending-moment diagram does not change abruptly at these points.

Maximum bending moment. The maximum moment occurs where the shear force equals zero. This point can be found by setting the shear force V (from Eq. 4-23a) equal to zero and solving for the value of x, which we will denote by x_1. The result is

$$x_1 = a + \frac{b}{2L}(b + 2c) \tag{4-25}$$

Now we substitute x_1 into the expression for the bending moment (Eq. 4-23b) and solve for the maximum moment. The result is

$$M_{max} = \frac{qb}{8L^2}(b + 2c)(4aL + 2bc + b^2) \tag{4-26}$$

The maximum bending moment always occurs within the region of the uniform load, as shown by Eq. (4-25).

Special cases. If the uniform load is symmetrically placed on the beam ($a = c$), then we obtain the following simplified results from Eqs. (4-25) and (4-26):

$$x_1 = \frac{L}{2} \qquad M_{max} = \frac{qb(2L - b)}{8} \tag{4-27a,b}$$

If the uniform load extends over the entire span, then $b = L$ and $M_{max} = qL^2/8$, which agrees with Fig. 4-16 and Eq. (4-15).

Example 4-5

Draw the shear-force and bending-moment diagrams for a cantilever beam with two concentrated loads (Fig. 4-19a).

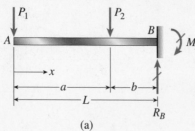

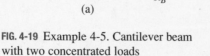

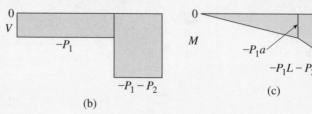

FIG. 4-19 Example 4-5. Cantilever beam with two concentrated loads

Solution

Reactions. From the free-body diagram of the entire beam we find the vertical reaction R_B (positive when upward) and the moment reaction M_B (positive when clockwise):

$$R_B = P_1 + P_2 \qquad M_B = P_1L + P_2b \qquad \text{(4-28a,b)}$$

Shear forces and bending moments. We obtain the shear forces and bending moments by cutting through the beam in each of the two segments, drawing the corresponding free-body diagrams, and solving the equations of equilibrium. Again measuring the distance x from the left-hand end of the beam, we get

$$V = -P_1 \qquad M = -P_1x \qquad (0 < x < a) \qquad \text{(4-29a,b)}$$

$$V = -P_1 - P_2 \qquad M = -P_1x - P_2(x - a) \qquad (a < x < L) \qquad \text{(4-30a,b)}$$

The corresponding shear-force and bending-moment diagrams are shown in Figs. 4-19b and c. The shear force is constant between the loads and reaches its maximum numerical value at the support, where it is equal numerically to the vertical reaction R_B (Eq. 4-28a).

The bending-moment diagram consists of two inclined straight lines, each having a slope equal to the shear force in the corresponding segment of the beam. The maximum bending moment occurs at the support and is equal numerically to the moment reaction M_B (Eq. 4-28b). It is also equal to the area of the entire shear-force diagram, as expected from Eq. (4-7).

Example 4-6

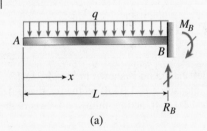

(a)

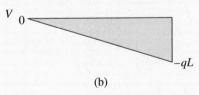

(b)

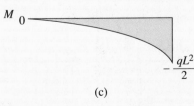

(c)

FIG. 4-20 Example 4-6. Cantilever beam with a uniform load

A cantilever beam supporting a uniform load of constant intensity q is shown in Fig. 4-20a. Draw the shear-force and bending-moment diagrams for this beam.

Solution

Reactions. The reactions R_B and M_B at the fixed support are obtained from equations of equilibrium for the entire beam; thus,

$$R_B = qL \qquad M_B = \frac{qL^2}{2} \qquad \text{(4-31a,b)}$$

Shear forces and bending moments. These quantities are found by cutting through the beam at distance x from the free end, drawing a free-body diagram of the left-hand part of the beam, and solving the equations of equilibrium. By this means we obtain

$$V = -qx \qquad M = -\frac{qx^2}{2} \qquad \text{(4-32a,b)}$$

The shear-force and bending-moment diagrams are obtained by plotting these equations (see Figs. 4-20b and c). Note that the slope of the shear-force diagram is equal to $-q$ (see Eq. 4-4) and the slope of the bending-moment diagram is equal to V (see Eq. 4-6).

continued

The maximum values of the shear force and bending moment occur at the fixed support where $x = L$:

$$V_{max} = -ql \qquad M_{max} = -\frac{qL^2}{2} \qquad \text{(4-33a,b)}$$

These values are consistent with the values of the reactions R_B and M_B (Eqs. 4-31a and b).

Alternative solution. Instead of using free-body diagrams and equations of equilibrium, we can determine the shear forces and bending moments by integrating the differential relationships between load, shear force, and bending moment. The shear force V at distance x from the free end A is obtained from the load by integrating Eq. (4-5), as follows:

$$V - V_A = V - 0 = V = -\int_0^x q \, dx = -qx \qquad \text{(a)}$$

which agrees with the previous result (Eq. 4-32a).

The bending moment M at distance x from the end is obtained from the shear force by integrating Eq. (4-7):

$$M - M_A = M - 0 = M = \int_0^x V \, dx = \int_0^x -qx \, dx = -\frac{qx^2}{2} \qquad \text{(b)}$$

which agrees with Eq. 4-32b.

Integrating the differential relationships is quite simple in this example because the loading pattern is continuous and there are no concentrated loads or couples in the regions of integration. If concentrated loads or couples were present, discontinuities in the V and M diagrams would exist, and we cannot integrate Eq. (4-5) through a concentrated load nor can we integrate Eq. (4-7) through a couple (see Section 4.4).

Example 4-7

A beam ABC with an overhang at the left-hand end is shown in Fig. 4-21a. The beam is subjected to a uniform load of intensity $q = 1.0$ k/ft on the overhang AB and a counterclockwise couple $M_0 = 12.0$ k-ft acting midway between the supports at B and C.

Draw the shear-force and bending-moment diagrams for this beam.

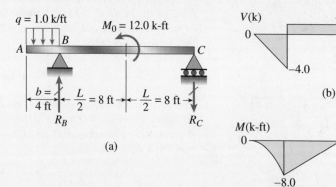

FIG. 4-21 Example 4-7. Beam with an overhang

Solution

Reactions. We can readily calculate the reactions R_B and R_C from a free-body diagram of the entire beam (Fig. 4-21a). In so doing, we find that R_B is upward and R_C is downward, as shown in the figure. Their numerical values are

$$R_B = 5.25 \text{ k} \qquad R_C = 1.25 \text{ k}$$

Shear forces. The shear force equals zero at the free end of the beam and equals $-qb$ (or -4.0 k) just to the left of support B. Since the load is uniformly distributed (that is, q is constant), the slope of the shear diagram is constant and equal to $-q$ (from Eq. 4-4). Therefore, the shear diagram is an inclined straight line with negative slope in the region from A to B (Fig. 4-21b).

Because there are no concentrated or distributed loads between the supports, the shear-force diagram is horizontal in this region. The shear force is equal to the reaction R_C, or 1.25 k, as shown in the figure. (Note that the shear force does not change at the point of application of the couple M_0.)

The numerically largest shear force occurs just to the left of support B and equals -4.0 k.

Bending moments. The bending moment is zero at the free end and decreases algebraically (but increases numerically) as we move to the right until support B is reached. The slope of the moment diagram, equal to the value of the shear force (from Eq. 4-6), is zero at the free end and -4.0 k just to the left of support B. The diagram is parabolic (second degree) in this region, with the vertex at the end of the beam. The moment at point B is

$$M_B = -\frac{qb^2}{2} = -\frac{1}{2}(1.0 \text{ k/ft})(4.0 \text{ ft})^2 = -8.0 \text{ k-ft}$$

which is also equal to the area of the shear-force diagram between A and B (see Eq. 4-7).

The slope of the bending-moment diagram from B to C is equal to the shear force, or 1.25 k. Therefore, the bending moment just to the left of the couple M_0 is

$$-8.0 \text{ k-ft} + (1.25 \text{ k})(8.0 \text{ ft}) = 2.0 \text{ k-ft}$$

as shown on the diagram. Of course, we can get this same result by cutting through the beam just to the left of the couple, drawing a free-body diagram, and solving the equation of moment equilibrium.

The bending moment changes abruptly at the point of application of the couple M_0, as explained earlier in connection with Eq. (4-9). Because the couple acts counterclockwise, the moment decreases by an amount equal to M_0. Thus, the moment just to the right of the couple M_0 is

$$2.0 \text{ k-ft} - 12.0 \text{ k-ft} = -10.0 \text{ k-ft}$$

From that point to support C the diagram is again a straight line with slope equal to 1.25 k. Therefore, the bending moment at the support is

$$-10.0 \text{ k-ft} + (1.25 \text{ k})(8.0 \text{ ft}) = 0$$

as expected.

Maximum and minimum values of the bending moment occur where the shear force changes sign and where the couple is applied. Comparing the various high and low points on the moment diagram, we see that the numerically largest bending moment equals -10.0 k-ft and occurs just to the right of the couple M_0.

If a roller support is now added at joint A and a shear release is inserted just to the left of joint B (Fig. 4-21d), the support reactions must be recomputed. The

continued

beam is broken into two free-body diagrams, AB and BC, by cutting through the shear release (where $V = 0$), and reaction R_A is found to be 4 kips by summing vertical forces in the left free-body diagram. Then by summing moments and forces in the entire structure, $R_B = -R_C = 0.25$ kips. Finally, shear and moment diagrams can be plotted for the modified structure.

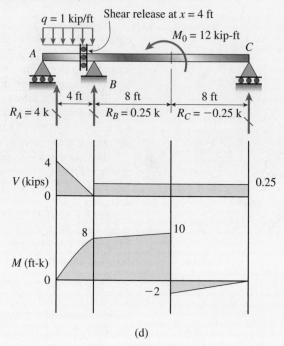

FIG. 4-21 Example 4-7. Modified beam with overhang—add shear release

(d)

In Chapter 4, we reviewed the analysis of statically determinate beams and simple frames to find support reactions and internal stress resultants (N, V, and M), then plotted axial force, shear, and bending-moment diagrams to show the variation of these quantities throughout the structure. We considered clamped, sliding, pinned and roller supports, and both concentrated and distributed loadings in assembling models of a variety of structures with different support conditions. In some cases, internal releases were included in the model to represent known locations of zero values of N, V, or M. Some of the major concepts presented in this chapter are as follows:

1. If the structure is **statically determinate** and stable, the laws of statics alone are sufficient to solve for all values of support reaction forces and moments, as well as the magnitude of the internal axial force (N), shear force (V), and bending moment (M) at any location in the structure.

2. If axial, shear, or moment **releases** are present in the structure model, the structure should be broken into separate free-body diagrams (FBD) by cutting through the release; an additional equation of equilibrium is then available for use in solving for the unknown support reactions shown in that FBD.

3. Graphical displays or **diagrams** showing the variation of N, V, and M over a structure are useful in design because they readily show the location of maximum values of N, V, and M needed in **design** (to be considered for beams in Chapter 5).

4. The **rules for drawing shear and bending moment diagrams** may be summarized as follows:

   a. The ordinate on the distributed load curve (q) is equal to the negative of the slope on the shear diagram.

$$\frac{dV}{dx} = -q$$

   b. The difference in shear values between any two points on the shear diagram is equal to the ($-$) area under the distributed load curve between those same two points.

$$\int_A^B dV = -\int_A^B q \, dx$$

$$V_B - V_A = -\int_A^B q \, dx$$

$$= -(\text{area of the loading diagram between } A \text{ and } B)$$

   c. The ordinate on the shear diagram (V) is equal to the slope on the bending moment diagram.

$$\frac{dM}{dx} = V$$

   d. The difference in values between any two points on the moment diagram is equal to the area under the shear diagram between those same two points;

$$\int_A^B dM = \int_A^B V \, dx$$

$$M_B - M_A = \int_A^B V \, dx$$

$$= (\text{area of the shear-force diagram between } A \text{ and } B)$$

   e. At those points at which the shear curve crosses the reference axis (i.e., $V = 0$), the value of the moment on the moment diagram is a local maximum or minimum.

   f. The ordinate on the axial force diagram (N) is equal to zero at an axial force release; the ordinate on the shear diagram (V) is zero at a shear release; and the ordinate on the moment diagram (M) is zero at a moment release.

PROBLEMS CHAPTER 4

Shear Forces and Bending Moments

4.3-1 Calculate the shear force V and bending moment M at a cross section just to the left of the 1600-1b load acting on the simple beam AB shown in the figure.

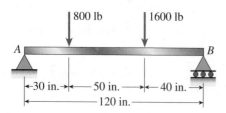

PROB. 4.3-1

4.3-2 Determine the shear force V and bending moment M at the midpoint C of the simple beam AB shown in the figure.

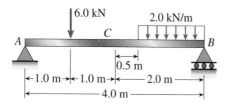

PROB. 4.3-2

4.3-3 Determine the shear force V and bending moment M at the midpoint of the beam with overhangs (see figure). Note that one load acts downward and the other upward, and clockwise moments Pb are applied at each support.

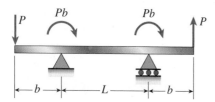

PROB. 4.3-3

4.3-4 Calculate the shear force V and bending moment M at a cross section located 0.5 m from the fixed support of the cantilever beam AB shown in the figure.

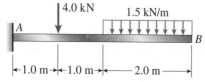

PROB. 4.3-4

4.3-5 Determine the shear force V and bending moment M at a cross section located 18 ft from the left-hand end A of the beam with an overhang shown in the figure.

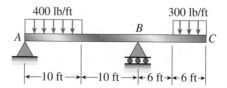

PROB. 4.3-5

4.3-6 The beam ABC shown in the figure is simply supported at A and B and has an overhang from B to C. The loads consist of a horizontal force $P_1 = 4.0$ kN acting at the end of a vertical arm and a vertical force $P_2 = 8.0$ kN acting at the end of the overhang.

Determine the shear force V and bending moment M at a cross section located 3.0 m from the left-hand support. (*Note:* Disregard the widths of the beam and vertical arm and use centerline dimensions when making calculations.)

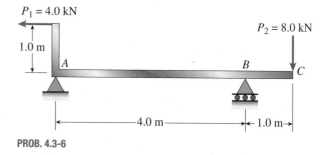

PROB. 4.3-6

4.3-7 The beam ABCD shown in the figure has overhangs at each end and carries a uniform load of intensity q.

For what ratio b/L will the bending moment at the midpoint of the beam be zero?

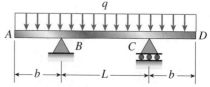

PROB. 4.3-7

4.3-8 At full draw, an archer applies a pull of 130 N to the bowstring of the bow shown in the figure. Determine the bending moment at the midpoint of the bow.

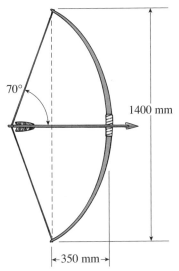

70°

1400 mm

←350 mm→

PROB. 4.3-8

4.3-9 A curved bar *ABC* is subjected to loads in the form of two equal and opposite forces *P*, as shown in the figure. The axis of the bar forms a semicircle of radius *r*.

Determine the axial force *N*, shear force *V*, and bending moment *M* acting at a cross section defined by the angle θ.

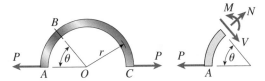

PROB. 4.3-9

4.3-10 Under cruising conditions the distributed load acting on the wing of a small airplane has the idealized variation shown in the figure.

Calculate the shear force *V* and bending moment *M* at the inboard end of the wing.

Wings of small airplane have distributed uplift loads
(Thomas Gulla/Shutterstock)

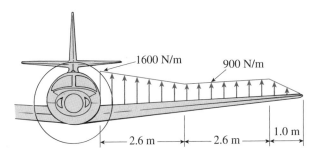

1600 N/m   900 N/m

←2.6 m→←2.6 m→|1.0 m|

PROB. 4.3-10

4.3-11 A beam *ABCD* with a vertical arm *CE* is supported as a simple beam at *A* and *D* (see figure). A cable passes over a small pulley that is attached to the arm at *E*. One end of the cable is attached to the beam at point *B*.

What is the force *P* in the cable if the bending moment in the beam just to the left of point *C* is equal numerically to 640 lb-ft? (*Note:* Disregard the widths of the beam and vertical arm and use centerline dimensions when making calculations.)

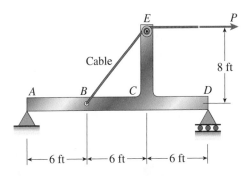

E   *P*

Cable

8 ft

A   *B*   *C*   *D*

←6 ft→←6 ft→←6 ft→

PROB. 4.3-11

4.3-12 A simply supported beam *AB* supports a trapezoidally distributed load (see figure). The intensity of the load varies linearly from 50 kN/m at support *A* to 25 kN/m at support *B*.

Calculate the shear force *V* and bending moment *M* at the midpoint of the beam.

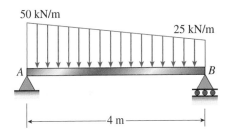

50 kN/m

25 kN/m

A   *B*

←4 m→

PROB. 4.3-12

4.3-13 Beam *ABCD* represents a reinforced-concrete foundation beam that supports a uniform load of intensity $q_1 = 3500$ lb/ft (see figure). Assume that the soil pressure on the underside of the beam is uniformly distributed with intensity q_2.

(a) Find the shear force V_B and bending moment M_B at point *B*.

(b) Find the shear force V_m and bending moment M_m at the midpoint of the beam.

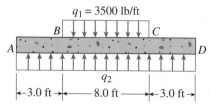

PROB. 4.3-13

4.3-14 The simply supported beam *ABCD* is loaded by a weight $W = 27$ kN through the arrangement shown in the figure. The cable passes over a small frictionless pulley at *B* and is attached at *E* to the end of the vertical arm.

Calculate the axial force *N*, shear force *V*, and bending moment *M* at section *C*, which is just to the left of the vertical arm. (*Note:* Disregard the widths of the beam and vertical arm and use centerline dimensions when making calculations.)

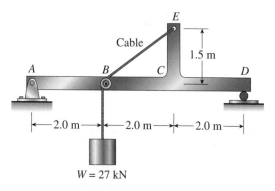

PROB. 4.3-14

4.3-15 The centrifuge shown in the figure rotates in a horizontal plane (the *xy* plane) on a smooth surface about the *z* axis (which is vertical) with an angular acceleration α. Each of the two arms has weight *w* per unit length and supports a weight $W = 2.0wL$ at its end.

Derive formulas for the maximum shear force and maximum bending moment in the arms, assuming $b = L/9$ and $c = L/10$.

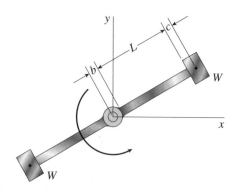

PROB. 4.3-15

Shear-Force and Bending-Moment Diagrams

When solving the problems for Section 4.5, draw the shear-force and bending-moment diagrams approximately to scale and label all critical ordinates, including the maximum and minimum values.

Probs 4.5-1 through 4.5-10 are symbolic problems and Probs. 4.5-11 through 4.5-24 are numerical problems. The remaining problems (4.5-25 through 4.5-30) involve specialized topics, such as optimization, beams with hinges, and moving loads.

4.5-1 Draw the shear-force and bending-moment diagrams for a simple beam *AB* supporting two equal concentrated loads *P* (see figure).

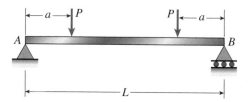

PROB. 4.5-1

4.5-2 A simple beam *AB* is subjected to a counterclockwise couple of moment M_0 acting at distance *a* from the left-hand support (see figure).

Draw the shear-force and bending-moment diagrams for this beam.

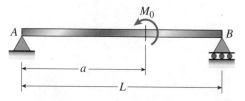

PROB. 4.5-2

4.5-3 Draw the shear-force and bending-moment diagrams for a cantilever beam AB carrying a uniform load of intensity q over one-half of its length (see figure).

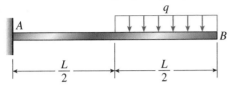

PROB. 4.5-3

4.5-4 The cantilever beam AB shown in the figure is subjected to a concentrated load P at the midpoint and a counterclockwise couple of moment $M_1 = PL/4$ at the free end.

Draw the shear-force and bending-moment diagrams for this beam.

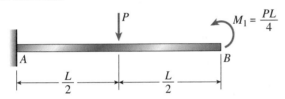

PROB. 4.5-4

4.5-5 The simple beam AB shown in the figure is subjected to a concentrated load P and a clockwise couple $M_1 = PL/3$ acting at the third points.

Draw the shear-force and bending-moment diagrams for this beam.

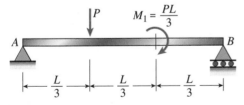

PROB. 4.5-5

4.5-6 A simple beam AB subjected to couples M_1 and $3M_1$ acting at the third points is shown in the figure.

Draw the shear-force and bending-moment diagrams for this beam.

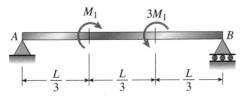

PROB. 4.5-6

4.5-7 A simply supported beam ABC is loaded by a vertical load P acting at the end of a bracket BDE (see figure).

Draw the shear-force and bending-moment diagrams for beam ABC.

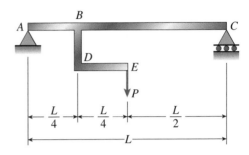

PROB. 4.5-7

4.5-8 A beam ABC is simply supported at A and B and has an overhang BC (see figure). The beam is loaded by two forces P and a clockwise couple of moment Pa that act through the arrangement shown.

Draw the shear-force and bending-moment diagrams for beam ABC.

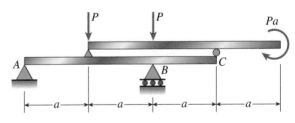

PROB. 4.5-8

4.5-9 Beam $ABCD$ is simply supported at B and C and has overhangs at each end (see figure). The span length is L and each overhang has length $L/3$. A uniform load of intensity q acts along the entire length of the beam.

Draw the shear-force and bending-moment diagrams for this beam.

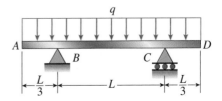

PROB. 4.5-9

4.5-10 Draw the shear-force and bending-moment diagrams for a cantilever beam AB supporting a linearly varying load of maximum intensity q_0 (see figure).

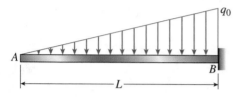

PROB. 4.5-10

4.5-11 The simple beam AB supports a triangular load of maximum intensity $q_0 = 10$ lb/in. acting over one-half of the span and a concentrated load $P = 80$ lb acting at midspan (see figure). Draw the shear-force and bending-moment diagrams for this beam.

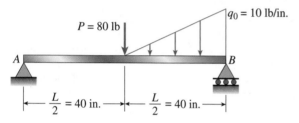

PROB. 4.5-11

4.5-12 The beam AB shown in the figure supports a uniform load of intensity 3000 N/m acting over half the length of the beam. The beam rests on a foundation that produces a uniformly distributed load over the entire length. Draw the shear-force and bending-moment diagrams for this beam.

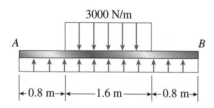

PROB. 4.5-12

4.5-13 A cantilever beam AB supports a couple and a concentrated load, as shown in the figure. Draw the shear-force and bending-moment diagrams for this beam.

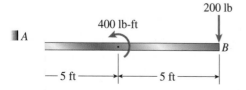

4.5-14 The cantilever beam AB shown in the figure is subjected to a triangular load acting throughout one-half of its length and a concentrated load acting at the free end.

Draw the shear-force and bending-moment diagrams for this beam.

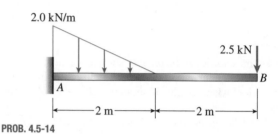

PROB. 4.5-14

4.5-15 The uniformly loaded beam ABC has simple supports at A and B and an overhang BC (see figure).

Draw the shear-force and bending-moment diagrams for this beam.

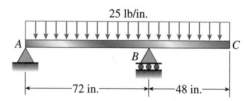

PROB. 4.5-15

4.5-16 A beam ABC with an overhang at one end supports a uniform load of intensity 12 kN/m and a concentrated moment of magnitude 3 kN · m at C (see figure).

Draw the shear-force and bending-moment diagrams for this beam.

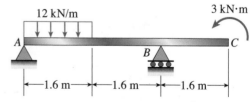

PROB. 4.5-16

4.5-17 Consider two beams, which are loaded the same but have different support conditions. Which beam has the larger maximum moment?

First, find support reactions, then plot axial force (N), shear (V), and moment (M) diagrams for all three beams. *Label* all critical N, V, and M values and also the *distance* to points where N, V, and/or M is zero.

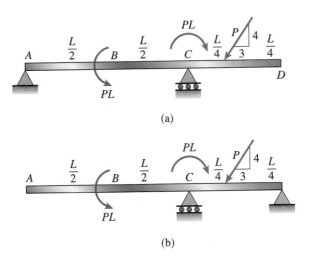

(a)

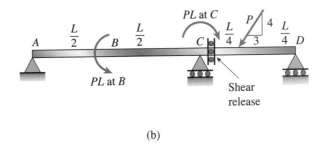

(b)

4.5-18 The three beams below are loaded the same and have the same support conditions. However, one has a *moment release* just to the left of C, the second has a *shear release* just to the right of C and the third has an axial release just to the left of C. Which beam has the largest maximum moment?

First, find support reactions, then plot axial force (N), shear (V), and moment (M) diagrams for all three beams. *Label* all critical N, V, and M values and also the *distance* to points where N, V, and/or M is zero.

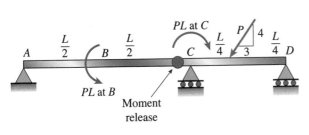

(a)

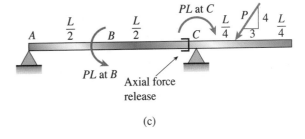

(c)

4.5-19 The beam *ABC* shown in the figure is simply supported at *A* and *B* and has an overhang from *B* to *C*. The loads consist of a horizontal force $P_1 = 400$ lb acting at the end of the vertical arm and a vertical force $P_2 = 900$ lb acting at the end of the overhang.

Draw the shear-force and bending-moment diagrams for this beam. (*Note:* Disregard the widths of the beam and vertical arm and use centerline dimensions when making calculations.)

4.5-20 A simple beam *AB* is loaded by two segments of uniform load and two horizontal forces acting at the ends of a vertical arm (see figure).

Draw the shear-force and bending-moment diagrams for this beam.

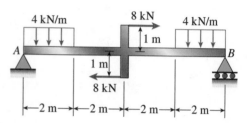

PROB. 4.5-20

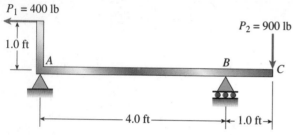

PROB. 4.5-19

4.5-21 The two beams below are loaded the same and have the same support conditions. However, the location of internal *axial, shear,* and *moment releases* is different for each beam (see figures). Which beam has the larger maximum moment?

First, find support reactions, then plot axial force (*N*), shear (*V*), and moment (*M*) diagrams for both beams. *Label* all critical *N*, *V*, and *M* values and also the *distance* to points where *N*, *V*, and/or *M* is zero.

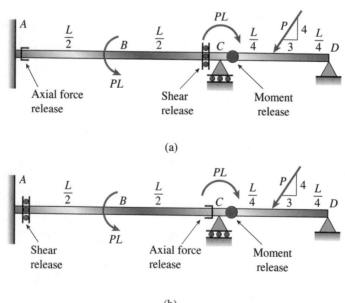

PROB. 4.5-21

4.5-22 The beam *ABCD* shown in the figure has overhangs that extend in both directions for a distance of 4.2 m from the supports at *B* and *C*, which are 1.2 m apart.

Draw the shear-force and bending-moment diagrams for this overhanging beam.

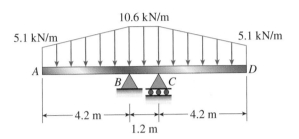

PROB. 4.5-22

4.5-23 A beam *ABCD* with a vertical arm *CE* is supported as a simple beam at *A* and *D* (see figure). A cable passes over a small pulley that is attached to the arm at *E*. One end of the cable is attached to the beam at point *B*. The tensile force in the cable is 1800 lb.

Draw the shear-force and bending-moment diagrams for beam *ABCD*. (*Note:* Disregard the widths of the beam and vertical arm and use centerline dimensions when making calculations.)

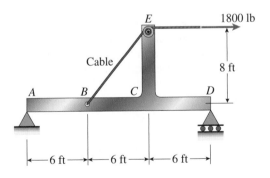

PROB. 4.5-23

4.5-24 Beams *ABC* and *CD* are supported at *A*, *C*, and *D* and are joined by a hinge (or *moment release*) just to the left of *C*. The support at *A* is a sliding support (hence reaction $A_y = 0$ for the loading shown below). Find all support reactions then plot

shear (V) and moment (M) diagrams. *Label* all critical V and M values and also the *distance* to points where either V and/or M is zero.

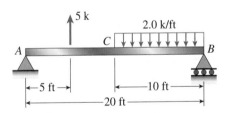

PROB. 4.5-24

4.5-25 The simple beam *AB* shown in the figure supports a concentrated load and a segment of uniform load.

Draw the shear-force and bending-moment diagrams for this beam.

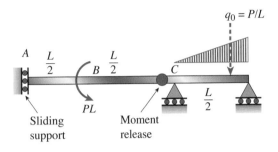

PROB. 4.5-25

4.5-26 The cantilever beam shown in the figure supports a concentrated load and a segment of uniform load.

Draw the shear-force and bending-moment diagrams for this cantilever beam.

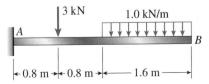

PROB. 4.5-26

4.5-27 The simple beam *ACB* shown in the figure is subjected to a triangular load of maximum intensity 180 lb/ft and a concentrated moment of 300 lb-ft at *A*.

Draw the shear-force and bending-moment diagrams for this beam.

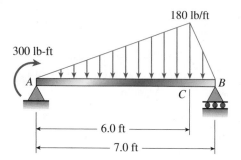

PROB. 4.5-27

4.5-28 A beam with simple supports is subjected to a trapezoidally distributed load (see figure). The intensity of the load varies from 1.0 kN/m at support *A* to 3.0 kN/m at support *B*.

Draw the shear-force and bending-moment diagrams for this beam.

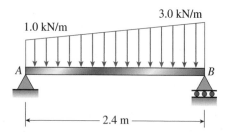

PROB. 4.5-28

4.5-29 A beam of length *L* is being designed to support a uniform load of intensity *q* (see figure). If the supports of the beam are placed at the ends, creating a simple beam, the maximum bending moment in the beam is $qL^2/8$. However, if the supports of the beam are moved symmetrically toward the middle of the beam (as pictured), the maximum bending moment is reduced.

Determine the distance *a* between the supports so that the maximum bending moment in the beam has the smallest possible numerical value. Draw the shear-force and bending-moment diagrams for this condition.

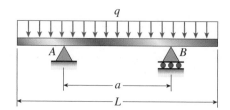

PROB. 4.5-29

4.5-30 The compound beam *ABCDE* shown in the figure consists of two beams (*AD* and *DE*) joined by a hinged connection at *D*. The hinge can transmit a shear force but not a bending moment. The loads on the beam consist of a 4-kN force at the end of a bracket attached at point *B* and a 2-kN force at the midpoint of beam *DE*. Draw the shear-force and bending-moment diagrams for this compound beam.

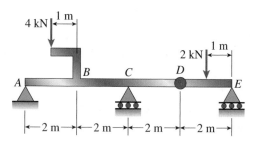

PROB. 4.5-30

4.5-31 The beam shown below has a sliding support at *A* and an elastic support with spring constant *k* at *B*. A distributed load *q(x)* is applied over the entire beam. Find all support reactions, then plot shear (*V*) and moment (*M*) diagrams for beam *AB*; *label* all critical *V* and *M* values and also the *distance* to points where any critical ordinates are zero.

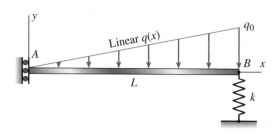

PROB. 4.5-31

4.5-32 The shear-force diagram for a simple beam is shown in the figure.

Determine the loading on the beam and draw the bending-moment diagram, assuming that no couples act as loads on the beam.

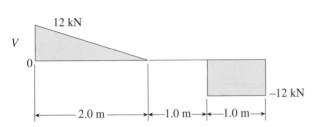

PROB. 4.5-32

4.5-33 The shear-force diagram for a beam is shown in the figure. Assuming that no couples act as loads on the beam, determine the forces acting on the beam and draw the bending-moment diagram.

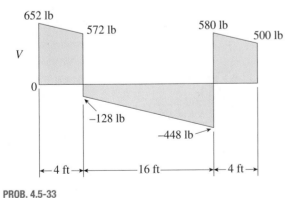

PROB. 4.5-33

4.5-34 The compound beam below has an internal *moment release* just to the left of *B* and a *shear release* just to the right of *C*. Reactions have been computed at *A*, *C*, and *D* and are shown in the figure.

First, confirm the reaction expressions using statics, then plot shear (V) and moment (M) diagrams. *Label* all critical V and M values and also the *distance* to points where either V and/or M is zero.

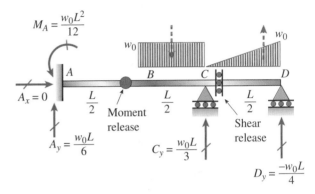

PROB. 4.5-34

4.5-35 The compound beam below has a *shear release* just to the left of *C* and a *moment release* just to the right of *C*. A plot of the moment diagram is provided below for applied load *P* at *B* and triangular distributed loads $w(x)$ on segments *BC* and *CD*.

First, solve for reactions using statics, then plot axial force (N) and shear (V) diagrams. Confirm that the moment diagram is that shown below. *Label* all critical N, V, and M values and also the *distance* to points where N, V, and/or M is zero.

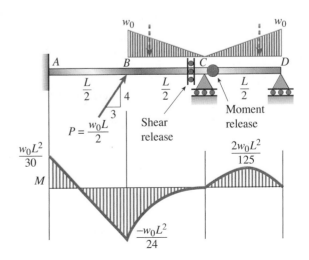

PROB. 4.5-35

4.5-36 A simple beam AB supports two connected wheel loads P and $2P$ that are distance d apart (see figure). The wheels may be placed at any distance x from the left-hand support of the beam.

(a) Determine the distance x that will produce the maximum shear force in the beam, and also determine the maximum shear force V_{max}.

(b) Determine the distance x that will produce the maximum bending moment in the beam, and also draw the corresponding bending-moment diagram. (Assume $P = 10$ kN, $d = 2.4$ m, and $L = 12$ m.)

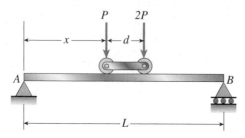

PROB. 4.5-36

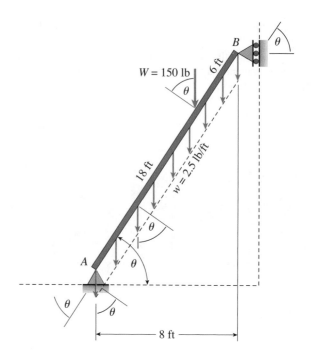

PROB. 4.5-37

4.5-37 The inclined beam represents a ladder with the following applied loads: the weight (W) of the house painter and the distributed weight (w) of the ladder itself. Find support reactions at A and B, then plot axial force (N), shear (V), and moment (M) diagrams. *Label* all critical N, V, and M values and also the *distance* to points where any critical ordinates are zero. Plot N, V, and M diagrams normal to the inclined ladder.

4.5-38 Beam ABC is supported by a tie rod CD as shown. Two configurations are possible: pin support at A and downward triangular load on AB or pin at B and upward load on AB. Which has the larger maximum moment?

First, find all support reactions, then plot axial force (N), shear (V), and moment (M) diagrams for ABC only and *label* all critical N, V, and M values. Label the *distance* to points where any critical ordinates are zero.

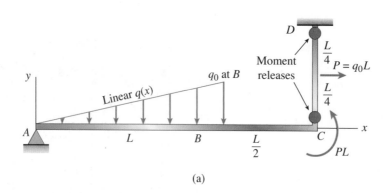

(a)

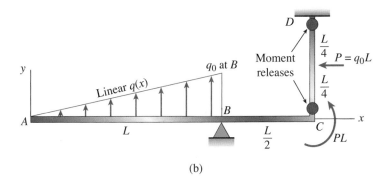

(b)

PROB. 4.5-38

4.5-39 The plane frame below consists of column AB and beam BC which carries a triangular distributed load. Support A is fixed, and there is a roller support at C. Column AB has a *moment release* just below joint B.

Find support reactions at A and C, then plot axial force (N), shear (V), and moment (M) diagrams for both members. *Label* all critical N, V, and M values and also the *distance* to points where any critical ordinates are zero.

4.5-40 The plane frame shown below is part of an elevated freeway system. Supports at A and D are fixed but there are *moment releases* at the base of both columns (AB and DE), as well as in column BC and at the end of beam BE.

Find all support reactions, then plot axial force (N), shear (V), and moment (M) diagrams for all beam and column members. *Label* all critical N, V, and M values and also the *distance* to points where any critical ordinates are zero.

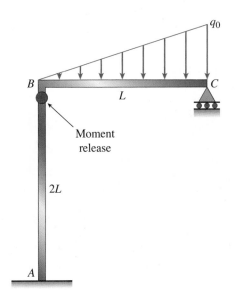

PROB. 4.5-39

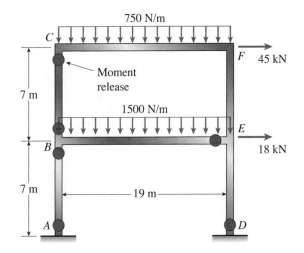

PROB. 4.5-40

Beams are essential load carrying components in modern building and bridge construction. (Lester Lefkowitz/Getty Images)

5

Stresses in Beams

CHAPTER OVERVIEW

Chapter 5 is concerned with stresses and strains in beams which have loads applied in the xy plane, a plane of symmetry of the cross section, resulting in beam deflection in that same plane, known as the **plane of bending**. Both **pure bending** (beam flexure under constant bending moment) and **nonuniform bending** (flexure in the presence of shear forces) are discussed (Section 5.2). We will see that strains and stresses in the beam are directly related to the **curvature** κ of the deflection curve (Section 5.3). A **strain-curvature relation** will be developed from consideration of longitudinal strains developed in the beam during bending; these strains vary linearly with distance from the neutral surface of the beam (Section 5.4). When Hooke's law (which applies for linearly elastic materials) is combined with the strain-curvature relation, we find that the neutral axis passes through the centroid of the cross section. As a result, x and y axes are seen to be **principal centroidal axes**. By consideration of the moment resultant of the normal stresses acting over the cross section, we next derive the **moment-curvature relation** which relates curvature (κ) to moment (M) and flexural rigidity (EI). This will lead to the differential equation of the beam elastic curve, a topic for consideration in Chapter 8 when we will discuss beam deflections in detail. Of immediate interest here, however, are beam stresses, and the moment-curvature relation is next used to develop the **flexure formula** (Section 5.5). The **flexure formula** shows that normal stresses (σ_x) vary linearly with distance (y) from the neutral surface and depend on bending moment (M) and moment of inertia (I) of the cross section. Next, the section modulus (S) of the beam cross section is defined and then used in **design** of beams in Section 5.6. In beam design, we use the maximum bending moment (M_{max}), obtained from the bending moment diagram (Section 4.5) and the allowable normal stress for the material (σ_{allow}) to compute the required section modulus, then select an appropriate beam of steel or wood from the tables in Appendices F and G (available online).

For beams in nonuniform bending, both normal and shear stresses are developed and must be considered in beam analysis and design. Normal stresses are computed using the **flexure formula**, as noted above, and the

shear formula must be used to calculate shear stresses (τ) which vary over the height of the beam (Sections 5.7 and 5.8). Maximum normal and shear stresses do not occur at the same location along a beam, but in most cases, maximum normal stresses control the design of the beam. Special consideration is given to shear stresses in beams with flanges (e.g., W and C shapes) (Section 5.9).

Finally, stresses and strains in **composite beams**, that is beams fabricated of more than one material, is discussed in Section 5.10. First, we locate the neutral axis then find the flexure formula for a composite beam made up of two different materials. Lastly, we study the **transformed-section method** as an alternative procedure for analyzing the bending stresses in a composite beam.

Chapter 5 is organized as follows:

5.1 INTRODUCTION

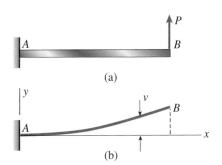

FIG. 5-1 Bending of a cantilever beam: (a) beam with load, and (b) deflection curve

In the preceding chapter we saw how the loads acting on a beam create internal actions (or *stress resultants*) in the form of shear forces and bending moments. In this chapter we go one step further and investigate the *stresses* and *strains* associated with those shear forces and bending moments. Knowing the stresses and strains, we will be able to analyze and design beams subjected to a variety of loading conditions.

The loads acting on a beam cause the beam to bend (or *flex*), thereby deforming its axis into a curve. As an example, consider a cantilever beam *AB* subjected to a load *P* at the free end (Fig. 5-1a). The initially straight axis is bent into a curve (Fig. 5-1b), called the **deflection curve** of the beam.

For reference purposes, we construct a system of **coordinate axes** (Fig. 5-1b) with the origin located at a suitable point on the longitudinal axis of the beam. In this illustration, we place the origin at the fixed support. The positive *x* axis is directed to the right, and the positive *y* axis is directed upward. The *z* axis, not shown in the figure, is directed outward (that is, toward the viewer), so that the three axes form a right-handed coordinate system.

The beams considered in this chapter (like those discussed in Chapter 4) are assumed to be symmetric about the *xy* plane, which means that the *y* axis is an axis of symmetry of the cross section. In addition, all loads must act in the *xy* plane. As a consequence, the bending deflections occur in this same plane, known as the **plane of bending**. Thus, the deflection curve shown in Fig. 5-1b is a plane curve lying in the plane of bending.

The **deflection** of the beam at any point along its axis is the *displacement* of that point from its original position, measured in the *y* direction. We denote the deflection by the letter *v* to distinguish it from the coordinate *y* itself (see Fig. 5-1b).[*]

5.2 PURE BENDING AND NONUNIFORM BENDING

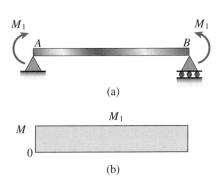

FIG. 5-2 Simple beam in pure bending ($M = M_1$)

When analyzing beams, it is often necessary to distinguish between pure bending and nonuniform bending. **Pure bending** refers to flexure of a beam under a constant bending moment. Therefore, pure bending occurs only in regions of a beam where the shear force is zero (because $V = dM/dx$; see Eq. 4-6). In contrast, **nonuniform bending** refers to flexure in the presence of shear forces, which means that the bending moment changes as we move along the axis of the beam.

As an example of pure bending, consider a simple beam *AB* loaded by two couples M_1 having the same magnitude but acting in opposite directions (Fig. 5-2a). These loads produce a constant bending moment $M = M_1$ throughout the length of the beam, as shown by the bending

---

[*]In applied mechanics, the traditional symbols for displacements in the *x*, *y*, and *z* directions are *u*, *v*, and *w*, respectively.

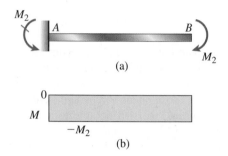

FIG. 5-3 Cantilever beam in pure bending $(M = -M_2)$

moment diagram in part (b) of the figure. Note that the shear force V is zero at all cross sections of the beam.

Another illustration of pure bending is given in Fig. 5-3a, where the cantilever beam AB is subjected to a clockwise couple M_2 at the free end. There are no shear forces in this beam, and the bending moment M is constant throughout its length. The bending moment is negative ($M = -M_2$), as shown by the bending moment diagram in part (b) of Fig. 5-3.

The symmetrically loaded simple beam of Fig. 5-4a is an example of a beam that is partly in pure bending and partly in nonuniform bending, as seen from the shear-force and bending-moment diagrams (Figs. 5-4b and c). The central region of the beam is in pure bending because the shear force is zero and the bending moment is constant. The parts of the beam near the ends are in nonuniform bending because shear forces are present and the bending moments vary.

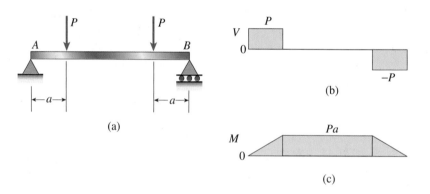

FIG. 5-4 Simple beam with central region in pure bending and end regions in nonuniform bending

In the following two sections we will investigate the strains and stresses in beams subjected only to pure bending. Fortunately, we can often use the results obtained for pure bending even when shear forces are present, as explained later (see the last paragraph in Section 5.7).

5.3 CURVATURE OF A BEAM

When loads are applied to a beam, its longitudinal axis is deformed into a curve, as illustrated previously in Fig. 5-1. The resulting strains and stresses in the beam are directly related to the **curvature** of the deflection curve.

To illustrate the concept of curvature, consider again a cantilever beam subjected to a load P acting at the free end (see Fig. 5-5a on the next page). The deflection curve of this beam is shown in Fig. 5-5b. For purposes of analysis, we identify two points m_1 and m_2 on the deflection curve. Point m_1 is selected at an arbitrary distance x from the y axis and point m_2 is located a small distance ds further along the curve. At each of these points we draw a line normal to the *tangent* to the deflection curve,

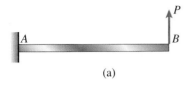

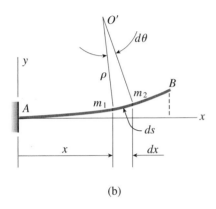

(a)

(b)

FIG. 5-5 Curvature of a bent beam:
(a) beam with load, and (b) deflection
curve

that is, normal to the curve itself. These normals intersect at point O', which is the **center of curvature** of the deflection curve. Because most beams have very small deflections and nearly flat deflection curves, point O' is usually located much farther from the beam than is indicated in the figure.

The distance $m_1 O'$ from the curve to the center of curvature is called the **radius of curvature** ρ (Greek letter rho), and the **curvature** κ (Greek letter kappa) is defined as the reciprocal of the radius of curvature. Thus,

$$\kappa = \frac{1}{\rho} \tag{5-1}$$

Curvature is a measure of how sharply a beam is bent. If the load on a beam is small, the beam will be nearly straight, the radius of curvature will be very large, and the curvature will be very small. If the load is increased, the amount of bending will increase—the radius of curvature will become smaller, and the curvature will become larger.

From the geometry of triangle $O' m_1 m_2$ (Fig. 5-5b) we obtain

$$\rho \, d\theta = ds \tag{a}$$

in which $d\theta$ (measured in radians) is the infinitesimal angle between the normals and ds is the infinitesimal distance along the curve between points m_1 and m_2. Combining Eq. (a) with Eq. (5-1), we get

$$\kappa = \frac{1}{\rho} = \frac{d\theta}{ds} \tag{5-2}$$

This equation for **curvature** is derived in textbooks on calculus and holds for any curve, regardless of the amount of curvature. If the curvature is *constant* throughout the length of a curve, the radius of curvature will also be constant and the curve will be an arc of a circle.

The deflections of a beam are usually very small compared to its length (consider, for instance, the deflections of the structural frame of an automobile or a beam in a building). Small deflections mean that the deflection curve is nearly flat. Consequently, the distance ds along the curve may be set equal to its horizontal projection dx (see Fig. 5-5b). Under these special conditions of **small deflections**, the equation for the curvature becomes

$$\kappa = \frac{1}{\rho} = \frac{d\theta}{dx} \tag{5-3}$$

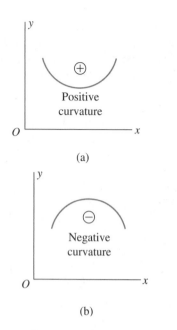

FIG. 5-6 Sign convention for curvature

Both the curvature and the radius of curvature are functions of the distance x measured along the x axis. It follows that the position O' of the center of curvature also depends upon the distance x.

In Section 5.5 we will see that the curvature at a particular point on the axis of a beam depends upon the bending moment at that point and upon the properties of the beam itself (shape of cross section and type of material). Therefore, if the beam is prismatic and the material is homogeneous, the curvature will vary only with the bending moment. Consequently, a beam in *pure bending* will have constant curvature and a beam in *nonuniform bending* will have varying curvature.

The **sign convention for curvature** depends upon the orientation of the coordinate axes. If the x axis is positive to the right and the y axis is positive upward, as shown in Fig. 5-6, then the curvature is positive when the beam is bent concave upward and the center of curvature is above the beam. Conversely, the curvature is negative when the beam is bent concave downward and the center of curvature is below the beam.

In the next section we will see how the longitudinal strains in a bent beam are determined from its curvature, and in Chapter 8 we will see how curvature is related to the deflections of beams.

5.4 LONGITUDINAL STRAINS IN BEAMS

The longitudinal strains in a beam can be found by analyzing the curvature of the beam and the associated deformations. For this purpose, let us consider a portion AB of a beam in pure bending subjected to positive bending moments M (Fig. 5-7a). We assume that the beam initially has a straight longitudinal axis (the x axis in the figure) and that its cross section is symmetric about the y axis, as shown in Fig. 5-7b.

Under the action of the bending moments, the beam deflects in the xy plane (the plane of bending) and its longitudinal axis is bent into a circular curve (curve ss in Fig. 5-7c). The beam is bent concave upward, which is positive curvature (Fig. 5-6a).

Cross sections of the beam, such as sections mn and pq in Fig. 5-7a, remain plane and normal to the longitudinal axis (Fig. 5-7c). The fact that cross sections of a beam in pure bending remain plane is so fundamental to beam theory that it is often called an assumption. However, we could also call it a theorem, because it can be proved rigorously using only rational arguments based upon symmetry (Ref. 5-1). The basic point is that the symmetry of the beam and its loading (Figs. 5-7a and b) means that all elements of the beam (such as element $mpqn$) must deform in an identical manner, which is possible only if cross sections remain plane during bending (Fig. 5-7c). This conclusion is valid for beams of any material, whether the material is elastic or inelastic, linear or nonlinear. Of course, the material properties, like the dimensions, must

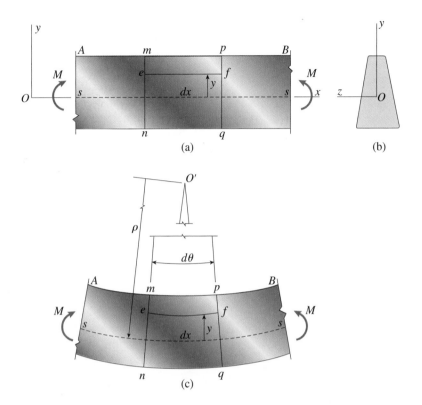

FIG. 5-7 Deformations of a beam in pure bending: (a) side view of beam, (b) cross section of beam, and (c) deformed beam

be symmetric about the plane of bending. (*Note:* Even though a plane cross section in pure bending remains plane, there still may be deformations in the plane itself. Such deformations are due to the effects of Poisson's ratio, as explained at the end of this discussion.)

Because of the bending deformations shown in Fig. 5-7c, cross sections *mn* and *pq* rotate with respect to each other about axes perpendicular to the *xy* plane. Longitudinal lines on the lower part of the beam are elongated, whereas those on the upper part are shortened. Thus, the lower part of the beam is in tension and the upper part is in compression. Somewhere between the top and bottom of the beam is a surface in which longitudinal lines do not change in length. This surface, indicated by the dashed line *ss* in Figs. 5-7a and c, is called the **neutral surface** of the beam. Its intersection with any cross-sectional plane is called the **neutral axis** of the cross section; for instance, the *z* axis is the neutral axis for the cross section of Fig. 5-7b.

The planes containing cross sections *mn* and *pq* in the deformed beam (Fig. 5-7c) intersect in a line through the center of curvature O'. The angle between these planes is denoted $d\theta$, and the distance from O' to the neutral surface *ss* is the radius of curvature ρ. The initial distance dx between the two planes (Fig. 5-7a) is unchanged at the neutral surface (Fig. 5-7c), hence $\rho\,d\theta = dx$. However, all other longitudinal lines between the two planes either lengthen or shorten, thereby creating **normal strains** ϵ_x.

To evaluate these normal strains, consider a typical longitudinal line *ef* located within the beam between planes *mn* and *pq* (Fig. 5-7a). We identify line *ef* by its distance *y* from the neutral surface in the initially straight beam. Thus, we are now assuming that the *x* axis lies along the neutral surface of the *undeformed* beam. Of course, when the beam deflects, the neutral surface moves with the beam, but the *x* axis remains fixed in position. Nevertheless, the longitudinal line *ef* in the deflected beam (Fig. 5-7c) is still located at the same distance *y* from the neutral surface. Thus, the length L_1 of line *ef* after bending takes place is

$$L_1 = (\rho - y)\, d\theta = dx - \frac{y}{\rho}dx$$

in which we have substituted $d\theta = dx/\rho$.

Since the original length of line *ef* is *dx*, it follows that its elongation is $L_1 - dx$, or $-y\,dx/\rho$. The corresponding *longitudinal strain* is equal to the elongation divided by the initial length *dx*; therefore, the **strain-curvature relation** is

$$\epsilon_x = -\frac{y}{\rho} = -\kappa y \tag{5-4}$$

where κ is the curvature (see Eq. 5-1).

The preceding equation shows that the longitudinal strains in the beam are proportional to the curvature and vary linearly with the distance *y* from the neutral surface. When the point under consideration is above the neutral surface, the distance *y* is positive. If the curvature is also positive (as in Fig. 5-7c), then ϵ_x will be a negative strain, representing a shortening. By contrast, if the point under consideration is below the neutral surface, the distance *y* will be negative and, if the curvature is positive, the strain ϵ_x will also be positive, representing an elongation. Note that the **sign convention** for ϵ_x is the same as that used for normal strains in earlier chapters, namely, elongation is positive and shortening is negative.

Equation (5-4) for the normal strains in a beam was derived solely from the geometry of the deformed beam—the properties of the material did not enter into the discussion. Therefore, *the strains in a beam in pure bending vary linearly with distance from the neutral surface regardless of the shape of the stress-strain curve of the material.*

The next step in our analysis, namely, finding the stresses from the strains, requires the use of the *stress-strain curve*. This step is described in the next section for linearly elastic materials.

The longitudinal strains in a beam are accompanied by *transverse strains* (that is, normal strains in the *y* and *z* directions) because of the effects of Poisson's ratio. However, there are no accompanying transverse stresses because beams are free to deform laterally. This stress condition is analogous to that of a prismatic bar in tension or compression, and therefore *longitudinal elements in a beam in pure bending are in a state of uniaxial stress.*

Example 5-1

A simply supported steel beam *AB* (Fig. 5-8a) of length $L = 8.0$ ft and height $h = 6.0$ in. is bent by couples M_0 into a circular arc with a downward deflection δ at the midpoint (Fig. 5-8b). The longitudinal normal strain (elongation) on the bottom surface of the beam is 0.00125, and the distance from the neutral surface to the bottom surface of the beam is 3.0 in.

Determine the radius of curvature ρ, the curvature κ, and the deflection δ of the beam.

Note: This beam has a relatively large deflection because its length is large compared to its height ($L/h = 16$) and the strain of 0.00125 is also large. (It is about the same as the yield strain for ordinary structural steel.)

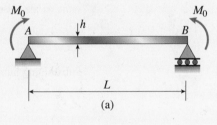

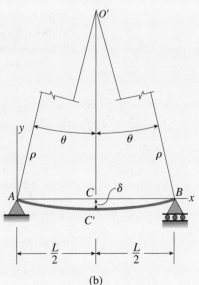

FIG. 5-8 Example 5-1. Beam in pure bending: (a) beam with loads, and (b) deflection curve

continued

Solution

Curvature. Since we know the longitudinal strain at the bottom surface of the beam ($\epsilon_x = 0.00125$), and since we also know the distance from the neutral surface to the bottom surface ($y = -3.0$ in.), we can use Eq. (5-4) to calculate both the radius of curvature and the curvature. Rearranging Eq. (5-4) and substituting numerical values, we get

$$\rho = -\frac{y}{\epsilon_x} = -\frac{-3.0 \text{ in.}}{0.00125} = 2400 \text{ in.} = 200 \text{ ft} \qquad \kappa = \frac{1}{\rho} = 0.0050 \text{ ft}^{-1} \quad \Longleftarrow$$

These results show that the radius of curvature is extremely large compared to the length of the beam even when the strain in the material is large. If, as usual, the strain is less, the radius of curvature is even larger.

Deflection. As pointed out in Section 5.3, a constant bending moment (pure bending) produces constant curvature throughout the length of a beam. Therefore, the deflection curve is a circular arc. From Fig. 5-8b we see that the distance from the center of curvature O' to the midpoint C' of the deflected beam is the radius of curvature ρ, and the distance from O' to point C on the x axis is $\rho \cos \theta$, where θ is angle $BO'C$. This leads to the following expression for the deflection at the midpoint of the beam:

$$\delta = \rho(1 - \cos \theta) \tag{5-5}$$

For a nearly flat curve, we can assume that the distance between supports is the same as the length of the beam itself. Therefore, from triangle $BO'C$ we get

$$\sin \theta = \frac{L/2}{\rho} \tag{5-6}$$

Substituting numerical values, we obtain

$$\sin \theta = \frac{(8.0 \text{ ft})(12 \text{ in./ft})}{2(2400 \text{ in.})} = 0.0200$$

and $\qquad\qquad\qquad \theta = 0.0200 \text{ rad} = 1.146°$

Note that for practical purposes we may consider $\sin \theta$ and θ (radians) to be equal numerically because θ is a very small angle.

Now we substitute into Eq. (5-5) for the deflection and obtain

$$\delta = \rho(1 - \cos \theta) = (2400 \text{ in.})(1 - 0.999800) = 0.480 \text{ in.} \quad \Longleftarrow$$

This deflection is very small compared to the length of the beam, as shown by the ratio of the span length to the deflection:

$$\frac{L}{\delta} = \frac{(8.0 \text{ ft})(12 \text{ in./ft})}{0.480 \text{ in.}} = 200$$

Thus, we have confirmed that the deflection curve is nearly flat in spite of the large strains. Of course, in Fig. 5-8b the deflection of the beam is highly exaggerated for clarity.

Note: The purpose of this example is to show the relative magnitudes of the radius of curvature, length of the beam, and deflection of the beam. However, the method used for finding the deflection has little practical value because it is limited to pure bending, which produces a circular deflected shape. More useful methods for finding beam deflections are presented later in Chapter 8.

5.5 NORMAL STRESSES IN BEAMS (LINEARLY ELASTIC MATERIALS)

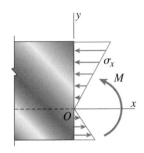

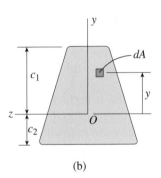

FIG. 5-9 Normal stresses in a beam of linearly elastic material: (a) side view of beam showing distribution of normal stresses, and (b) cross section of beam showing the z axis as the neutral axis of the cross section

In the preceding section we investigated the longitudinal strains ϵ_x in a beam in pure bending (see Eq. 5-4 and Fig. 5-7). Since longitudinal elements of a beam are subjected only to tension or compression, we can use the **stress-strain curve** for the material to determine the stresses from the strains. The stresses act over the entire cross section of the beam and vary in intensity depending upon the shape of the stress-strain diagram and the dimensions of the cross section. Since the x direction is longitudinal (Fig. 5-7a), we use the symbol σ_x to denote these stresses.

The most common stress-strain relationship encountered in engineering is the equation for a **linearly elastic material**. For such materials we substitute Hooke's law for uniaxial stress ($\sigma = E\epsilon$) into Eq. (5-4) and obtain

$$\sigma_x = E\epsilon_x = -\frac{Ey}{\rho} = -E\kappa y \tag{5-7}$$

This equation shows that the normal stresses acting on the cross section vary linearly with the distance y from the neutral surface. This stress distribution is pictured in Fig. 5-9a for the case in which the bending moment M is positive and the beam bends with positive curvature.

When the curvature is positive, the stresses σ_x are negative (compression) above the neutral surface and positive (tension) below it. In the figure, compressive stresses are indicated by arrows pointing *toward* the cross section and tensile stresses are indicated by arrows pointing *away* from the cross section.

In order for Eq. (5-7) to be of practical value, we must locate the origin of coordinates so that we can determine the distance y. In other words, we must locate the neutral axis of the cross section. We also need to obtain a relationship between the curvature and the bending moment—so that we can substitute into Eq. (5-7) and obtain an equation relating the stresses to the bending moment. These two objectives can be accomplished by determining the resultant of the stresses σ_x acting on the cross section.

In general, the **resultant of the normal stresses** consists of two stress resultants: (1) a force acting in the x direction, and (2) a bending couple acting about the z axis. However, the axial force is zero when a beam is in pure bending. Therefore, we can write the following equations of statics: (1) The resultant force in the x direction is equal to zero, and (2) the resultant moment is equal to the bending moment M. The first equation gives the location of the neutral axis and the second gives the moment-curvature relationship.

Location of Neutral Axis

To obtain the first equation of statics, we consider an element of area dA in the cross section (Fig. 5-9b). The element is located at distance y from

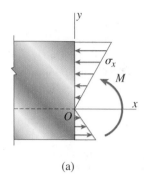

(a)

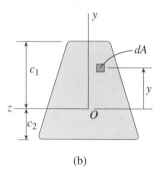

(b)

FIG. 5-9 (Repeated)

the neutral axis, and therefore the stress σ_x acting on the element is given by Eq. (5-7). The *force* acting on the element is equal to $\sigma_x dA$ and is compressive when y is positive. Because there is no resultant force acting on the cross section, the integral of $\sigma_x dA$ over the area A of the entire cross section must vanish; thus, the *first equation of statics* is

$$\int_A \sigma_x dA = -\int_A E\kappa y\, dA = 0 \qquad (a)$$

Because the curvature κ and modulus of elasticity E are nonzero constants at any given cross section of a bent beam, they are not involved in the integration over the cross-sectional area. Therefore, we can drop them from the equation and obtain

$$\int_A y\, dA = 0 \qquad (5-8)$$

This equation states that the first moment of the area of the cross section, evaluated with respect to the z axis, is zero. In other words, the z axis must pass through the centroid of the cross section.[*]

Since the z axis is also the neutral axis, we have arrived at the following important conclusion: *The neutral axis passes through the centroid of the cross-sectional area when the material follows Hooke's law and there is no axial force acting on the cross section.* This observation makes it relatively simple to determine the position of the neutral axis.

As explained in Section 5.1, our discussion is limited to beams for which the y axis is an axis of symmetry. Consequently, the y axis also passes through the centroid. Therefore, we have the following additional conclusion: *The origin O of coordinates* (Fig. 5-9b) *is located at the centroid of the cross-sectional area.*

Because the y axis is an axis of symmetry of the cross section, it follows that the y axis is a *principal axis* (see Chapter 10, Section 10.9 available online for a discussion of principal axes). Since the z axis is perpendicular to the y axis, it too is a principal axis. Thus, when a beam of linearly elastic material is subjected to pure bending, *the y and z axes are principal centroidal axes.*

Moment-Curvature Relationship

The *second equation of statics* expresses the fact that the moment resultant of the normal stresses σ_x acting over the cross section is equal to the bending moment M (Fig. 5-9a). The element of force $\sigma_x dA$ acting on the element of area dA (Fig. 5-9b) is in the positive direction of the x axis when σ_x is positive and in the negative direction when σ_x is negative. Since the

---

[*]Centroids and first moments of areas are discussed in Chapter 10, Sections 10.2 and 10.3, which is available online.

element dA is located above the neutral axis, a positive stress σ_x acting on that element produces an element of moment equal to $\sigma_x y \, dA$. This element of moment acts opposite in direction to the positive bending moment M shown in Fig. 5-9a. Therefore, the elemental moment is

$$dM = -\sigma_x y \, dA$$

The integral of all such elemental moments over the entire cross-sectional area A must equal the bending moment:

$$M = -\int_A \sigma_x y \, dA \tag{b}$$

or, upon substituting for σ_x from Eq. (5-7),

$$M = \int_A \kappa E y^2 \, dA = \kappa E \int_A y^2 \, dA \tag{5-9}$$

This equation relates the curvature of the beam to the bending moment M.

Since the integral in the preceding equation is a property of the cross-sectional area, it is convenient to rewrite the equation as follows:

$$M = \kappa E I \tag{5-10}$$

in which

$$I = \int_A y^2 \, dA \tag{5-11}$$

This integral is the **moment of inertia** of the cross-sectional area with respect to the z axis (that is, with respect to the neutral axis). Moments of inertia are always positive and have dimensions of length to the fourth power; for instance, typical USCS units are in.[4] and typical SI units are mm[4] when performing beam calculations.*

Equation (5-10) can now be rearranged to express the *curvature* in terms of the bending moment in the beam:

$$\kappa = \frac{1}{\rho} = \frac{M}{EI} \tag{5-12}$$

Known as the **moment-curvature equation**, Eq. (5-12) shows that the curvature is directly proportional to the bending moment M and inversely proportional to the quantity EI, which is called the **flexural rigidity** of the beam. Flexural rigidity is a measure of the resistance of a beam to bending, that is, the larger the flexural rigidity, the smaller the curvature for a given bending moment.

Comparing the **sign convention** for bending moments (Fig. 4-5) with that for curvature (Fig. 5-6), we see that *a positive bending moment produces positive curvature and a negative bending moment produces negative curvature* (see Fig. 5-10).

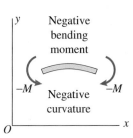

FIG. 5-10 Relationships between signs of bending moments and signs of curvatures

---

*Moments of inertia of areas are discussed in Chapter 10, Section 10.4 available online.

Flexure Formula

Now that we have located the neutral axis and derived the moment-curvature relationship, we can determine the stresses in terms of the bending moment. Substituting the expression for curvature (Eq. 5-12) into the expression for the stress σ_x (Eq. 5-7), we get

$$\sigma_x = - \frac{My}{I} \tag{5-13}$$

This equation, called the **flexure formula**, shows that the stresses are directly proportional to the bending moment M and inversely proportional to the moment of inertia I of the cross section. Also, the stresses vary linearly with the distance y from the neutral axis, as previously observed. Stresses calculated from the flexure formula are called **bending stresses** or **flexural stresses**.

If the bending moment in the beam is positive, the bending stresses will be positive (tension) over the part of the cross section where y is negative, that is, over the lower part of the beam. The stresses in the upper part of the beam will be negative (compression). If the bending moment is negative, the stresses will be reversed. These relationships are shown in Fig. 5-11.

Maximum Stresses at a Cross Section

The maximum tensile and compressive bending stresses acting at any given cross section occur at points located farthest from the neutral axis. Let us denote by c_1 and c_2 the distances from the neutral axis to the extreme elements in the positive and negative y directions, respectively (see Fig. 5-9b and Fig. 5-11). Then the corresponding **maximum normal stresses** σ_1 and σ_2 (from the flexure formula) are

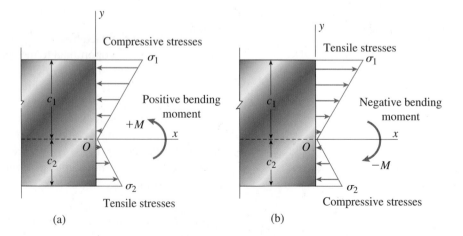

FIG. 5-11 Relationships between signs of bending moments and directions of normal stresses: (a) positive bending moment, and (b) negative bending moment

$$\sigma_1 = -\frac{Mc_1}{I} = -\frac{M}{S_1} \qquad \sigma_2 = \frac{Mc_2}{I} = \frac{M}{S_2} \qquad \text{(5-14a,b)}$$

in which

$$S_1 = \frac{I}{c_1} \qquad S_2 = \frac{I}{c_2} \qquad \text{(5-15a,b)}$$

The quantities S_1 and S_2 are known as the **section moduli** of the cross-sectional area. From Eqs. (5-15a and b) we see that each section modulus has dimensions of length to the third power (for example, in.^{3} or mm^3). Note that the distances c_1 and c_2 to the top and bottom of the beam are always taken as positive quantities.

The advantage of expressing the maximum stresses in terms of section moduli arises from the fact that each section modulus combines the beam's relevant cross-sectional properties into a single quantity. Then this quantity can be listed in tables and handbooks as a property of the beam, which is a convenience to designers. (Design of beams using section moduli is explained in the next section.)

Doubly Symmetric Shapes

If the cross section of a beam is symmetric with respect to the z axis as well as the y axis (*doubly symmetric cross section*), then $c_1 = c_2 = c$ and the maximum tensile and compressive stresses are equal numerically:

$$\sigma_1 = -\sigma_2 = -\frac{Mc}{I} = -\frac{M}{S} \qquad \text{or} \qquad \sigma_{\max} = \frac{M}{S} \qquad \text{(5-16a,b)}$$

in which

$$S = \frac{I}{c} \qquad \text{(5-17)}$$

is the only section modulus for the cross section.

For a beam of **rectangular cross section** with width b and height h (Fig. 5-12a), the moment of inertia and section modulus are

$$I = \frac{bh^3}{12} \qquad S = \frac{bh^2}{6} \qquad \text{(5-18a,b)}$$

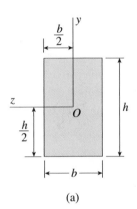

(a)

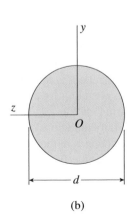

(b)

FIG. 5-12 Doubly symmetric cross-sectional shapes

For a **circular cross section** of diameter d (Fig. 5-12b), these properties are

$$I = \frac{\pi d^4}{64} \qquad S = \frac{\pi d^3}{32} \qquad\qquad (5\text{-}19a,b)$$

Properties of other doubly symmetric shapes, such as hollow tubes (either rectangular or circular) and wide-flange shapes, can be readily obtained from the preceding formulas.

Properties of Beam Cross Sections

Moments of inertia of many plane figures are listed in Appendix E (available online) for convenient reference. Also, the dimensions and properties of standard sizes of steel and wood beams are listed in Appendixes F and G (available online) and in many engineering handbooks, as explained in more detail in the next section.

For other cross-sectional shapes, we can determine the location of the neutral axis, the moment of inertia, and the section moduli by direct calculation, using the techniques described in Chapter 10 (available online). This procedure is illustrated later in Example 5-4.

Limitations

The analysis presented in this section is for pure bending of prismatic beams composed of homogeneous, linearly elastic materials. If a beam is subjected to nonuniform bending, the shear forces will produce warping (or out-of-plane distortion) of the cross sections. Thus, a cross section that was plane before bending is no longer plane after bending. Warping due to shear deformations greatly complicates the behavior of the beam. However, detailed investigations show that the normal stresses calculated from the flexure formula are not significantly altered by the presence of shear stresses and the associated warping (Ref. 2-1, pp. 42 and 48; a list of references is available online). Thus, we may justifiably use the theory of pure bending for calculating normal stresses in beams subjected to nonuniform bending.[*]

The flexure formula gives results that are accurate only in regions of the beam where the stress distribution is not disrupted by changes in the shape of the beam or by discontinuities in loading. For instance, the flexure formula is not applicable near the supports of a beam or close to a concentrated load. Such irregularities produce localized stresses, or *stress concentrations*, that are much greater than the stresses obtained from the flexure formula.

---

[*]Beam theory began with Galileo Galilei (1564–1642), who investigated the behavior of various types of beams. His work in mechanics of materials is described in his famous book *Two New Sciences*, first published in 1638 (Ref. 5-2; a list of references is available online). Although Galileo made many important discoveries regarding beams, he did not obtain the stress distribution that we use today. Further progress in beam theory was made by Mariotte, Jacob Bernoulli, Euler, Parent, Saint-Venant, and others (Ref. 5-3).

Example 5-2

A high-strength steel wire of diameter d is bent around a cylindrical drum of radius R_0 (Fig. 5-13).

Determine the bending moment M and maximum bending stress σ_{max} in the wire, assuming $d = 4$ mm and $R_0 = 0.5$ m. (The steel wire has modulus of elasticity $E = 200$ GPa and proportional limit $\sigma_{pl} = 1200$ MPa.)

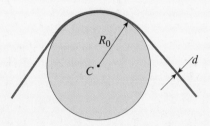

FIG. 5-13 Example 5-2. Wire bent around a drum

Solution

The first step in this example is to determine the radius of curvature ρ of the bent wire. Then, knowing ρ, we can find the bending moment and maximum stresses.

Radius of curvature. The radius of curvature of the bent wire is the distance from the center of the drum to the neutral axis of the cross section of the wire:

$$\rho = R_0 + \frac{d}{2} \tag{5-20}$$

Bending moment. The bending moment in the wire may be found from the moment-curvature relationship (Eq. 5-12):

$$M = \frac{EI}{\rho} = \frac{2EI}{2R_0 + d} \tag{5-21}$$

in which I is the moment of inertia of the cross-sectional area of the wire. Substituting for I in terms of the diameter d of the wire (Eq. 5-19a), we get

$$M = \frac{\pi E d^4}{32(2R_0 + d)} \tag{5-22}$$

continued

This result was obtained without regard to the *sign* of the bending moment, since the direction of bending is obvious from the figure.

Maximum bending stresses. The maximum tensile and compressive stresses, which are equal numerically, are obtained from the flexure formula as given by Eq. (5-16b):

$$\sigma_{max} = \frac{M}{S}$$

in which S is the section modulus for a circular cross section. Substituting for M from Eq. (5-22) and for S from Eq. (5-19b), we get

$$\sigma_{max} = \frac{Ed}{2R_0 + d} \tag{5-23}$$

This same result can be obtained directly from Eq. (5-7) by replacing y with $d/2$ and substituting for ρ from Eq. (5-20).

We see by inspection of Fig. 5-13 that the stress is compressive on the lower (or inner) part of the wire and tensile on the upper (or outer) part.

Numerical results. We now substitute the given numerical data into Eqs. (5-22) and (5-23) and obtain the following results:

$$M = \frac{\pi E d^4}{32(2R_0 + d)} = \frac{\pi (200 \text{ GPa})(4 \text{ mm})^4}{32[2(0.5 \text{ m}) + 4 \text{ mm}]} = 5.01 \text{ N·m} \quad \Longleftarrow$$

$$\sigma_{max} = \frac{Ed}{2R_0 + d} = \frac{(200 \text{ GPa})(4 \text{ mm})}{2(0.5 \text{ m}) + 4 \text{ mm}} = 797 \text{ MPa} \quad \Longleftarrow$$

Note that σ_{max} is less than the proportional limit of the steel wire, and therefore the calculations are valid.

Note: Because the radius of the drum is large compared to the diameter of the wire, we can safely disregard d in comparison with $2R_0$ in the denominators of the expressions for M and σ_{max}. Then Eqs. (5-22) and (5-23) yield the following results:

$$M = 5.03 \text{ N·m} \qquad \sigma_{max} = 800 \text{ MPa}$$

These results are on the conservative side and differ by less than 1% from the more precise values.

Example 5-3

A simple beam AB of span length $L = 22$ ft (Fig. 5-14a) supports a uniform load of intensity $q = 1.5$ k/ft and a concentrated load $P = 12$ k. The uniform load includes an allowance for the weight of the beam. The concentrated load acts at a point 9.0 ft from the left-hand end of the beam. The beam is constructed of glued laminated wood and has a cross section of width $b = 8.75$ in. and height $h = 27$ in. (Fig. 5-14b).

Determine the maximum tensile and compressive stresses in the beam due to bending.

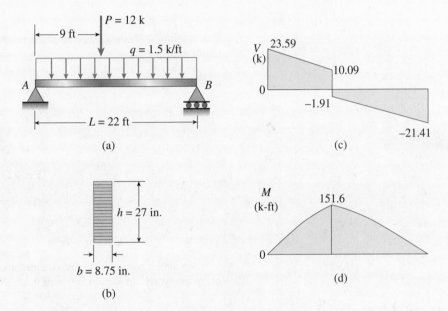

FIG. 5-14 Example 5-3. Stresses in a simple beam

Solution

Reactions, shear forces, and bending moments. We begin the analysis by calculating the reactions at supports A and B, using the techniques described in Chapter 4. The results are

$$R_A = 23.59 \text{ k} \qquad R_B = 21.41 \text{ k}$$

Knowing the reactions, we can construct the shear-force diagram, shown in Fig. 5-14c. Note that the shear force changes from positive to negative under the concentrated load P, which is at a distance of 9 ft from the left-hand support.

continued

Next, we draw the bending-moment diagram (Fig. 5-14d) and determine the maximum bending moment, which occurs under the concentrated load where the shear force changes sign. The maximum moment is

$$M_{max} = 151.6 \text{ k-ft}$$

The maximum bending stresses in the beam occur at the cross section of maximum moment.

Section modulus. The section modulus of the cross-sectional area is calculated from Eq. (5-18b), as follows:

$$S = \frac{bh^2}{6} = \frac{1}{6}(8.75 \text{ in.})(27 \text{ in.})^2 = 1063 \text{ in.}^3$$

Maximum stresses. The maximum tensile and compressive stresses σ_t and σ_c, respectively, are obtained from Eq. (5-16a):

$$\sigma_t = \sigma_2 = \frac{M_{max}}{S} = \frac{(151.6 \text{ k-ft})(12 \text{ in./ft})}{1063 \text{ in.}^3} = 1710 \text{ psi} \quad \Longleftarrow$$

$$\sigma_c = \sigma_1 = -\frac{M_{max}}{S} = -1710 \text{ psi} \quad \Longleftarrow$$

Because the bending moment is positive, the maximum tensile stress occurs at the bottom of the beam and the maximum compressive stress occurs at the top.

Example 5-4

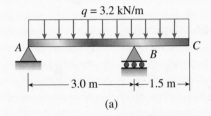

(a)

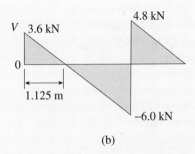

(b)

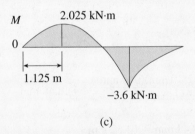

(c)

FIG. 5-15 Example 5-4. Stresses in a beam with an overhang

The beam ABC shown in Fig 5-15a has simple supports at A and B and an overhang from B to C. The length of the span is 3.0 m and the length of the overhang is 1.5 m. A uniform load of intensity $q = 3.2$ kN/m acts throughout the entire length of the beam (4.5 m).

The beam has a cross section of channel shape with width $b = 300$ mm and height $h = 80$ mm (Fig. 5-16a). The web thickness is $t = 12$ mm, and the average thickness of the sloping flanges is the same. For the purpose of calculating the properties of the cross section, assume that the cross section consists of three rectangles, as shown in Fig. 5-16b.

Determine the maximum tensile and compressive stresses in the beam due to the uniform load.

Solution

Reactions, shear forces, and bending moments. We begin the analysis of this beam by calculating the reactions at supports A and B, using the techniques described in Chapter 4. The results are

$$R_A = 3.6 \text{ kN} \qquad R_B = 10.8 \text{ kN}$$

From these values, we construct the shear-force diagram (Fig. 5-15b). Note that the shear force changes sign and is equal to zero at two locations: (1) at a distance of 1.125 m from the left-hand support, and (2) at the right-hand reaction.

Next, we draw the bending-moment diagram, shown in Fig. 5-15c. Both the maximum positive and maximum negative bending moments occur at the cross sections where the shear force changes sign. These maximum moments are

$$M_{\text{pos}} = 2.025 \text{ kN·m} \qquad M_{\text{neg}} = -3.6 \text{ kN·m}$$

respectively.

Neutral axis of the cross section (Fig. 5-16b). The origin O of the yz coordinates is placed at the centroid of the cross-sectional area, and therefore the z axis becomes the neutral axis of the cross section. The centroid is located by using the techniques described in Chapter 10, Section 10.3 (available online), as follows.

First, we divide the area into three rectangles (A_1, A_2, and A_3). Second, we establish a reference axis Z-Z across the upper edge of the cross section, and we let y_1 and y_2 be the distances from the Z-Z axis to the centroids of areas A_1 and

continued

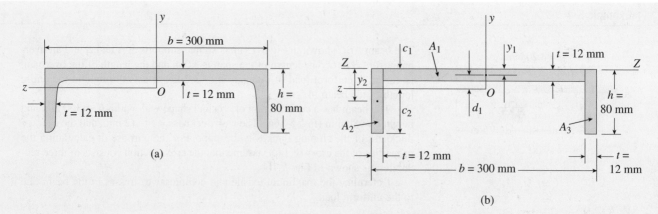

FIG. 5-16 Cross section of beam discussed in Example 5-4. (a) Actual shape, and (b) idealized shape for use in analysis (the thickness of the beam is exaggerated for clarity)

A_2, respectively. Then the calculations for locating the centroid of the entire channel section (distances c_1 and c_2) are as follows:

Area 1:
$$y_1 = t/2 = 6 \text{ mm}$$
$$A_1 = (b - 2t)(t) = (276 \text{ mm})(12 \text{ mm}) = 3312 \text{ mm}^2$$

Area 2:
$$y_2 = h/2 = 40 \text{ mm}$$
$$A_2 = ht = (80 \text{ mm})(12 \text{ mm}) = 960 \text{ mm}^2$$

Area 3:
$$y_3 = y_2 \qquad A_3 = A_2$$

$$c_1 = \frac{\sum y_i A_i}{\sum A_i} = \frac{y_1 A_1 + 2y_2 A_2}{A_1 + 2A_2}$$

$$= \frac{(6 \text{ mm})(3312 \text{ mm}^2) + 2(40 \text{ mm})(960 \text{ mm}^2)}{3312 \text{ mm}^2 + 2(960 \text{ mm}^2)} = 18.48 \text{ mm}$$

$$c_2 = h - c_1 = 80 \text{ mm} - 18.48 \text{ mm} = 61.52 \text{ mm}$$

Thus, the position of the neutral axis (the z axis) is determined.

Moment of inertia. In order to calculate the stresses from the flexure formula, we must determine the moment of inertia of the cross-sectional area with respect to the neutral axis. These calculations require the use of the parallel-axis theorem (see Chapter 10, Section 10.5 available online).

Beginning with area A_1, we obtain its moment of inertia $(I_z)_1$ about the z axis from the equation

$$(I_z)_1 = (I_c)_1 + A_1 d_1^2 \tag{c}$$

In this equation, $(I_c)_1$ is the moment of inertia of area A_1 about its own centroidal axis:

$$(I_c)_1 = \frac{1}{12}(b - 2t)(t)^3 = \frac{1}{12}(276 \text{ mm})(12 \text{ mm})^3 = 39{,}744 \text{ mm}^4$$

and d_1 is the distance from the centroidal axis of area A_1 to the z axis:

$$d_1 = c_1 - t/2 = 18.48 \text{ mm} - 6 \text{ mm} = 12.48 \text{ mm}$$

Therefore, the moment of inertia of area A_1 about the z axis (from Eq. c) is

$$(I_z)_1 = 39{,}744 \text{ mm}^4 + (3312 \text{ mm}^2)(12.48 \text{ mm}^2) = 555{,}600 \text{ mm}^4$$

Proceeding in the same manner for areas A_2 and A_3, we get

$$(I_z)_2 = (I_z)_3 = 956{,}600 \text{ mm}^4$$

Thus, the centroidal moment of inertia I_z of the entire cross-sectional area is

$$I_z = (I_z)_1 + (I_z)_2 + (I_z)_3 = 2.469 \times 10^6 \text{ mm}^4$$

Section moduli. The section moduli for the top and bottom of the beam, respectively, are

$$S_1 = \frac{I_z}{c_1} = 133{,}600 \text{ mm}^3 \qquad S_2 = \frac{I_z}{c_2} = 40{,}100 \text{ mm}^3$$

(see Eqs. 5-15a and b). With the cross-sectional properties determined, we can now proceed to calculate the maximum stresses from Eqs. (5-14a and b).

Maximum stresses. At the cross section of maximum positive bending moment, the largest tensile stress occurs at the bottom of the beam (σ_2) and the largest compressive stress occurs at the top (σ_1). Thus, from Eqs. (5-14b) and (5-14a), respectively, we get

$$\sigma_t = \sigma_2 = \frac{M_{\text{pos}}}{S_2} = \frac{2.025 \text{ kN·m}}{40{,}100 \text{ mm}^3} = 50.5 \text{ MPa}$$

$$\sigma_c = \sigma_1 = -\frac{M_{\text{pos}}}{S_1} = -\frac{2.025 \text{ kN·m}}{133{,}600 \text{ mm}^3} = -15.2 \text{ MPa}$$

Similarly, the largest stresses at the section of maximum negative moment are

$$\sigma_t = \sigma_1 = -\frac{M_{\text{neg}}}{S_1} = -\frac{-3.6 \text{ kN·m}}{133{,}600 \text{ mm}^3} = 26.9 \text{ MPa}$$

$$\sigma_c = \sigma_2 = \frac{M_{\text{neg}}}{S_2} = \frac{-3.6 \text{ kN·m}}{40{,}100 \text{ mm}^3} = -89.8 \text{ MPa}$$

A comparison of these four stresses shows that the largest tensile stress in the beam is 50.5 MPa and occurs at the bottom of the beam at the cross section of maximum positive bending moment; thus,

$$(\sigma_t)_{\text{max}} = 50.5 \text{ MPa}$$

The largest compressive stress is -89.8 MPa and occurs at the bottom of the beam at the section of maximum negative moment:

$$(\sigma_c)_{\text{max}} = -89.8 \text{ MPa}$$

Thus, we have determined the maximum bending stresses due to the uniform load acting on the beam.

5.6 DESIGN OF BEAMS FOR BENDING STRESSES

The process of designing a beam requires that many factors be considered, including the type of structure (airplane, automobile, bridge, building, or whatever), the materials to be used, the loads to be supported, the environmental conditions to be encountered, and the costs to be paid. However, from the standpoint of strength, the task eventually reduces to selecting a shape and size of beam such that the actual stresses in the beam do not exceed the allowable stresses for the material. In this section, we will consider only the bending stresses (that is, the stresses obtained from the flexure formula, Eq. 5-13). Later, we will consider the effects of shear stresses (Sections 5.7, 5.8, and 5.9).

When designing a beam to resist bending stresses, we usually begin by calculating the **required section modulus**. For instance, if the beam has a doubly symmetric cross section and the allowable stresses are the same for both tension and compression, we can calculate the required modulus by dividing the maximum bending moment by the allowable bending stress for the material (see Eq. 5-16):

$$S = \frac{M_{\max}}{\sigma_{\text{allow}}} \qquad (5\text{-}24)$$

The allowable stress is based upon the properties of the material and the desired factor of safety. To ensure that this stress is not exceeded, we must choose a beam that provides a section modulus at least as large as that obtained from Eq. (5-24).

If the cross section is not doubly symmetric, or if the allowable stresses are different for tension and compression, we usually need to determine two required section moduli—one based upon tension and the other based upon compression. Then we must provide a beam that satisfies both criteria.

To minimize weight and save material, we usually select a beam that has the least cross-sectional area while still providing the required section moduli (and also meeting any other design requirements that may be imposed).

Beams are constructed in a great variety of shapes and sizes to suit a myriad of purposes. For instance, very large steel beams are fabricated by welding (Fig. 5-17), aluminum beams are extruded as round or rectangular tubes, wood beams are cut and glued to fit special requirements, and reinforced concrete beams are cast in any desired shape by proper construction of the forms.

In addition, beams of steel, aluminum, plastic, and wood can be ordered in **standard shapes and sizes** from catalogs supplied by dealers and manufacturers. Readily available shapes include wide-flange beams, I-beams, angles, channels, rectangular beams, and tubes.

Beams of Standardized Shapes and Sizes

The dimensions and properties of many kinds of beams are listed in engineering handbooks. For instance, in the United States the shapes and sizes of structural-steel beams are standardized by the American Institute of Steel Construction (AISC), which publishes manuals giving their properties in both USCS and SI units (Ref. 5-4; a list of references is available online). The tables in these manuals list cross-sectional dimensions and properties such as weight, cross-sectional area, moment of inertia, and section modulus.

Properties of aluminum and wood beams are tabulated in a similar manner and are available in publications of the Aluminum Association (Ref. 5-5) and the American Forest and Paper Association (Ref. 5-6).

Abridged tables of steel beams and wood beams are given later in this book for use in solving problems using both USCS and SI units (see Appendixes F and G available online).

Structural-steel sections are given a designation such as W 30×211 in USCS units, which means that the section is of W shape (also called a wide-flange shape) with a nominal depth of 30 in. and a weight of 211 lb per ft of length (see Table F-1(a), Appendix F). The corresponding properties for each W shape are also given in SI units in Table F-1(b). For example, in SI units, the W 30×211 is listed as W 760×314 with a nominal depth of 760 millimeters and mass of 314 kilograms per meter of length.

Similar designations are used for S shapes (also called I-beams) and C shapes (also called channels), as shown in Tables F-2(a) and F-3(a) in USCS units and in Tables F-2(b) and F-3(b) in SI units. Angle sections, or L shapes, are designated by the lengths of the two legs and the thickness (see Tables F-4 and F-5). For example, L $8 \times 6 \times 1$ [see Table F-5(a)] denotes an angle with unequal legs, one of length 8 in. and the other of length 6 in., with a thickness of 1 in. The corresponding label in SI units for this unequal leg angle is L $203 \times 152 \times 25.4$ [see Table F-5(b)].

The standardized steel sections described above are manufactured by *rolling*, a process in which a billet of hot steel is passed back and forth between rolls until it is formed into the desired shape.

Aluminum structural sections are usually made by the process of *extrusion*, in which a hot billet is pushed, or extruded, through a shaped die. Since dies are relatively easy to make and the material is workable, aluminum beams can be extruded in almost any desired shape. Standard shapes of wide-flange beams, I-beams, channels, angles, tubes, and other sections are listed in the *Aluminum Design Manual* (Ref. 5-5). In addition, custom-made shapes can be ordered.

Most **wood beams** have rectangular cross sections and are designated by nominal dimensions, such as 4×8 inches. These dimensions represent the rough-cut size of the lumber. The net dimensions (or actual dimensions) of a wood beam are smaller than the nominal dimensions if the sides of the rough lumber have been planed, or *surfaced*, to make them smooth. Thus, a 4 × 8 wood beam has actual dimensions 3.5 × 7.25 in. after it has been surfaced. Of course, the net dimensions of surfaced lumber should be used in all engineering computations. Therefore, net dimensions and the corresponding properties (in USCS units) are given in Appendix G available online. Similar tables are available in SI units.

Relative Efficiency of Various Beam Shapes

One of the objectives in designing a beam is to use the material as efficiently as possible within the constraints imposed by function, appearance, manufacturing costs, and the like. From the standpoint of strength alone, efficiency in bending depends primarily upon the shape of the cross section. In particular, the most efficient beam is one in which the material is located as far as practical from the neutral axis. The farther a given amount of material is from the neutral axis, the larger the section modulus becomes—and the larger the section modulus, the larger the bending moment that can be resisted (for a given allowable stress).

As an illustration, consider a cross section in the form of a **rectangle** of width b and height h (Fig. 5-18a). The section modulus (from Eq. 5-18b) is

$$S = \frac{bh^2}{6} = \frac{Ah}{6} = 0.167Ah \tag{5-25}$$

where A denotes the cross-sectional area. This equation shows that a rectangular cross section of given area becomes more efficient as the height h is increased (and the width b is decreased to keep the area constant). Of course, there is a practical limit to the increase in height, because the beam becomes laterally unstable when the ratio of height to width becomes too large. Thus, a beam of very narrow rectangular section will fail due to lateral (sideways) buckling rather than to insufficient strength of the material.

Next, let us compare a **solid circular cross section** of diameter d (Fig. 5-18b) with a square cross section of the same area. The side h of a square having the same area as the circle is $h = (d/2)\sqrt{\pi}$. The corresponding section moduli (from Eqs. 5-18b and 5-19b) are

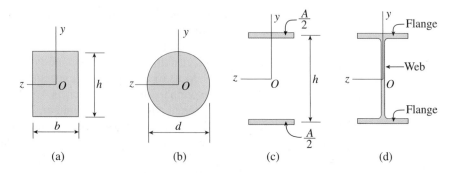

FIG. 5-18 Cross-sectional shapes of beams

(a)   (b)   (c)   (d)

$$S_{\text{square}} = \frac{h^3}{6} = \frac{\pi\sqrt{\pi}d^3}{48} = 0.1160d^3 \qquad (5\text{-}26\text{a})$$

$$S_{\text{circle}} = \frac{\pi d^3}{32} = 0.0982d^3 \qquad (5\text{-}26\text{b})$$

from which we get

$$\frac{S_{\text{square}}}{S_{\text{circle}}} = 1.18 \qquad (5\text{-}27)$$

This result shows that a beam of square cross section is more efficient in resisting bending than is a circular beam of the same area. The reason, of course, is that a circle has a relatively larger amount of material located near the neutral axis. This material is less highly stressed, and therefore it does not contribute as much to the strength of the beam.

The **ideal cross-sectional shape** for a beam of given cross-sectional area A and height h would be obtained by placing one-half of the area at a distance $h/2$ above the neutral axis and the other half at distance $h/2$ below the neutral axis, as shown in Fig. 5-18c. For this ideal shape, we obtain

$$I = 2\left(\frac{A}{2}\right)\left(\frac{h}{2}\right)^2 = \frac{Ah^2}{4} \qquad S = \frac{I}{h/2} = 0.5Ah \qquad (5\text{-}28\text{a,b})$$

These theoretical limits are approached in practice by wide-flange sections and I-sections, which have most of their material in the flanges (Fig. 5-18d). For standard wide-flange beams, the section modulus is approximately

$$S \approx 0.35Ah \qquad (5\text{-}29)$$

which is less than the ideal but much larger than the section modulus for a rectangular cross section of the same area and height (see Eq. 5-25).

Another desirable feature of a wide-flange beam is its greater width, and hence greater stability with respect to sideways buckling, when compared to a rectangular beam of the same height and section modulus. On the other hand, there are practical limits to how thin we can make the web of a wide-flange beam. If the web is too thin, it will be susceptible to localized buckling or it may be overstressed in shear, a topic that is discussed in Section 5.9.

The following four examples illustrate the process of selecting a beam on the basis of the allowable stresses. In these examples, only the effects of bending stresses (obtained from the flexure formula) are considered.

Note: When solving examples and problems that require the selection of a steel or wood beam from the tables in the appendix, we use the following rule: *If several choices are available in a table, select the lightest beam that will provide the required section modulus.*

Example 5-5

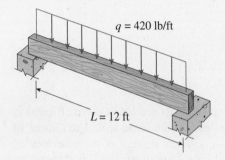

$q = 420$ lb/ft

$L = 12$ ft

FIG. 5-19 Example 5-5. Design of a simply supported wood beam

A simply supported wood beam having a span length $L = 12$ ft carries a uniform load $q = 420$ lb/ft (Fig. 5-19). The allowable bending stress is 1800 psi, the wood weighs 35 lb/ft³, and the beam is supported laterally against sideways buckling and tipping.

Select a suitable size for the beam from the table in Appendix G available online.

Solution

Since we do not know in advance how much the beam weighs, we will proceed by trial-and-error as follows: (1) Calculate the required section modulus based upon the given uniform load. (2) Select a trial size for the beam. (3) Add the weight of the beam to the uniform load and calculate a new required section modulus. (4) Check to see that the selected beam is still satisfactory. If it is not, select a larger beam and repeat the process.

(1) The maximum bending moment in the beam occurs at the midpoint (see Eq. 4-15):

$$M_{max} = \frac{qL^2}{8} = \frac{(420 \text{ lb/ft})(12 \text{ ft})^2(12 \text{ in./ft})}{8} = 90{,}720 \text{ lb-in.}$$

The required section modulus (Eq. 5-24) is

$$S = \frac{M_{max}}{\sigma_{allow}} = \frac{90{,}720 \text{ lb-in.}}{1800 \text{ psi}} = 50.40 \text{ in.}^3$$

(2) From the table in Appendix G we see that the lightest beam that supplies a section modulus of at least 50.40 in.³ about axis 1-1 is a 3 × 12 in. beam (nominal dimensions). This beam has a section modulus equal to 52.73 in.³ and weighs 6.8 lb/ft. (Note that Appendix G available online gives weights of beams based upon a density of 35 lb/ft³.)

(3) The uniform load on the beam now becomes 426.8 lb/ft, and the corresponding required section modulus is

$$S = (50.40 \text{ in.}^3)\left(\frac{426.8 \text{ lb/ft}}{420 \text{ lb/ft}}\right) = 51.22 \text{ in.}^3$$

(4) The previously selected beam has a section modulus of 52.73 in.³, which is larger than the required modulus of 51.22 in.³

Therefore, a 3 × 12 in. beam is satisfactory.   ⬅

Note: If the weight density of the wood is other than 35 lb/ft³, we can obtain the weight of the beam per linear foot by multiplying the value in the last column in Appendix G by the ratio of the actual weight density to 35 lb/ft³.

Example 5-6

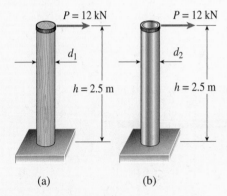

FIG. 5-20 Example 5-6. (a) Solid wood
post, and (b) aluminum tube

A vertical post 2.5-meters high must support a lateral load $P = 12$ kN at its upper
end (Fig. 5-20). Two plans are proposed—a solid wood post and a hollow
aluminum tube.

(a) What is the minimum required diameter d_1 of the wood post if the allowable bending stress in the wood is 15 MPa?

(b) What is the minimum required outer diameter d_2 of the aluminum tube
if its wall thickness is to be one-eighth of the outer diameter and the allowable
bending stress in the aluminum is 50 MPa?

Solution

Maximum bending moment. The maximum moment occurs at the base of the
post and is equal to the load P times the height h; thus,

$$M_{max} = Ph = (12 \text{ kN})(2.5 \text{ m}) = 30 \text{ kN·m}$$

(a) *Wood post.* The required section modulus S_1 for the wood post (see
Eqs. 5-19b and 5-24) is

$$S_1 = \frac{\pi d_1^3}{32} = \frac{M_{max}}{\sigma_{allow}} = \frac{30 \text{ kN·m}}{15 \text{ MPa}} = 0.0020 \text{ m}^3 = 2 \times 10^6 \text{ mm}^3$$

Solving for the diameter, we get

$$d_1 = 273 \text{ mm}$$

The diameter selected for the wood post must be equal to or larger than 273 mm
if the allowable stress is not to be exceeded.

(b) *Aluminum tube.* To determine the section modulus S_2 for the tube, we
first must find the moment of inertia I_2 of the cross section. The wall thickness
of the tube is $d_2/8$, and therefore the inner diameter is $d_2 - d_2/4$, or $0.75d_2$. Thus,
the moment of inertia (see Eq. 5-19a) is

$$I_2 = \frac{\pi}{64}\left[d_2^4 - (0.75d_2)^4\right] = 0.03356d_2^4$$

The section modulus of the tube is now obtained from Eq. (5-17) as follows:

$$S_2 = \frac{I_2}{c} = \frac{0.03356d_2^4}{d_2/2} = 0.06712d_2^3$$

The required section modulus is obtained from Eq. (5-24):

$$S_2 = \frac{M_{max}}{\sigma_{allow}} = \frac{30 \text{ kN·m}}{50 \text{ MPa}} = 0.0006 \text{ m}^3 = 600 \times 10^3 \text{ mm}^3$$

By equating the two preceding expressions for the section modulus, we can solve
for the required outer diameter:

$$d_2 = \left(\frac{600 \times 10^3 \text{ mm}^3}{0.06712}\right)^{1/3} = 208 \text{ mm}$$

The corresponding inner diameter is 0.75(208 mm), or 156 mm.

Example 5-7

A simple beam AB of span length 21 ft must support a uniform load $q = 2000$ lb/ft distributed along the beam in the manner shown in Fig. 5-21a.

Considering both the uniform load and the weight of the beam, and also using an allowable bending stress of 18,000 psi, select a structural steel beam of wide-flange shape to support the loads.

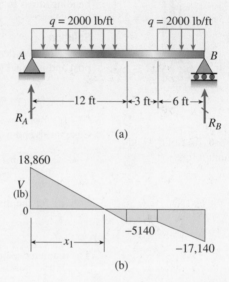

(a)

(b)

FIG. 5-21 Example 5-7. Design of a simple beam with partial uniform loads

Solution

In this example, we will proceed as follows: (1) Find the maximum bending moment in the beam due to the uniform load. (2) Knowing the maximum moment, find the required section modulus. (3) Select a trial wide-flange beam from Table F-1 in Appendix F (available online) and obtain the weight of the beam. (4) With the weight known, calculate a new value of the bending moment and a new value of the section modulus. (5) Determine whether the selected beam is still satisfactory. If it is not, select a new beam size and repeat the process until a satisfactory size of beam has been found.

Maximum bending moment. To assist in locating the cross section of maximum bending moment, we construct the shear-force diagram (Fig. 5-21b) using the methods described in Chapter 4. As part of that process, we determine the reactions at the supports:

$$R_A = 18,860 \text{ lb} \qquad R_B = 17,140 \text{ lb}$$

The distance x_1 from the left-hand support to the cross section of zero shear force is obtained from the equation

$$V = R_A - qx_1 = 0$$

which is valid in the range $0 \leq x \leq 12$ ft. Solving for x_1, we get

$$x_1 = \frac{R_A}{q} = \frac{18{,}860 \text{ lb}}{2000 \text{ lb/ft}} = 9.430 \text{ ft}$$

which is less than 12 ft, and therefore the calculation is valid.

The maximum bending moment occurs at the cross section where the shear force is zero; therefore,

$$M_{\max} = R_A x_1 - \frac{q x_1^2}{2} = 88{,}920 \text{ lb-ft}$$

Required section modulus. The required section modulus (based only upon the load q) is obtained from Eq. (5-24):

$$S = \frac{M_{\max}}{\sigma_{\text{allow}}} = \frac{(88{,}920 \text{ lb-ft})(12 \text{ in./ft})}{18{,}000 \text{ psi}} = 59.3 \text{ in.}^3$$

Trial beam. We now turn to Table F-1 and select the lightest wide-flange beam having a section modulus greater than 59.3 in.^{3} The lightest beam that provides this section modulus is W 12×50 with $S = 64.7$ in.^{3} This beam weighs 50 lb/ft. (Recall that the tables in Appendix F are abridged, and therefore a lighter beam may actually be available.)

We now recalculate the reactions, maximum bending moment, and required section modulus with the beam loaded by both the uniform load q and its own weight. Under these combined loads the reactions are

$$R_A = 19{,}380 \text{ lb} \qquad R_B = 17{,}670 \text{ lb}$$

and the distance to the cross section of zero shear becomes

$$x_1 = \frac{19{,}380 \text{ lb}}{2050 \text{ lb/ft}} = 9.454 \text{ ft}$$

The maximum bending moment increases to 91,610 lb-ft, and the new required section modulus is

$$S = \frac{M_{\max}}{\sigma_{\text{allow}}} = \frac{(91{,}610 \text{ lb-ft})(12 \text{ in./ft})}{18{,}000 \text{ psi}} = 61.1 \text{ in.}^3$$

Thus, we see that the W 12×50 beam with section modulus $S = 64.7$ in.^{3} is still satisfactory.

Note: If the new required section modulus exceeded that of the W 12×50 beam, a new beam with a larger section modulus would be selected and the process repeated.

Example 5-8

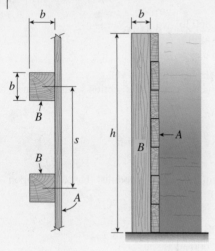

(a) Top view   (b) Side view

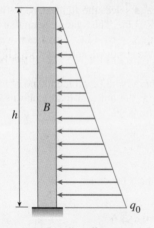

(c) Loading diagram

FIG. 5-22 Example 5-8. Wood dam with horizontal planks A supported by vertical posts B

A temporary wood dam is constructed of horizontal planks A supported by vertical wood posts B that are sunk into the ground so that they act as cantilever beams (Fig. 5-22). The posts are of square cross section (dimensions $b \times b$) and spaced at distance $s = 0.8$ m, center to center. Assume that the water level behind the dam is at its full height $h = 2.0$ m.

Determine the minimum required dimension b of the posts if the allowable bending stress in the wood is $\sigma_{allow} = 8.0$ MPa.

Solution

Loading diagram. Each post is subjected to a triangularly distributed load produced by the water pressure acting against the planks. Consequently, the loading diagram for each post is triangular (Fig. 5-22c). The maximum intensity q_0 of the load on the posts is equal to the water pressure at depth h times the spacing s of the posts:

$$q_0 = \gamma h s \tag{a}$$

in which γ is the specific weight of water. Note that q_0 has units of force per unit distance, γ has units of force per unit volume, and both h and s have units of length.

Section modulus. Since each post is a cantilever beam, the maximum bending moment occurs at the base and is given by the following expression:

$$M_{max} = \frac{q_0 h}{2}\left(\frac{h}{3}\right) = \frac{\gamma h^3 s}{6} \tag{b}$$

Therefore, the required section modulus (Eq. 5-24) is

$$S = \frac{M_{max}}{\sigma_{allow}} = \frac{\gamma h^3 s}{6\sigma_{allow}} \tag{c}$$

For a beam of square cross section, the section modulus is $S = b^3/6$ (see Eq. 5-18b). Substituting this expression for S into Eq. (c), we get a formula for the cube of the minimum dimension b of the posts:

$$b^3 = \frac{\gamma h^3 s}{\sigma_{allow}} \tag{d}$$

Numerical values. We now substitute numerical values into Eq. (d) and obtain

$$b^3 = \frac{(9.81 \text{ kN/m}^3)(2.0 \text{ m})^3(0.8 \text{ m})}{8.0 \text{ MPa}} = 0.007848 \text{ m}^3 = 7.848 \times 10^6 \text{ mm}^3$$

from which

$$b = 199 \text{ mm}$$

Thus, the minimum required dimension b of the posts is 199 mm. Any larger dimension, such as 200 mm, will ensure that the actual bending stress is less than the allowable stress.

5.7 SHEAR STRESSES IN BEAMS OF RECTANGULAR CROSS SECTION

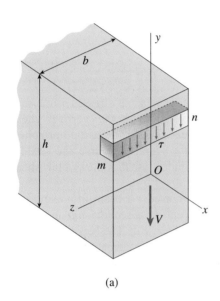

(a)

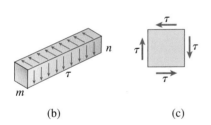

(b)   (c)

FIG. 5-23 Shear stresses in a beam of rectangular cross section

When a beam is in *pure bending*, the only stress resultants are the bending moments and the only stresses are the normal stresses acting on the cross sections. However, most beams are subjected to loads that produce both bending moments and shear forces (*nonuniform bending*). In these cases, both normal and shear stresses are developed in the beam. The normal stresses are calculated from the flexure formula (see Section 5.5), provided the beam is constructed of a linearly elastic material. The shear stresses are discussed in this and the following two sections.

Vertical and Horizontal Shear Stresses

Consider a beam of rectangular cross section (width b and height h) subjected to a positive shear force V (Fig. 5-23a). It is reasonable to assume that the shear stresses τ acting on the cross section are parallel to the shear force, that is, parallel to the vertical sides of the cross section. It is also reasonable to assume that the shear stresses are uniformly distributed across the width of the beam, although they may vary over the height. Using these two assumptions, we can determine the intensity of the shear stress at any point on the cross section.

For purposes of analysis, we isolate a small element mn of the beam (Fig. 5-23a) by cutting between two adjacent cross sections and between two horizontal planes. According to our assumptions, the shear stresses τ acting on the front face of this element are vertical and uniformly distributed from one side of the beam to the other. Also, from the discussion of shear stresses in Section 1.6, we know that shear stresses acting on one side of an element are accompanied by shear stresses of equal magnitude acting on perpendicular faces of the element (see Figs. 5-23b and c). Thus, there are horizontal shear stresses acting between horizontal layers of the beam as well as vertical shear stresses acting on the cross sections. At any point in the beam, these complementary shear stresses are equal in magnitude.

The equality of the horizontal and vertical shear stresses acting on an element leads to an important conclusion regarding the shear stresses at the top and bottom of the beam. If we imagine that the element mn (Fig. 5-23a) is located at either the top or the bottom, we see that the horizontal shear stresses must vanish, because there are no stresses on the outer surfaces of the beam. It follows that the vertical shear stresses must also vanish at those locations; in other words, $\tau = 0$ where $y = \pm h/2$.

The existence of horizontal shear stresses in a beam can be demonstrated by a simple experiment. Place two identical rectangular beams on simple supports and load them by a force P, as shown in Fig. 5-24a. If friction between the beams is small, the beams will bend independently (Fig. 5-24b). Each beam will be in compression above its own neutral axis and in tension below its neutral axis, and therefore the bottom surface of the upper beam will slide with respect to the top surface of the lower beam.

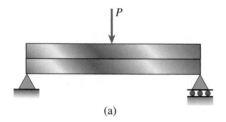

(a)

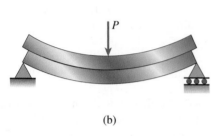

(b)

FIG. 5-24 Bending of two separate beams

Now suppose that the two beams are glued along the contact surface, so that they become a single solid beam. When this beam is loaded, horizontal shear stresses must develop along the glued surface in order to prevent the sliding shown in Fig. 5-24b. Because of the presence of these shear stresses, the single solid beam is much stiffer and stronger than the two separate beams.

Derivation of Shear Formula

We are now ready to derive a formula for the shear stresses τ in a rectangular beam. However, instead of evaluating the vertical shear stresses acting on a cross section, it is easier to evaluate the horizontal shear stresses acting between layers of the beam. Of course, the vertical shear stresses have the same magnitudes as the horizontal shear stresses.

With this procedure in mind, let us consider a beam in nonuniform bending (Fig. 5-25a). We take two adjacent cross sections mn and m_1n_1, distance dx apart, and consider the **element** mm_1n_1n. The bending moment and shear force acting on the left-hand face of this element are denoted M and V, respectively. Since both the bending moment and shear force may change as we move along the axis of the beam, the corresponding quantities on the right-hand face (Fig. 5-25a) are denoted $M + dM$ and $V + dV$.

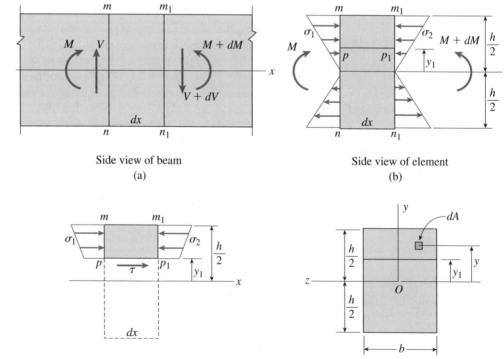

Side view of beam
(a)

Side view of element
(b)

Side view of subelement
(c)

Cross section of beam at subelement
(d)

FIG. 5-25 Shear stresses in a beam of rectangular cross section

Because of the presence of the bending moments and shear forces, the element shown in Fig. 5-25a is subjected to normal and shear stresses on both cross-sectional faces. However, only the normal stresses are needed in the following derivation, and therefore only the normal stresses are shown in Fig. 5-25b. On cross sections mn and m_1n_1 the normal stresses are, respectively,

$$\sigma_1 = -\frac{My}{I} \quad \text{and} \quad \sigma_2 = -\frac{(M+dM)y}{I} \tag{a,b}$$

as given by the flexure formula (Eq. 5-13). In these expressions, y is the distance from the neutral axis and I is the moment of inertia of the cross-sectional area about the neutral axis.

Next, we isolate a **subelement** mm_1p_1p by passing a horizontal plane pp_1 through element mm_1n_1n (Fig. 5-25b). The plane pp_1 is at distance y_1 from the neutral surface of the beam. The subelement is shown separately in Fig. 5-25c. We note that its top face is part of the upper surface of the beam and thus is free from stress. Its bottom face (which is parallel to the neutral surface and distance y_1 from it) is acted upon by the horizontal shear stresses τ existing at this level in the beam. Its cross-sectional faces mp and m_1p_1 are acted upon by the bending stresses σ_1 and σ_2, respectively, produced by the bending moments. Vertical shear stresses also act on the cross-sectional faces; however, these stresses do not affect the equilibrium of the subelement in the horizontal direction (the x direction), so they are not shown in Fig. 5-25c.

If the bending moments at cross sections mn and m_1n_1 (Fig. 5-25b) are equal (that is, if the beam is in pure bending), the normal stresses σ_1 and σ_2 acting over the sides mp and m_1p_1 of the subelement (Fig. 5-25c) also will be equal. Under these conditions, the subelement will be in equilibrium under the action of the normal stresses alone, and therefore the shear stresses τ acting on the bottom face pp_1 will vanish. This conclusion is obvious inasmuch as a beam in pure bending has no shear force and hence no shear stresses.

If the bending moments vary along the x axis (nonuniform bending), we can determine the shear stress τ acting on the bottom face of the subelement (Fig. 5-25c) by considering the equilibrium of the subelement in the x direction.

We begin by identifying an element of area dA in the *cross section* at distance y from the neutral axis (Fig. 5-25d). The force acting on this element is σdA, in which σ is the normal stress obtained from the flexure formula. If the element of area is located on the left-hand face mp of the subelement (where the bending moment is M), the normal stress is given by Eq. (a), and therefore the element of force is

$$\sigma_1 dA = \frac{My}{I}\,dA$$

Note that we are using only absolute values in this equation because the directions of the stresses are obvious from the figure. Summing these elements of force over the area of face mp of the subelement (Fig. 5-25c) gives the total horizontal force F_1 acting on that face:

$$F_1 = \int \sigma_1 \, dA = \int \frac{My}{I} \, dA \qquad (c)$$

Note that this integration is performed over the area of the shaded part of the cross section shown in Fig. 5-25d, that is, over the area of the cross section from $y = y_1$ to $y = h/2$.

The force F_1 is shown in Fig. 5-26 on a partial free-body diagram of the subelement (vertical forces have been omitted).

In a similar manner, we find that the total force F_2 acting on the right-hand face m_1p_1 of the subelement (Fig. 5-26 and Fig. 5-25c) is

$$F_2 = \int \sigma_2 \, dA = \int \frac{(M+dM)y}{I} \, dA \qquad (d)$$

Knowing the forces F_1 and F_2, we can now determine the horizontal force F_3 acting on the bottom face of the subelement.

Since the subelement is in equilibrium, we can sum forces in the x direction and obtain

$$F_3 = F_2 - F_1 \qquad (e)$$

or

$$F_3 = \int \frac{(M + dM)y}{I} \, dA - \int \frac{My}{I} \, dA = \int \frac{(dM)y}{I} \, dA$$

The quantities dM and I in the last term can be moved outside the integral sign because they are constants at any given cross section and are not involved in the integration. Thus, the expression for the force F_3 becomes

$$F_3 = \frac{dM}{I} \int y \, dA \qquad (5\text{-}30)$$

If the shear stresses τ are uniformly distributed across the width b of the beam, the force F_3 is also equal to the following:

$$F_3 = \tau b \, dx \qquad (5\text{-}31)$$

in which $b \, dx$ is the area of the bottom face of the subelement.

Combining Eqs. (5-30) and (5-31) and solving for the shear stress τ, we get

$$\tau = \frac{dM}{dx} \left(\frac{1}{Ib} \right) \int y \, dA \qquad (5\text{-}32)$$

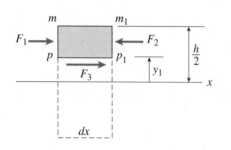

FIG. 5-26 Partial free-body diagram of subelement showing all horizontal forces (compare with Fig. 5-25c)

The quantity dM/dx is equal to the shear force V (see Eq. 4-6), and therefore the preceding expression becomes

$$\tau = \frac{V}{Ib} \int y\, dA \qquad (5\text{-}33)$$

The integral in this equation is evaluated over the shaded part of the cross section (Fig. 5-25d), as already explained. Thus, the integral is the first moment of the shaded area with respect to the neutral axis (the z axis). In other words, *the integral is the first moment of the cross-sectional area above the level at which the shear stress τ is being evaluated.* This first moment is usually denoted by the symbol Q:

$$Q = \int y\, dA \qquad (5\text{-}34)$$

With this notation, the equation for the shear stress becomes

$$\tau = \frac{VQ}{Ib} \qquad (5\text{-}35)$$

This equation, known as the **shear formula**, can be used to determine the shear stress τ at any point in the cross section of a rectangular beam. Note that for a specific cross section, the shear force V, moment of inertia I, and width b are constants. However, the first moment Q (and hence the shear stress τ) varies with the distance y_1 from the neutral axis.

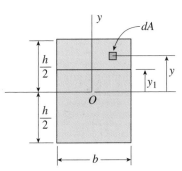

Cross section of beam at subelement
(d)

FIG. 5-25d   (Repeated)

Calculation of the First Moment Q

If the level at which the shear stress is to be determined is above the neutral axis, as shown in Fig. 5-25d, it is natural to obtain Q by calculating the first moment of the cross-sectional area *above* that level (the shaded area in the figure). However, as an alternative, we could calculate the first moment of the remaining cross-sectional area, that is, the area *below* the shaded area. Its first moment is equal to the negative of Q.

The explanation lies in the fact that the first moment of the entire cross-sectional area with respect to the neutral axis is equal to zero (because the neutral axis passes through the centroid). Therefore, the value of Q for the area below the level y_1 is the negative of Q for the area above that level. As a matter of convenience, we usually use the area above the level y_1 when the point where we are finding the shear stress is in the upper part of the beam, and we use the area below the level y_1 when the point is in the lower part of the beam.

Furthermore, we usually don't bother with sign conventions for V and Q. Instead, we treat all terms in the shear formula as positive quantities and determine the direction of the shear stresses by inspection, since the stresses act in the same direction as the shear force V itself. This procedure for determining shear stresses is illustrated later in Example 5-9.

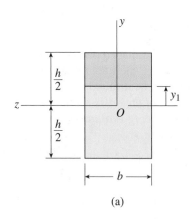

(a)

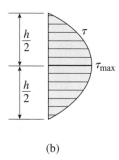

(b)

FIG. 5-27 Distribution of shear stresses in a beam of rectangular cross section: (a) cross section of beam, and (b) diagram showing the parabolic distribution of shear stresses over the height of the beam

Distribution of Shear Stresses in a Rectangular Beam

We are now ready to determine the distribution of the shear stresses in a beam of rectangular cross section (Fig. 5-27a). The first moment Q of the shaded part of the cross-sectional area is obtained by multiplying the area by the distance from its own centroid to the neutral axis:

$$Q = b\left(\frac{h}{2} - y_1\right)\left(y_1 + \frac{h/2 - y_1}{2}\right) = \frac{b}{2}\left(\frac{h^2}{4} - y_1^2\right) \qquad \text{(f)}$$

Of course, this same result can be obtained by integration using Eq. (5-34):

$$Q = \int y\, dA = \int_{y_1}^{h/2} yb\, dy = \frac{b}{2}\left(\frac{h^2}{4} - y_1^2\right) \qquad \text{(g)}$$

Substituting the expression for Q into the shear formula (Eq. 5-35), we get

$$\tau = \frac{V}{2I}\left(\frac{h^2}{4} - y_1^2\right) \qquad \text{(5-36)}$$

This equation shows that the shear stresses in a rectangular beam vary quadratically with the distance y_1 from the neutral axis. Thus, when plotted along the height of the beam, τ varies as shown in Fig. 5-27b. Note that the shear stress is zero when $y_1 = \pm h/2$.

The maximum value of the shear stress occurs at the neutral axis ($y_1 = 0$) where the first moment Q has its maximum value. Substituting $y_1 = 0$ into Eq. (5-36), we get

$$\tau_{max} = \frac{Vh^2}{8I} = \frac{3V}{2A} \qquad \text{(5-37)}$$

in which $A = bh$ is the cross-sectional area. Thus, the maximum shear stress in a beam of rectangular cross section is 50% larger than the average shear stress V/A.

Note again that the preceding equations for the shear stresses can be used to calculate either the vertical shear stresses acting on the cross sections or the horizontal shear stresses acting between horizontal layers of the beam.[*]

---

[*]The shear-stress analysis presented in this section was developed by the Russian engineer D. J. Jourawski; see Refs. 5-7 and 5-8; a list of references is available online.

Limitations

The formulas for shear stresses presented in this section are subject to the same restrictions as the flexure formula from which they are derived. Thus, they are valid only for beams of linearly elastic materials with small deflections.

In the case of rectangular beams, the accuracy of the shear formula depends upon the height-to-width ratio of the cross section. The formula may be considered as exact for very narrow beams (height h much larger than the width b). However, it becomes less accurate as b increases relative to h. For instance, when the beam is square ($b = h$), the true maximum shear stress is about 13% larger than the value given by Eq. (5-37). (For a more complete discussion of the limitations of the shear formula, see Ref. 5-9 available online.)

A common error is to apply the shear formula (Eq. 5-35) to cross-sectional shapes for which it is not applicable. For instance, it is not applicable to sections of triangular or semicircular shape. To avoid misusing the formula, we must keep in mind the following assumptions that underlie the derivation: (1) The edges of the cross section must be parallel to the y axis (so that the shear stresses act parallel to the y axis), and (2) the shear stresses must be uniform across the width of the cross section. These assumptions are fulfilled only in certain cases, such as those discussed in this and the next two sections.

Finally, the shear formula applies only to prismatic beams. If a beam is nonprismatic (for instance, if the beam is tapered), the shear stresses are quite different from those predicted by the formulas given here (see Refs. 5-9 and 5-10; a list of references is available online).

Effects of Shear Strains

Because the shear stress τ varies parabolically over the height of a rectangular beam, it follows that the shear strain $\gamma = \tau/G$ also varies parabolically. As a result of these shear strains, cross sections of the beam that were originally plane surfaces become warped. This warping is shown in Fig. 5-28, where cross sections mn and pq, originally plane, have become curved surfaces m_1n_1 and p_1q_1, with the maximum shear strain occurring at the neutral surface. At points m_1, p_1, n_1, and q_1 the shear strain is zero, and therefore the curves m_1n_1 and p_1q_1 are perpendicular to the upper and lower surfaces of the beam.

If the shear force V is constant along the axis of the beam, warping is the same at every cross section. Therefore, stretching and shortening of longitudinal elements due to the bending moments is unaffected by the shear strains, and the distribution of the normal stresses is the same as in pure bending. Moreover, detailed investigations using advanced methods of analysis show that warping of cross sections due to shear strains does not substantially affect the longitudinal strains even when the shear force varies continuously along the length. Thus, under most conditions it is justifiable to use the flexure formula (Eq. 5-13) for nonuniform bending, even though the formula was derived for pure bending.

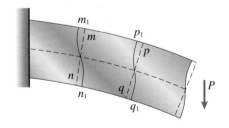

FIG. 5-28 Warping of the cross sections of a beam due to shear strains

Example 5-9

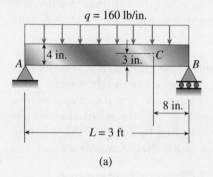

$q = 160$ lb/in.

4 in.

3 in. C

A

B

8 in.

$L = 3$ ft

(a)

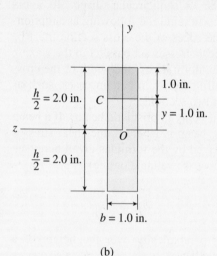

y

1.0 in.

$\dfrac{h}{2} = 2.0$ in. C

$y = 1.0$ in.

z

O

$\dfrac{h}{2} = 2.0$ in.

$b = 1.0$ in.

(b)

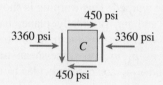

450 psi

3360 psi C 3360 psi

450 psi

FIG. 5-29 Example 5-9. (a) Simple beam with uniform load, (b) cross section of beam, and (c) stress element showing the normal and shear stresses at point C

A metal beam with span $L = 3$ ft is simply supported at points A and B (Fig. 5-29a). The uniform load on the beam (including its own weight) is $q = 160$ lb/in. The cross section of the beam is rectangular (Fig. 5-29b) with width $b = 1$ in. and height $h = 4$ in. The beam is adequately supported against sideways buckling.

Determine the normal stress σ_C and shear stress τ_C at point C, which is located 1 in. below the top of the beam and 8 in. from the right-hand support. Show these stresses on a sketch of a stress element at point C.

Solution

Shear force and bending moment. The shear force V_C and bending moment M_C at the cross section through point C are found by the methods described in Chapter 4. The results are

$$M_C = 17{,}920 \text{ lb-in.} \qquad V_C = -1600 \text{ lb}$$

The signs of these quantities are based upon the standard sign conventions for bending moments and shear forces (see Fig. 4-5).

Moment of inertia. The moment of inertia of the cross-sectional area about the neutral axis (the z axis in Fig. 5-29b) is

$$I = \frac{bh^3}{12} = \frac{1}{12}(1.0 \text{ in.})(4.0 \text{ in.})^3 = 5.333 \text{ in.}^4$$

Normal stress at point C. The normal stress at point C is found from the flexure formula (Eq. 5-13) with the distance y from the neutral axis equal to 1.0 in.; thus,

$$\sigma_C = -\frac{My}{I} = -\frac{(17{,}920 \text{ lb-in.})(1.0 \text{ in.})}{5.333 \text{ in.}^4} = -3360 \text{ psi} \quad \Longleftarrow$$

The minus sign indicates that the stress is compressive, as expected.

Shear stress at point C. To obtain the shear stress at point C, we need to evaluate the first moment Q_C of the cross-sectional area above point C (Fig. 5-29b). This first moment is equal to the product of the area and its centroidal distance (denoted y_C) from the z axis; thus,

$$A_C = (1.0 \text{ in.})(1.0 \text{ in.}) = 1.0 \text{ in.}^2 \qquad y_C = 1.5 \text{ in.} \qquad Q_C = A_C y_C = 1.5 \text{ in.}^3$$

Now we substitute numerical values into the shear formula (Eq. 5-35) and obtain the magnitude of the shear stress:

$$\tau_C = \frac{V_C Q_C}{Ib} = \frac{(1600 \text{ lb})(1.5 \text{ in.}^3)}{(5.333 \text{ in.}^4)(1.0 \text{ in.})} = 450 \text{ psi} \quad \Longleftarrow$$

The direction of this stress can be established by inspection, because it acts in the same direction as the shear force. In this example, the shear force acts upward on the part of the beam to the left of point C and downward on the part of the beam to the right of point C. The best way to show the directions of both the normal and shear stresses is to draw a stress element, as follows.

Stress element at point C. The stress element, shown in Fig. 5-29c, is cut from the side of the beam at point C (Fig. 5-29a). Compressive stresses $\sigma_C = 3360$ psi act on the cross-sectional faces of the element and shear stresses $\tau_C = 450$ psi act on the top and bottom faces as well as the cross-sectional faces.

Example 5-10

A wood beam AB supporting two concentrated loads P (Fig. 5-30a) has a rectangular cross section of width $b = 100$ mm and height $h = 150$ mm (Fig. 5-30b). The distance from each end of the beam to the nearest load is $a = 0.5$ m.

Determine the maximum permissible value P_{max} of the loads if the allowable stress in bending is $\sigma_{allow} = 11$ MPa (for both tension and compression) and the allowable stress in horizontal shear is $\tau_{allow} = 1.2$ MPa. (Disregard the weight of the beam itself.)

Note: Wood beams are much weaker in *horizontal shear* (shear parallel to the longitudinal fibers in the wood) than in *cross-grain shear* (shear on the cross sections). Consequently, the allowable stress in horizontal shear is usually considered in design.

Solution

The maximum shear force occurs at the supports and the maximum bending moment occurs throughout the region between the loads. Their values are

$$V_{max} = P \qquad M_{max} = Pa$$

Also, the section modulus S and cross-sectional area A are

$$S = \frac{bh^2}{6} \qquad A = bh$$

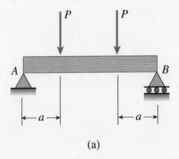

(a)

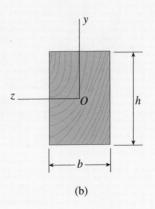

(b)

FIG. 5-30 Example 5-10. Wood beam with concentrated loads

continued

The maximum normal and shear stresses in the beam are obtained from the flexure and shear formulas (Eqs. 5-16 and 5-37):

$$\sigma_{max} = \frac{M_{max}}{S} = \frac{6Pa}{bh^2} \qquad \tau_{max} = \frac{3V_{max}}{2A} = \frac{3P}{2bh}$$

Therefore, the maximum permissible values of the load P in bending and shear, respectively, are

$$P_{bending} = \frac{\sigma_{allow}bh^2}{6a} \qquad P_{shear} = \frac{2\tau_{allow}bh}{3}$$

Substituting numerical values into these formulas, we get

$$P_{bending} = \frac{(11 \text{ MPa})(100 \text{ mm})(150 \text{ mm})^2}{6(0.5 \text{ m})} = 8.25 \text{ kN}$$

$$P_{shear} = \frac{2(1.2 \text{ MPa})(100 \text{ mm})(150 \text{ mm})}{3} = 12.0 \text{ kN}$$

Thus, the bending stress governs the design, and the maximum permissible load is

$$P_{max} = 8.25 \text{ kN} \qquad \Longleftarrow$$

A more complete analysis of this beam would require that the weight of the beam be taken into account, thus reducing the permissible load.

Notes:

(1) In this example, the maximum normal stresses and maximum shear stresses do not occur at the same locations in the beam—the normal stress is maximum in the middle region of the beam at the top and bottom of the cross section, and the shear stress is maximum near the supports at the neutral axis of the cross section.

(2) For most beams, the bending stresses (not the shear stresses) control the allowable load, as in this example.

(3) Although wood is not a homogeneous material and often departs from linearly elastic behavior, we can still obtain approximate results from the flexure and shear formulas. These approximate results are usually adequate for designing wood beams.

5.8 SHEAR STRESSES IN BEAMS OF CIRCULAR CROSS SECTION

When a beam has a **circular cross section** (Fig. 5-31), we can no longer assume that the shear stresses act parallel to the y axis. For instance, we can easily prove that at point m (on the boundary of the cross section) the shear stress τ must act *tangent* to the boundary. This observation follows from the fact that the outer surface of the beam is free of stress, and therefore the shear stress acting on the cross section can have no component in the radial direction.

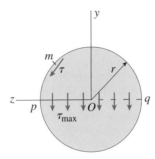

FIG. 5-31 Shear stresses acting on the cross section of a circular beam

Although there is no simple way to find the shear stresses acting throughout the entire cross section, we can readily determine the shear stresses at the neutral axis (where the stresses are the largest) by making some reasonable assumptions about the stress distribution. We assume that the stresses act parallel to the y axis and have constant intensity across the width of the beam (from point p to point q in Fig. 5-31). Since these assumptions are the same as those used in deriving the shear formula $\tau = VQ/Ib$ (Eq. 5-35), we can use the shear formula to calculate the stresses at the neutral axis.

For use in the shear formula, we need the following properties pertaining to a circular cross section having radius r:

$$I = \frac{\pi r^4}{4} \qquad Q = A\bar{y} = \left(\frac{\pi r^2}{2}\right)\left(\frac{4r}{3\pi}\right) = \frac{2r^3}{3} \qquad b = 2r \qquad \text{(5-38a,b)}$$

The expression for the moment of inertia I is taken from Case 9 of Appendix E (available online), and the expression for the first moment Q is based upon the formulas for a semicircle (Case 10, Appendix E). Substituting these expressions into the shear formula, we obtain

$$\tau_{max} = \frac{VQ}{Ib} = \frac{V(2r^3/3)}{(\pi r^4/4)(2r)} = \frac{4V}{3\pi r^2} = \frac{4V}{3A} \qquad \text{(5-39)}$$

in which $A = \pi r^2$ is the area of the cross section. This equation shows that the maximum shear stress in a circular beam is equal to 4/3 times the average vertical shear stress V/A.

If a beam has a **hollow circular cross section** (Fig. 5-32), we may again assume with reasonable accuracy that the shear stresses at the neutral axis are parallel to the y axis and uniformly distributed across the section. Consequently, we may again use the shear formula to find the maximum stresses. The required properties for a hollow circular section are

$$I = \frac{\pi}{4}\left(r_2^4 - r_1^4\right) \qquad Q = \frac{2}{3}\left(r_2^3 - r_1^3\right) \qquad b = 2(r_2 - r_1) \qquad \text{(5-40a,b,c)}$$

in which r_1 and r_2 are the inner and outer radii of the cross section. Therefore, the maximum stress is

$$\tau_{max} = \frac{VQ}{Ib} = \frac{4V}{3A}\left(\frac{r_2^2 + r_2 r_1 + r_1^2}{r_2^2 + r_1^2}\right) \qquad \text{(5-41)}$$

in which

$$A = \pi\left(r_2^2 - r_1^2\right)$$

is the area of the cross section. Note that if $r_1 = 0$, Eq. (5-41) reduces to Eq. (5-39) for a solid circular beam.

Although the preceding theory for shear stresses in beams of circular cross section is approximate, it gives results differing by only a few percent from those obtained using the exact theory of elasticity (Ref. 5-9 available online). Consequently, Eqs. (5-39) and (5-41) can be used to determine the maximum shear stresses in circular beams under ordinary circumstances.

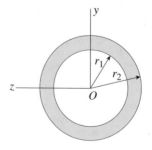

FIG. 5-32 Hollow circular cross section

Example 5-11

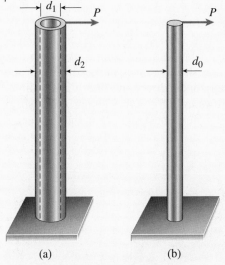

(a)

(b)

FIG. 5-33 Example 5-11. Shear stresses in beams of circular cross section

A vertical pole consisting of a circular tube of outer diameter $d_2 = 4.0$ in. and inner diameter $d_1 = 3.2$ in. is loaded by a horizontal force $P = 1500$ lb (Fig. 5-33a).

(a) Determine the maximum shear stress in the pole.

(b) For the same load P and the same maximum shear stress, what is the diameter d_0 of a solid circular pole (Fig. 5-33b)?

Solution

(a) *Maximum shear stress.* For the pole having a hollow circular cross section (Fig. 5-33a), we use Eq. (5-41) with the shear force V replaced by the load P and the cross-sectional area A replaced by the expression $\pi(r_2^2 - r_1^2)$; thus,

$$\tau_{max} = \frac{4P}{3\pi}\left(\frac{r_2^2 + r_2 r_1 + r_1^2}{r_2^4 - r_1^4}\right) \tag{a}$$

Next, we substitute numerical values, namely,

$$P = 1500 \text{ lb} \qquad r_2 = d_2/2 = 2.0 \text{ in.} \qquad r_1 = d_1/2 = 1.6 \text{ in.}$$

and obtain

$$\tau_{max} = 658 \text{ psi} \qquad \Longleftarrow$$

which is the maximum shear stress in the pole.

(b) *Diameter of solid circular pole.* For the pole having a solid circular cross section (Fig. 5-33b), we use Eq. (5-39) with V replaced by P and r replaced by $d_0/2$:

$$\tau_{max} = \frac{4P}{3\pi(d_0/2)^2} \tag{b}$$

Solving for d_0, we obtain

$$d_0^2 = \frac{16P}{3\pi \tau_{max}} = \frac{16(1500 \text{ lb})}{3\pi(658 \text{ psi})} = 3.87 \text{ in.}^2$$

from which we get

$$d_0 = 1.97 \text{ in.} \qquad \Longleftarrow$$

In this particular example, the solid circular pole has a diameter approximately one-half that of the tubular pole.

Note: Shear stresses rarely govern the design of either circular or rectangular beams made of metals such as steel and aluminum. In these kinds of materials, the allowable shear stress is usually in the range 25 to 50% of the allowable tensile stress. In the case of the tubular pole in this example, the maximum shear stress is only 658 psi. In contrast, the maximum bending stress obtained from the flexure formula is 9700 psi for a relatively short pole of length 24 in. Thus, as the load increases, the allowable tensile stress will be reached long before the allowable shear stress is reached.

The situation is quite different for materials that are weak in shear, such as wood. For a typical wood beam, the allowable stress in horizontal shear is in the range 4 to 10% of the allowable bending stress. Consequently, even though the maximum shear stress is relatively low in value, it sometimes governs the design.

5.9 SHEAR STRESSES IN THE WEBS OF BEAMS WITH FLANGES

When a beam of wide-flange shape (Fig. 5-34a) is subjected to shear forces as well as bending moments (nonuniform bending), both normal and shear stresses are developed on the cross sections. The distribution of the shear stresses in a wide-flange beam is more complicated than in a rectangular beam. For instance, the shear stresses in the flanges of the beam act in both vertical and horizontal directions (the y and z directions), as shown by the small arrows in Fig. 5-34b. The horizontal shear stresses are much larger than the vertical shear stresses in the flanges.

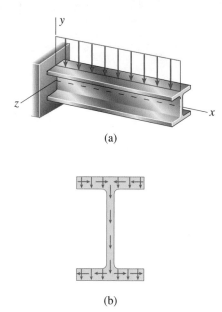

(a)

(b)

FIG. 5-34 (a) Beam of wide-flange shape, and (b) directions of the shear stresses acting on a cross section

The shear stresses in the web of a wide-flange beam act only in the vertical direction and are larger than the stresses in the flanges. These stresses can be found by the same techniques we used for finding shear stresses in rectangular beams.

Shear Stresses in the Web

Let us begin the analysis by determining the shear stresses at line ef in the web of a wide-flange beam (Fig. 5-35a). We will make the same assumptions as those we made for a rectangular beam; that is, we assume that the shear stresses act parallel to the y axis and are uniformly distributed across the thickness of the web. Then the shear formula $\tau = VQ/Ib$ will still apply. However, the width b is now the thickness t of the web, and the area used in calculating the first moment Q is the area between line ef and the top edge of the cross section (indicated by the shaded area of Fig. 5-35a).

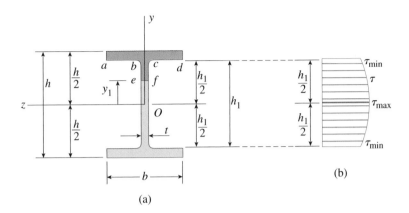

FIG. 5-35 Shear stresses in the web of a wide-flange beam. (a) Cross section of beam, and (b) distribution of vertical shear stresses in the web

When finding the first moment Q of the shaded area, we will disregard the effects of the small fillets at the juncture of the web and flange (points b and c in Fig. 5-35a). The error in ignoring the areas of these fillets is very small. Then we will divide the shaded area into two rectangles. The first rectangle is the upper flange itself, which has area

$$A_1 = b\left(\frac{h}{2} - \frac{h_1}{2}\right) \tag{a}$$

in which b is the width of the flange, h is the overall height of the beam, and h_1 is the distance between the insides of the flanges. The second rectangle is the part of the web between ef and the flange, that is, rectangle $efcb$, which has area

$$A_2 = t\left(\frac{h_1}{2} - y_1\right) \tag{b}$$

in which t is the thickness of the web and y_1 is the distance from the neutral axis to line ef.

The first moments of areas A_1 and A_2, evaluated about the neutral axis, are obtained by multiplying these areas by the distances from their respective centroids to the z axis. Adding these first moments gives the first moment Q of the combined area:

$$Q = A_1\left(\frac{h_1}{2} + \frac{h/2 - h_1/2}{2}\right) + A_2\left(y_1 + \frac{h_1/2 - y_1}{2}\right)$$

Upon substituting for A_1 and A_2 from Eqs. (a) and (b) and then simplifying, we get

$$Q = \frac{b}{8}(h^2 - h_1^2) + \frac{t}{8}(h_1^2 - 4y_1^2) \tag{5-42}$$

Therefore, the shear stress τ in the web of the beam at distance y_1 from the neutral axis is

$$\tau = \frac{VQ}{It} = \frac{V}{8It}\left[b(h^2 - h_1^2) + t(h_1^2 - 4y_1^2)\right] \qquad (5\text{-}43)$$

in which the moment of inertia of the cross section is

$$I = \frac{bh^3}{12} - \frac{(b - t)h_1^3}{12} = \frac{1}{12}(bh^3 - bh_1^3 + th_1^3) \qquad (5\text{-}44)$$

Since all quantities in Eq. (5-43) are constants except y_1, we see immediately that τ varies quadratically throughout the height of the web, as shown by the graph in Fig. 5-35b. Note that the graph is drawn only for the web and does not include the flanges. The reason is simple enough— Eq. (5-43) cannot be used to determine the vertical shear stresses in the flanges of the beam (see the discussion titled "Limitations" later in this section).

Maximum and Minimum Shear Stresses

The maximum shear stress in the web of a wide-flange beam occurs at the neutral axis, where $y_1 = 0$. The minimum shear stress occurs where the web meets the flanges ($y_1 = \pm\, h_1/2$). These stresses, found from Eq. (5-43), are

$$\tau_{\max} = \frac{V}{8It}(bh^2 - bh_1^2 + th_1^2) \qquad \tau_{\min} = \frac{Vb}{8It}(h^2 - h_1^2) \qquad (5\text{-}45a,b)$$

Both $\tau_{\max}$ and $\tau_{\min}$ are labeled on the graph of Fig. 5-35b. For typical wide-flange beams, the maximum stress in the web is from 10 to 60% greater than the minimum stress.

Although it may not be apparent from the preceding discussion, the stress $\tau_{\max}$ given by Eq. (5-45a) not only is the largest shear stress in the web but also is the largest shear stress anywhere in the cross section.

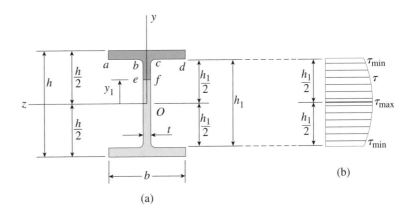

FIG. 5-35 (Repeated) Shear stresses in the web of a wide-flange beam. (a) Cross section of beam, and (b) distribution of vertical shear stresses in the web

Shear Force in the Web

The vertical shear force carried by the web alone may be determined by multiplying the area of the shear-stress diagram (Fig. 5-35b) by the thickness t of the web. The shear-stress diagram consists of two parts, a rectangle of area $h_1 \tau_{min}$ and a parabolic segment of area

$$\frac{2}{3}(h_1)(\tau_{max} - \tau_{min})$$

By adding these two areas, multiplying by the thickness t of the web, and then combining terms, we get the total shear force in the web:

$$V_{web} = \frac{th_1}{3}(2\tau_{max} + \tau_{min}) \tag{5-46}$$

For beams of typical proportions, the shear force in the web is 90 to 98% of the total shear force V acting on the cross section; the remainder is carried by shear in the flanges.

Since the web resists most of the shear force, designers often calculate an approximate value of the maximum shear stress by dividing the total shear force by the area of the web. The result is the average shear stress in the web, assuming that the web carries *all* of the shear force:

$$\tau_{aver} = \frac{V}{th_1} \tag{5-47}$$

For typical wide-flange beams, the average stress calculated in this manner is within 10% (plus or minus) of the maximum shear stress calculated from Eq. (5-45a). Thus, Eq. (5-47) provides a simple way to estimate the maximum shear stress.

Limitations

The elementary shear theory presented in this section is suitable for determining the vertical shear stresses in the web of a wide-flange beam. However, when investigating vertical shear stresses in the flanges, we can no longer assume that the shear stresses are constant across the width of the section, that is, across the width b of the flanges (Fig. 5-35a). Hence, we cannot use the shear formula to determine these stresses.

To emphasize this point, consider the junction of the web and upper flange ($y_1 = h_1/2$), where the width of the section changes abruptly from t to b. The shear stresses on the free surfaces ab and cd (Fig. 5-35a) must be zero, whereas the shear stress across the web at line bc is τ_{min}. These observations indicate that the distribution of shear stresses at the junction of the web and the flange is quite complex and cannot be investigated by elementary methods. The stress analysis is further complicated by the use of fillets at the re-entrant corners (corners b and c). The fillets are necessary to prevent the stresses from becoming dangerously large, but they also alter the stress distribution across the web.

Thus, we conclude that the shear formula cannot be used to determine the vertical shear stresses in the flanges. However, the shear formula does give good results for the shear stresses acting *horizontally* in the flanges (Fig. 5-34b).

The method described above for determining shear stresses in the webs of wide-flange beams can also be used for other sections having thin webs. For instance, Example 5-13 illustrates the procedure for a T-beam.

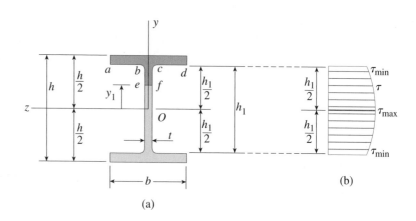

FIG. 5-35 (Repeated) Shear stresses in the web of a wide-flange beam. (a) Cross section of beam, and (b) distribution of vertical shear stresses in the web

Example 5-12

A beam of wide-flange shape (Fig. 5-36a) is subjected to a vertical shear force $V = 45$ kN. The cross-sectional dimensions of the beam are $b = 165$ mm, $t = 7.5$ mm, $h = 320$ mm, and $h_1 = 290$ mm.

Determine the maximum shear stress, minimum shear stress, and total shear force in the web. (Disregard the areas of the fillets when making calculations.)

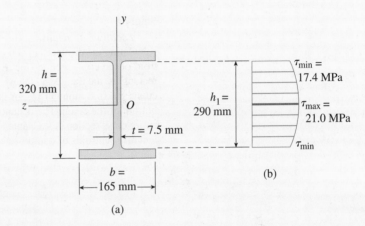

FIG. 5-36 Example 5-12. Shear stresses in the web of a wide-flange beam

Solution

Maximum and minimum shear stresses. The maximum and minimum shear stresses in the web of the beam are given by Eqs. (5-45a) and (5-45b). Before substituting into those equations, we calculate the moment of inertia of the cross-sectional area from Eq. (5-44):

$$I = \frac{1}{12}(bh^3 - bh_1^3 + th_1^3) = 130.45 \times 10^6 \text{ mm}^4$$

Now we substitute this value for I, as well as the numerical values for the shear force V and the cross-sectional dimensions, into Eqs. (5-45a) and (5-45b):

$$\tau_{max} = \frac{V}{8It}(bh^2 - bh_1^2 + th_1^2) = 21.0 \text{ MPa} \qquad \Longleftarrow$$

$$\tau_{min} = \frac{Vb}{8It}(h^2 - h_1^2) = 17.4 \text{ MPa} \qquad \Longleftarrow$$

In this case, the ratio of τ_{max} to τ_{min} is 1.21, that is, the maximum stress in the web is 21% larger than the minimum stress. The variation of the shear stresses over the height h_1 of the web is shown in Fig. 5-36b.

Total shear force. The shear force in the web is calculated from Eq. (5-46) as follows:

$$V_{web} = \frac{th_1}{3}(2\tau_{max} + \tau_{min}) = 43.0 \text{ kN} \qquad \Longleftarrow$$

From this result we see that the web of this particular beam resists 96% of the total shear force.

Note: The average shear stress in the web of the beam (from Eq. 5-47) is

$$\tau_{aver} = \frac{V}{th_1} = 20.7 \text{ MPa}$$

which is only 1% less than the maximum stress.

Example 5-13

A beam having a T-shaped cross section (Fig. 5-37a) is subjected to a vertical shear force $V = 10{,}000$ lb. The cross-sectional dimensions are $b = 4$ in., $t = 1.0$ in., $h = 8.0$ in., and $h_1 = 7.0$ in.

Determine the shear stress τ_1 at the top of the web (level nn) and the maximum shear stress τ_{max}. (Disregard the areas of the fillets.)

Solution

Location of neutral axis. The neutral axis of the T-beam is located by calculating the distances c_1 and c_2 from the top and bottom of the beam to the centroid of the cross section (Fig. 5-37a). First, we divide the cross section into two rectangles, the flange and the web (see the dashed line in Fig. 5-37a). Then we calculate the first moment Q_{aa} of these two rectangles with respect to line aa at the bottom of the beam. The distance c_2 is equal to Q_{aa} divided by the area A of the entire cross section (see Chapter 10, Section 10.3 available online, for methods for locating centroids of composite areas). The calculations are as follows:

$$A = \Sigma A_i = b(h - h_1) + th_1 = 11.0 \text{ in.}^2$$

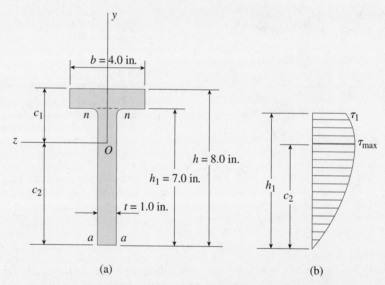

FIG. 5-37 Example 5-13. Shear stresses in web of T-shaped beam

(a)

(b)

$$Q_{aa} = \Sigma y_i A_i = \left(\frac{h + h_1}{2}\right)(b)(h - h_1) + \frac{h_1}{2}(th_1) = 54.5 \text{ in.}^3$$

$$c_2 = \frac{Q_{aa}}{A} = \frac{54.5 \text{ in.}^3}{11.0 \text{ in.}^2} = 4.955 \text{ in.} \qquad c_1 = h - c_2 = 3.045 \text{ in.}$$

Moment of inertia. The moment of inertia I of the entire cross-sectional area (with respect to the neutral axis) can be found by determining the moment of inertia I_{aa} about line aa at the bottom of the beam and then using the parallel-axis theorem (see Section 10.5 available online):

$$I = I_{aa} - Ac_2^2$$

The calculations are as follows:

$$I_{aa} = \frac{bh^3}{3} - \frac{(b-t)h_1^3}{3} = 339.67 \text{ in.}^4 \qquad Ac_2^2 = 270.02 \text{ in.}^4 \qquad I = 69.65 \text{ in.}^4$$

Shear stress at top of web. To find the shear stress τ_1 at the top of the web (along line *nn*) we need to calculate the first moment Q_1 of the area above level *nn*. This first moment is equal to the area of the flange times the distance from the neutral axis to the centroid of the flange:

$$Q_1 = b(h - h_1)\left(c_1 - \frac{h - h_1}{2}\right)$$

$$= (4 \text{ in.})(1 \text{ in.})(3.045 \text{ in.} - 0.5 \text{ in.}) = 10.18 \text{ in.}^3$$

Of course, we get the same result if we calculate the first moment of the area *below* level *nn*:

$$Q_1 = th_1\left(c_2 - \frac{h_1}{2}\right) = (1 \text{ in.})(7 \text{ in.})(4.955 \text{ in.} - 3.5 \text{ in.}) = 10.18 \text{ in.}^3$$

Substituting into the shear formula, we find

$$\tau_1 = \frac{VQ_1}{It} = \frac{(10,000 \text{ lb})(10.18 \text{ in.}^3)}{(69.65 \text{ in.}^4)(1 \text{ in.})} = 1460 \text{ psi} \qquad \longleftarrow$$

This stress exists both as a vertical shear stress acting on the cross section and as a horizontal shear stress acting on the horizontal plane between the flange and the web.

Maximum shear stress. The maximum shear stress occurs in the web at the neutral axis. Therefore, we calculate the first moment Q_{max} of the cross-sectional area below the neutral axis:

$$Q_{max} = tc_2\left(\frac{c_2}{2}\right) = (1 \text{ in.})(4.955 \text{ in.})\left(\frac{4.955 \text{ in.}}{2}\right) = 12.28 \text{ in.}^3$$

As previously indicated, we would get the same result if we calculated the first moment of the area above the neutral axis, but those calculations would be slightly longer.

Substituting into the shear formula, we obtain

$$\tau_{max} = \frac{VQ_{max}}{It} = \frac{(10,000 \text{ lb})(12.28 \text{ in.}^3)}{(69.65 \text{ in.}^4)(1 \text{ in.})} = 1760 \text{ psi} \qquad \longleftarrow$$

which is the maximum shear stress in the beam.

The parabolic distribution of shear stresses in the web is shown in Fig. 5-37b.

5.10 COMPOSITE BEAMS

Beams that are fabricated of more than one material are called **composite beams**. Examples are bimetallic beams (such as those used in thermostats), plastic coated pipes, and wood beams with steel reinforcing plates (see Fig. 5-38). Many other types of composite beams have been developed in recent years, primarily to save material and reduce weight. For instance, **sandwich beams** are widely used in the aviation and aerospace industries, where light weight plus high strength and rigidity are required. Such familiar objects as skis, doors, wall panels, book shelves, and cardboard boxes are also manufactured in sandwich style.

A typical sandwich beam (Fig. 5-39) consists of two thin *faces* of relatively high-strength material (such as aluminum) separated by a thick *core* of lightweight, low-strength material. Since the faces are at the greatest distance from the neutral axis (where the bending stresses are highest), they function somewhat like the flanges of an I-beam. The core serves as a filler and provides support for the faces, stabilizing them against wrinkling or buckling. Lightweight plastics and foams, as well as honeycombs and corrugations, are often used for cores.

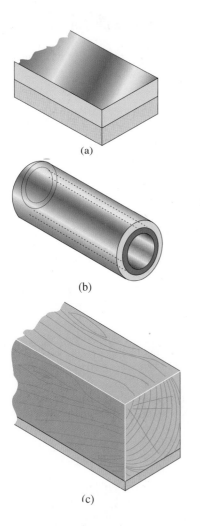

General Theory for Composite Beams

In this section, we will study the flexure of composite beams made up of two different materials. First the general theory of flexure developed in Sections 5.2–5.5 will be expanded for the case of composite beams. Then an alternative approach, known as the transformed-section method, will be discussed. In the transformed-section method, bending of a composite beam is analyzed by converting the composite beam into an equivalent beam of one material only. Examples of both procedures are provided in the following section.

Strains and Stresses

The strains in composite beams are determined from the same basic axiom that we used for finding the strains in beams of one material, namely, cross sections remain plane during bending. This axiom is valid for pure bending regardless of the nature of the material (see Section 5.4). Therefore, the longitudinal strains ϵ_x in a composite beam vary linearly from top to bottom of the beam, as expressed by Eq. (5-4), which is repeated here:

$$\epsilon_x = -\frac{y}{\rho} = -\kappa y \tag{5-48}$$

FIG. 5-38 Examples of composite beams: (a) bimetallic beam, (b) plastic-coated steel pipe, and (c) wood beam reinforced with a steel plate

In this equation, y is the distance from the neutral axis, ρ is the radius of curvature, and κ is the curvature.

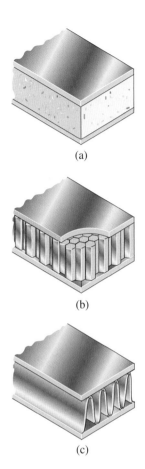

FIG. 5-39 Sandwich beams with:
(a) plastic core, (b) honeycomb core,
and (c) corrugated core

Beginning with the linear strain distribution represented by Eq. (5-48), we can determine the strains and stresses in any composite beam. To show how this is accomplished, consider the composite beam shown in Fig. 5-40. This beam consists of two materials, labeled 1 and 2 in the figure, which are securely bonded so that they act as a single solid beam.

As in our previous discussions of beams, we assume that the xy plane is a plane of symmetry and that the xz plane is the neutral plane of the beam. However, the neutral axis (the z axis in Fig. 5-40b) does *not* pass through the centroid of the cross-sectional area when the beam is made of two different materials.

If the beam is bent with positive curvature, the strains ϵ_x will vary as shown in Fig. 5-40c, where ϵ_A is the compressive strain at the top of the beam, ϵ_B is the tensile strain at the bottom, and ϵ_C is the strain at the contact surface of the two materials. Of course, the strain is zero at the neutral axis (the z axis).

The normal stresses acting on the cross section can be obtained from the strains by using the stress-strain relationships for the two materials. Let us assume that both materials behave in a linearly elastic manner so that Hooke's law for uniaxial stress is valid. Then the stresses in the materials are obtained by multiplying the strains by the appropriate modulus of elasticity.

Denoting the moduli of elasticity for materials 1 and 2 as E_1 and E_2, respectively, and also assuming that $E_2 > E_1$, we obtain the stress diagram shown in Fig. 5-40d. The compressive stress at the top of the beam is $\sigma_A = E_1\epsilon_A$ and the tensile stress at the bottom is $\sigma_B = E_2\epsilon_B$.

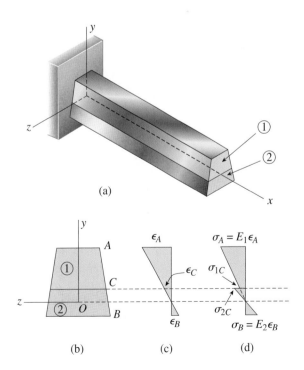

FIG. 5-40 (a) Composite beam of two materials, (b) cross section of beam, (c) distribution of strains ϵ_x throughout the height of the beam, and (d) distribution of stresses σ_x in the beam for the case where $E_2 > E_1$

At the contact surface (C) the stresses in the two materials are different because their moduli are different. In material 1 the stress is $\sigma_{1C} = E_1\epsilon_C$ and in material 2 it is $\sigma_{2C} = E_2\epsilon_C$.

Using Hooke's law and Eq. (5-48), we can express the normal stresses at distance y from the neutral axis in terms of the curvature:

$$\sigma_{x1} = -E_1\kappa y \qquad \sigma_{x2} = -E_2\kappa y \qquad \text{(5-49a,b)}$$

in which σ_{x1} is the stress in material 1 and σ_{x2} is the stress in material 2. With the aid of these equations, we can locate the neutral axis and obtain the moment-curvature relationship.

Neutral Axis

The position of the neutral axis (the z axis) is found from the condition that the resultant axial force acting on the cross section is zero (see Section 5.5); therefore,

$$\int_1 \sigma_{x1}\,dA + \int_2 \sigma_{x2}\,dA = 0 \qquad \text{(a)}$$

where it is understood that the first integral is evaluated over the cross-sectional area of material 1 and the second integral is evaluated over the cross-sectional area of material 2. Replacing σ_{x1} and σ_{x2} in the preceding equation by their expressions from Eqs. (5-49a) and (4-49b), we get

$$-\int_1 E_1\kappa y\,dA - \int_2 E_2\kappa y\,dA = 0$$

Since the curvature is a constant at any given cross section, it is not involved in the integrations and can be cancelled from the equation; thus, the equation for locating the **neutral axis** becomes

$$E_1\int_1 y\,dA + E_2\int_2 y\,dA = 0 \qquad \text{(5-50)}$$

The integrals in this equation represent the first moments of the two parts of the cross-sectional area with respect to the neutral axis. (If there are more than two materials—a rare condition—additional terms are required in the equation.)

Equation (5-50) is a generalized form of the analogous equation for a beam of one material (Eq. 5-8). The details of the procedure for locating the neutral axis with the aid of Eq. (5-50) are illustrated later in Example 5-14.

If the cross section of a beam is **doubly symmetric,** as in the case of a wood beam with steel cover plates on the top and bottom (Fig. 5-41), the neutral axis is located at the midheight of the cross section and Eq. (5-50) is not needed.

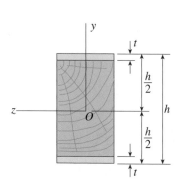

FIG. 5-41 Doubly symmetric cross section

Moment-Curvature Relationship

The moment-curvature relationship for a composite beam of two materials (Fig. 5-40) may be determined from the condition that the moment resultant of the bending stresses is equal to the bending moment M acting at the cross section. Following the same steps as for a beam of one material (see Eqs. 5-9 through 5-12), and also using Eqs. (5-49a) and (5-49b), we obtain

$$M = -\int_A \sigma_x y\, dA = -\int_1 \sigma_{x1} y\, dA - \int_2 \sigma_{x2} y\, dA$$

$$= \kappa E_1 \int_1 y^2\, dA + \kappa E_2 \int_2 y^2\, dA \qquad \text{(b)}$$

This equation can be written in the simpler form

$$M = \kappa(E_1 I_1 + E_2 I_2) \qquad (5\text{-}51)$$

in which I_1 and I_2 are the moments of inertia about the neutral axis (the z axis) of the cross-sectional areas of materials 1 and 2, respectively. Note that $I = I_1 + I_2$, where I is the moment of inertia of the *entire* cross-sectional area about the neutral axis.

Equation (5-51) can now be solved for the curvature in terms of the bending moment:

$$\kappa = \frac{1}{\rho} = \frac{M}{E_1 I_1 + E_2 I_2} \qquad (5\text{-}52)$$

This equation is the **moment-curvature relationship** for a beam of two materials (compare with Eq. 5-12 for a beam of one material). The denominator on the right-hand side is the **flexural rigidity** of the composite beam.

Normal Stresses (Flexure Formulas)

The normal stresses (or bending stresses) in the beam are obtained by substituting the expression for curvature (Eq. 5-52) into the expressions for σ_{x1} and σ_{x2} (Eqs. 5-49a and 5-49b); thus,

$$\sigma_{x1} = -\frac{M y E_1}{E_1 I_1 + E_2 I_2} \qquad \sigma_{x2} = -\frac{M y E_2}{E_1 I_1 + E_2 I_2} \qquad (5\text{-}53\text{a,b})$$

These expressions, known as the **flexure formulas for a composite beam**, give the normal stresses in materials 1 and 2, respectively. If the two materials have the same modulus of elasticity ($E_1 = E_2 = E$), then both equations reduce to the flexure formula for a beam of one material (Eq. 5-13).

The analysis of composite beams, using Eqs. (5-50) through (5-53), is illustrated in Examples 5-14 and 5-15 at the end of this section.

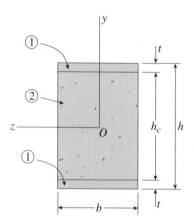

FIG. 5-42 Cross section of a sandwich beam having two axes of symmetry (doubly symmetric cross section)

Approximate Theory for Bending of Sandwich Beams

Sandwich beams having doubly symmetric cross sections and composed of two linearly elastic materials (Fig. 5-42) can be analyzed for bending using Eqs. (5-52) and (5-53), as described previously. However, we can also develop an approximate theory for bending of sandwich beams by introducing some simplifying assumptions.

If the material of the faces (material 1) has a much larger modulus of elasticity than does the material of the core (material 2), it is reasonable to disregard the normal stresses in the core and assume that the faces resist all of the longitudinal bending stresses. This assumption is equivalent to saying that the modulus of elasticity E_2 of the core is zero. Under these conditions the flexure formula for material 2 (Eq. 5-53b) gives $\sigma_{x2} = 0$ (as expected), and the flexure formula for material 1 (Eq. 5-53a) gives

$$\sigma_{x1} = -\frac{My}{I_1} \tag{5-54}$$

which is similar to the ordinary flexure formula (Eq. 5-13). The quantity I_1 is the moment of inertia of the two faces evaluated with respect to the neutral axis; thus,

$$I_1 = \frac{b}{12}\left(h^3 - h_c^3\right) \tag{5-55}$$

in which b is the width of the beam, h is the overall height of the beam, and h_c is the height of the core. Note that $h_c = h - 2t$ where t is the thickness of the faces.

The maximum normal stresses in the sandwich beam occur at the top and bottom of the cross section where $y = h/2$ and $-h/2$, respectively. Thus, from Eq. (5-54), we obtain

$$\sigma_{\text{top}} = -\frac{Mh}{2I_1} \qquad \sigma_{\text{bottom}} = \frac{Mh}{2I_1} \tag{5-56a,b}$$

If the bending moment M is positive, the upper face is in compression and the lower face is in tension. (These equations are conservative because they give stresses in the faces that are higher than those obtained from Eqs. 5-53a and 5-53b.)

If the faces are thin compared to the thickness of the core (that is, if t is small compared to h_c), we can disregard the shear stresses in the faces and assume that the core carries all of the shear stresses. Under these conditions the average shear stress and average shear strain in the core are, respectively,

$$\tau_{\text{aver}} = \frac{V}{bh_c} \qquad \gamma_{\text{aver}} = \frac{V}{bh_c G_c} \qquad (5\text{-}57\text{a,b})$$

in which V is the shear force acting on the cross section and G_c is the shear modulus of elasticity for the core material. (Although the maximum shear stress and maximum shear strain are larger than the average values, the average values are often used for design purposes.)

Limitations

Throughout the preceding discussion of composite beams, we assumed that both materials followed Hooke's law and that the two parts of the beam were adequately bonded so that they acted as a single unit. Thus, our analysis is highly idealized and represents only a first step in understanding the behavior of composite beams and composite materials. Methods for dealing with nonhomogeneous and nonlinear materials, bond stresses between the parts, shear stresses on the cross sections, buckling of the faces, and other such matters are treated in reference books dealing specifically with composite construction.

Reinforced concrete beams are one of the most complex types of composite construction (Fig. 5-43), and their behavior differs significantly from that of the composite beams discussed in this section. Concrete is strong in compression but extremely weak in tension. Consequently, its tensile strength is usually disregarded entirely. Under those conditions, *the formulas given in this section do not apply*.

Furthermore, most reinforced concrete beams are not designed on the basis of linearly elastic behavior—instead, more realistic design methods (based upon load-carrying capacity instead of allowable stresses) are used. The design of reinforced concrete members is a highly specialized subject that is presented in courses and textbooks devoted solely to that subject.

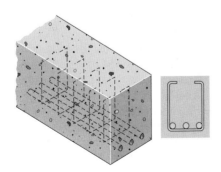

FIG. 5-43 Reinforced concrete beam with longitudinal reinforcing bars and vertical stirrups

Example 5-14

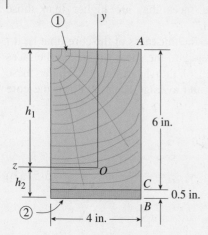

FIG. 5-44 Example 5-14. Cross section of a composite beam of wood and steel

A composite beam (Fig. 5-44) is constructed from a wood beam (4.0 in. × 6.0 in. actual dimensions) and a steel reinforcing plate (4.0 in. wide and 0.5 in. thick). The wood and steel are securely fastened to act as a single beam. The beam is subjected to a positive bending moment $M = 60$ k-in.

Calculate the largest tensile and compressive stresses in the wood (material 1) and the maximum and minimum tensile stresses in the steel (material 2) if $E_1 = 1500$ ksi and $E_2 = 30,000$ ksi.

Solution

Neutral axis. The first step in the analysis is to locate the neutral axis of the cross section. For that purpose, let us denote the distances from the neutral axis to the top and bottom of the beam as h_1 and h_2, respectively. To obtain these distances, we use Eq. (5-50). The integrals in that equation are evaluated by taking the first moments of areas 1 and 2 about the z axis, as follows:

$$\int_1 y\, dA = \bar{y}_1 A_1 = (h_1 - 3 \text{ in.})(4 \text{ in.} \times 6 \text{ in.}) = (h_1 - 3 \text{ in.})(24 \text{ in.}^2)$$

$$\int_2 y\, dA = \bar{y}_2 A_2 = -(6.25 \text{ in.} - h_1)(4 \text{ in.} \times 0.5 \text{ in.}) = (h_1 - 6.25 \text{ in.})(2 \text{ in.}^2)$$

in which A_1 and A_2 are the areas of parts 1 and 2 of the cross section, $\bar{y}_1$ and $\bar{y}_2$ are the y coordinates of the centroids of the respective areas, and h_1 has units of inches.

Substituting the preceding expressions into Eq. (5-50) gives the equation for locating the neutral axis, as follows:

$$E_1 \int_1 y\, dA + E_2 \int_2 y\, dA = 0$$

or

$$(1500 \text{ ksi})(h_1 - 3 \text{ in.})(24 \text{ in.}^2) + (30,000 \text{ ksi})(h_1 - 6.25 \text{ in.})(2 \text{ in.}^2) = 0$$

Solving this equation, we obtain the distance h_1 from the neutral axis to the top of the beam:

$$h_1 = 5.031 \text{ in.}$$

Also, the distance h_2 from the neutral axis to the bottom of the beam is

$$h_2 = 6.5 \text{ in.} - h_1 = 1.469 \text{ in.}$$

Thus, the position of the neutral axis is established.

Moments of inertia. The moments of inertia I_1 and I_2 of areas A_1 and A_2 with respect to the neutral axis can be found by using the parallel-axis theorem (see Section 10.5 of Chapter 10, available online). Beginning with area 1 (Fig. 5-44), we get

$$I_1 = \frac{1}{12}(4 \text{ in.})(6 \text{ in.})^3 + (4 \text{ in.})(6 \text{ in.})(h_1 - 3 \text{ in.})^2 = 171.0 \text{ in.}^4$$

Similarly, for area 2 we get

$$I_2 = \frac{1}{12}(4 \text{ in.})(0.5 \text{ in.})^3 + (4 \text{ in.})(0.5 \text{ in.})(h_2 - 0.25 \text{ in.})^2 = 3.01 \text{ in.}^4$$

To check these calculations, we can determine the moment of inertia I of the entire cross-sectional area about the z axis as follows:

$$I = \frac{1}{3}(4 \text{ in.})h_1^3 + \frac{1}{3}(4 \text{ in.})h_2^3 = 169.8 + 4.2 = 174.0 \text{ in.}^4$$

which agrees with the sum of I_1 and I_2.

Normal stresses. The stresses in materials 1 and 2 are calculated from the flexure formulas for composite beams (Eqs. 5-53a and b). The largest compressive stress in material 1 occurs at the top of the beam (A) where $y = h_1 = 5.031$ in. Denoting this stress by σ_{1A} and using Eq. (5-53a), we get

$$\sigma_{1A} = -\frac{Mh_1E_1}{E_1I_1 + E_2I_2}$$

$$= -\frac{(60 \text{ k-in.})(5.031 \text{ in.})(1500 \text{ ksi})}{(1500 \text{ ksi})(171.0 \text{ in.}^4) + (30,000 \text{ ksi})(3.01 \text{ in.}^4)} = -1310 \text{ psi} \quad \Longleftarrow$$

The largest tensile stress in material 1 occurs at the contact plane between the two materials (C) where $y = -(h_2 - 0.5 \text{ in.}) = -0.969$ in. Proceeding as in the previous calculation, we get

$$\sigma_{1C} = -\frac{(60 \text{ k-in.})(-0.969 \text{ in.})(1500 \text{ ksi})}{(1500 \text{ ksi})(171.0 \text{ in.}^4) + (30,000 \text{ ksi})(3.01 \text{ in.}^4)} = 251 \text{ psi} \quad \Longleftarrow$$

Thus, we have found the largest compressive and tensile stresses in the wood.

The steel plate (material 2) is located below the neutral axis, and therefore it is entirely in tension. The maximum tensile stress occurs at the bottom of the beam (B) where $y = -h_2 = -1.469$ in. Hence, from Eq. (5-53b) we get

$$\sigma_{2B} = -\frac{M(-h_2)E_2}{E_1I_1 + E_2I_2}$$

$$= -\frac{(60 \text{ k-in.})(-1.469 \text{ in.})(30,000 \text{ ksi})}{(1500 \text{ ksi})(171.0 \text{ in.}^4) + (30,000 \text{ ksi})(3.01 \text{ in.}^4)} = 7620 \text{ psi} \quad \Longleftarrow$$

The minimum tensile stress in material 2 occurs at the contact plane (C) where $y = -0.969$ in. Thus,

$$\sigma_{2C} = -\frac{(60 \text{ k-in.})(-0.969 \text{ in.})(30,000 \text{ ksi})}{(1500 \text{ ksi})(171.0 \text{ in.}^4) + (30,000 \text{ ksi})(3.01 \text{ in.}^4)} = 5030 \text{ psi} \quad \Longleftarrow$$

These stresses are the maximum and minimum tensile stresses in the steel.

Note: At the contact plane the ratio of the stress in the steel to the stress in the wood is

$$\sigma_{2C}/\sigma_{1C} = 5030 \text{ psi}/251 \text{ psi} = 20$$

which is equal to the ratio E_2/E_1 of the moduli of elasticity (as expected). Although the strains in the steel and wood are equal at the contact plane, the stresses are different because of the different moduli.

Example 5-15

A sandwich beam having aluminum-alloy faces enclosing a plastic core (Fig. 5-45) is subjected to a bending moment $M = 3.0$ kN·m. The thickness of the faces is $t = 5$ mm and their modulus of elasticity is $E_1 = 72$ GPa. The height of the plastic core is $h_c = 150$ mm and its modulus of elasticity is $E_2 = 800$ MPa. The overall dimensions of the beam are $h = 160$ mm and $b = 200$ mm.

Determine the maximum tensile and compressive stresses in the faces and the core using: (a) the general theory for composite beams, and (b) the approximate theory for sandwich beams.

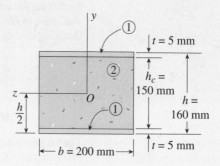

FIG. 5-45 Example 5-15. Cross section of sandwich beam having aluminum-alloy faces and a plastic core

Solution

Neutral axis. Because the cross section is doubly symmetric, the neutral axis (the z axis in Fig. 5-45) is located at midheight.

Moments of inertia. The moment of inertia I_1 of the cross-sectional areas of the faces (about the z axis) is

$$I_1 = \frac{b}{12}(h^3 - h_c^3) = \frac{200 \text{ mm}}{12}\left[(160 \text{ mm})^3 - (150 \text{ mm})^3\right] = 12.017 \times 10^6 \text{ mm}^4$$

and the moment of inertia I_2 of the plastic core is

$$I_2 = \frac{b}{12}(h_c^3) = \frac{200 \text{ mm}}{12}(150 \text{ mm})^3 = 56.250 \times 10^6 \text{ mm}^4$$

As a check on these results, note that the moment of inertia of the entire cross-sectional area about the z axis ($I = bh^3/12$) is equal to the sum of I_1 and I_2.

(a) *Normal stresses calculated from the general theory for composite beams.* To calculate these stresses, we use Eqs. (5-53a) and (5-53b). As a preliminary matter, we will evaluate the term in the denominator of those equations (that is, the flexural rigidity of the composite beam):

$$E_1I_1 + E_2I_2 = (72 \text{ GPa})(12.017 \times 10^6 \text{ mm}^4) + (800 \text{ MPa})(56.250 \times 10^6 \text{ mm}^4)$$

$$= 910,200 \text{ N·m}^2$$

The maximum tensile and compressive stresses in the aluminum faces are found from Eq. (5-53a):

$$(\sigma_1)_{max} = \pm \frac{M(h/2)(E_1)}{E_1 I_1 + E_2 I_2}$$

$$= \pm \frac{(3.0 \text{ kN·m})(80 \text{ mm})(72 \text{ GPa})}{910{,}200 \text{ N·m}^2} = \pm 19.0 \text{ MPa} \qquad \Longleftarrow$$

The corresponding quantities for the plastic core (from Eq. 5-53b) are

$$(\sigma_2)_{max} = \pm \frac{M(h_c/2)(E_2)}{E_1 I_1 + E_2 I_2}$$

$$= \pm \frac{(3.0 \text{ kN·m})(75 \text{ mm})(800 \text{ MPa})}{910{,}200 \text{ N·m}^2} = \pm 0.198 \text{ MPa} \qquad \Longleftarrow$$

The maximum stresses in the faces are 96 times greater than the maximum stresses in the core, primarily because the modulus of elasticity of the aluminum is 90 times greater than that of the plastic.

(b) *Normal stresses calculated from the approximate theory for sandwich beams.* In the approximate theory we disregard the normal stresses in the core and assume that the faces transmit the entire bending moment. Then the maximum tensile and compressive stresses in the faces can be found from Eqs. (5-56a) and (5-56b), as follows:

$$(\sigma_1)_{max} = \pm \frac{Mh}{2I_1} = \pm \frac{(3.0 \text{ kN·m})(80 \text{ mm})}{12.017 \times 10^6 \text{ mm}^4} = \pm 20.0 \text{ MPa} \qquad \Longleftarrow$$

As expected, the approximate theory gives slightly higher stresses in the faces than does the general theory for composite beams.

Transformed-Section Method for Composite Beams

The general theory for flexure of composite beams of two materials was presented above. Now an alternative procedure known as the transformed-section method is presented for finding the bending stresses in a composite beam. The method is based upon the theories and equations developed in the preceding section, and therefore it is subject to the same limitations (for instance, it is valid only for linearly elastic materials) and gives the same results. Although the transformed-section method does not reduce the calculating effort, many designers find that it provides a convenient way to visualize and organize the calculations.

The method consists of transforming the cross section of a composite beam into an equivalent cross section of an imaginary beam that is composed of only one material. This new cross section is called the transformed section. Then the imaginary beam with the transformed section is analyzed in the customary manner for a beam of one material. As a final step, the stresses in the transformed beam are converted to those in the original beam.

Neutral Axis and Transformed Section

If the transformed beam is to be equivalent to the original beam, *its neutral axis must be located in the same place and its moment-resisting capacity must be the same*. To show how these two requirements are met, consider again a composite beam of two materials (Fig. 5-46a). The **neutral axis** of the cross section is obtained from Eq. (5-50), which is repeated here:

$$E_1 \int_1 y\,dA + E_2 \int_2 y\,dA = 0 \tag{5-58}$$

In this equation, the integrals represent the first moments of the two parts of the cross section with respect to the neutral axis.

Let us now introduce the notation

$$n = \frac{E_2}{E_1} \tag{5-59}$$

where n is the **modular ratio**. With this notation, we can rewrite Eq. (5-58) in the form

$$\int_1 y\,dA + \int_2 yn\,dA = 0 \tag{5-60}$$

Since Eqs. (5-58) and (5-60) are equivalent, the preceding equation shows that the neutral axis is unchanged if each element of area dA in material 2 is multiplied by the factor n, provided that the y coordinate for each such element of area is not changed.

Therefore, we can create a new cross section consisting of two parts: (1) area 1 with its dimensions unchanged, and (2) area 2 with its *width* (that is, its dimension parallel to the neutral axis) multiplied by n. This new cross section (the transformed section) is shown in Fig. 5-46b for the case where $E_2 > E_1$ (and therefore $n > 1$). Its neutral axis is in the same position as the neutral axis of the original beam. (Note that all dimensions perpendicular to the neutral axis remain the same.)

Since the stress in the material (for a given strain) is proportional to the modulus of elasticity ($\sigma = E\epsilon$), we see that multiplying the width of material 2 by $n = E_2/E_1$ is equivalent to transforming it to material 1. For instance, suppose that $n = 10$. Then the area of part 2 of the cross section is now 10 times wider than before. If we imagine that this part of

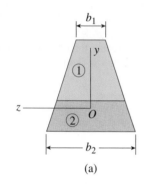

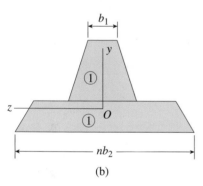

FIG. 5-46 Composite beam of two materials: (a) actual cross section, and (b) transformed section consisting only of material 1

the beam is now material 1, we see that it will carry the same force as before because its modulus is *reduced* by a factor of 10 (from E_2 to E_1) at the same time that its area is *increased* by a factor of 10. Thus, the new section (the transformed section) consists only of material 1.

Moment-Curvature Relationship

The *moment-curvature relationship* for the transformed beam must be the same as for the original beam. To show that this is indeed the case, we note that the stresses in the transformed beam (since it consists only of material 1) are given by Eq. (5-7) of Section 5.5:

$$\sigma_x = -E_1 \kappa y$$

Using this equation, and also following the same procedure as for a beam of one material (see Section 5.5), we can obtain the moment-curvature relation for the transformed beam:

$$M = -\int_A \sigma_x y \, dA = -\int_1 \sigma_x y \, dA - \int_2 \sigma_x y \, dA$$

$$= E_1 \kappa \int_1 y^2 dA + E_1 \kappa \int_2 y^2 dA = \kappa(E_1 I_1 + E_1 n I_2)$$

or

$$M = \kappa(E_1 I_1 + E_2 I_2) \tag{5-61}$$

This equation is the same as Eq. (5-51), thereby demonstrating that the moment-curvature relationship for the transformed beam is the same as for the original beam.

Normal Stresses

Since the transformed beam consists of only one material, the *normal stresses* (or *bending stresses*) can be found from the standard flexure formula (Eq. 5-13). Thus, the normal stresses in the beam transformed to material 1 (Fig. 5-46b) are

$$\sigma_{x1} = -\frac{My}{I_T} \tag{5-62}$$

where I_T is the moment of inertia of the transformed section with respect to the neutral axis. By substituting into this equation, we can calculate the stresses at any point in the *transformed* beam. (As explained later, the stresses in the transformed beam match those in the original beam in the part of the original beam consisting of material 1; however, in the part of the original beam consisting of material 2, the stresses are different from those in the transformed beam.)

We can easily verify Eq. (5-62) by noting that the moment of inertia of the transformed section (Fig. 5-46b) is related to the moment of inertia of the original section (Fig. 5-46a) by the following relation:

$$I_T = I_1 + nI_2 = I_1 + \frac{E_2}{E_1} I_2 \tag{5-63}$$

Substituting this expression for I_T into Eq. (5-62) gives

$$\sigma_{x1} = -\frac{MyE_1}{E_1I_1 + E_2I_2} \tag{a}$$

which is the same as Eq. (5-53a), thus demonstrating that the stresses in material 1 in the original beam are the same as the stresses in the corresponding part of the transformed beam.

As mentioned previously, the stresses in material 2 in the original beam are *not* the same as the stresses in the corresponding part of the transformed beam. Instead, the stresses in the transformed beam (Eq. 5-62) must be multiplied by the modular ratio n to obtain the stresses in material 2 of the original beam:

$$\sigma_{x2} = -\frac{My}{I_T} n \tag{5-64}$$

We can verify this formula by noting that when Eq. (5-63) for I_T is substituted into Eq. (5-64), we get

$$\sigma_{x2} = -\frac{MynE_1}{E_1I_1 + E_2I_2} = -\frac{MyE_2}{E_1I_1 + E_2I_2} \tag{b}$$

which is the same as Eq. (5-53b).

General Comments

In this discussion of the transformed-section method we chose to transform the original beam to a beam consisting entirely of material 1. It is also possible to transform the beam to material 2. In that case the stresses in the original beam in material 2 will be the same as the stresses in the corresponding part of the transformed beam. However, the stresses in material 1 in the original beam must be obtained by multiplying the stresses in the corresponding part of the transformed beam by the modular ratio n, which in this case is defined as $n = E_1/E_2$.

It is also possible to transform the original beam into a material having any arbitrary modulus of elasticity E, in which case all parts of the beam must be transformed to the fictitious material. Of course, the calculations are simpler if we transform to one of the original materials. Finally, with a little ingenuity it is possible to extend the transformed-section method to composite beams of more than two materials.

Example 5-16

The composite beam shown in Fig. 5-47a is formed of a wood beam (4.0 in. × 6.0 in. actual dimensions) and a steel reinforcing plate (4.0 in. wide and 0.5 in. thick). The beam is subjected to a positive bending moment $M = 60$ k-in.

Using the transformed-section method, calculate the largest tensile and compressive stresses in the wood (material 1) and the maximum and minimum tensile stresses in the steel (material 2) if $E_1 = 1500$ ksi and $E_2 = 30,000$ ksi.

Note: This same beam was analyzed previously in Example 5-14 above.

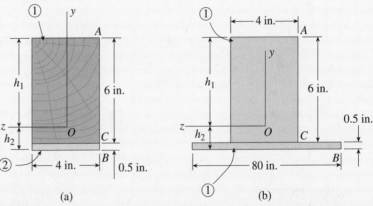

FIG. 5-47 Example 5-16. Composite beam of Example 5-14 analyzed by the transformed-section method: (a) cross section of original beam, and (b) transformed section (material 1)

Solution

Transformed section. We will transform the original beam into a beam of material 1, which means that the modular ratio is defined as

$$n = \frac{E_2}{E_1} = \frac{30,000 \text{ ksi}}{1,500 \text{ ksi}} = 20$$

The part of the beam made of wood (material 1) is not altered but the part made of steel (material 2) has its width multiplied by the modular ratio. Thus, the width of this part of the beam becomes

$$n(4 \text{ in.}) = 20(4 \text{ in.}) = 80 \text{ in.}$$

in the transformed section (Fig. 5-47b).

Neutral axis. Because the transformed beam consists of only one material, the neutral axis passes through the centroid of the cross-sectional area. Therefore, with the top edge of the cross section serving as a reference line, and with the distance y_i measured positive downward, we can calculate the distance h_1 to the centroid as follows:

$$h_1 = \frac{\sum y_i A_i}{\sum A_i} = \frac{(3 \text{ in.})(4 \text{ in.})(6 \text{ in.}) + (6.25 \text{ in.})(80 \text{ in.})(0.5 \text{ in.})}{(4 \text{ in.})(6 \text{ in.}) + (80 \text{ in.})(0.5 \text{ in.})}$$

$$= \frac{322.0 \text{ in.}^3}{64.0 \text{ in.}^2} = 5.031 \text{ in.}$$

continued

Also, the distance h_2 from the lower edge of the section to the centroid is

$$h_2 = 6.5 \text{ in.} - h_1 = 1.469 \text{ in.}$$

Thus, the location of the neutral axis is determined.

Moment of inertia of the transformed section. Using the parallel-axis theorem (see Section 10.5 of Chapter 10, available online), we can calculate the moment of inertia I_T of the entire cross-sectional area with respect to the neutral axis as follows:

$$I_T = \frac{1}{12}(4 \text{ in.})(6 \text{ in.})^3 + (4 \text{ in.})(6 \text{ in.})(h_1 - 3 \text{ in.})^2$$

$$+ \frac{1}{12}(80 \text{ in.})(0.5 \text{ in.})^3 + (80 \text{ in.})(0.5 \text{ in.})(h_2 - 0.25 \text{ in.})^2$$

$$= 171.0 \text{ in.}^4 + 60.3 \text{ in.}^4 = 231.3 \text{ in.}^4$$

Normal stresses in the wood (material 1). The stresses in the transformed beam (Fig. 5-47b) at the top of the cross section (A) and at the contact plane between the two parts (C) are the same as in the original beam (Fig. 5-47a). These stresses can be found from the flexure formula (Eq. 5-62), as follows:

$$\sigma_{1A} = -\frac{My}{I_T} = -\frac{(60 \text{ k-in.})(5.031 \text{ in.})}{231.3 \text{ in.}^4} = -1310 \text{ psi} \quad \Leftarrow$$

$$\sigma_{1C} = -\frac{My}{I_T} = -\frac{(60 \text{ k-in.})(-0.969 \text{ in.})}{231.3 \text{ in.}^4} = 251 \text{ psi} \quad \Leftarrow$$

These are the largest tensile and compressive stresses in the wood (material 1) in the original beam. The stress σ_{1A} is compressive and the stress σ_{1C} is tensile.

Normal stresses in the steel (material 2). The maximum and minimum stresses in the steel plate are found by multiplying the corresponding stresses in the transformed beam by the modular ratio n (Eq. 5-64). The maximum stress occurs at the lower edge of the cross section (B) and the minimum stress occurs at the contact plane (C):

$$\sigma_{2B} = -\frac{My}{I_T}n = -\frac{(60 \text{ k-in.})(-1.469 \text{ in.})}{231.3 \text{ in.}^4}(20) = 7620 \text{ psi} \quad \Leftarrow$$

$$\sigma_{2C} = -\frac{My}{I_T}n = -\frac{(60 \text{ k-in.})(-0.969 \text{ in.})}{231.3 \text{ in.}^4}(20) = 5030 \text{ psi} \quad \Leftarrow$$

Both of these stresses are tensile.

Note that the stresses calculated by the transformed-section method agree with those found in Example 5-14 by direct application of the formulas for a composite beam.

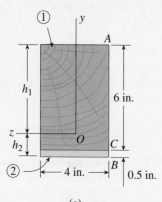

(a)

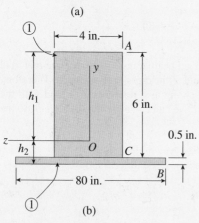

(b)

FIG. 5-47 (Repeated)

CHAPTER SUMMARY & REVIEW

In Chapter 5, we investigated the behavior of beams with loads applied and bending occurring in the *x-y* plane: a plane of symmetry in the beam cross section. Both pure bending and nonuniform bending were considered. The normal stresses were seen to vary linearly from the neutral surface in accordance with the **flexure formula**, which showed that the stresses are directly proportional to the bending moment M and inversely proportional to the moment of inertia I of the cross section. Next, the relevant properties of the beam cross section were combined into a single quantity known as the **section modulus** S of the beam: a useful property in **beam design** once the maximum moment (M_{max}) and allowable normal stress (σ_{allow}) are known. Next, horizontal and vertical shear stresses (τ) were computed using the **shear formula** for the case of nonuniform bending of beams with either rectangular or circular cross sections. The special case of shear in beams with flanges also was considered. Finally, the analysis of composite beams (that is, beams of more than one material) was discussed.

Some of the major concepts, formulas and findings presented in this chapter are as follows:

1. If the *xy* plane is a plane of symmetry of a beam cross section and applied loads act in the *xy* plane, the bending deflections occur in this same plane, known as the **plane of bending**.

2. A beam in pure bending has constant curvature κ, and a beam in nonuniform bending has varying curvature. Longitudinal strains (ϵ_x) in a bent beam are proportional to its curvature, and the strains in a beam in pure bending vary linearly with distance from the neutral surface, regardless of the shape of the stress-strain curve of the material in accordance with Eq. (5-4):

$$\epsilon_x = -\kappa y$$

3. The neutral axis passes through the centroid of the cross-sectional area when the material follows Hooke's law and there is no axial force acting on the cross section. When a beam of linearly elastic material is subjected to pure bending, the *y* and *z* axes are **principal centroidal axes**.

4. If the material of a beam is linearly elastic and follows Hooke's law, the **moment-curvature equation** shows that the curvature is directly proportional to the bending moment M and inversely proportional to the quantity EI, referred to as the **flexural rigidity** of the beam. The moment curvature relation was given in Eq. (5-12):

$$\kappa = \frac{M}{EI}$$

5. The **flexure formula** shows that the normal stresses σ_x are directly proportional to the bending moment M and inversely proportional to the moment of inertia I of the cross section as given in Eq. (5-13):

$$\sigma_x = -\frac{My}{I}.$$

The maximum tensile and compressive bending stresses acting at any given cross section occur at points located farthest from the neutral axis

$$(y = c_1, y = -c_2).$$

continued

6. The normal stresses calculated from the flexure formula are not significantly altered by the presence of shear stresses and the associated warping of the cross section for the case of nonuniform bending. However, the flexure formula is not applicable near the supports of a beam or close to a concentrated load, because such irregularities produce **stress concentrations** that are much greater than the stresses obtained from the flexure formula.

7. To **design** a beam to resist bending stresses, we calculate the required **section modulus** S from the maximum moment and allowable normal stress as follows:

$$S = \frac{M_{max}}{\sigma_{allow}}$$

To minimize weight and save material, we usually select a beam from a material design manual (e.g., see sample tables in Appendices F and G for steel and wood, available online) that has the least cross-sectional area while still providing the required section modulus; wide-flange sections, and I-sections have most of their material in the flanges and the width of their flanges helps to reduce the likelihood of sideways buckling.

8. Beams subjected to loads that produce both bending moments (M) and shear forces (V) (**nonuniform bending**) develop both normal and shear stresses in the beam. Normal stresses are calculated from the **flexure formula** (provided the beam is constructed of a linearly elastic material), and shear stresses are computed using the **shear formula** as follows:

$$\tau = \frac{VQ}{Ib}$$

Shear stress varies parabolically over the height of a rectangular beam, and shear strain also varies parabolically; these shear strains cause cross sections of the beam that were originally plane surfaces to become warped. The maximum values of the shear stress and strain (τ_{max}, γ_{max}) occur at the neutral axis, and the shear stress and strain are zero on the top and bottom surfaces of the beam.

9. The shear formula applies only to prismatic beams and is valid only for beams of linearly elastic materials with small deflections; also, the edges of the cross section must be **parallel** to the y axis. For rectangular beams, the accuracy of the shear formula depends upon the height-to-width ratio of the cross section: the formula may be considered as exact for very narrow beams but becomes less accurate as width b increases relative to height h. Note that we can use the shear formula to calculate the shear stresses only at the neutral axis of a beam of **circular** cross section.

For rectangular cross sections,

$$\tau_{max} = \frac{3}{2} \frac{V}{A}$$

and for solid circular cross sections

$$\tau_{max} = \frac{4}{3} \frac{V}{A}$$

10. Shear stresses rarely govern the design of either circular or rectangular beams made of metals such as steel and aluminum for which the allowable shear stress is usually in the range 25 to 50% of the allowable tensile stress. However, for **materials that are weak in shear**, such as wood, the allowable stress in horizontal shear is in the range of 4 to 10% of the allowable bending stress and so may govern the design.

11. Shear stresses in the flanges of **wide-flange beams** act in both vertical and horizontal directions. The horizontal shear stresses are much larger than the vertical shear stresses in the flanges. The shear stresses in the **web of a wide-flange beam** act only in the vertical direction, are larger than the stresses in the flanges, and may be computed using the shear formula. The maximum shear stress in the web of a wide-flange beam occurs at the neutral axis, and the minimum shear stress occurs where the web meets the flanges. For beams of typical proportions, the shear force in the web is 90 to 98% of the total shear force V acting on the cross section; the remainder is carried by shear in the flanges.

12. In the introduction to **composite beams**, specialized moment-curvature relationship and flexure formulas for composite beams of two materials were developed:

$$\kappa = \frac{1}{\rho} = \frac{M}{E_1 I_1 + E_2 I_2} \quad \sigma_{x1} = -\frac{M y E_1}{E_1 I_1 + E_2 I_2} \quad \sigma_{x2} = -\frac{M y E_2}{E_1 I_1 + E_2 I_2}$$

We assumed that both materials follow Hooke's law and that the two parts of the beam are adequately bonded so that they act as a single unit. Advanced topics such as nonhomogeneous and nonlinear materials, bond stresses between the parts, shear stresses on the cross sections, buckling of the faces, and other such matters are not considered. In particular, the formulas presented herein do not apply to reinforced concrete beams which are not designed on the basis of linearly elastic behavior.

13. The **transformed-section method** offers a convenient way of transforming the cross section of a composite beam into an equivalent cross section of an imaginary beam that is composed of only one material. The ratio of the modulus of elasticity of material 2 to that of material 1 is known as the **modular ratio**, $n = E_2/E_1$. The neutral axis of the transformed beam is located in the same place, and its moment-resisting capacity is the same as that of the original composite beam. The moment of inertia of the transformed section is defined as follows:

$$I_T = I_1 + n I_2 = I_1 + \frac{E_2}{E_1} I_2$$

Normal stresses in the beam transformed to material 1 are computed using the simplified flexure formula

$$\sigma_{x1} = -\frac{My}{I_T}$$

while those in material 2 are computed as follows:

$$\sigma_{x2} = -\frac{My}{I_T} n$$

PROBLEMS CHAPTER 5

Longitudinal Strains in Beams

5.4-1 Determine the maximum normal strain ϵ_{max} produced in a steel wire of diameter $d = 1/16$ in. when it is bent around a cylindrical drum of radius $R = 24$ in. (see figure).

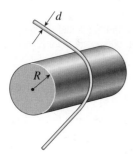

PROB. 5.4-1

5.4-2 A copper wire having diameter $d = 3$ mm is bent into a circle and held with the ends just touching (see figure). If the maximum permissible strain in the copper is $\epsilon_{max} = 0.0024$, what is the shortest length L of wire that can be used?

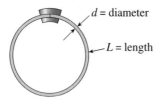

PROB. 5.4-2

5.4-3 A 4.5 in. outside diameter polyethylene pipe designed to carry chemical wastes is placed in a trench and bent around a quarter-circular 90° bend (see figure). The bent section of the pipe is 46 ft long.

Determine the maximum compressive strain ϵ_{max} in the pipe.

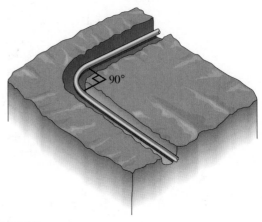

PROB. 5.4-3

5.4-4 A cantilever beam AB is loaded by a couple M_0 at its free end (see figure). The length of the beam is $L = 2.0$ m, and the longitudinal normal strain at the top surface is 0.0012. The distance from the top surface of the beam to the neutral surface is 82.5 mm.

Calculate the radius of curvature ρ, the curvature κ, and the vertical deflection δ at the end of the beam.

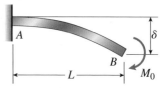

PROB. 5.4-4

5.4-5 A thin strip of steel of length $L = 28$ in. and thickness $t = 0.25$ in. is bent by couples M_0 (see figure). The deflection at the midpoint of the strip (measured from a line joining its end points) is found to be 0.20 in.

Determine the longitudinal normal strain ε at the top surface of the strip.

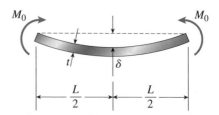

PROB. 5.4-5

5.4-6 A bar of rectangular cross section is loaded and supported as shown in the figure. The distance between supports is $L = 1.5$ m, and the height of the bar is $h = 120$ mm. The deflection at the midpoint is measured as 3.0 mm.

What is the maximum normal strain ε at the top and bottom of the bar?

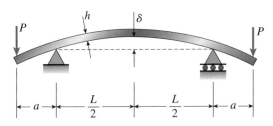

PROB. 5.4-6

Normal Stresses in Beams

5.5-1 A thin strip of hard copper ($E = 16,000$ ksi) having length $L = 90$ in. and thickness $t = 3/32$ in. is bent into a circle and held with the ends just touching (see figure).

(a) Calculate the maximum bending stress σ_{max} in the strip.

(b) By what percent does the stress increase or decrease if the thickness of the strip is increased by 1/32 in.?

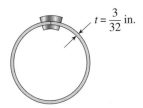

PROB. 5.5-1

5.5-2 A steel wire ($E = 200$ GPa) of diameter $d = 1.25$ mm is bent around a pulley of radius $R_0 = 500$ mm (see figure).

(a) What is the maximum stress σ_{max} in the wire?

(b) By what percent does the stress increase or decrease if the radius of the pulley is increased by 25%?

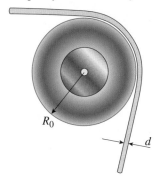

PROB. 5.5-2

5.5-3 A thin, high-strength steel rule ($E = 30 \times 10^6$ psi) having thickness $t = 0.175$ in. and length $L = 48$ in. is bent by couples M_0 into a circular arc subtending a central angle $\alpha = 40°$ (see figure).

(a) What is the maximum bending stress σ_{max} in the rule?

(b) By what percent does the stress increase or decrease if the central angle is increased by 10%?

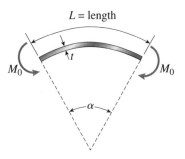

PROB. 5.5-3

5.5-4 A simply supported wood beam AB with span length $L = 4$ m carries a uniform load of intensity $q = 5.8$ kN/m (see figure).

(a) Calculate the maximum bending stress σ_{max} due to the load q if the beam has a rectangular cross section with width $b = 140$ mm and height $h = 240$ mm.

(b) Repeat (a) but use the trapezoidal distributed load shown in the figure part (b).

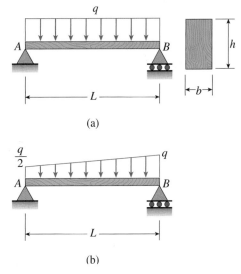

PROB. 5.5-4

5.5-5 Each girder of the lift bridge (see figure) is 180 ft long and simply supported at the ends. The design load for each girder is a uniform load of intensity 1.6 k/ft. The girders are fabricated by welding three steel plates so as to form an I-shaped cross section (see figure) having section modulus $S = 3600 \text{ in}^3$.

What is the maximum bending stress σ_{max} in a girder due to the uniform load?

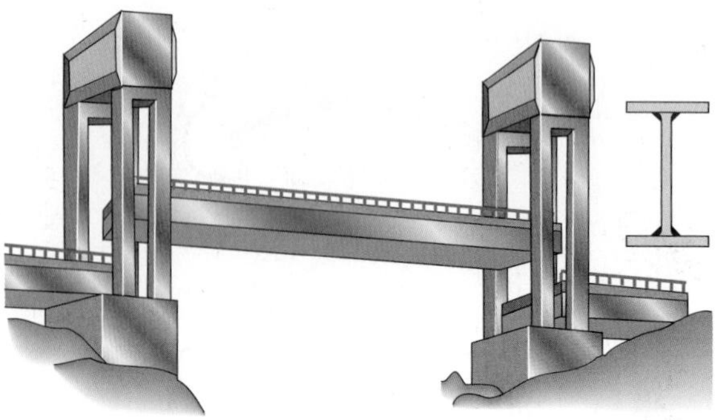

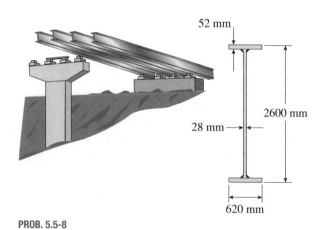

PROB. 5.5-5

5.5-6 A freight-car axle AB is loaded approximately as shown in the figure, with the forces P representing the car loads (transmitted to the axle through the axle boxes) and the forces R representing the rail loads (transmitted to the axle through the wheels). The diameter of the axle is $d = 82$ mm, the distance between centers of the rails is L, and the distance between the forces P and R is $b = 220$ mm.

Calculate the maximum bending stress σ_{max} in the axle if $P = 50$ kN.

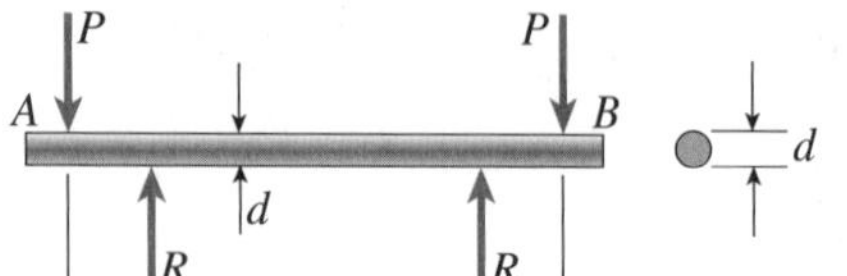

PROB. 5.5-6

5.5-7 A seesaw weighing 3 lb/ft of length is occupied by two children, each weighing 90 lb (see figure). The center of gravity of each child is 8 ft from the fulcrum. The board is 19 ft long, 8 in. wide, and 1.5 in. thick.

What is the maximum bending stress in the board?

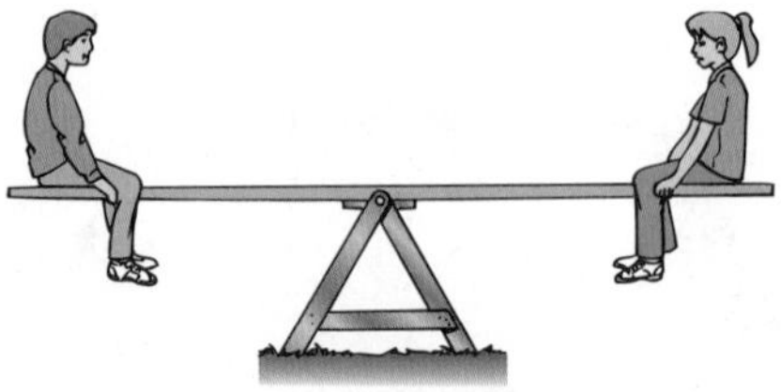

PROB. 5.5-7

5.5-8 During construction of a highway bridge, the main girders are cantilevered outward from one pier toward the next (see figure). Each girder has a cantilever length of 48 m and an I-shaped cross section with dimensions shown in the figure. The load on each girder (during construction) is assumed to be 9.5 kN/m, which includes the weight of the girder.

Determine the maximum bending stress in a girder due to this load.

PROB. 5.5-8

5.5-9 The horizontal beam ABC of an oil-well pump has the cross section shown in the figure. If the vertical pumping force acting at end C is 9 k and if the distance from the line of action of that force to point B is 16 ft, what is the maximum bending stress in the beam due to the pumping force?

Horizontal beam transfers loads as part of oil well pump
(Gabriel M. Covian/Getty Images)

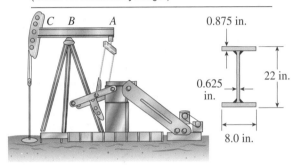

PROB. 5.5-9

5.5-10 A railroad tie (or *sleeper*) is subjected to two rail loads, each of magnitude $P = 175$ kN, acting as shown in the figure. The reaction q of the ballast is assumed to be uniformly distributed over the length of the tie, which has cross-sectional dimensions $b = 300$ mm and $h = 250$ mm.

Calculate the maximum bending stress σ_{max} in the tie due to the loads P, assuming the distance $L = 1500$ mm and the overhang length $a = 500$ mm.

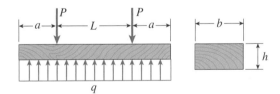

PROB. 5.5-10

5.5-11 A fiberglass pipe is lifted by a sling, as shown in the figure. The outer diameter of the pipe is 6.0 in., its thickness is 0.25 in., and its weight density is 0.053 lb/in.^{3} The length of the pipe is $L = 36$ ft and the distance between lifting points is $s = 11$ ft.

Determine the maximum bending stress in the pipe due to its own weight.

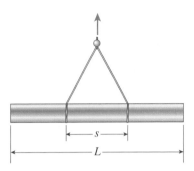

PROB. 5.5-11

5.5-12 A small dam of height $h = 2.0$ m is constructed of vertical wood beams AB of thickness $t = 120$ mm, as shown in the figure. Consider the beams to be simply supported at the top and bottom.

Determine the maximum bending stress σ_{max} in the beams, assuming that the weight density of water is $\gamma = 9.81$ kN/m^3.

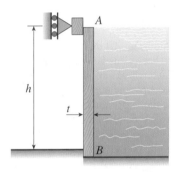

PROB. 5.5-12

5.5-13 Determine the maximum tensile stress σ_t (due to pure bending about a horizontal axis through C by positive bending moments M) for beams having cross sections as follows (see figure).

(a) A semicircle of diameter d
(b) An isosceles trapezoid with bases $b_1 = b$ and $b_2 = 4b/3$, and altitude h
(c) A circular sector with $\alpha = \pi/3$ and $r = d/2$

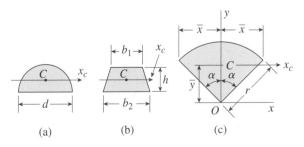

(a)   (b)   (c)

PROB. 5.5-13

5.5-14 Determine the maximum bending stress σ_{max} (due to pure bending by a moment M) for a beam having a cross section in the form of a circular core (see figure). The circle has diameter d and the angle $\beta = 60°$. (*Hint:* Use the formulas given in Appendix E, Cases 9 and 15 available online.)

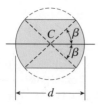

PROB. 5.5-14

5.5-15 A simple beam AB of span length $L = 24$ ft is subjected to two wheel loads acting at distance $d = 5$ ft apart (see figure). Each wheel transmits a load $P = 3.0$ k, and the carriage may occupy any position on the beam.

Determine the maximum bending stress σ_{max} due to the wheel loads if the beam is an I-beam having section modulus $S = 16.2$ in.^{3}

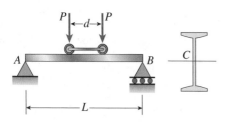

PROB. 5.5-15

5.5-16 Determine the maximum tensile stress σ_t and maximum compressive stress σ_c due to the load P acting on the simple beam AB (see figure).

Data are as follows: $P = 6.2$ kN, $L = 3.2$ m, $d = 1.25$ m, $b = 80$ mm, $t = 25$ mm, $h = 120$ mm, and $h_1 = 90$ mm.

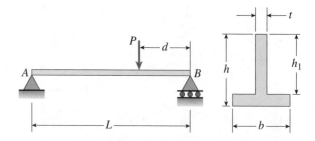

PROB. 5.5-16

5.5-17 A cantilever beam AB, loaded by a uniform load and a concentrated load (see figure), is constructed of a channel section.

Find the maximum tensile stress σ_t and maximum compressive stress σ_c if the cross section has the dimensions indicated and the moment of inertia about the z axis (the neutral axis) is $I = 3.36$ in.^{4} (*Note:* The uniform load represents the weight of the beam.)

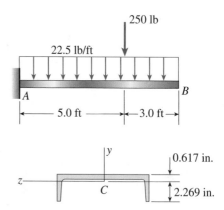

PROB. 5.5-17

5.5-18 A cantilever beam AB of isosceles trapezoidal cross section has length $L = 0.8$ m, dimensions $b_1 = 80$ mm, $b_2 = 90$ mm, and height $h = 110$ mm (see figure). The beam is made of brass weighing 85 kN/m^3.

(a) Determine the maximum tensile stress σ_t and maximum compressive stress σ_c due to the beam's own weight.

(b) If the width b_1 is doubled, what happens to the stresses?

(c) If the height h is doubled, what happens to the stresses?

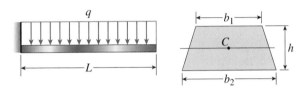

PROB. 5.5-18

5.5-19 A beam ABC with an overhang from B to C supports a uniform load of 200 lb/ft throughout its length (see figure). The beam is a channel section with dimensions as shown in the figure. The moment of inertia about the z axis (the neutral axis) equals 8.13 in.[4]

Calculate the maximum tensile stress σ_t and maximum compressive stress σ_c due to the uniform load.

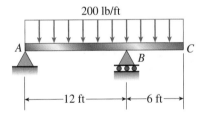

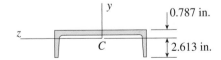

PROB. 5.5-19

5.5-20 A frame ABC travels horizontally with an acceleration a_0 (see figure). Obtain a formula for the maximum

stress σ_{max} in the vertical arm AB, which has length L, thickness t, and mass density ρ.

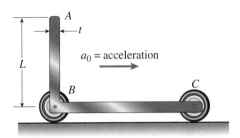

PROB. 5.5-20

5.5-21 A beam of T-section is supported and loaded as shown in the figure. The cross section has width $b = 2\ 1/2$ in., height $h = 3$ in., and thickness $t = 3/8$ in.

Determine the maximum tensile and compressive stresses in the beam.

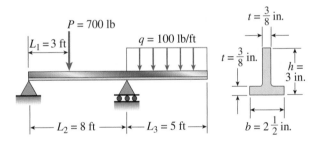

PROB. 5.5-21

5.5-22 A cantilever beam AB with a rectangular cross section has a longitudinal hole drilled throughout its length (see figure). The beam supports a load $P = 600$ N. The cross section is 25 mm wide and 50 mm high, and the hole has a diameter of 10 mm.

Find the bending stresses at the top of the beam, at the top of the hole, and at the bottom of the beam.

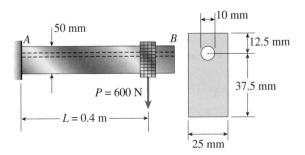

PROB. 5.5-22

5.5-23 A small dam of height $h = 6$ ft is constructed of vertical wood beams AB, as shown in the figure. The wood beams, which have thickness $t = 2.5$ in., are simply supported by horizontal steel beams at A and B.

Construct a graph showing the maximum bending stress σ_{max} in the wood beams versus the depth d of the water above the lower support at B. Plot the stress σ_{max} (psi) as the ordinate and the depth d (ft) as the abscissa. (*Note:* The weight density γ of water equals 62.4 lb/ft^3.)

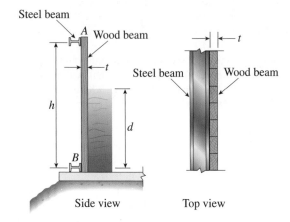

Side view   Top view

PROB. 5.5-23

5.5-24 Consider the nonprismatic *cantilever beam* of circular cross section shown. The beam has an internal cylindrical hole in segment 1; the bar is solid (radius r) in segment 2. The beam is loaded by a downward triangular load with maximum intensity q_0 as shown.

Find expressions for maximum tensile and compressive flexural stresses at joint 1.

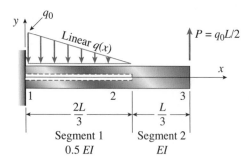

PROB. 5.5-24

5.5-25 A steel post ($E = 30 \times 10^6$ psi) having thickness $t = 1/8$ in. and height $L = 72$ in. supports a stop sign (see figure: $s = 12.5$ in.). The height of the post L is measured from the base to the centroid of the sign. The stop sign is subjected to wind pressure $p = 20$ lb/ft^2 normal to its surface. Assume that the post is fixed at its base.

(a) What is the resultant load on the sign? (See Appendix E, Case 25, available online for properties of an octagon, $n = 8$).

(b) What is the maximum bending stress σ_{max} in the post?

Design of Beams

5.6-1 The cross section of a narrow-gage railway bridge is shown in part (a) of the figure. The bridge is constructed with longitudinal steel girders that support the wood cross ties. The girders are restrained against lateral buckling by diagonal bracing, as indicated by the dashed lines.

The spacing of the girders is $s_1 = 50$ in. and the spacing of the rails is $s_2 = 30$ in. The load transmitted by each rail to a single tie is $P = 1500$ lb. The cross section of a tie, shown in part (b) of the figure, has width $b = 5.0$ in. and depth d.

Determine the minimum value of d based upon an allowable bending stress of 1125 psi in the wood tie. (Disregard the weight of the tie itself.)

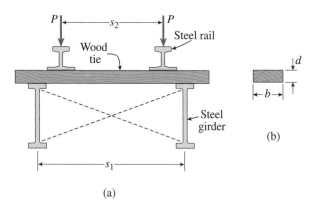

PROB. 5.6-1

5.6-2 A fiberglass bracket $ABCD$ of solid circular cross section has the shape and dimensions shown in the figure. A vertical load $P = 40$ N acts at the free end D.

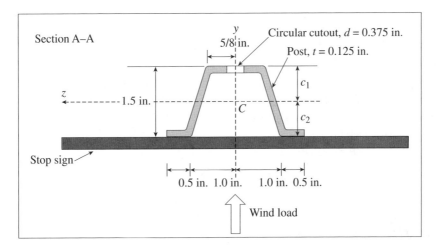

Section A–A

y

5/8 in.

Circular cutout, $d = 0.375$ in.

Post, $t = 0.125$ in.

c_1

z

1.5 in.

C

c_2

Stop sign

0.5 in. 1.0 in.   1.0 in. 0.5 in.

Wind load

Numerical properties of post

$A = 0.578$ in.^{2}, $c_1 = 0.769$ in., $c_2 = 0.731$ in., $I_y = 0.44867$ in.^{4}, $I_z = 0.16101$ in.^{4}

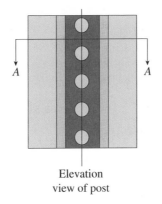

Elevation
view of post

PROB. 5.5-25

Determine the minimum permissible diameter d_{min} of the bracket if the allowable bending stress in the material is 30 MPa and $b = 37$ mm. (*Note:* Disregard the weight of the bracket itself.)

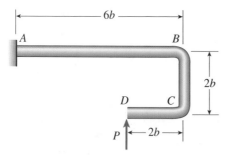

PROB. 5.6-2

5.6-3 A cantilever beam of length $L = 7.5$ ft supports a uniform load of intensity $q = 225$ lb/ft and a concentrated load $P = 2750$ lb (see figure).

Calculate the required section modulus S if $\sigma_{allow} = 17,000$ psi. Then select a suitable wide-flange beam (W shape) from Table F-1(a), Appendix F available online, and recalculate S taking into account the weight of the beam. Select a new beam size if necessary.

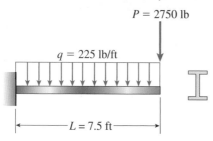

PROB. 5.6-3

5.6-4 A simple beam of length $L = 5$ m carries a uniform load of intensity $q = 5.8 \dfrac{\text{kN}}{\text{m}}$ and a concentrated load 22.5 kN (see figure).

Assuming $\sigma_{allow} = 110$ MPa, calculate the required section modulus S. Then select an 200 mm wide-flange beam (W shape) from Table F-1(b) Appendix F available online, and recalculate S taking into account the weight of the beam. Select a new 200 mm beam if necessary.

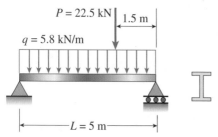

PROB. 5.6-4

5.6-5 A simple beam AB is loaded as shown in the figure.

Calculate the required section modulus S if $\sigma_{allow} = 17,000$ psi, $L = 28$ ft, $P = 2200$ lb, and $q = 425$ lb/ft. Then select a suitable I-beam (S shape) from Table F-2(a), Appendix F available online, and recalculate S taking into account the weight of the beam. Select a new beam size if necessary.

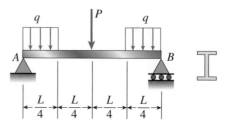

PROB. 5.6-5

5.6-6 A pontoon bridge (see figure) is constructed of two longitudinal wood beams, known as *balks*, that span between adjacent pontoons and support the transverse floor beams, which are called *chesses*.

For purposes of design, assume that a uniform floor load of 8.0 kPa acts over the chesses. (This load includes an allowance for the weights of the chesses and balks.) Also, assume that the chesses are 2.0 m long and that the balks are simply supported with a span of 3.0 m. The allowable bending stress in the wood is 16 MPa.

If the balks have a square cross section, what is their minimum required width b_{min}?

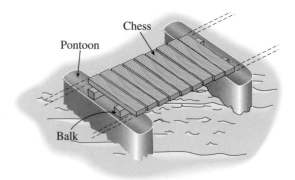

PROB. 5.6-6

5.6-7 A floor system in a small building consists of wood planks supported by 2 in. (nominal width) joists spaced at distance s, measured from center to center (see figure). The span length L of each joist is 10.5 ft, the spacing s of the joists is 16 in., and the allowable bending stress in the wood is 1350 psi. The uniform floor load is 120 lb/ft², which includes an allowance for the weight of the floor system itself.

Calculate the required section modulus S for the joists, and then select a suitable joist size (surfaced lumber) from Appendix G, assuming that each joist may be represented as a simple beam carrying a uniform load.

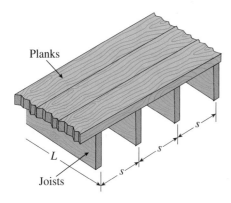

PROBS. 5.6-7 and 5.6-8

5.6-8 The wood joists supporting a plank floor (see figure) are 40 mm × 180 mm in cross section (actual dimensions) and have a span length $L = 4.0$ m. The floor load is 3.6 kPa, which includes the weight of the joists and the floor.

Calculate the maximum permissible spacing s of the joists if the allowable bending stress is 15 MPa. (Assume that each joist may be represented as a simple beam carrying a uniform load.)

5.6-9 A beam ABC with an overhang from B to C is constructed of a C 10 × 30 channel section (see figure). The beam supports its own weight (30 lb/ft) plus a *triangular* load of maximum intensity q_0 acting on the overhang. The allowable stresses in tension and compression are 20 ksi and 11 ksi, respectively.

Determine the allowable *triangular* load intensity $q_{0,\text{allow}}$ if the distance L equals 3.5 ft.

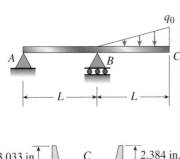

PROB. 5.6-9

5.6-10 A so-called "trapeze bar" in a hospital room provides a means for patients to exercise while in bed (see figure). The bar is 2.1 m long and has a cross section in the shape of a regular octagon. The design load is 1.2 kN applied at the midpoint of the bar, and the allowable bending stress is 200 MPa.

Determine the minimum height h of the bar. (Assume that the ends of the bar are simply supported and that the weight of the bar is negligible.)

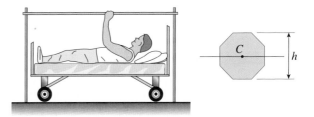

PROB. 5.6-10

5.6-11 A two-axle carriage that is part of an overhead traveling crane in a testing laboratory moves slowly across a simple beam AB (see figure). The load transmitted to the beam from the front axle is 2200 lb and from the rear axle is 3800 lb. The weight of the beam itself may be disregarded.

(a) Determine the minimum required section modulus S for the beam if the allowable bending stress is 17.0 ksi, the length of the beam is 18 ft, and the wheelbase of the carriage is 5 ft.

(b) Select the most economical I-beam (S shape) from Table F-2(a), Appendix F available online.

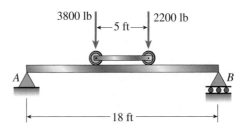

PROB. 5.6-11

5.6-12 A cantilever beam AB of circular cross section and length $L = 450$ mm supports a load $P = 400$ N acting at the free end (see figure). The beam is made of steel with an allowable bending stress of 60 MPa.

Determine the required diameter d_{min} of the beam, considering the effect of the beam's own weight.

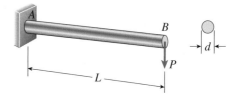

PROB. 5.6-12

5.6-13 A compound beam *ABCD* (see figure) is supported at points *A*, *B*, and *D* and has a splice at point *C*. The distance $a = 6.25$ ft, and the beam is a S 18×70 wide-flange shape with an allowable bending stress of 12,800 psi.

(a) If the splice is a *moment release*, find the allowable uniform load q_{allow} that may be placed on top of the beam, taking into account the weight of the beam itself. [See figure part (a).]

(b) Repeat assuming now that the splice is a *shear release*, as in figure part (b).

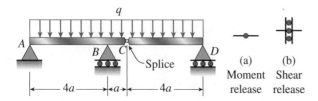

(a) Moment release (b) Shear release

PROB. 5.6-13

5.6-14 A small balcony constructed of wood is supported by three identical cantilever beams (see figure). Each beam has length $L_1 = 2.1$ m, width *b*, and height $h = 4b/3$. The dimensions of the balcony floor are $L_1 \times L_2$, with $L_2 = 2.5$ m. The design load is 5.5 kPa acting over the entire floor area. (This load accounts for all loads except the weights of the cantilever beams, which have a weight density $\gamma = 5.5$ kN/m³.) The allowable bending stress in the cantilevers is 15 MPa.

Assuming that the middle cantilever supports 50% of the load and each outer cantilever supports 25% of the load, determine the required dimensions *b* and *h*.

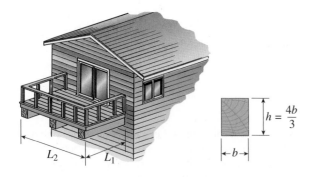

PROB. 5.6-14

5.6-15 A beam having a cross section in the form of an unsymmetric wide-flange shape (see figure) is subjected to a negative bending moment acting about the *z* axis.

Determine the width *b* of the top flange in order that the stresses at the top and bottom of the beam will be in the ratio 4:3, respectively.

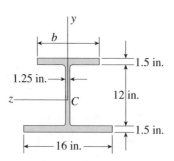

PROB. 5.6-15

5.6-16 A beam having a cross section in the form of a channel (see figure) is subjected to a bending moment acting about the *z* axis.

Calculate the thickness *t* of the channel in order that the bending stresses at the top and bottom of the beam will be in the ratio 7:3, respectively.

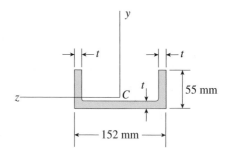

PROB. 5.6-16

5.6-17 Determine the ratios of the weights of three beams that have the same length, are made of the same material, are subjected to the same maximum bending moment, and have the same maximum bending stress if their cross sections are (1) a rectangle with height equal to twice the width, (2) a square, and (3) a circle (see figures).

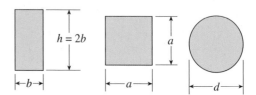

PROB. 5.6-17

5.6-18 A horizontal shelf *AD* of length $L = 915$ mm, width $b = 305$ mm, and thickness $t = 22$ mm is supported by brackets at *B* and *C* [see part (a) of the figure]. The brackets are adjustable and may be placed in any desired positions between the ends of the shelf. A uniform load of intensity q, which includes the weight of the shelf itself, acts on the shelf [see part (b) of the figure].

Determine the maximum permissible value of the load q if the allowable bending stress in the shelf is $\sigma_{allow} = 7.5$ MPa and the position of the supports is adjusted for maximum load-carrying capacity.

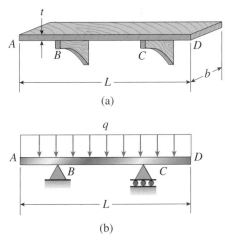

(a)

(b)

PROB. 5.6-18

5.6-19 A steel plate (called a *cover plate*) having cross sectional dimensions 6.0 in. $\times$ 0.5 in. is welded along the full length of the bottom flange of a W12 $\times$ 50 wide-flange beam (see figure, which shows the beam cross section).

What is the percent increase in the smaller section modulus (as compared to the wide-flange beam alone)?

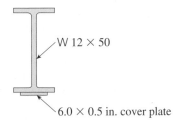

W 12 $\times$ 50

6.0 $\times$ 0.5 in. cover plate

PROB. 5.6-19

5.6-20 A steel beam ABC is simply supported at A and B and has an overhang BC of length $L = 150$ mm (see figure). The beam supports a uniform load of intensity $q = 4.0$ kN/m over its entire span AB and $1.5q$ over BC. The cross section of the beam is rectangular with width b and height $2b$. The allowable bending stress in the steel is $\sigma_{allow} = 60$ MPa, and its weight density is $\gamma = 77.0$ kN/m^3.

(a) Disregarding the weight of the beam, calculate the required width b of the rectangular cross section.

(b) Taking into account the weight of the beam, calculate the required width b.

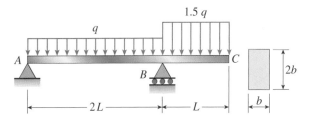

PROB. 5.6-20

5.6-21 A retaining wall 5 ft high is constructed of horizontal wood planks 3 in. thick (actual dimension) that are supported by vertical wood piles of 12 in. diameter (actual dimension), as shown in the figure. The lateral earth pressure is $p_1 = 100$ lb/ft^2 at the top of the wall and $p_2 = 400$ lb/ft^2 at the bottom.

Assuming that the allowable stress in the wood is 1200 psi, calculate the maximum permissible spacing s of the piles.

(*Hint:* Observe that the spacing of the piles may be governed by the load-carrying capacity of either the planks or the piles. Consider the piles to act as cantilever beams subjected to a trapezoidal distribution of load, and consider the planks to act as simple beams between the piles. To be on the safe side, assume that the pressure on the bottom plank is uniform and equal to the maximum pressure.)

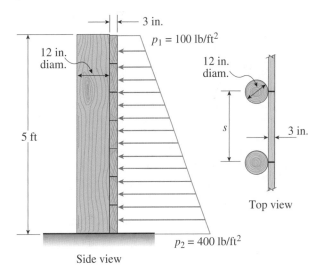

Side view

PROB. 5.6-21

5.6-22 A beam of square cross section ($a =$ length of each side) is bent in the plane of a diagonal (see figure). By removing a small amount of material at the top and bottom corners, as shown by the shaded triangles in the figure, we can increase the section modulus and obtain a stronger beam, even though the area of the cross section is reduced.

(a) Determine the ratio β defining the areas that should be removed in order to obtain the strongest cross section in bending.

(b) By what percent is the section modulus increased when the areas are removed?

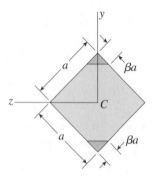

PROB. 5.6-22

5.6-23 The cross section of a rectangular beam having width b and height h is shown in part (a) of the figure. For reasons unknown to the beam designer, it is planned to add structural projections of width $b/9$ and height d to the top and bottom of the beam [see part (b) of the figure].

For what values of d is the bending-moment capacity of the beam increased? For what values is it decreased?

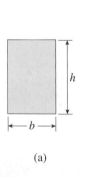 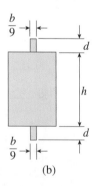

(a)   (b)

PROB. 5.6-23

Shear Stresses in Rectangular Beams

5.7-1 The shear stresses τ in a rectangular beam are given by Eq. (5-36):

$$\tau = \frac{V}{2I}\left(\frac{h^2}{4} - y_1^2\right)$$

in which V is the shear force, I is the moment of inertia of the cross-sectional area, h is the height of the beam, and y_1 is the distance from the neutral axis to the point where the shear stress is being determined (Fig. 5-27).

By integrating over the cross-sectional area, show that the resultant of the shear stresses is equal to the shear force V.

5.7-2 Calculate the maximum shear stress τ_{max} and the maximum bending stress σ_{max} in a wood beam (see figure) carrying a uniform load of 22.5 kN/m (which includes the weight of the beam) if the length is 1.95 m and the cross section is rectangular with width 150 mm and height 300 mm, and the beam is (a) simply supported as in the figure part (a) and (b) has a sliding support at right as in the figure part (b).

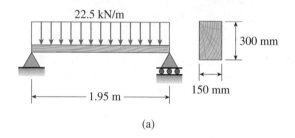

(a)

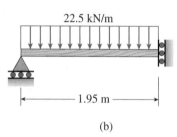

(b)

PROB. 5.7-2

5.7-3 Two wood beams, each of rectangular cross section (3.0 in. × 4.0 in., actual dimensions) are glued together to form a solid beam of dimensions 6.0 in. × 4.0 in. (see figure). The beam is simply supported with a span of 8 ft.

What is the maximum moment M_{max} that may be applied at the left support if the allowable shear stress in the glued joint is 200 psi? (Include the effects of the beam's own weight, assuming that the wood weighs 35 lb/ft³.)

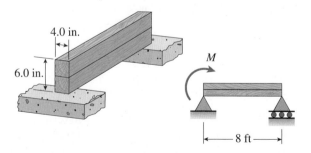

PROB. 5.7-3

5.7-4 A cantilever beam of length $L = 2$ m supports a load $P = 8.0$ kN (see figure). The beam is made of wood with cross-sectional dimensions 120 mm × 200 mm.

Calculate the shear stresses due to the load P at points located 25 mm, 50 mm, 75 mm, and 100 mm from the top surface of the beam. From these results, plot a graph showing the distribution of shear stresses from top to bottom of the beam.

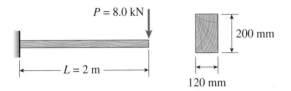

PROB. 5.7-4

5.7-5 A steel beam of length $L = 16$ in. and cross-sectional dimensions $b = 0.6$ in. and $h = 2$ in. (see figure) supports a uniform load of intensity $q = 240$ lb/in., which includes the weight of the beam.

Calculate the shear stresses in the beam (at the cross section of maximum shear force) at points located 1/4 in., 1/2 in., 3/4 in., and 1 in. from the top surface of the beam. From these calculations, plot a graph showing the distribution of shear stresses from top to bottom of the beam.

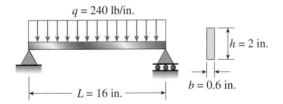

PROB. 5.7-5

5.7-6 A beam of rectangular cross section (width b and height h) supports a uniformly distributed load along its entire length L. The allowable stresses in bending and shear are σ_{allow} and τ_{allow}, respectively.

(a) If the beam is simply supported, what is the span length L_0 below which the shear stress governs the allowable load and above which the bending stress governs?

(b) If the beam is supported as a cantilever, what is the length L_0 below which the shear stress governs the allowable load and above which the bending stress governs?

5.7-7 A laminated wood beam on simple supports is built up by gluing together four 2 in. × 4 in. boards (actual dimensions) to form a solid beam 4 in. × 8 in. in cross section, as shown in the figure. The allowable shear stress in the glued joints is 65 psi, and the allowable bending stress in the wood is 1800 psi.

If the beam is 9 ft long, what is the allowable load P acting at the one-third point along the beam as shown? (Include the effects of the beam's own weight, assuming that the wood weighs 35 lb/ft^3.)

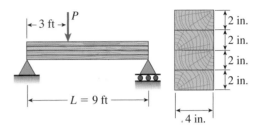

PROB. 5.7-7

5.7-8 A laminated plastic beam of square cross section is built up by gluing together three strips, each 10 mm × 30 mm in cross section (see figure). The beam has a total weight of 3.6 N and is simply supported with span length $L = 360$ mm.

Considering the weight of the beam (q) calculate the maximum permissible CCW moment M that may be placed at the right support.

(a) If the allowable shear stress in the glued joints is 0.3 MPa.

(b) If the allowable bending stress in the plastic is 8 MPa.

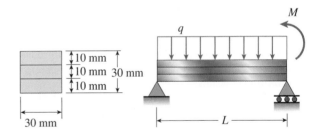

PROB. 5.7-8

5.7-9 A wood beam AB on simple supports with span length equal to 10 ft is subjected to a uniform load of intensity 125 lb/ft acting along the entire length of the beam, a concentrated load of magnitude 7500 lb acting at a point 3 ft from the right-hand support, and a moment at A of 18,500 ft-lb (see figure on the next page). The allowable stresses in bending and shear, respectively, are 2250 psi and 160 psi.

(a) From the table in Appendix G available online, select the lightest beam that will support the loads (disregard the weight of the beam).

(b) Taking into account the weight of the beam (weight density = 35 lb/ft^3), verify that the selected beam is satisfactory, or if it is not, select a new beam.

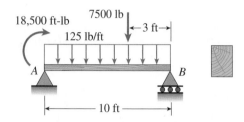

PROB. 5.7-9

5.7-10 A simply supported wood beam of rectangular cross section and span length 1.2 m carries a concentrated load P at midspan in addition to its own weight (see figure). The cross section has width 140 mm and height 240 mm. The weight density of the wood is 5.4 kN/m^3.

Calculate the maximum permissible value of the load P if (a) the allowable bending stress is 8.5 MPa, and (b) the allowable shear stress is 0.8 MPa.

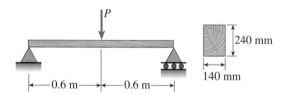

PROB. 5.7-10

5.7-11 A square wood platform, 8 ft × 8 ft in area, rests on masonry walls (see figure). The deck of the platform is constructed of 2 in. nominal thickness tongue-and-groove planks (actual thickness 1.5 in.; see Appendix G) supported on two 8-ft long beams. The beams have 4 in. × 6 in. nominal dimensions (actual dimensions 3.5 in. × 5.5 in.).

The planks are designed to support a uniformly distributed load w (lb/ft^2) acting over the entire top surface of the platform. The allowable bending stress for the planks is 2400 psi and the allowable shear stress is 100 psi. When analyzing the planks, disregard their weights and assume that their reactions are uniformly distributed over the top surfaces of the supporting beams.

(a) Determine the allowable platform load w_1 (lb/ft^2) based upon the bending stress in the planks.

(b) Determine the allowable platform load w_2 (lb/ft^2) based upon the shear stress in the planks.

(c) Which of the preceding values becomes the allowable load w_{allow} on the platform?

(*Hints:* Use care in constructing the loading diagram for the planks, noting especially that the reactions are distributed loads instead of concentrated loads. Also, note that the maximum shear forces occur at the inside faces of the supporting beams.)

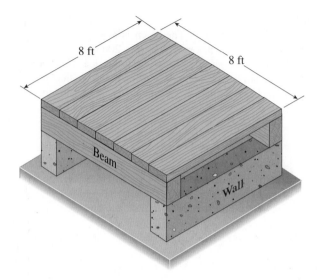

PROB. 5.7-11

5.7-12 A wood beam ABC with simple supports at A and B and an overhang BC has height $h = 300$ mm (see figure). The length of the main span of the beam is $L = 3.6$ m and the length of the overhang is $L/3 = 1.2$ m. The beam supports a concentrated load $3P = 18$ kN at the midpoint of the main span and a moment $PL/2 = 10.8$ kN · m at the free end of the overhang. The wood has weight density $\gamma = 5.5$ kN/m^3.

(a) Determine the required width b of the beam based upon an allowable bending stress of 8.2 MPa.

(b) Determine the required width based upon an allowable shear stress of 0.7 MPa.

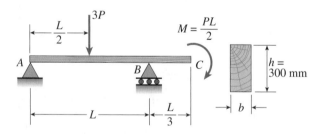

PROB. 5.7-12

Shear Stresses in Circular Beams

5.8-1 A wood pole of solid circular cross section ($d = $ diameter) is subjected to a triangular distributed horizontal force of peak intensity $q_0 = 20$ lb/in. (see figure). The length of the pole is $L = 6$ ft, and the allowable stresses in the wood are 1900 psi in bending and 120 psi in shear.

Determine the minimum required diameter of the pole based upon (a) the allowable bending stress and (b) the allowable shear stress.

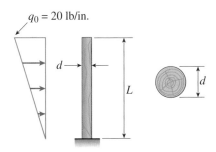

PROB. 5.8-1

5.8-2 A simple log bridge in a remote area consists of two parallel logs with planks across them (see figure). The logs are Douglas fir with average diameter 300 mm. A truck moves slowly across the bridge, which spans 2.5 m. Assume that the weight of the truck is equally distributed between the two logs.

Because the wheelbase of the truck is greater than 2.5 m, only one set of wheels is on the bridge at a time. Thus, the wheel load on one log is equivalent to a concentrated load W acting at any position along the span. In addition, the weight of one log and the planks it supports is equivalent to a uniform load of 850 N/m acting on the log.

Determine the maximum permissible wheel load W based upon (a) an allowable bending stress of 7.0 MPa, and (b) an allowable shear stress of 0.75 MPa.

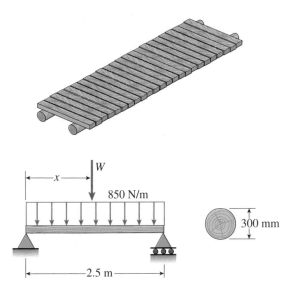

PROB. 5.8-2

5.8-3 A sign for an automobile service station is supported by two aluminum poles of hollow circular cross section, as shown in the figure. The poles are being designed to resist a wind pressure of 75 lb/ft² against the full area of the sign. The dimensions of the poles and sign are $h_1 = 20$ ft, $h_2 = 5$ ft, and $b = 10$ ft. To prevent buckling of the walls of the poles, the thickness t is specified as one-tenth the outside diameter d.

(a) Determine the minimum required diameter of the poles based upon an allowable bending stress of 7500 psi in the aluminum.

(b) Determine the minimum required diameter based upon an allowable shear stress of 2000 psi.

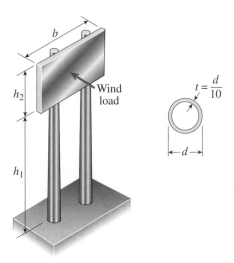

PROBS. 5.8-3 and 5.8-4

5.8-4 Solve the preceding problem for a sign and poles having the following dimensions: $h_1 = 6.0$ m, $h_2 = 1.5$ m, $b = 3.0$ m, and $t = d/10$. The design wind pressure is 3.6 kPa, and the allowable stresses in the aluminum are 50 MPa in bending and 14 MPa in shear.

Shear Stresses in Beams with Flanges

5.9-1 through 5.9-6 A wide-flange beam (see figure) having the cross section described below is subjected to a shear force V. Using the dimensions of the cross section, calculate the moment of inertia and then determine the following quantities:

(a) The maximum shear stress τ_{max} in the web.

(b) The minimum shear stress τ_{min} in the web.

(c) The average shear stress τ_{aver} (obtained by dividing the shear force by the area of the web) and the ratio τ_{max}/τ_{aver}.

(d) The shear force V_{web} carried in the web and the ratio V_{web}/V.

(*Note:* Disregard the fillets at the junctions of the web and flanges and determine all quantities, including the moment of inertia, by considering the cross section to consist of three rectangles.)

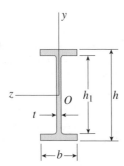

PROBS. 5.9-1 through 5.9-6

5.9-1 Dimensions of cross section: $b = 6$ in., $t = 0.5$ in., $h = 12$ in., $h_1 = 10.5$ in., and $V = 30$ k.

5.9-2 Dimensions of cross section: $b = 180$ mm, $t = 12$ mm, $h = 420$ mm, $h_1 = 380$ mm, and $V = 125$ kN.

5.9-3 Wide-flange shape, W 8 × 28 (see Table F-1, Appendix F available online); $V = 10$ k.

5.9-4 Dimensions of cross section: $b = 220$ mm, $t = 12$ mm, $h = 600$ mm, $h_1 = 570$ mm, and $V = 200$ kN.

5.9-5 Wide-flange shape, W 18 × 71 (see Table F-1, Appendix F available online); $V = 21$ k.

5.9-6 Dimensions of cross section: $b = 120$ mm, $t = 7$ mm, $h = 350$ mm, $h_1 = 330$ mm, and $V = 60$ kN.

5.9-7 A cantilever beam AB of length $L = 6.5$ ft supports a trapezoidal distributed load of peak intensity q, and minimum intensity $q/2$, that includes the weight of the beam (see figure). The beam is a steel W 12 × 14 wide-flange shape (see Table F-1(a), Appendix F available online).

Calculate the maximum permissible load q based upon (a) an allowable bending stress $\sigma_{allow} = 18$ ksi and (b) an allowable shear stress $\tau_{allow} = 7.5$ ksi. (*Note:* Obtain the moment of inertia and section modulus of the beam from Table F-1(a))

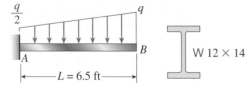

PROB. 5.9-7

5.9-8 A bridge girder AB on a simple span of length $L = 14$ m supports a distributed load of maximum intensity q at midspan and minimum intensity $q/2$ at supports A and B that includes the weight of the girder (see figure). The girder is constructed of three plates welded to form the cross section shown.

Determine the maximum permissible load q based upon (a) an allowable bending stress $\sigma_{allow} = 110$ MPa and (b) an allowable shear stress $\tau_{allow} = 50$ MPa.

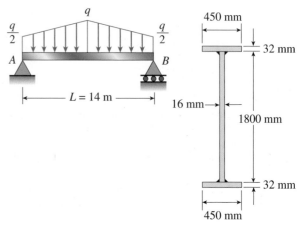

PROB. 5.9-8

5.9-9 A simple beam with an overhang supports a uniform load of intensity $q = 1200$ lb/ft and a concentrated load $P = 3000$ lb at 8 ft to the right of A and also at C (see figure). The uniform load includes an allowance for the weight of the beam. The allowable stresses in bending and shear are 18 ksi and 11 ksi, respectively.

Select from Table F-2(a), Appendix F available online, the lightest I-beam (S shape) that will support the given loads.

(*Hint:* Select a beam based upon the bending stress and then calculate the maximum shear stress. If the beam is overstressed in shear, select a heavier beam and repeat.)

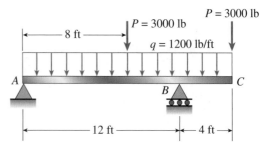

PROB. 5.9-9

5.9-10 A hollow steel box beam has the rectangular cross section shown in the figure. Determine the maximum allowable shear force V that may act on the beam if the allowable shear stress is 36 MPa.

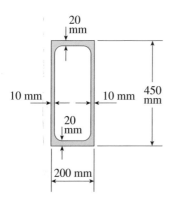

PROB. 5.9-10

5.9-11 A hollow aluminum box beam has the square cross section shown in the figure. Calculate the maximum and minimum shear stresses τ_{max} and τ_{min} in the webs of the beam due to a shear force $V = 28$ k.

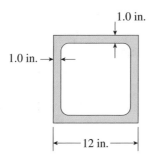

PROB. 5.9-11

5.9-12 The T-beam shown in the figure has cross-sectional dimensions as follows: $b = 210$ mm, $t = 16$ mm, $h = 300$ mm, and $h_1 = 280$ mm. The beam is subjected to a shear force $V = 68$ kN.

Determine the maximum shear stress τ_{max} in the web of the beam.

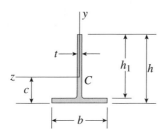

PROBS. 5.9-12 and 5.9-13

5.9-13 Calculate the maximum shear stress τ_{max} in the web of the T-beam shown in the figure if $b = 10$ in., $t = 0.5$ in., $h = 7$ in., $h_1 = 6.2$ in., and the shear force $V = 5300$ lb.

Composite Beams

When solving the problems for Section 5.10, assume that the component parts of the beams are securely bonded by adhesives or connected by fasteners. Also, be sure to use the general theory for composite beams described in Section 5.10.

5.10-1 A composite beam consisting of fiberglass faces and a core of particle board has the cross section shown in the figure. The width of the beam is 2.0 in., the thickness of the faces is 0.10 in., and the thickness of the core is 0.50 in. The beam is subjected to a bending moment of 250 lb-in. acting about the z axis.

Find the maximum bending stresses σ_{face} and σ_{core} in the faces and the core, respectively, if their respective moduli of elasticity are 4×10^6 psi and 1.5×10^6 psi.

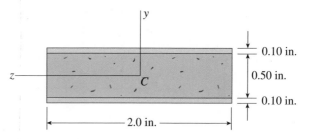

PROB. 5.10-1

5.10-2 A wood beam with cross-sectional dimensions 200 mm × 300 mm is reinforced on its sides by steel plates 12 mm thick (see figure). The moduli of elasticity for the steel and wood are $E_s = 190$ GPa and $E_w = 11$ GPa, respectively. Also, the corresponding allowable stresses are $\sigma_s = 110$ MPa and $\sigma_w = 7.5$ MPa.

(a) Calculate the maximum permissible bending moment M_{max} when the beam is bent about the z axis.

(b) Repeat part a if the beam is now bent about its y axis.

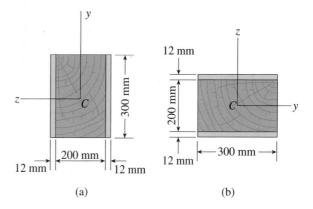

(a)   (b)

PROB. 5.10-2

5.10-3 A hollow box beam is constructed with webs of Douglas-fir plywood and flanges of pine, as shown in the figure in a cross-sectional view. The plywood is 1 in. thick and 12 in. wide; the flanges are 2 in. × 4 in. (nominal size). The modulus of elasticity for the plywood is 1,800,000 psi and for the pine is 1,400,000 psi.

(a) If the allowable stresses are 2000 psi for the plywood and 1750 psi for the pine, find the allowable bending moment M_{max} when the beam is bent about the z axis.

(b) Repeat part (a) if the beam is now bent about its y axis.

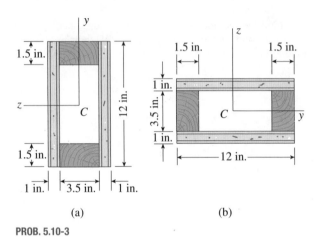

(a)   (b)

PROB. 5.10-3

5.10-4 A round steel tube of outside diameter d and a brass core of diameter $2d/3$ are bonded to form a composite beam, as shown in the figure.

Derive a formula for the allowable bending moment M that can be carried by the beam based upon an allowable stress σ_s in the steel. (Assume that the moduli of elasticity for the steel and brass are E_s and E_b, respectively.)

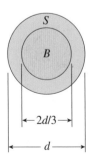

PROB. 5.10-4

5.10-5 A beam with a guided support and 10 ft span supports a distributed load of intensity $q = 660$ lb/ft over its first half

[see figure part (a)] and a moment $M_0 = 300$ ft-lb at joint B. The beam consists of a wood member (nominal dimensions 6 in. × 12 in., actual dimensions 5.5 in. × 11.5 in. in cross section, as shown in the figure part (b) that is reinforced by 0.25-in.-thick steel plates on top and bottom. The moduli of elasticity for the steel and wood are $E_s = 30 \times 10^6$ psi and $E_w = 1.5 \times 10^6$ psi, respectively.

(a) Calculate the maximum bending stresses σ_s in the steel plates and σ_w in the wood member due to the applied loads.

(b) If the allowable bending stress in the steel plates is $\sigma_{as} = 14,000$ psi and that in the wood is $\sigma_{aw} = 900$ psi, find q_{max}. (Assume that the moment at B, M_0, remains at 300 ft-lb.)

(c) If $q = 660$ lb/ft and allowable stress values in (b) apply, what is $M_{0,max}$ at B?

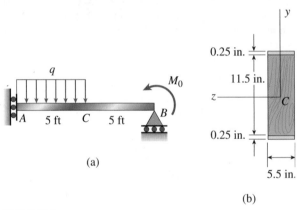

(a)

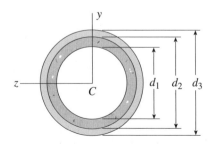

(b)

PROB. 5.10-5

5.10-6 A plastic-lined steel pipe has the cross-sectional shape shown in the figure. The steel pipe has outer diameter $d_3 = 100$ mm and inner diameter $d_2 = 94$ mm. The plastic liner has inner diameter $d_1 = 82$ mm. The modulus of elasticity of the steel is 75 times the modulus of the plastic.

Determine the allowable bending moment M_{allow} if the allowable stress in the steel is 35 MPa and in the plastic is 600 kPa.

PROB. 5.10-6

5.10-7 The cross section of a sandwich beam consisting of aluminum alloy faces and a foam core is shown in the figure. The width b of the beam is 8.0 in., the thickness t of the faces is 0.25 in., and the height h_c of the core is 5.5 in. (total height $h = 6.0$ in.). The moduli of elasticity are 10.5×10^6 psi for the aluminum faces and 12,000 psi for the foam core. A bending moment $M = 40$ k-in. acts about the z axis.

Determine the maximum stresses in the faces and the core using (a) the general theory for composite beams, and (b) the approximate theory for sandwich beams.

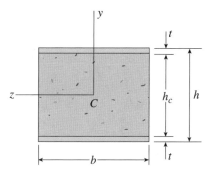

PROBS. 5.10-7 and 5.10-8

5.10-8 The cross section of a sandwich beam consisting of fiberglass faces and a lightweight plastic core is shown in the figure. The width b of the beam is 50 mm, the thickness t of the faces is 4 mm, and the height h_c of the core is 92 mm (total height $h = 100$ mm). The moduli of elasticity are 75 GPa for the fiberglass and 1.2 GPa for the plastic. A bending moment $M = 275$ N·m acts about the z axis.

Determine the maximum stresses in the faces and the core using (a) the general theory for composite beams, and (b) the approximate theory for sandwich beams.

5.10-9 A bimetallic beam used in a temperature-control switch consists of strips of aluminum and copper bonded together as shown in the figure, which is a cross-sectional view. The width of the beam is 1.0 in., and each strip has a thickness of 1/16 in.

Under the action of a bending moment $M = 12$ lb-in. acting about the z axis, what are the maximum stresses σ_a and σ_c in the aluminum and copper, respectively? (Assume $E_a = 10.5 \times 10^6$ psi and $E_c = 16.8 \times 10^6$ psi.)

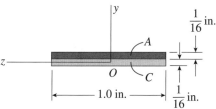

PROB. 5.10-9

5.10-10 A simply supported composite beam 3 m long carries a uniformly distributed load of intensity $q = 3.0$ kN/m (see figure). The beam is constructed of a wood member, 100 mm wide by 150 mm deep, reinforced on its lower side by a steel plate 8 mm thick and 100 mm wide.

Find the maximum bending stresses σ_w and σ_s in the wood and steel, respectively, due to the uniform load if the moduli of elasticity are $E_w = 10$ GPa for the wood and $E_s = 210$ GPa for the steel.

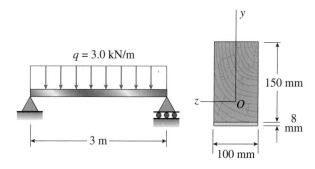

PROB. 5.10-10

5.10-11 A simply supported wooden I-beam with a 12 ft span supports a distributed load of intensity $q = 90$ lb/ft over its length [see figure part (a)]. The beam is constructed with a web of Douglas-fir plywood and flanges of pine glued to the web as shown in the figure part (b). The plywood is 3/8 in. thick; the flanges are 2 in. × 2 in. (actual size). The modulus

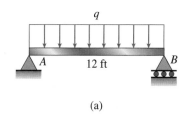

(a)

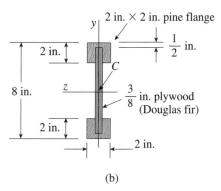

(b)

PROB. 5.10-11

of elasticity for the plywood is 1,600,000 psi and for the pine is 1,200,000 psi.

(a) Calculate the maximum bending stresses in the pine flanges and in the plywood web.

(b) What is q_{max} if allowable stresses are 1600 psi in the flanges and 1200 psi in the web?

5.10-12 A simply supported composite beam with a 3.6 m span supports a triangularly distributed load of peak intensity q_0 at midspan [see figure part (a)]. The beam is constructed of two wood joists, each 50 mm × 280 mm, fastened to two steel plates, one of dimensions 6 mm × 80 mm and the lower plate of dimensions 6 mm × 120 mm [see figure part (b)]. The modulus of elasticity for the wood is 11 GPa and for the steel is 210 GPa.

If the allowable stresses are 7 MPa for the wood and 120 MPa for the steel, find the allowable peak load intensity $q_{0,max}$ when the beam is bent about the z axis. Neglect the weight of the beam.

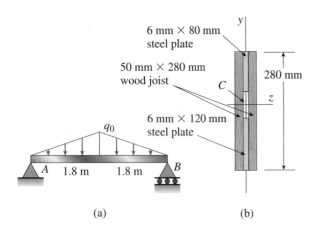

(a)   (b)

PROB. 5.10-12

Transformed-Section Method

When solving the problems in this section, assume that the component parts of the beams are securely bonded by adhesives or connected by fasteners. Also, be sure to use the transformed-section method in the solutions.

5.10-13 A wood beam 8 in. wide and 12 in. deep (nominal dimensions) is reinforced on top and bottom by 0.25-in.-thick steel plates [see figure part (a)].

(a) Find the allowable bending moment M_{max} about the z axis if the allowable stress in the wood is 1,100 psi and in the steel is 15,000 psi. (Assume that the ratio of the moduli of elasticity of steel and wood is 20.)

(b) Compare the moment capacity of the beam in part (a) with that shown in the figure part (b) which has two 4 in. × 12 in. joists (nominal dimensions) attached to a 1/4 in. × 11.0 in. steel plate.

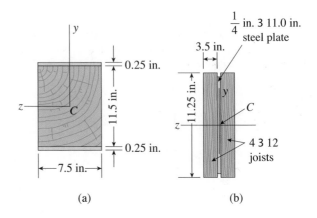

(a)   (b)

PROB. 5.10-13

5.10-14 A simple beam of span length 3.2 m carries a uniform load of intensity 48 kN/m. The cross section of the beam is a hollow box with wood flanges and steel side plates, as shown in the figure. The wood flanges are 75 mm by 100 mm in cross section, and the steel plates are 300 mm deep.

What is the required thickness t of the steel plates if the allowable stresses are 120 MPa for the steel and 6.5 MPa for the wood? (Assume that the moduli of elasticity for the steel and wood are 210 GPa and 10 GPa, respectively, and disregard the weight of the beam.)

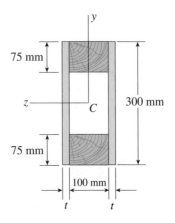

PROB. 5.10-14

5.10-15 A simple beam that is 18 ft long supports a uniform load of intensity q. The beam is constructed of two C 8 × 11.5 sections (channel sections or C shapes) on either side of a 4 × 8 (actual dimensions) wood beam (see the cross section shown in the figure part (a). The modulus of elasticity of the steel (E_s = 30,000 ksi) is 20 times that of the wood (E_w).

(a) If the allowable stresses in the steel and wood are 12,000 psi and 900 psi, respectively, what is the allowable load q_{allow}? (*Note*: Disregard the weight of the beam, and see Table F-3a of Appendix F available online for the dimensions and properties of the C-shape beam.)

(b) If the beam is rotated 90° to bend about its y axis [see figure part (b)], and uniform load q = 250 lb/ft is applied, find the maximum stresses σ_s and σ_w in the steel and wood, respectively. Include the weight of the beam. (Assume weight densities of 35 lb/ft³ and 490 lb/ft³ for the wood and steel, respectively.)

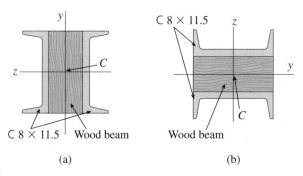

(a)   (b)

PROB. 5.10-15

5.10-16 The composite beam shown in the figure is simply supported and carries a total uniform load of 50 kN/m on a span length of 4.0 m. The beam is built of a wood member having cross-sectional dimensions 150 mm × 250 mm and two steel plates of cross-sectional dimensions 50 mm × 150 mm.

Determine the maximum stresses σ_s and σ_w in the steel and wood, respectively, if the moduli of elasticity are E_s = 209 GPa and E_w = 11 GPa. (Disregard the weight of the beam.)

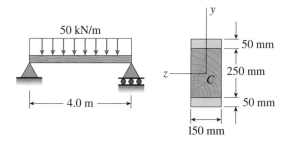

PROB. 5.10-16

5.10-17 The cross section of a beam made of thin strips of aluminum separated by a lightweight plastic is shown in the figure. The beam has width b = 3.0 in., the aluminum strips have thickness t = 0.1 in., and the plastic segments have heights d = 1.2 in. and $3d$ = 3.6 in. The total height of the beam is h = 6.4 in.

The moduli of elasticity for the aluminum and plastic are E_a = 11 × 10⁶ psi and E_p = 440 × 10³ psi, respectively.

Determine the maximum stresses σ_a and σ_p in the aluminum and plastic, respectively, due to a bending moment of 6.0 k-in.

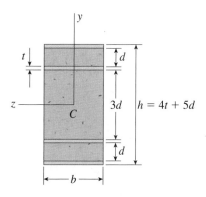

PROBS. 5.10-17 and 5.10-18

5.10-18 Consider the preceding problem if the beam has width b = 75 mm, the aluminum strips have thickness t = 3 mm, the plastic segments have heights d = 40 mm and $3d$ = 120 mm, and the total height of the beam is h = 212 mm. Also, the moduli of elasticity are E_a = 75 GPa and E_p = 3 GPa, respectively.

Determine the maximum stresses σ_a and σ_p in the aluminum and plastic, respectively, due to a bending moment of 1.0 kN·m.

5.10-19 A simple beam that is 18 ft long supports a uniform load of intensity q. The beam is constructed of two angle sections, each L 6 × 4 × 1/2, on either side of a 2 in. × 8 in. (actual dimensions) wood beam [see the cross section shown in the figure part (a) on the next page]. The modulus of elasticity of the steel is 20 times that of the wood.

(a) If the allowable stresses in the steel and wood are 12,000 psi and 900 psi, respectively, what is the allowable load q_{allow}? (*Note:* Disregard the weight of the beam, and see Table F-5a of Appendix F available online for the dimensions and properties of the angles.)

(b) Repeat part (a) if a 1 in. × 10 in. wood flange (actual dimensions) is added [see figure part (b)].

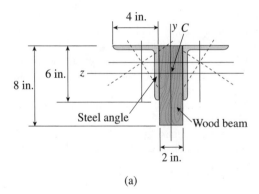

(a)

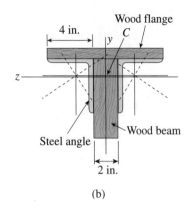

(b)

PROB. 5.10-19

5.10-20 The cross section of a composite beam made of aluminum and steel is shown in the figure. The moduli of elasticity are $E_a = 75$ GPa and $E_s = 200$ GPa.

Under the action of a bending moment that produces a maximum stress of 50 MPa in the aluminum, what is the maximum stress σ_s in the steel?

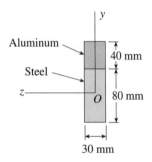

PROB. 5.10-20

5.10-21 A beam is constructed of two angle sections, each L 5 × 3 × 1/2, which reinforce a 2 × 8 (actual dimensions) wood plank (see the cross section shown in the figure). The modulus of elasticity for the wood is $E_w = 1.2 \times 10^6$ psi and for the steel is $E_s = 30 \times 10^6$ psi.

Find the allowable bending moment M_{allow} for the beam if the allowable stress in the wood is $\sigma_w = 1100$ psi and in the steel is $\sigma_s = 12,000$ psi. (*Note:* Disregard the weight of the beam, and see Table F-5a of Appendix F available online for the dimensions and properties of the angles.)

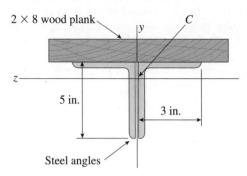

PROB. 5.10-21

5.10-22 The cross section of a bimetallic strip is shown in the figure. Assuming that the moduli of elasticity for metals A and B are $E_A = 168$ GPa and $E_B = 90$ GPa, respectively, determine the smaller of the two section moduli for the beam. (Recall that section modulus is equal to bending moment divided by maximum bending stress.) In which material does the maximum stress occur?

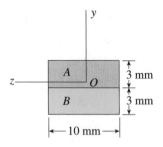

PROB. 5.10-22

5.10-23 A W 12 × 50 steel wide-flange beam and a segment of a 4-inch-thick-concrete slab (see figure) jointly resist a positive bending moment of 95 k-ft. The beam and slab are joined by shear connectors that are welded to the steel beam. (These connectors resist the horizontal shear at the contact surface.) The moduli of elasticity of the steel and the concrete are in the ratio 12 to 1.

Determine the maximum stresses σ_s and σ_c in the steel and concrete, respectively. (*Note:* See Table F-1a of Appendix F available online for the dimensions and properties of the steel beam.)

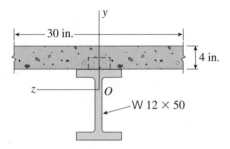

PROB. 5.10-23

5.10-24 A wood beam reinforced by an aluminum channel section is shown in the figure. The beam has a cross section of dimensions 150 mm by 250 mm, and the channel has a uniform thickness of 6 mm.

If the allowable stresses in the wood and aluminum are 8.0 MPa and 38 MPa, respectively, and if their moduli of elasticity are in the ratio 1 to 6, what is the maximum allowable bending moment for the beam?

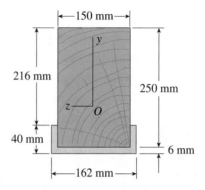

PROB. 5.10-24

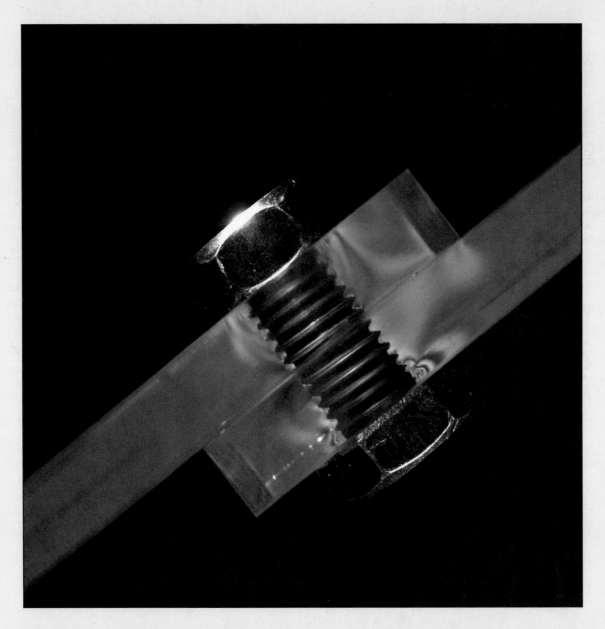

Photoelasticity is an experimental method that can be used to find the complex state of stress near a bolt connecting two plates. (Alfred Pasieka/Peter Arnold, Inc.)

Analysis of Stress and Strain

CHAPTER OVERVIEW

Chapter 6 is concerned with finding normal and shear stresses acting on inclined sections cut through a member, because these stresses may be larger than those on a stress element aligned with the cross section. In two dimensions, a stress element displays the state of **plane stress** at a point (normal stresses σ_x, σ_y, and shear stress τ_{xy}) (Section 6.2), and transformation equations (Section 6.3) are needed to find the stresses acting on an element rotated by some angle θ from that position. The resulting expressions for normal and shear stresses can be reduced to those examined in Section 2.6 for uniaxial stress ($\sigma_x \neq 0$, $\sigma_y = 0$, $\tau_{xy} = 0$) and in Section 3.5 for pure shear ($\sigma_x = 0$, $\sigma_y = 0$, $\tau_{xy} \neq 0$). Maximum values of stress are needed for design, and the transformation equations can be used to find these **principal stresses** and the planes on which they act (Section 6.3). There are no shear stresses acting on the principal planes, but a separate analysis can be made to find the maximum shear stress (τ_{max}) and the inclined plane on which it acts. **Maximum shear stress** is shown to be equal to one-half of the difference between the principal normal stresses (σ_1, σ_2). A graphical representation of the transformation equations for plane stress, known as **Mohr's Circle**, provides a convenient way of calculating stresses on any inclined plane of interest and those on principal planes, in particular (Section 6.4). In Section 6.5, normal and shear strains (ϵ_x, ϵ_y, γ_{xy}) are studied, and **Hooke's law for plane stress** is derived, which relates elastic moduli E and G and Poisson's ratio ν for homogeneous and isotropic materials. The general expressions for Hooke's law can be simplified to the stress-strain relationships for biaxial stress, uniaxial stress, and pure shear. Further examination of strains leads to an expression for unit volume change (or *dilatation e*) in plane stress (Section 6.5). Finally, **triaxial stress** is discussed (Section 6.6). Special cases of triaxial stress, known as **spherical stress** and **hydrostatic stress** are explained: for spherical stress, the three normal stresses are equal and tensile, while for hydrostatic stress, they are equal and compressive.

The discussions in Chapter 6 are organized as follows:

6.1 INTRODUCTION

Normal and shear stresses in beams, shafts, and bars can be calculated from the basic formulas discussed in the preceding chapters. For instance, the stresses in a beam are given by the flexure and shear formulas ($\sigma = My/I$ and $\tau = VQ/Ib$), and the stresses in a shaft are given by the torsion formula ($\tau = T\rho/I_P$). The stresses calculated from these formulas act on cross sections of the members, but larger stresses may occur on **inclined sections**. Therefore, we will begin our analysis of stresses and strains by discussing methods for finding the normal and shear stresses acting on inclined sections cut through a member.

We have already derived expressions for the normal and shear stresses acting on inclined sections in both *uniaxial stress* and *pure shear* (see Sections 2.6 and 3.5, respectively). In the case of uniaxial stress, we found that the maximum shear stresses occur on planes inclined at 45° to the axis, whereas the maximum normal stresses occur on the cross sections. In the case of pure shear, we found that the maximum tensile and compressive stresses occur on 45° planes. In an analogous manner, the stresses on inclined sections cut through a beam may be larger than the stresses acting on a cross section. To calculate such stresses, we need to determine the stresses acting on inclined planes under a more general stress state known as **plane stress** (Section 6.2).

In our discussions of plane stress we will use **stress elements** to represent the state of stress at a point in a body. Stress elements were discussed previously in a specialized context (see Sections 2.6 and 3.5), but now we will use them in a more formalized manner. We will begin our analysis by considering an element on which the stresses are known, and then we will derive the **transformation equations** that give the stresses acting on the sides of an element oriented in a different direction.

When working with stress elements, we must always keep in mind that only one intrinsic **state of stress** exists at a point in a stressed body, regardless of the orientation of the element being used to portray that state of stress. When we have two elements with different orientations at the same point in a body, the stresses acting on the faces of the two elements are different, but they still represent the same state of stress, namely, the stress at the point under consideration. This situation is analogous to the representation of a force vector by its components—although the components are different when the coordinate axes are rotated to a new position, the force itself is the same.

Furthermore, we must always keep in mind that stresses are *not* vectors. This fact can sometimes be confusing, because we customarily represent stresses by arrows just as we represent force vectors by arrows. *Although the arrows used to represent stresses have magnitude and direction, they are not vectors because they do not combine according to the parallelogram law of addition.* Instead, stresses are much more complex quantities than are vectors, and in mathematics they are called **tensors**. Other tensor quantities in mechanics are strains and moments of inertia.

6.2 PLANE STRESS

The stress conditions that we encountered in earlier chapters when analyzing bars in tension and compression, shafts in torsion, and beams in bending are examples of a state of stress called **plane stress**. To explain plane stress, we will consider the stress element shown in Fig. 6-1a. This element is infinitesimal in size and can be sketched either as a cube or as a rectangular parallelepiped. The *xyz* axes are parallel to the edges of the element, and the faces of the element are designated by the directions of their outward normals, as explained previously in Section 1.6. For instance, the right-hand face of the element is referred to as the positive *x* face, and the left-hand face (hidden from the viewer) is referred to as the negative *x* face. Similarly, the top face is the positive *y* face, and the front face is the positive *z* face.

When the material is in plane stress in the *xy* plane, only the *x* and *y* faces of the element are subjected to stresses, and all stresses act parallel to the *x* and *y* axes, as shown in Fig. 6-1a. This stress condition is very common because it exists at the surface of any stressed body, except at points where external loads act on the surface. When the element shown in Fig. 6-1a is located at the free surface of a body, the *z* axis is normal to the surface and the *z* face is in the plane of the surface.

The symbols for the stresses shown in Fig. 6-1a have the following meanings. A **normal stress** σ has a subscript that identifies the face on which the stress acts; for instance, the stress σ_x acts on the *x* face of the element and the stress σ_y acts on the *y* face of the element. Since the element is infinitesimal in size, equal normal stresses act on the opposite faces. The **sign convention for normal stresses** is the familiar one, namely, tension is positive and compression is negative.

A **shear stress** τ has two subscripts—the first subscript denotes the face on which the stress acts, and the second gives the direction on that face. Thus, the stress τ_{xy} acts on the *x* face in the direction of the *y* axis (Fig. 6-1a), and the stress τ_{yx} acts on the *y* face in the direction of the *x* axis.

FIG. 6-1 Elements in plane stress: (a) three-dimensional view of an element oriented to the *xyz* axes, (b) two-dimensional view of the same element, and (c) two-dimensional view of an element oriented to the $x_1y_1z_1$ axes

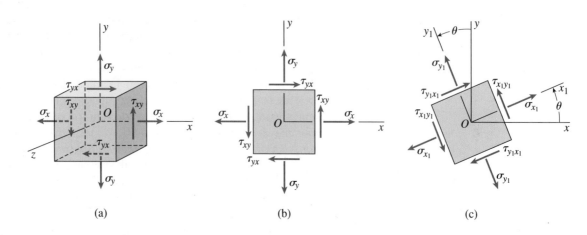

(a)   (b)   (c)

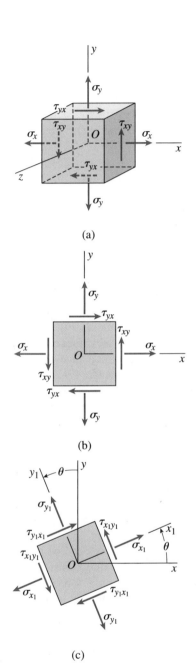

FIG. 6-1 (Repeated)

The **sign convention for shear stresses** is as follows. A shear stress is positive when it acts on a positive face of an element in the positive direction of an axis, and it is negative when it acts on a positive face of an element in the negative direction of an axis. Therefore, the stresses τ_{xy} and τ_{yx} shown on the positive x and y faces in Fig. 6-1a are positive shear stresses. Similarly, on a negative face of the element, a shear stress is positive when it acts in the negative direction of an axis. Hence, the stresses τ_{xy} and τ_{yx} shown on the negative x and y faces of the element are also positive.

This sign convention for shear stresses is easy to remember if we state it as follows:

A shear stress is positive when the directions associated with its subscripts are plus-plus or minus-minus; the stress is negative when the directions are plus-minus or minus-plus.

The preceding sign convention for shear stresses is consistent with the equilibrium of the element, because we know that shear stresses on opposite faces of an infinitesimal element must be equal in magnitude and opposite in direction. Hence, according to our sign convention, a positive stress τ_{xy} acts upward on the positive face (Fig. 6-1a) and downward on the negative face. In a similar manner, the stresses τ_{yx} acting on the top and bottom faces of the element are positive although they have opposite directions.

We also know that shear stresses on perpendicular planes are equal in magnitude and have directions such that both stresses point toward, or both point away from, the line of intersection of the faces. In as much as τ_{xy} and τ_{yx} are positive in the directions shown in the figure, they are consistent with this observation. Therefore, we note that

$$\tau_{xy} = \tau_{yx} \tag{6-1}$$

This relationship was derived previously from equilibrium of the element (see Section 1.6).

For convenience in sketching plane-stress elements, we usually draw only a two-dimensional view of the element, as shown in Fig. 6-1b. Although a figure of this kind is adequate for showing all stresses acting on the element, we must still keep in mind that the element is a solid body with a thickness perpendicular to the plane of the figure.

Stresses on Inclined Sections

We are now ready to consider the stresses acting on inclined sections, assuming that the stresses σ_x, σ_y, and τ_{xy} (Figs. 6-1a and b) are known. To portray the stresses acting on an inclined section, we consider a new stress element (Fig. 6-1c) that is located at the same point in the material as the original element (Fig. 6-1b). However, the new element has faces that are parallel and perpendicular to the inclined direction. Associated with this new element are axes x_1, y_1, and z_1, such that the z_1 axis coincides with the

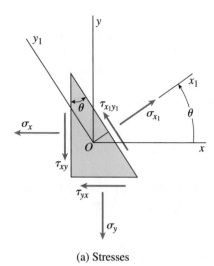

(a) Stresses

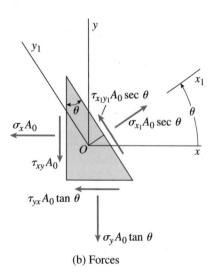

(b) Forces

FIG. 6-2 Wedge-shaped stress element in plane stress: (a) stresses acting on the element, and (b) forces acting on the element (free-body diagram)

z axis and the $x_1 y_1$ axes are rotated counterclockwise through an angle θ with respect to the *xy* axes.

The normal and shear stresses acting on this new element are denoted σ_{x_1}, σ_{y_1}, $\tau_{x_1 y_1}$, and $\tau_{y_1 x_1}$, using the same subscript designations and sign conventions described previously for the stresses acting on the *xy* element. The previous conclusions regarding the shear stresses still apply, so that

$$\tau_{x_1 y_1} = \tau_{y_1 x_1} \tag{6-2}$$

From this equation and the equilibrium of the element, we see that *the shear stresses acting on all four side faces of an element in plane stress are known if we determine the shear stress acting on any one of those faces.*

The stresses acting on the inclined $x_1 y_1$ element (Fig. 6-1c) can be expressed in terms of the stresses on the *xy* element (Fig. 6-1b) by using equations of equilibrium. For this purpose, we choose a **wedge-shaped stress element** (Fig. 6-2a) having an inclined face that is the same as the x_1 face of the inclined element shown in Fig. 6-1c. The other two side faces of the wedge are parallel to the *x* and *y* axes.

In order to write equations of equilibrium for the wedge, we need to construct a free-body diagram showing the forces acting on the faces. Let us denote the area of the left-hand side face (that is, the negative *x* face) as A_0. Then the normal and shear forces acting on that face are $\sigma_x A_0$ and $\tau_{xy} A_0$, as shown in the free-body diagram of Fig. 6-2b. The area of the bottom face (or negative *y* face) is $A_0 \tan \theta$, and the area of the inclined face (or positive x_1 face) is $A_0 \sec \theta$. Thus, the normal and shear forces acting on these faces have the magnitudes and directions shown in Fig. 6-2b.

The forces acting on the left-hand and bottom faces can be resolved into orthogonal components acting in the x_1 and y_1 directions. Then we can obtain two equations of equilibrium by summing forces in those directions. The first equation, obtained by summing forces in the x_1 direction, is

$$\sigma_{x_1} A_0 \sec \theta - \sigma_x A_0 \cos \theta - \tau_{xy} A_0 \sin \theta$$

$$- \sigma_y A_0 \tan \theta \sin \theta - \tau_{yx} A_0 \tan \theta \cos \theta = 0$$

In the same manner, summation of forces in the y_1 direction gives

$$\tau_{x1y1} A_0 \sec \theta + \sigma_x A_0 \sin \theta - \tau_{xy} A_0 \cos \theta$$

$$- \sigma_y A_0 \tan \theta \cos \theta + \tau_{yx} A_0 \tan \theta \sin \theta = 0$$

Using the relationship $\tau_{xy} = \tau_{yx}$, and also simplifying and rearranging, we obtain the following two equations:

$$\sigma_{x_1} = \sigma_x \cos^2 \theta + \sigma_y \sin^2 \theta + 2\tau_{xy} \sin \theta \cos \theta \qquad \text{(6-3a)}$$

$$\tau_{x_1 y_1} = -(\sigma_x - \sigma_y) \sin \theta \cos \theta + \tau_{xy} (\cos^2 \theta - \sin^2 \theta) \qquad \text{(6-3b)}$$

Equations (6-3a) and (6-3b) give the normal and shear stresses acting on the x_1 plane in terms of the angle θ and the stresses σ_x, σ_y, and τ_{xy} acting on the x and y planes.

For the special case when $\theta = 0$, we note that Eqs. (6-3a) and (6-3b) give $\sigma_{x_1} = \sigma_x$ and $\tau_{x_1 y_1} = \tau_{xy}$, as expected. Also, when $\theta = 90°$, the equations give $\sigma_{x_1} = \sigma_y$ and $\tau_{x_1 y_1} = -\tau_{xy} = -\tau_{yx}$. In the latter case, since the x_1 axis is vertical when $\theta = 90°$, the stress $\tau_{x_1 y_1}$ will be positive when it acts to the left. However, the stress τ_{yx} acts to the right, and therefore $\tau_{x_1 y_1} = -\tau_{yx}$.

Transformation Equations for Plane Stress

Equations (6-3a) and (6-3b) for the stresses on an inclined section can be expressed in a more convenient form by introducing the following trigonometric identities (see Appendix D available online):

$$\cos^2 \theta = \frac{1}{2}(1 + \cos 2\theta) \qquad \sin^2 \theta = \frac{1}{2}(1 - \cos 2\theta)$$

$$\sin \theta \cos \theta = \frac{1}{2} \sin 2\theta$$

When these substitutions are made, the equations become

$$\sigma_{x_1} = \frac{\sigma_x + \sigma_y}{2} + \frac{\sigma_x - \sigma_y}{2} \cos 2\theta + \tau_{xy} \sin 2\theta \qquad \text{(6-4a)}$$

$$\tau_{x_1 y_1} = -\frac{\sigma_x - \sigma_y}{2} \sin 2\theta + \tau_{xy} \cos 2\theta \qquad \text{(6-4b)}$$

These equations are usually called the **transformation equations for plane stress** because they transform the stress components from one set of axes to another. However, as explained previously, the intrinsic state of stress at the point under consideration is the same whether represented by stresses acting on the xy element (Fig. 6-1b) or by stresses acting on the inclined $x_1 y_1$ element (Fig. 6-1c).

Since the transformation equations were derived solely from equilibrium of an element, they are applicable to stresses in any kind of material, whether linear or nonlinear, elastic or inelastic.

An important observation concerning the normal stresses can be obtained from the transformation equations. As a preliminary matter, we note that the normal stress σ_{y_1} acting on the y_1 face of the inclined element (Fig. 6-1c) can be obtained from Eq. (6-4a) by substituting $\theta + 90°$ for θ. The result is the following equation for σ_{y_1}:

$$\sigma_{y_1} = \frac{\sigma_x + \sigma_y}{2} - \frac{\sigma_x - \sigma_y}{2} \cos 2\theta - \tau_{xy} \sin 2\theta \qquad (6\text{-}5)$$

Summing the expressions for σ_{x_1} and σ_{y_1} (Eqs. 6-4a and 6-5), we obtain the following equation for plane stress:

$$\sigma_{x_1} + \sigma_{y_1} = \sigma_x + \sigma_y \qquad (6\text{-}6)$$

This equation shows that the sum of the normal stresses acting on perpendicular faces of plane-stress elements (at a given point in a stressed body) is constant and independent of the angle θ.

The manner in which the normal and shear stresses vary is shown in Fig. 6-3, which is a graph of σ_{x_1} and $\tau_{x_1y_1}$ versus the angle θ (from Eqs. 6-4a and 6-4b). The graph is plotted for the particular case of $\sigma_y = 0.2\sigma_x$ and $\tau_{xy} = 0.8\sigma_x$. We see from the plot that the stresses vary continuously as the orientation of the element is changed. At certain angles, the normal stress reaches a maximum or minimum value; at other angles, it becomes zero. Similarly, the shear stress has maximum, minimum, and zero values at certain angles. A detailed investigation of these maximum and minimum values is made in Section 6.3.

Special Cases of Plane Stress

The general case of plane stress reduces to simpler states of stress under special conditions. For instance, if all stresses acting on the xy element (Fig. 6-1b) are zero except for the normal stress σ_x, then the element is

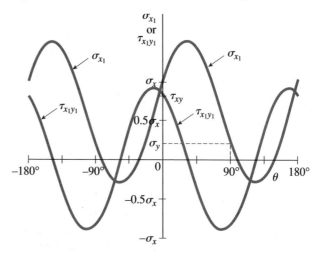

FIG. 6-3 Graph of normal stress σ_{x_1} and shear stress $\tau_{x_1y_1}$ versus the angle θ (for $\sigma_y = 0.2\sigma_x$ and $\tau_{xy} = 0.8\sigma_x$)

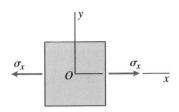

FIG. 6-4 Element in uniaxial stress

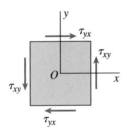

FIG. 6-5 Element in pure shear

in **uniaxial stress** (Fig. 6-4). The corresponding transformation equations, obtained by setting σ_y and τ_{xy} equal to zero in Eqs. (6-4a) and (6-4b), are

$$\sigma_{x_1} = \frac{\sigma_x}{2}(1 + \cos 2\theta) \qquad \tau_{x_1y_1} = -\frac{\sigma_x}{2}(\sin 2\theta) \qquad \text{(6-7a,b)}$$

These equations agree with the equations derived previously in Section 2.6 (see Eqs. 2-29a and 2-29b), except that now we are using a more generalized notation for the stresses acting on an inclined plane.

Another special case is **pure shear** (Fig. 6-5), for which the transformation equations are obtained by substituting $\sigma_x = 0$ and $\sigma_y = 0$ into Eqs. (6-4a) and (6-4b):

$$\sigma_{x_1} = \tau_{xy} \sin 2\theta \qquad \tau_{x_1y_1} = \tau_{xy} \cos 2\theta \qquad \text{(6-8a,b)}$$

Again, these equations correspond to those derived earlier (see Eqs. 3-30a and 3-30b in Section 3.5).

Finally, we note the special case of **biaxial stress**, in which the xy element is subjected to normal stresses in both the x and y directions but without any shear stresses (Fig. 6-6). The equations for biaxial stress are obtained from Eqs. (6-4a) and (6-4b) simply by dropping the terms containing τ_{xy}, as follows:

$$\sigma_{x_1} = \frac{\sigma_x + \sigma_y}{2} + \frac{\sigma_x - \sigma_y}{2}\cos 2\theta \qquad \text{(6-9a)}$$

$$\tau_{x_1y_1} = -\frac{\sigma_x - \sigma_y}{2}\sin 2\theta \qquad \text{(6-9b)}$$

Biaxial stress occurs in many kinds of structures, including thin-walled pressure vessels (see Sections 7.2 and 7.3).

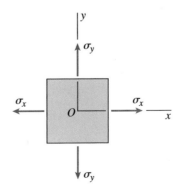

FIG. 6-6 Element in biaxial stress

Example 6-1

An element in plane stress is subjected to stresses σ_x = 16,000 psi, σ_y = 6,000 psi, and τ_{xy} = τ_{yx} = 4,000 psi, as shown in Fig. 6-7a.

Determine the stresses acting on an element inclined at an angle θ = 45°.

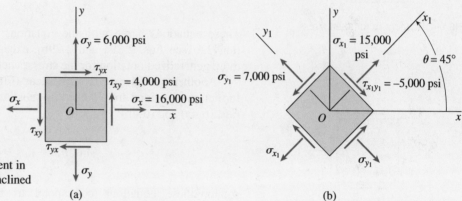

FIG. 6-7 Example 6-1. (a) Element in plane stress, and (b) element inclined at an angle θ = 45°

(a)   (b)

Solution

Transformation equations. To determine the stresses acting on an inclined element, we will use the transformation equations (Eqs. 6-4a and 6-4b). From the given numerical data, we obtain the following values for substitution into those equations:

$$\frac{\sigma_x + \sigma_y}{2} = 11{,}000 \text{ psi} \qquad \frac{\sigma_x - \sigma_y}{2} = 5{,}000 \text{ psi} \qquad \tau_{xy} = 4{,}000 \text{ psi}$$

$$\sin 2\theta = \sin 90° = 1 \qquad \cos 2\theta = \cos 90° = 0$$

Substituting these values into Eqs. (6-4a) and (6-4b), we get

$$\sigma_{x_1} = \frac{\sigma_x + \sigma_y}{2} + \frac{\sigma_x - \sigma_y}{2} \cos 2\theta + \tau_{xy} \sin 2\theta$$

$$= 11{,}000 \text{ psi} + (5{,}000 \text{ psi})(0) + (4{,}000 \text{ psi})(1) = 15{,}000 \text{ psi} \qquad \Longleftarrow$$

$$\tau_{x_1 y_1} = -\frac{\sigma_x - \sigma_y}{2} \sin 2\theta + \tau_{xy} \cos 2\theta$$

$$= -(5{,}000 \text{ psi})(1) + (4{,}000 \text{ psi})(0) = -5{,}000 \text{ psi} \qquad \Longleftarrow$$

In addition, the stress σ_{y_1} may be obtained from Eq. (6-5):

$$\sigma_{y_1} = \frac{\sigma_x + \sigma_y}{2} - \frac{\sigma_x - \sigma_y}{2} \cos 2\theta - \tau_{xy} \sin 2\theta$$

$$= 11{,}000 \text{ psi} - (5{,}000 \text{ psi})(0) - (4{,}000 \text{ psi})(1) = 7{,}000 \text{ psi} \qquad \Longleftarrow$$

Stress elements. From these results we can readily obtain the stresses acting on all sides of an element oriented at θ = 45°, as shown in Fig. 6-7b. The arrows show the true directions in which the stresses act. Note especially the directions of the shear stresses, all of which have the same magnitude. Also, observe that the sum of the normal stresses remains constant and equal to 22,000 psi (see Eq. 6-6).

Note: The stresses shown in Fig. 6-7b represent the same intrinsic state of stress as do the stresses shown in Fig. 6-7a. However, the stresses have different values because the elements on which they act have different orientations.

Example 6-2

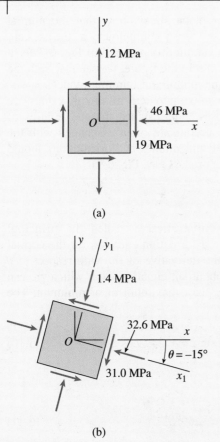

FIG. 6-8 Example 6-2. (a) Element in plane stress, and (b) element inclined at an angle $\theta = -15°$

A plane-stress condition exists at a point on the surface of a loaded structure, where the stresses have the magnitudes and directions shown on the stress element of Fig. 6-8a.

Determine the stresses acting on an element that is oriented at a clockwise angle of 15° with respect to the original element.

Solution

The stresses acting on the original element (Fig. 6-8a) have the following values:

$$\sigma_x = -46 \text{ MPa} \qquad \sigma_y = 12 \text{ MPa} \qquad \tau_{xy} = -19 \text{ MPa}$$

An element oriented at a clockwise angle of 15° is shown in Fig. 6-8b, where the x_1 axis is at an angle $\theta = -15°$ with respect to the x axis. (As an alternative, the x_1 axis could be placed at a positive angle $\theta = 75°$.)

Stress transformation equations. We can readily calculate the stresses on the x_1 face of the element oriented at $\theta = -15°$ by using the transformation equations (Eqs. 6-4a and 6-4b). The calculations proceed as follows:

$$\frac{\sigma_x + \sigma_y}{2} = -17 \text{ MPa} \qquad \frac{\sigma_x - \sigma_y}{2} = -29 \text{ MPa}$$

$$\sin 2\theta = \sin(-30°) = -0.5 \qquad \cos 2\theta = \cos(-30°) = 0.8660$$

Substituting into the transformation equations, we get

$$\sigma_{x_1} = \frac{\sigma_x + \sigma_y}{2} + \frac{\sigma_x - \sigma_y}{2} \cos 2\theta + \tau_{xy} \sin 2\theta$$

$$= -17 \text{ MPa} + (-29 \text{ MPa})(0.8660) + (-19 \text{ MPa})(-0.5)$$

$$= -32.6 \text{ MPa} \qquad \Longleftarrow$$

$$\tau_{x_1 y_1} = -\frac{\sigma_x - \sigma_y}{2} \sin 2\theta + \tau_{xy} \cos 2\theta$$

$$= -(-29 \text{ MPa})(-0.5) + (-19 \text{ MPa})(0.8660)$$

$$= -31.0 \text{ MPa} \qquad \Longleftarrow$$

The normal stress acting on the y_1 face (Eq. 6-5) is

$$\sigma_{y_1} = \frac{\sigma_x + \sigma_y}{2} - \frac{\sigma_x - \sigma_y}{2} \cos 2\theta - \tau_{xy} \sin 2\theta$$

$$= -17 \text{ MPa} - (-29 \text{ MPa})(0.8660) - (-19 \text{ MPa})(-0.5)$$

$$= -1.4 \text{ MPa} \qquad \Longleftarrow$$

This stress can be verified by substituting $\theta = 75°$ into Eq. (6-4a). As a further check on the results, we note that $\sigma_{x_1} + \sigma_{y_1} = \sigma_x + \sigma_y$.

The stresses acting on the inclined element are shown in Fig. 6-8b, where the arrows indicate the true directions of the stresses. Again we note that both stress elements shown in Fig. 6-8 represent the same state of stress.

6.3 PRINCIPAL STRESSES AND MAXIMUM SHEAR STRESSES

(a) Photo of a crane-hook
(Frans Lemmens/Getty Images)

(b) Photoelastic fringe pattern
(Courtesy of Eann Patterson)

FIG. 6-9 Photoelastic fringe pattern displays principal stresses in a model of a crane-hook

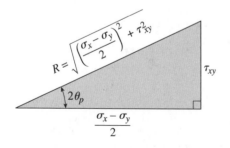

FIG. 6-10 Geometric representation of Eq. (6-11)

The transformation equations for plane stress show that the normal stresses σ_{x_1} and the shear stresses $\tau_{x_1y_1}$ vary continuously as the axes are rotated through the angle θ. This variation is pictured in Fig. 6-3 for a particular combination of stresses. From the figure, we see that both the normal and shear stresses reach maximum and minimum values at 90° intervals. Not surprisingly, these maximum and minimum values are usually needed for design purposes. For instance, fatigue failures of structures such as machines and aircraft are often associated with the maximum stresses, and hence their magnitudes and orientations should be determined as part of the design process (see Fig. 6-9).

Principal Stresses

The maximum and minimum normal stresses, called the **principal stresses**, can be found from the transformation equation for the normal stress σ_{x_1} (Eq. 6-4a). By taking the derivative of σ_{x_1} with respect to θ and setting it equal to zero, we obtain an equation from which we can find the values of θ at which σ_{x_1} is a maximum or a minimum. The equation for the derivative is

$$\frac{d\sigma_{x_1}}{d\theta} = -(\sigma_x - \sigma_y)\sin 2\theta + 2\tau_{xy}\cos 2\theta = 0 \qquad (6\text{-}10)$$

from which we get

$$\tan 2\theta_p = \frac{2\tau_{xy}}{\sigma_x - \sigma_y} \qquad (6\text{-}11)$$

The subscript p indicates that the angle θ_p defines the orientation of the **principal planes**, that is, the planes on which the principal stresses act.

Two values of the angle $2\theta_p$ in the range from 0 to 360° can be obtained from Eq. (6-11). These values differ by 180°, with one value between 0 and 180° and the other between 180° and 360°. Therefore, the angle θ_p has two values that differ by 90°, one value between 0 and 90° and the other between 90° and 180°. The two values of θ_p are known as the **principal angles**. For one of these angles, the normal stress σ_{x_1} is a *maximum* principal stress; for the other, it is a *minimum* principal stress. Because the principal angles differ by 90°, we see that *the principal stresses occur on mutually perpendicular planes*.

The principal stresses can be calculated by substituting each of the two values of θ_p into the first stress-transformation equation (Eq. 6-4a) and solving for σ_{x_1}. By determining the principal stresses in this manner, we not only obtain the values of the principal stresses but we also learn which principal stress is associated with which principal angle.

We can also obtain general formulas for the principal stresses. To do so, refer to the right triangle in Fig. 6-10, which is constructed from

Eq. (6-11). Note that the hypotenuse of the triangle, obtained from the Pythagorean theorem, is

$$R = \sqrt{\left(\frac{\sigma_x - \sigma_y}{2}\right)^2 + \tau_{xy}^2} \tag{6-12}$$

The quantity R is always a positive number and, like the other two sides of the triangle, has units of stress. From the triangle we obtain two additional relations:

$$\cos 2\theta_p = \frac{\sigma_x - \sigma_y}{2R} \qquad \sin 2\theta_p = \frac{\tau_{xy}}{R} \tag{6-13a,b}$$

Now we substitute these expressions for $\cos 2\theta_p$ and $\sin 2\theta_p$ into Eq. (6-4a) and obtain the algebraically larger of the two principal stresses, denoted by σ_1:

$$\sigma_1 = \sigma_{x_1} = \frac{\sigma_x + \sigma_y}{2} + \frac{\sigma_x - \sigma_y}{2} \cos 2\theta_p + \tau_{xy} \sin 2\theta_p$$

$$= \frac{\sigma_x + \sigma_y}{2} + \frac{\sigma_x - \sigma_y}{2}\left(\frac{\sigma_x - \sigma_y}{2R}\right) + \tau_{xy}\left(\frac{\tau_{xy}}{R}\right)$$

After substituting for R from Eq. (6-12) and performing some algebraic manipulations, we obtain

$$\sigma_1 = \frac{\sigma_x + \sigma_y}{2} + \sqrt{\left(\frac{\sigma_x - \sigma_y}{2}\right)^2 + \tau_{xy}^2} \tag{6-14}$$

The smaller of the principal stresses, denoted by σ_2, may be found from the condition that the sum of the normal stresses on perpendicular planes is constant (see Eq. 6-6):

$$\sigma_1 + \sigma_2 = \sigma_x + \sigma_y \tag{6-15}$$

Substituting the expression for σ_1 into Eq. (6-15) and solving for σ_2, we get

$$\sigma_2 = \sigma_x + \sigma_y - \sigma_1$$

$$= \frac{\sigma_x + \sigma_y}{2} - \sqrt{\left(\frac{\sigma_x - \sigma_y}{2}\right)^2 + \tau_{xy}^2} \tag{6-16}$$

This equation has the same form as the equation for σ_1 but differs by the presence of the minus sign before the square root.

The preceding formulas for σ_1 and σ_2 can be combined into a single formula for the **principal stresses**:

$$\sigma_{1,2} = \frac{\sigma_x + \sigma_y}{2} \pm \sqrt{\left(\frac{\sigma_x - \sigma_y}{2}\right)^2 + \tau_{xy}^2} \tag{6-17}$$

The plus sign gives the algebraically larger principal stress and the minus sign gives the algebraically smaller principal stress.

Principal Angles

Let us now denote the two angles defining the principal planes as θ_{p_1} and θ_{p_2}, corresponding to the principal stresses σ_1 and σ_2, respectively. Both angles can be determined from the equation for $\tan 2\theta_p$ (Eq. 6-11). However, we cannot tell from that equation which angle is θ_{p_1} and which is θ_{p_2}. A simple procedure for making this determination is to take one of the values and substitute it into the equation for σ_{x_1} (Eq. 6-4a). The resulting value of σ_{x_1} will be recognized as either σ_1 or σ_2 (assuming we have already found σ_1 and σ_2 from Eq. 6-17), thus correlating the two principal angles with the two principal stresses.

Another method for correlating the principal angles and principal stresses is to use Eqs. (6-13a) and (6-13b) to find θ_p, since the only angle that satisfies *both* of those equations is θ_{p_1}. Thus, we can rewrite those equations as follows:

$$\cos 2\theta_{p_1} = \frac{\sigma_x - \sigma_y}{2R} \qquad \sin 2\theta_{p_1} = \frac{\tau_{xy}}{R} \qquad (6\text{-}18\text{a,b})$$

Only one angle exists between 0 and 360° that satisfies both of these equations. Thus, the value of θ_{p_1} can be determined uniquely from Eqs. (6-18a) and (6-18b). The angle θ_{p_2}, corresponding to σ_2, defines a plane that is perpendicular to the plane defined by θ_{p_1}. Therefore, θ_{p_2} can be taken as 90° larger or 90° smaller than θ_{p_1}.

Shear Stresses on the Principal Planes

An important characteristic of the principal planes can be obtained from the transformation equation for the shear stresses (Eq. 6-4b). If we set the shear stress $\tau_{x_1y_1}$ equal to zero, we get an equation that is the same as Eq. (6-10). Therefore, if we solve that equation for the angle 2θ, we get the same expression for $\tan 2\theta$ as before (Eq. 6-11). In other words, the angles to the planes of zero shear stress are the same as the angles to the principal planes.

Thus, we can make the following important observation: *The shear stresses are zero on the principal planes.*

Special Cases

The principal planes for elements in **uniaxial stress** and **biaxial stress** are the x and y planes themselves (Fig. 6-11), because $\tan 2\theta_p = 0$ (see Eq. 6-11) and the two values of θ_p are 0 and 90°. We also know that the x and y planes are the principal planes from the fact that the shear stresses are zero on those planes.

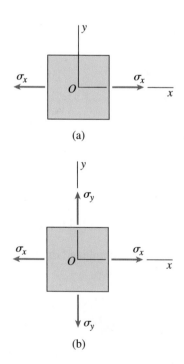

(a)

(b)

FIG. 6-11 Elements in uniaxial and biaxial stress

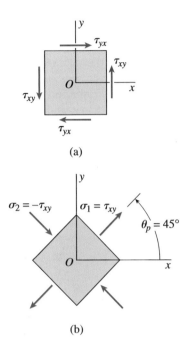

FIG. 6-12 (a) Element in pure shear, and (b) principal stresses

For an element in **pure shear** (Fig. 6-12a), the principal planes are oriented at 45° to the x axis (Fig. 6-12b), because tan $2\theta_p$ is infinite and the two values of θ_p are 45° and 135°. If τ_{xy} is positive, the principal stresses are $\sigma_1 = \tau_{xy}$ and $\sigma_2 = -\tau_{xy}$ (see Section 3.5 for a discussion of pure shear).

The Third Principal Stress

The preceding discussion of principal stresses refers only to rotation of axes in the xy plane, that is, rotation about the z axis (Fig. 6-13a). Therefore, the two principal stresses determined from Eq. (6-17) are called the **in-plane principal stresses**. However, we must not overlook the fact that the stress element is actually three-dimensional and has three (not two) principal stresses acting on three mutually perpendicular planes.

By making a more complete three-dimensional analysis, it can be shown that the three principal planes for a plane-stress element are the two principal planes already described plus the z face of the element. These principal planes are shown in Fig. 6-13b, where a stress element has been oriented at the principal angle θ_{p_1}, which corresponds to the principal stress σ_1. The principal stresses σ_1 and σ_2 are given by Eq. (6-17), and the third principal stress (σ_3) equals zero.

By definition, σ_1 is algebraically larger than σ_2, but σ_3 may be algebraically larger than, between, or smaller than σ_1 and σ_2. Of course, it is also possible for some or all of the principal stresses to be equal. Note again that there are no shear stresses on any of the principal planes.[*]

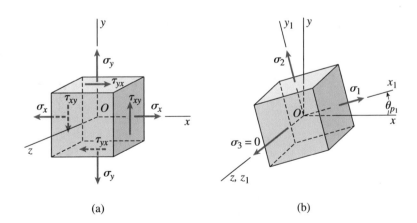

FIG. 6-13 Elements in plane stress: (a) original element, and (b) element oriented to the three principal planes and three principal stresses

---

[*]The determination of principal stresses is an example of a type of mathematical analysis known as *eigenvalue analysis*, which is described in books on matrix algebra. The stress-transformation equations and the concept of principal stresses are due to the French mathematicians A. L. Cauchy (1789–1857) and Barré de Saint-Venant (1797–1886) and to the Scottish scientist and engineer W. J. M. Rankine (1820–1872); see Refs. 6-1, 6-2, and 6-3 respectively. A list of references is available online.

Maximum Shear Stresses

Having found the principal stresses and their directions for an element in plane stress, we now consider the determination of the maximum shear stresses and the planes on which they act. The shear stresses $\tau_{x_1y_1}$ acting on inclined planes are given by the second transformation equation (Eq. 6-4b). Taking the derivative of $\tau_{x_1y_1}$ with respect to θ and setting it equal to zero, we obtain

$$\frac{d\tau_{x_1y_1}}{d\theta} = -(\sigma_x - \sigma_y)\cos 2\theta - 2\tau_{xy}\sin 2\theta = 0 \qquad (6\text{-}19)$$

from which

$$\tan 2\theta_s = -\frac{\sigma_x - \sigma_y}{2\tau_{xy}} \qquad (6\text{-}20)$$

The subscript s indicates that the angle θ_s defines the orientation of the planes of maximum positive and negative shear stresses.

Equation (6-20) yields one value of θ_s between 0 and 90° and another between 90° and 180°. Furthermore, these two values differ by 90°, and therefore the maximum shear stresses occur on perpendicular planes. Because shear stresses on perpendicular planes are equal in absolute value, the maximum positive and negative shear stresses differ only in sign.

Comparing Eq. (6-20) for θ_s with Eq. (6-11) for θ_p shows that

$$\tan 2\theta_s = -\frac{1}{\tan 2\theta_p} = -\cot 2\theta_p \qquad (6\text{-}21)$$

From this equation we can obtain a relationship between the angles θ_s and θ_p. First, we rewrite the preceding equation in the form

$$\frac{\sin 2\theta_s}{\cos 2\theta_s} + \frac{\cos 2\theta_p}{\sin 2\theta_p} = 0$$

Multiplying by the terms in the denominator, we get

$$\sin 2\theta_s \sin 2\theta_p + \cos 2\theta_s \cos 2\theta_p = 0$$

which is equivalent to the following expression (see Appendix D available online):

$$\cos(2\theta_s - 2\theta_p) = 0$$

Therefore,

$$2\theta_s - 2\theta_p = \pm 90°$$

and

$$\theta_s = \theta_p \pm 45° \qquad (6\text{-}22)$$

This equation shows that *the planes of maximum shear stress occur at 45° to the principal planes.*

The plane of the maximum positive shear stress $\tau_{\max}$ is defined by the angle θ_{s_1}, for which the following equations apply:

$$\cos 2\theta_{s_1} = \frac{\tau_{xy}}{R} \qquad \sin 2\theta_{s_1} = -\frac{\sigma_x - \sigma_y}{2R} \qquad \text{(6-23a,b)}$$

in which R is given by Eq. (6-12). Also, the angle θ_{s_1} is related to the angle θ_{p_1} (see Eqs. 6-18a and 6-18b) as follows:

$$\theta_{s_1} = \theta_{p_1}^{\cdot} - 45° \qquad \text{(6-24)}$$

The corresponding maximum shear stress is obtained by substituting the expressions for $\cos 2\theta_{s_1}$ and $\sin 2\theta_{s_1}$ into the second transformation equation (Eq. 6-4b), yielding

$$\tau_{\max} = \sqrt{\left(\frac{\sigma_x - \sigma_y}{2}\right)^2 + \tau_{xy}^2} \qquad \text{(6-25)}$$

The maximum negative shear stress $\tau_{\min}$ has the same magnitude but opposite sign.

Another expression for the maximum shear stress can be obtained from the principal stresses σ_1 and σ_2, both of which are given by Eq. (6-17). Subtracting the expression for σ_2 from that for σ_1, and then comparing with Eq. (6-25), we see that

$$\tau_{\max} = \frac{\sigma_1 - \sigma_2}{2} \qquad \text{(6-26)}$$

Thus, *the maximum shear stress is equal to one-half the difference of the principal stresses.*

The planes of maximum shear stress also contain normal stresses. The **normal stress** acting on the planes of maximum positive shear stress can be determined by substituting the expressions for the angle θ_{s_1} (Eqs. 6-23a and 6-23b) into the equation for σ_{x_1} (Eq. 6-4a). The resulting stress is equal to the average of the normal stresses on the x and y planes:

$$\sigma_{\text{aver}} = \frac{\sigma_x + \sigma_y}{2} \qquad \text{(6-27)}$$

This same normal stress acts on the planes of maximum negative shear stress.

In the particular cases of **uniaxial stress** and **biaxial stress** (Fig. 6-11), the planes of maximum shear stress occur at 45° to the x

and y axes. In the case of **pure shear** (Fig. 6-12), the maximum shear stresses occur on the x and y planes.

In-Plane and Out-of-Plane Shear Stresses

The preceding analysis of shear stresses has dealt only with **in-plane shear stresses**, that is, stresses acting in the xy plane. To obtain the maximum in-plane shear stresses (Eqs. 6-25 and 6-26), we considered elements that were obtained by rotating the xyz axes about the z axis, which is a principal axis (Fig. 6-13a). We found that the maximum shear stresses occur on planes at 45° to the principal planes. The principal planes for the element of Fig. 6-13a are shown in Fig. 6-13b, where σ_1 and σ_2 are the principal stresses. Therefore, the maximum in-plane shear stresses are found on an element obtained by rotating the $x_1 y_1 z_1$ axes (Fig. 6-13b) about the z_1 axis through an angle of 45°. These stresses are given by Eq. (6-25) or Eq. (6-26).

We can also obtain maximum shear stresses by 45° rotations about the other two principal axes (the x_1 and y_1 axes in Fig. 6-13b). As a result, we obtain three sets of **maximum positive and maximum negative shear stresses** (compare with Eq. 6-26):

$$(\tau_{\max})_{x_1} = \pm \frac{\sigma_2}{2} \quad (\tau_{\max})_{y_1} = \pm \frac{\sigma_1}{2} \quad (\tau_{\max})_{z_1} = \pm \frac{\sigma_1 - \sigma_2}{2} \quad \text{(6-28a,b,c)}$$

in which the subscripts indicate the principal axes about which the 45° rotations take place. The stresses obtained by rotations about the x_1 and y_1 axes are called **out-of-plane shear stresses**.

The algebraic values of σ_1 and σ_2 determine which of the preceding expressions gives the numerically largest shear stress. If σ_1 and σ_2 have the same sign, then one of the first two expressions is numerically largest; if they have opposite signs, the last expression is largest.

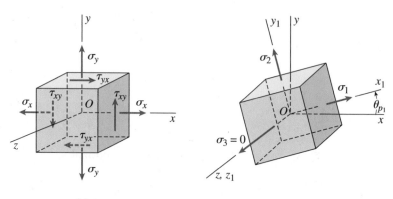

FIG. 6-13 (Repeated)

(a)   (b)

Example 6-3

An element in plane stress is subjected to stresses $\sigma_x = 12{,}300$ psi, $\sigma_y = -4{,}200$ psi, and $\tau_{xy} = -4{,}700$ psi, as shown in Fig. 6-14a.

(a) Determine the principal stresses and show them on a sketch of a properly oriented element.

(b) Determine the maximum shear stresses and show them on a sketch of a properly oriented element. (Consider only the in-plane stresses.)

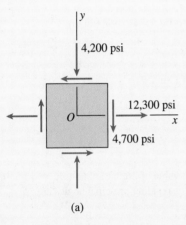

(a)

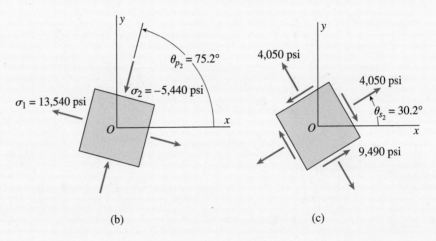

(b)   (c)

FIG. 6-14 Example 6-3. (a) Element in plane stress, (b) principal stresses, and (c) maximum shear stresses

Solution

(a) *Principal stresses.* The principal angles θ_p that locate the principal planes can be obtained from Eq. (6-11):

$$\tan 2\theta_p = \frac{2\tau_{xy}}{\sigma_x - \sigma_y} = \frac{2(-4{,}700 \text{ psi})}{12{,}300 \text{ psi} - (-4{,}200 \text{ psi})} = -0.5697$$

Solving for the angles, we get the following two sets of values:

$$2\theta_p = 150.3° \quad \text{and} \quad \theta_p = 75.2°$$

$$2\theta_p = 330.3° \quad \text{and} \quad \theta_p = 165.2°$$

The principal stresses may be obtained by substituting the two values of $2\theta_p$ into the transformation equation for σ_{x_1} (Eq. 6-4a). As a preliminary calculation, we determine the following quantities:

continued

$$\frac{\sigma_x + \sigma_y}{2} = \frac{12{,}300 \text{ psi} - 4{,}200 \text{ psi}}{2} = 4{,}050 \text{ psi}$$

$$\frac{\sigma_x - \sigma_y}{2} = \frac{12{,}300 \text{ psi} + 4{,}200 \text{ psi}}{2} = 8{,}250 \text{ psi}$$

Now we substitute the first value of $2\theta_p$ into Eq. (6-4a) and obtain

$$\sigma_{x_1} = \frac{\sigma_x + \sigma_y}{2} + \frac{\sigma_x - \sigma_y}{2} \cos 2\theta + \tau_{xy} \sin 2\theta$$
$$= 4{,}050 \text{ psi} + (8{,}250 \text{ psi})(\cos 150.3°) - (4{,}700 \text{ psi})(\sin 150.3°)$$
$$= -5{,}440 \text{ psi}$$

In a similar manner, we substitute the second value of $2\theta_p$ and obtain $\sigma_{x_1} = 13{,}540$ psi. Thus, the principal stresses and their corresponding principal angles are

$$\sigma_1 = 13{,}540 \text{ psi} \quad \text{and} \quad \theta_{p_1} = 165.2°$$

$$\sigma_2 = -5{,}440 \text{ psi} \quad \text{and} \quad \theta_{p_2} = 75.2°$$

Note that θ_{p_1} and θ_{p_2} differ by 90° and that $\sigma_1 + \sigma_2 = \sigma_x + \sigma_y$.

The principal stresses are shown on a properly oriented element in Fig. 6-14b. Of course, no shear stresses act on the principal planes.

Alternative solution for the principal stresses. The principal stresses may also be calculated directly from Eq. (6-17):

$$\sigma_{1,2} = \frac{\sigma_x + \sigma_y}{2} \pm \sqrt{\left(\frac{\sigma_x - \sigma_y}{2}\right)^2 + \tau_{xy}^2}$$
$$= 4{,}050 \text{ psi} \pm \sqrt{(8{,}250 \text{ psi})^2 + (-4{,}700 \text{ psi})^2}$$

$$\sigma_{1,2} = 4{,}050 \text{ psi} \pm 9{,}490 \text{ psi}$$

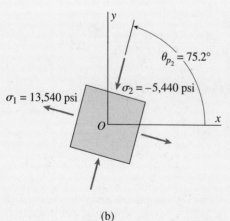

(b)

FIG. 6-14b (Repeated)

Therefore,

$$\sigma_1 = 13{,}540 \text{ psi} \qquad \sigma_2 = -5{,}440 \text{ psi}$$

The angle θ_{p_1} to the plane on which σ_1 acts is obtained from Eqs. (6-18a) and (6-18b):

$$\cos 2\theta_{p_1} = \frac{\sigma_x - \sigma_y}{2R} = \frac{8{,}250 \text{ psi}}{9{,}490 \text{ psi}} = 0.869$$

$$\sin 2\theta_{p_1} = \frac{\tau_{xy}}{R} = \frac{-4{,}700 \text{ psi}}{9{,}490 \text{ psi}} = -0.495$$

in which R is given by Eq. (6-12) and is equal to the square-root term in the preceding calculation for the principal stresses σ_1 and σ_2.

The only angle between 0 and 360° having the specified sine and cosine is $2\theta_{p_1} = 330.3°$; hence, $\theta_{p_1} = 165.2°$. This angle is associated with the algebraically larger principal stress $\sigma_1 = 13{,}540$ psi. The other angle is 90° larger or smaller than θ_{p_1}; hence, $\theta_{p_2} = 75.2°$. This angle corresponds to the smaller principal stress $\sigma_2 = -5{,}440$ psi. Note that these results for the principal stresses and principal angles agree with those found previously.

(b) *Maximum shear stresses.* The maximum in-plane shear stresses are given by Eq. (6-25):

$$\tau_{max} = \sqrt{\left(\frac{\sigma_x - \sigma_y}{2}\right)^2 + \tau_{xy}^2}$$

$$= \sqrt{(8{,}250 \text{ psi})^2 + (-4{,}700 \text{ psi})^2} = 9{,}490 \text{ psi} \qquad \Longleftarrow$$

The angle θ_{s_1} to the plane having the maximum positive shear stress is calculated from Eq. (6-24):

$$\theta_{s_1} = \theta_{p_1} - 45° = 165.2° - 45° = 120.2° \qquad \Longleftarrow$$

It follows that the maximum negative shear stress acts on the plane for which $\theta_{s_2} = 120.2° - 90° = 30.2°$.

The normal stresses acting on the planes of maximum shear stresses are calculated from Eq. (6-27):

$$\sigma_{aver} = \frac{\sigma_x + \sigma_y}{2} = 4{,}050 \text{ psi} \qquad \Longleftarrow$$

Finally, the maximum shear stresses and associated normal stresses are shown on the stress element of Fig. 6-14c.

As an alternative approach to finding the maximum shear stresses, we can use Eq. (6-20) to determine the two values of the angles θ_s, and then we can use the second transformation equation (Eq. 6-4b) to obtain the corresponding shear stresses.

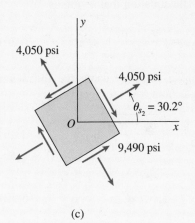

(c)

FIG. 6-14c (Repeated)

6.4 MOHR'S CIRCLE FOR PLANE STRESS

The transformation equations for plane stress can be represented in graphical form by a plot known as **Mohr's circle**. This graphical representation is extremely useful because it enables you to visualize the relationships between the normal and shear stresses acting on various inclined planes at a point in a stressed body. It also provides a means for calculating principal stresses, maximum shear stresses, and stresses on inclined planes. Furthermore, Mohr's circle is valid not only for stresses but also for other quantities of a similar mathematical nature, including strains and moments of inertia.[*]

Equations of Mohr's Circle

The equations of Mohr's circle can be derived from the transformation equations for plane stress (Eqs. 6-4a and 6-4b). The two equations are repeated here, but with a slight rearrangement of the first equation:

$$\sigma_{x_1} - \frac{\sigma_x + \sigma_y}{2} = \frac{\sigma_x - \sigma_y}{2} \cos 2\theta + \tau_{xy} \sin 2\theta \qquad (6\text{-}29a)$$

$$\tau_{x_1 y_1} = -\frac{\sigma_x - \sigma_y}{2} \sin 2\theta + \tau_{xy} \cos 2\theta \qquad (6\text{-}29b)$$

From analytic geometry, we might recognize that these two equations are the equations of a circle in parametric form. The angle 2θ is the parameter and the stresses σ_{x_1} and $\tau_{x_1 y_1}$ are the coordinates. However, it is not necessary to recognize the nature of the equations at this stage—if we eliminate the parameter, the significance of the equations will become apparent.

To eliminate the parameter 2θ, we square both sides of each equation and then add the two equations. The equation that results is

$$\left(\sigma_{x_1} - \frac{\sigma_x + \sigma_y}{2} \right)^2 + \tau_{x_1 y_1}^2 = \left(\frac{\sigma_x - \sigma_y}{2} \right)^2 + \tau_{xy}^2 \qquad (6\text{-}30)$$

This equation can be written in simpler form by using the following notation from Section 6.3 (see Eqs. 6-27 and 6-12, respectively):

$$\sigma_{\text{aver}} = \frac{\sigma_x + \sigma_y}{2} \qquad R = \sqrt{\left(\frac{\sigma_x - \sigma_y}{2} \right)^2 + \tau_{xy}^2} \qquad (6\text{-}31a,b)$$

Equation (6-30) now becomes

$$(\sigma_{x_1} - \sigma_{\text{aver}})^2 + \tau_{x_1 y_1}^2 = R^2 \qquad (6\text{-}32)$$

---

[*]Mohr's circle is named after the famous German civil engineer Otto Christian Mohr (1835–1918), who developed the circle in 1882 (Ref. 6-4 available online).

which is the equation of a circle in standard algebraic form. The coordinates are σ_{x_1} and $\tau_{x_1y_1}$, the radius is R, and the center of the circle has coordinates $\sigma_{x_1} = \sigma_{aver}$ and $\tau_{x_1y_1} = 0$.

Two Forms of Mohr's Circle

Mohr's circle can be plotted from Eqs. (6-29) and (6-32) in either of two forms. In the first form of Mohr's circle, we plot the normal stress σ_{x_1} positive to the right and the shear stress $\tau_{x_1y_1}$ positive downward, as shown in Fig. 6-15a. The advantage of plotting shear stresses positive downward is that the angle 2θ on Mohr's circle will be positive when counterclockwise, which agrees with the positive direction of 2θ in the derivation of the transformation equations (see Figs. 6-1 and 6-2).

In the second form of Mohr's circle, $\tau_{x_1y_1}$ is plotted positive upward but the angle 2θ is now positive clockwise (Fig. 6-15b), which is opposite to its usual positive direction.

Both forms of Mohr's circle are mathematically correct, and either one can be used. However, it is easier to visualize the orientation of the stress element if the positive direction of the angle 2θ is the same in Mohr's circle as it is for the element itself. Furthermore, a counterclockwise rotation agrees with the customary right-hand rule for rotation.

Therefore, we will choose the first form of Mohr's circle (Fig. 6-15a) in which *positive shear stress is plotted downward and a positive angle 2θ is plotted counterclockwise.*

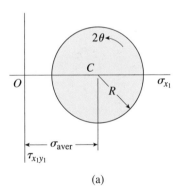

(a)

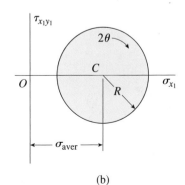

FIG. 6-15 Two forms of Mohr's circle: (a) $\tau_{x_1y_1}$ is positive downward and the angle 2θ is positive counterclockwise, and (b) $\tau_{x_1y_1}$ is positive upward and the angle 2θ is positive clockwise. (*Note:* The first form is used in this book.)

(b)

Construction of Mohr's Circle

Mohr's circle can be constructed in a variety of ways, depending upon which stresses are known and which are to be found. For our immediate purpose, which is to show the basic properties of the circle, let us assume that we know the stresses σ_x, σ_y, and τ_{xy} acting on the x and y planes of an element in plane stress (Fig. 6-16a). As we will see, this information is sufficient to construct the circle. Then, with the circle drawn, we can determine the stresses σ_{x_1}, σ_{y_1}, and $\tau_{x_1y_1}$ acting on an inclined element (Fig. 6-16b). We can also obtain the principal stresses and maximum shear stresses from the circle.

With σ_x, σ_y, and τ_{xy} known, the **procedure for constructing Mohr's circle** is as follows (see Fig. 6-16c):

1. Draw a set of coordinate axes with σ_{x_1} as abscissa (positive to the right) and $\tau_{x_1y_1}$ as ordinate (positive downward).
2. Locate the center C of the circle at the point having coordinates $\sigma_{x_1} = \sigma_{\mathrm{aver}}$ and $\tau_{x_1y_1} = 0$ (see Eqs. 6-31a and 6-32).
3. Locate point A, representing the stress conditions on the x face of the element shown in Fig. 6-16a, by plotting its coordinates $\sigma_{x_1} = \sigma_x$ and $\tau_{x_1y_1} = \tau_{xy}$. Note that point A on the circle corresponds to $\theta = 0$. Also, note that the x face of the element (Fig. 6-16a) is labeled "A" to show its correspondence with point A on the circle.
4. Locate point B, representing the stress conditions on the y face of the element shown in Fig. 6-16a, by plotting its coordinates

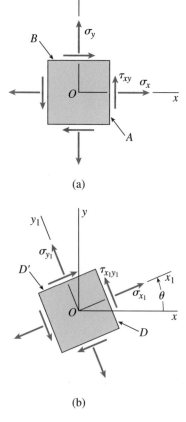

(a)

(b)

FIG. 6-16 Construction of Mohr's circle for plane stress

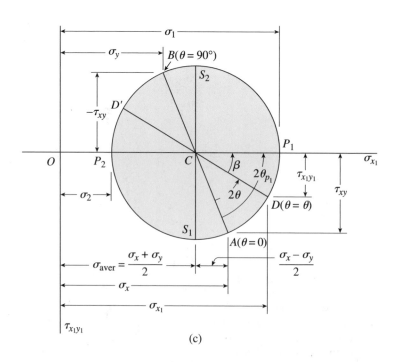

(c)

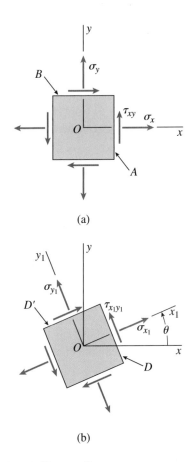

(a)

(b)

FIG. 6-16 (Repeated)

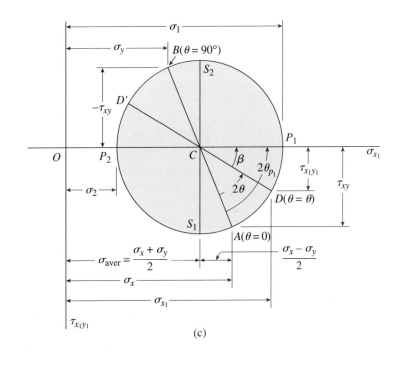

(c)

$\sigma_{x_1} = \sigma_y$ and $\tau_{x_1y_1} = -\tau_{xy}$. Note that point B on the circle corresponds to $\theta = 90°$. In addition, the y face of the element (Fig. 6-16a) is labeled "B" to show its correspondence with point B on the circle.

5. Draw a line from point A to point B. This line is a diameter of the circle and passes through the center C. Points A and B, representing the stresses on planes at 90° to each other (Fig. 6-16a), are at opposite ends of the diameter (and therefore are 180° apart on the circle).

6. Using point C as the center, draw Mohr's circle through points A and B. The circle drawn in this manner has radius R (Eq. 6-31b), as shown in the next paragraph.

Now that we have drawn the circle, we can verify by geometry that lines CA and CB are radii and have lengths equal to R. We note that the abscissas of points C and A are $(\sigma_x + \sigma_y)/2$ and σ_x, respectively. The difference in these abscissas is $(\sigma_x - \sigma_y)/2$, as dimensioned in the figure. Also, the ordinate to point A is τ_{xy}. Therefore, line CA is the hypotenuse of a right triangle having one side of length $(\sigma_x - \sigma_y)/2$ and the other side of length τ_{xy}. Taking the square root of the sum of the squares of these two sides gives the radius R:

$$R = \sqrt{\left(\frac{\sigma_x - \sigma_y}{2}\right)^2 + \tau_{xy}^2}$$

which is the same as Eq. (6-31b). By a similar procedure, we can show that the length of line CB is also equal to the radius R of the circle.

Stresses on an Inclined Element

Now we will consider the stresses σ_{x_1}, σ_{y_1}, and $\tau_{x_1y_1}$ acting on the faces of a plane-stress element oriented at an angle θ from the x axis (Fig. 6-16b). If the angle θ is known, these stresses can be determined from Mohr's circle. The procedure is as follows.

On the circle (Fig. 6-16c), we measure an angle 2θ counterclockwise from radius CA, because point A corresponds to $\theta = 0$ and is the reference point from which we measure angles. The angle 2θ locates point D on the circle, which (as shown in the next paragraph) has coordinates σ_{x_1} and $\tau_{x_1y_1}$. Therefore, point D represents the stresses on the x_1 face of the element of Fig. 6-16b. Consequently, this face of the element is labeled "D" in Fig. 6-16b.

Note that an angle 2θ on Mohr's circle corresponds to an angle θ on a stress element. For instance, point D on the circle is at an angle 2θ from point A, but the x_1 face of the element shown in Fig. 6-16b (the face labeled "D") is at an angle θ from the x face of the element shown in Fig. 6-16a (the face labeled "A"). Similarly, points A and B are 180° apart on the circle, but the corresponding faces of the element (Fig. 6-16a) are 90° apart.

To show that the coordinates σ_{x_1} and $\tau_{x_1y_1}$ of point D on the circle are indeed given by the stress-transformation equations (Eqs. 6-4a and 6-4b),

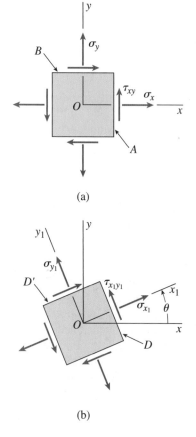

(a)

(b)

FIG. 6-16 (Repeated)

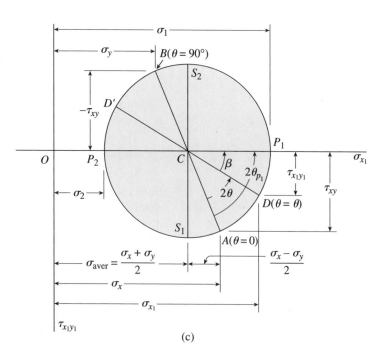

(c)

we again use the geometry of the circle. Let β be the angle between the radial line CD and the σ_{x_1} axis. Then, from the geometry of the figure, we obtain the following expressions for the coordinates of point D:

$$\sigma_{x_1} = \frac{\sigma_x + \sigma_y}{2} + R\cos\beta \qquad \tau_{x_1y_1} = R\sin\beta \qquad \text{(6-33a,b)}$$

Noting that the angle between the radius CA and the horizontal axis is $2\theta + \beta$, we get

$$\cos(2\theta + \beta) = \frac{\sigma_x - \sigma_y}{2R} \qquad \sin(2\theta + \beta) = \frac{\tau_{xy}}{R}$$

Expanding the cosine and sine expressions (see Appendix D available online) gives

$$\cos 2\theta \cos\beta - \sin 2\theta \sin\beta = \frac{\sigma_x - \sigma_y}{2R} \qquad \text{(a)}$$

$$\sin 2\theta \cos\beta + \cos 2\theta \sin\beta = \frac{\tau_{xy}}{R} \qquad \text{(b)}$$

Multiplying the first of these equations by $\cos 2\theta$ and the second by $\sin 2\theta$ and then adding, we obtain

$$\cos\beta = \frac{1}{R}\left(\frac{\sigma_x - \sigma_y}{2}\cos 2\theta + \tau_{xy}\sin 2\theta\right) \qquad \text{(c)}$$

Also, multiplying Eq. (a) by $\sin 2\theta$ and Eq. (b) by $\cos 2\theta$ and then subtracting, we get

$$\sin\beta = \frac{1}{R}\left(-\frac{\sigma_x - \sigma_y}{2}\sin 2\theta + \tau_{xy}\cos 2\theta\right) \qquad \text{(d)}$$

When these expressions for $\cos\beta$ and $\sin\beta$ are substituted into Eqs. (6-33a) and (6-33b), we obtain the stress-transformation equations for σ_{x_1} and $\tau_{x_1y_1}$ (Eqs. 6-4a and 6-4b). Thus, we have shown that point D on Mohr's circle, defined by the angle 2θ, represents the stress conditions on the x_1 face of the stress element defined by the angle θ (Fig. 6-16b).

Point D', which is diametrically opposite point D on the circle, is located by an angle 2θ (measured from line CA) that is 180° greater than the angle 2θ to point D. Therefore, point D' on the circle represents the stresses on a face of the stress element (Fig. 6-16b) at 90° from the face represented by point D. Thus, point D' on the circle gives the stresses σ_{y_1} and $-\tau_{x_1y_1}$ on the y_1 face of the stress element (the face labeled "D'" in Fig. 6-16b).

From this discussion we see how the stresses represented by points on Mohr's circle are related to the stresses acting on an element. The stresses on an inclined plane defined by the angle θ (Fig. 6-16b) are found on the circle at the point where the angle from the reference point (point A) is 2θ. Thus, as we rotate the x_1y_1 axes counterclockwise through an angle θ (Fig. 6-16b), the point on Mohr's circle corresponding to the x_1 face moves counterclockwise through an angle 2θ. Similarly, if we rotate the axes clockwise through an angle, the point on the circle moves clockwise through an angle twice as large.

Principal Stresses

The determination of principal stresses is probably the most important application of Mohr's circle. Note that as we move around Mohr's circle (Fig. 6-16c), we encounter point P_1 where the normal stress reaches its algebraically largest value and the shear stress is zero. Hence, point P_1 represents a **principal stress** and a **principal plane**. The abscissa σ_1 of point P_1 gives the algebraically larger principal stress and its angle $2\theta_{p_1}$ from the reference point A (where $\theta = 0$) gives the orientation of the principal plane. The other principal plane, associated with the algebraically smallest normal stress, is represented by point P_2, diametrically opposite point P_1.

From the geometry of the circle, we see that the algebraically larger principal stress is

$$\sigma_1 = \overline{OC} + \overline{CP_1} = \frac{\sigma_x + \sigma_y}{2} + R$$

which, upon substitution of the expression for R (Eq. 6-31b), agrees with the earlier equation for this stress (Eq. 6-14). In a similar manner, we can verify the expression for the algebraically smaller principal stress σ_2.

The principal angle θ_{p_1} between the x axis (Fig. 6-16a) and the plane of the algebraically larger principal stress is one-half the angle $2\theta_{p_1}$, which is the angle on Mohr's circle between radii CA and CP_1. The cosine and sine of the angle $2\theta_{p_1}$ can be obtained by inspection from the circle:

$$\cos 2\theta_{p_1} = \frac{\sigma_x - \sigma_y}{2R} \qquad \sin 2\theta_{p_1} = \frac{\tau_{xy}}{R}$$

These equations agree with Eqs. (6-18a) and (6-18b), and so once again we see that the geometry of the circle matches the equations derived earlier. On the circle, the angle $2\theta_{p_2}$ to the other principal point (point P_2) is 180° larger than $2\theta_{p_1}$; hence, $\theta_{p_2} = \theta_{p_1} + 90°$, as expected.

Maximum Shear Stresses

Points S_1 and S_2, representing the planes of maximum positive and maximum negative shear stresses, respectively, are located at the bottom and top of Mohr's circle (Fig. 6-16c). These points are at angles $2\theta = 90°$ from points P_1 and P_2, which agrees with the fact that the planes of maximum shear stress are oriented at 45° to the principal planes.

The maximum shear stresses are numerically equal to the radius R of the circle (compare Eq. 6-31b for R with Eq. 6-25 for τ_{max}). Also, the normal stresses on the planes of maximum shear stress are equal to the abscissa of point C, which is the average normal stress σ_{aver} (see Eq. 6-31a).

Alternative Sign Convention for Shear Stresses

An alternative sign convention for shear stresses is sometimes used when constructing Mohr's circle. In this convention, the direction of a shear stress acting on an element of the material is indicated by the sense of the rotation that it tends to produce (Figs. 6-17a and b). If the shear stress τ tends to rotate the stress element clockwise, it is called a *clockwise shear stress*, and if it tends to rotate it counterclockwise, it is called a *counterclockwise stress*. Then, when constructing Mohr's circle, clockwise shear stresses are plotted upward and counterclockwise shear stresses are plotted downward (Fig. 6-17c).

It is important to realize that *the alternative sign convention produces a circle that is identical to the circle already described* (Fig. 6-16c). The reason is that a positive shear stress $\tau_{x_1y_1}$ is also a counterclockwise shear stress, and both are plotted downward. Also, a negative shear stress $\tau_{x_1y_1}$ is a clockwise shear stress, and both are plotted upward.

Thus, the alternative sign convention merely provides a different point of view. Instead of thinking of the vertical axis as having negative shear stresses plotted upward and positive shear stresses plotted downward (which is a bit awkward), we can think of the vertical axis as having clockwise shear stresses plotted upward and counterclockwise shear stresses plotted downward (Fig. 6-17c).

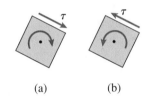

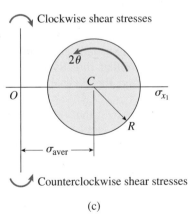

FIG. 6-17 Alternative sign convention for shear stresses: (a) clockwise shear stress, (b) counterclockwise shear stress, and (c) axes for Mohr's circle. (Note that clockwise shear stresses are plotted upward and counterclockwise shear stresses are plotted downward.)

General Comments about the Circle

From the preceding discussions in this section, it is apparent that we can find the stresses acting on any inclined plane, as well as the principal stresses and maximum shear stresses, from Mohr's circle. However, only rotations of axes in the xy plane (that is, rotations about the z axis) are considered, and therefore *all stresses on Mohr's circle are in-plane stresses*.

For convenience, the circle of Fig. 6-16 was drawn with σ_x, σ_y, and τ_{xy} as positive stresses, but the same procedures may be followed if one or more of the stresses is negative. If one of the normal stresses is negative, part or all of the circle will be located to the left of the origin, as illustrated in Example 6-6 that follows.

Point A in Fig. 6-16c, representing the stresses on the plane $\theta = 0$, may be situated anywhere around the circle. However, the angle 2θ is always measured counterclockwise from the radius CA, regardless of where point A is located.

In the special cases of *uniaxial stress*, *biaxial stress*, and *pure shear*, the construction of Mohr's circle is simpler than in the general case of plane stress. These special cases are illustrated in Example 6-4 and in Problems 6.4-1 through 6.4-9.

Besides using Mohr's circle to obtain the stresses on inclined planes when the stresses on the x and y planes are known, we can also use the circle in the opposite manner. If we know the stresses σ_{x_1}, σ_{y_1}, and $\tau_{x_1y_1}$ acting on an inclined element oriented at a known angle θ, we can easily construct the circle and determine the stresses σ_x, σ_y, and τ_{xy} for the angle $\theta = 0$. The procedure is to locate points D and D' from the known stresses and then draw the circle using line DD' as a diameter. By measuring the angle 2θ in a negative sense from radius CD, we can locate point A, corresponding to the x face of the element. Then we can locate point B by constructing a diameter from A. Finally, we can determine the coordinates of points A and B and thereby obtain the stresses acting on the element for which $\theta = 0$.

If desired, we can construct Mohr's circle to scale and measure values of stress from the drawing. However, it is usually preferable to obtain the stresses by numerical calculations, either directly from the various equations or by using trigonometry and the geometry of the circle.

Mohr's circle makes it possible to visualize the relationships between stresses acting on planes at various angles, and it also serves as a simple memory device for calculating stresses. Although many graphical techniques are no longer used in engineering work, Mohr's circle remains valuable because it provides a simple and clear picture of an otherwise complicated analysis.

Mohr's circle is also applicable to the transformations for plain strain and moments of inertia of plane areas, because these quantities follow the same transformation laws as do stresses (see Sections 10.8, and 10.9, available online).

Example 6-4

At a point on the surface of a pressurized cylinder, the material is subjected to biaxial stresses $\sigma_x = 90$ MPa and $\sigma_y = 20$ MPa, as shown on the stress element of Fig. 6-18a.

Using Mohr's circle, determine the stresses acting on an element inclined at an angle $\theta = 30°$. (Consider only the in-plane stresses, and show the results on a sketch of a properly oriented element.)

Solution

Construction of Mohr's circle. We begin by setting up the axes for the normal and shear stresses, with σ_{x_1} positive to the right and $\tau_{x_1y_1}$ positive downward, as shown in Fig. 6-18b. Then we place the center C of the circle on the σ_{x_1} axis at the point where the stress equals the average normal stress (Eq. 6-31a):

$$\sigma_{\text{aver}} = \frac{\sigma_x + \sigma_y}{2} = \frac{90 \text{ MPa} + 20 \text{ MPa}}{2} = 55 \text{ MPa}$$

Point A, representing the stresses on the x face of the element ($\theta = 0$), has coordinates

$$\sigma_{x_1} = 90 \text{ MPa} \qquad \tau_{x_1y_1} = 0$$

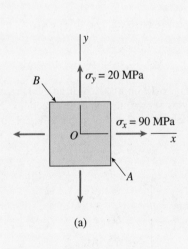

(a)

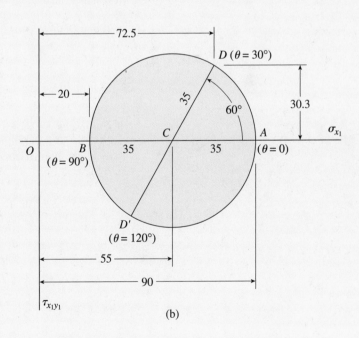

(b)

FIG. 6-18 Example 6-4. (a) Element in plane stress, and (b) the corresponding Mohr's circle. (*Note:* All stresses on the circle have units of MPa.)

continued

Similarly, the coordinates of point B, representing the stresses on the y face ($\theta = 90°$), are

$$\sigma_{x_1} = 20 \text{ MPa} \qquad \tau_{x_1 y_1} = 0$$

Now we draw the circle through points A and B with center at C and radius R (see Eq. 6-31b) equal to

$$R = \sqrt{\left(\frac{\sigma_x - \sigma_y}{2}\right)^2 + \tau_{xy}^2} = \sqrt{\left(\frac{90 \text{ MPa} - 20 \text{ MPa}}{2}\right)^2 + 0} = 35 \text{ MPa}$$

Stresses on an element inclined at $\theta = 30°$. The stresses acting on a plane oriented at an angle $\theta = 30°$ are given by the coordinates of point D, which is at an angle $2\theta = 60°$ from point A (Fig. 6-18b). By inspection of the circle, we see that the coordinates of point D are

(Point D) $\sigma_{x_1} = \sigma_{aver} + R \cos 60°$

$$= 55 \text{ MPa} + (35 \text{ MPa})(\cos 60°) = 72.5 \text{ MPa} \qquad \Longleftarrow$$

$$\tau_{x_1 y_1} = -R \sin 60° = -(35 \text{ MPa})(\sin 60°) = -30.3 \text{ MPa} \qquad \Longleftarrow$$

In a similar manner, we can find the stresses represented by point D', which corresponds to an angle $\theta = 120°$ (or $2\theta = 240°$):

(Point D') $\sigma_{x_1} = \sigma_{aver} - R \cos 60°$

$$= 55 \text{ MPa} - (35 \text{ MPa})(\cos 60°) = 37.5 \text{ MPa} \qquad \Longleftarrow$$

$$\tau_{x_1 y_1} = R \sin 60° = (35 \text{ MPa})(\sin 60°) = 30.3 \text{ MPa} \qquad \Longleftarrow$$

These results are shown in Fig. 6-19 on a sketch of an element oriented at an angle $\theta = 30°$, with all stresses shown in their true directions. Note that the sum of the normal stresses on the inclined element is equal to $\sigma_x + \sigma_y$, or 110 MPa.

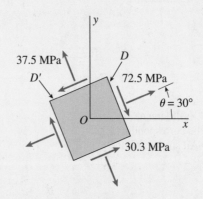

FIG. 6-19 Example 6-4 (continued). Stresses acting on an element oriented at an angle $\theta = 30°$

Example 6-5

An element in plane stress at the surface of a large machine is subjected to stresses $\sigma_x = 15{,}000$ psi, $\sigma_y = 5{,}000$ psi, and $\tau_{xy} = 4{,}000$ psi, as shown in Fig. 6-20a.

Using Mohr's circle, determine the following quantities: (a) the stresses acting on an element inclined at an angle $\theta = 40°$, (b) the principal stresses, and (c) the maximum shear stresses. (Consider only the in-plane stresses, and show all results on sketches of properly oriented elements.)

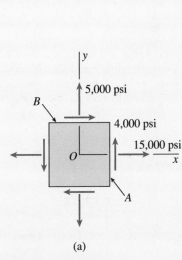

(a)

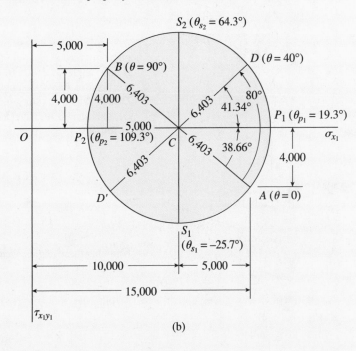

(b)

FIG. 6-20 Example 6-5. (a) Element in plane stress, and (b) the corresponding Mohr's circle. (*Note:* All stresses on the circle have units of psi.)

Solution

Construction of Mohr's circle. The first step in the solution is to set up the axes for Mohr's circle, with σ_{x_1} positive to the right and $\tau_{x_1 y_1}$ positive downward (Fig. 6-20b). The center C of the circle is located on the σ_{x_1} axis at the point where σ_{x_1} equals the average normal stress (Eq. 6-31a):

$$\sigma_{aver} = \frac{\sigma_x + \sigma_y}{2} = \frac{15{,}000 \text{ psi} + 5{,}000 \text{ psi}}{2} = 10{,}000 \text{ psi}$$

Point A, representing the stresses on the x face of the element ($\theta = 0$), has coordinates

$$\sigma_{x_1} = 15{,}000 \text{ psi} \qquad \tau_{x_1 y_1} = 4{,}000 \text{ psi}$$

Similarly, the coordinates of point B, representing the stresses on the y face ($\theta = 90°$) are

$$\sigma_{x_1} = 5{,}000 \text{ psi} \qquad \tau_{x_1 y_1} = -4{,}000 \text{ psi}$$

continued

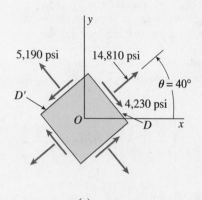

(a)

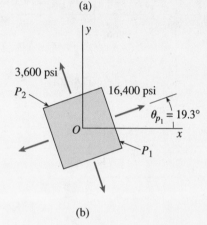

(b)

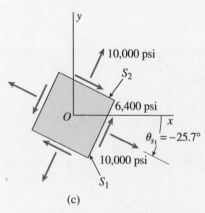

(c)

FIG. 6-21 Example 6-5 (continued).
(a) Stresses acting on an element
oriented at $\theta = 40°$, (b) principal stresses,
and (c) maximum shear stresses

The circle is now drawn through points A and B with center at C. The radius of the circle, from Eq. (6-31b), is

$$R = \sqrt{\left(\frac{\sigma_x - \sigma_y}{2}\right)^2 + \tau_{xy}^2}$$

$$= \sqrt{\left(\frac{15{,}000 \text{ psi} - 5{,}000 \text{ psi}}{2}\right)^2 + (4{,}000 \text{ psi})^2} = 6{,}403 \text{ psi}$$

(a) *Stresses on an element inclined at $\theta = 40°$.* The stresses acting on a plane oriented at an angle $\theta = 40°$ are given by the coordinates of point D, which is at an angle $2\theta = 80°$ from point A (Fig. 6-20b). To evaluate these coordinates, we need to know the angle between line CD and the σ_{x_1} axis (that is, angle DCP_1), which in turn requires that we know the angle between line CA and the σ_{x_1} axis (angle ACP_1). These angles are found from the geometry of the circle, as follows:

$$\tan \overline{ACP_1} = \frac{4{,}000 \text{ psi}}{5{,}000 \text{ psi}} = 0.8 \qquad \overline{ACP_1} = 38.66°$$

$$\overline{DCP_1} = 80° - \overline{ACP_1} = 80° - 38.66° = 41.34°$$

Knowing these angles, we can determine the coordinates of point D directly from the Figure 6-21a:

(Point D) $\sigma_{x_1} = 10{,}000 \text{ psi} + (6{,}403 \text{ psi})(\cos 41.34°) = 14{,}810 \text{ psi}$ ⬅

$\tau_{x_1 y_1} = -(6{,}403 \text{ psi})(\sin 41.34°) = -4{,}230 \text{ psi}$ ⬅

In an analogous manner, we can find the stresses represented by point D', which corresponds to a plane inclined at an angle $\theta = 130°$ (or $2\theta = 260°$):

(Point D') $\sigma_{x_1} = 10{,}000 \text{ psi} - (6{,}403 \text{ psi})(\cos 41.34°) = 5{,}190 \text{ psi}$ ⬅

$\tau_{x_1 y_1} = (6{,}403 \text{ psi})(\sin 41.34°) = 4{,}230 \text{ psi}$ ⬅

These stresses are shown in Fig. 6-21a on a sketch of an element oriented at an angle $\theta = 40°$ (all stresses are shown in their true directions). Also, note that the sum of the normal stresses is equal to $\sigma_x + \sigma_y$, or 20,000 psi.

(b) *Principal stresses.* The principal stresses are represented by points P_1 and P_2 on Mohr's circle (Fig. 6-20b). The algebraically larger principal stress (point P_1) is

$$\sigma_1 = 10{,}000 \text{ psi} + 6{,}400 \text{ psi} = 16{,}400 \text{ psi}$$ ⬅

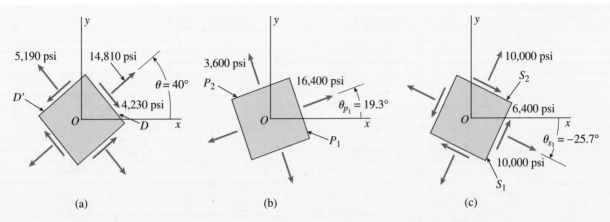

FIG. 6-21 (Repeated)

as seen by inspection of the circle. The angle $2\theta_{p_1}$ to point P_1 from point A is the angle ACP_1 on the circle, that is,

$$\overline{ACP_1} = 2\theta_{p_1} = 38.66° \qquad \theta_{p_1} = 19.3°$$

Thus, the plane of the algebraically larger principal stress is oriented at an angle $\theta_{p_1} = 19.3°$, as shown in Fig. 6-21b.

The algebraically smaller principal stress (represented by point P_2) is obtained from the circle in a similar manner:

$$\sigma_2 = 10,000 \text{ psi} - 6,400 \text{ psi} = 3,600 \text{ psi}$$

The angle $2\theta_{p_2}$ to point P_2 on the circle is $38.66° + 180° = 218.66°$; thus, the second principal plane is defined by the angle $\theta_{p_2} = 109.3°$. The principal stresses and principal planes are shown in Fig. 6-21b, and again we note that the sum of the normal stresses is equal to 20,000 psi.

(c) *Maximum shear stresses.* The maximum shear stresses are represented by points S_1 and S_2 on Mohr's circle; therefore, the maximum in-plane shear stress (equal to the radius of the circle) is

$$\tau_{max} = 6,400 \text{ psi}$$

The angle ACS_1 from point A to point S_1 is $90° - 38.66° = 51.34°$, and therefore the angle $2\theta_{s_1}$ for point S_1 is

$$2\theta_{s_1} = -51.34°$$

This angle is negative because it is measured clockwise on the circle. The corresponding angle θ_{s_1} to the plane of the maximum positive shear stress is one-half that value, or $\theta_{s_1} = -25.7°$, as shown in Figs. 6-20b and 6-21c. The maximum negative shear stress (point S_2 on the circle) has the same numerical value as the maximum positive stress (6,400 psi).

The normal stresses acting on the planes of maximum shear stress are equal to σ_{aver}, which is the abscissa of the center C of the circle (10,000 psi). These stresses are also shown in Fig. 6-21c. Note that the planes of maximum shear stress are oriented at 45° to the principal planes.

Example 6-6

At a point on the surface of a generator shaft the stresses are $\sigma_x = -50$ MPa, $\sigma_y = 10$ MPa, and $\tau_{xy} = -40$ MPa, as shown in Fig. 6-22a.

Using Mohr's circle, determine the following quantities: (a) the stresses acting on an element inclined at an angle $\theta = 45°$, (b) the principal stresses, and (c) the maximum shear stresses. (Consider only the in-plane stresses, and show all results on sketches of properly oriented elements.)

Solution

Construction of Mohr's circle. The axes for the normal and shear stresses are shown in Fig. 6-22b, with σ_{x_1} positive to the right and $\tau_{x_1y_1}$ positive downward. The center C of the circle is located on the σ_{x_1} axis at the point where the stress equals the average normal stress (Eq. 6-31a):

$$\sigma_{\text{aver}} = \frac{\sigma_x + \sigma_y}{2} = \frac{-50 \text{ MPa} + 10 \text{ MPa}}{2} = -20 \text{ MPa}$$

Point A, representing the stresses on the x face of the element ($\theta = 0$), has coordinates

$$\sigma_{x_1} = -50 \text{ MPa} \qquad \tau_{x_1y_1} = -40 \text{ MPa}$$

Similarly, the coordinates of point B, representing the stresses on the y face ($\theta = 90°$), are

$$\sigma_{x_1} = 10 \text{ MPa} \qquad \tau_{x_1y_1} = 40 \text{ MPa}$$

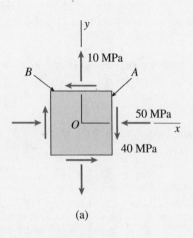

(a)

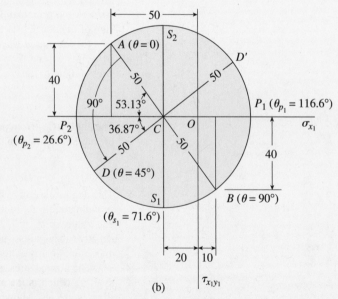

(b)

FIG. 6-22 Example 6-6. (a) Element in plane stress, and (b) the corresponding Mohr's circle. (*Note:* All stresses on the circle have units of MPa.)

The circle is now drawn through points A and B with center at C and radius R (from Eq. 6-31b) equal to

$$R = \sqrt{\left(\frac{\sigma_x - \sigma_y}{2}\right)^2 + \tau_{xy}^2}$$

$$= \sqrt{\left(\frac{-50 \text{ MPa} - 10 \text{ MPa}}{2}\right)^2 + (-40 \text{ MPa})^2} = 50 \text{ MPa}$$

(a) *Stresses on an element inclined at $\theta = 45°$.* The stresses acting on a plane oriented at an angle $\theta = 45°$ are given by the coordinates of point D, which is at an angle $2\theta = 90°$ from point A (Fig. 6-22b). To evaluate these coordinates, we need to know the angle between line CD and the negative σ_{x_1} axis (that is, angle DCP_2), which in turn requires that we know the angle between line CA and the negative σ_{x_1} axis (angle ACP_2). These angles are found from the geometry of the circle as follows:

$$\tan \overline{ACP_2} = \frac{40 \text{ MPa}}{30 \text{ MPa}} = \frac{4}{3} \qquad \overline{ACP_2} = 53.13°$$

$$\overline{DCP_2} = 90° - \overline{ACP_2} = 90° - 53.13° = 36.87°$$

Knowing these angles, we can obtain the coordinates of point D directly from Figure 6-23a:

(Point D) $\sigma_{x_1} = -20 \text{ MPa} - (50 \text{ MPa})(\cos 36.87°) = -60 \text{ MPa}$ $\Longleftarrow$

$\tau_{x_1 y_1} = (50 \text{ MPa})(\sin 36.87°) = 30 \text{ MPa}$ $\Longleftarrow$

In an analogous manner, we can find the stresses represented by point D', which corresponds to a plane inclined at an angle $\theta = 135°$ (or $2\theta = 270°$):

(Point D') $\sigma_{x_1} = -20 \text{ MPa} + (50 \text{ MPa})(\cos 36.87°) = 20 \text{ MPa}$ $\Longleftarrow$

$\tau_{x_1 y_1} = (-50 \text{ MPa})(\sin 36.87°) = -30 \text{ MPa}$ $\Longleftarrow$

These stresses are shown in Fig. 6-23a on a sketch of an element oriented at an angle $\theta = 45°$ (all stresses are shown in their true directions). Also, note that the sum of the normal stresses is equal to $\sigma_x + \sigma_y$, or -40 MPa.

(b) *Principal stresses.* The principal stresses are represented by points P_1 and P_2 on Mohr's circle. The algebraically larger principal stress (represented by point P_1) is

$$\sigma_1 = -20 \text{ MPa} + 50 \text{ MPa} = 30 \text{ MPa} \qquad \Longleftarrow$$

continued

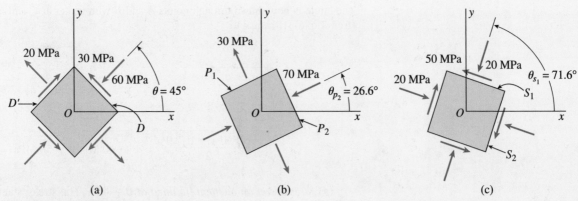

FIG. 6-23 Example 6-6 (continued). (a) Stresses acting on an element oriented at $\theta = 45°$, (b) principal stresses, and (c) maximum shear stresses

as seen by inspection of the circle. The angle $2\theta_{p_1}$ to point P_1 from point A is the angle ACP_1 measured counterclockwise on the circle, that is,

$$\overline{ACP_1} = 2\theta_{p_1} = 53.13° + 180° = 233.13° \qquad \theta_{p_1} = 116.6°$$

Thus, the plane of the algebraically larger principal stress is oriented at an angle $\theta_{p_1} = 116.6°$.

The algebraically smaller principal stress (point P_2) is obtained from the circle in a similar manner:

$$\sigma_2 = -20 \text{ MPa} - 50 \text{ MPa} = -70 \text{ MPa}$$

The angle $2\theta_{p_2}$ to point P_2 on the circle is $53.13°$; thus, the second principal plane is defined by the angle $\theta_{p_2} = 26.6°$.

The principal stresses and principal planes are shown in Fig. 6-23b, and again we note that the sum of the normal stresses is equal to $\sigma_x + \sigma_y$, or -40 MPa.

(c) *Maximum shear stresses.* The maximum positive and negative shear stresses are represented by points S_1 and S_2 on Mohr's circle (Fig. 6-22b). Their magnitudes, equal to the radius of the circle, are

$$\tau_{\max} = 50 \text{ MPa}$$

The angle ACS_1 from point A to point S_1 is $90° + 53.13° = 143.13°$, and therefore the angle $2\theta_{s_1}$ for point S_1 is

$$2\theta_{s_1} = 143.13°$$

The corresponding angle θ_{s_1} to the plane of the maximum positive shear stress is one-half that value, or $\theta_{s_1} = 71.6°$, as shown in Fig. 6-23c. The maximum negative shear stress (point S_2 on the circle) has the same numerical value as the positive stress (50 MPa).

The normal stresses acting on the planes of maximum shear stress are equal to σ_{aver}, which is the coordinate of the center C of the circle (-20 MPa). These stresses are also shown in Fig. 6-23c. Note that the planes of maximum shear stress are oriented at $45°$ to the principal planes.

6.5 HOOKE'S LAW FOR PLANE STRESS

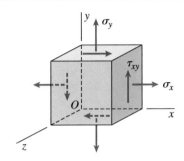

FIG. 6-24 Element of material in plane stress ($\sigma_z = 0$)

The stresses acting on inclined planes when the material is subjected to plane stress (Fig. 6-24) were discussed in Sections 6.2, 6.3, and 6.4. The stress-transformation equations derived in those discussions were obtained solely from equilibrium, and therefore the properties of the materials were not needed. Now, in this section, we will investigate the *strains* in the material, which means that the material properties must be considered. However, we will limit our discussion to materials that meet two important conditions: first, *the material is uniform throughout the body and has the same properties in all directions* (homogeneous and isotropic material), and second, *the material follows Hooke's law* (linearly elastic material). Under these conditions, we can readily obtain the relationships between the stresses and strains in the body.

Let us begin by considering the **normal strains** ϵ_x, ϵ_y, and ϵ_z in plane stress. The effects of these strains are pictured in Fig. 6-25, which shows the changes in dimensions of a small element having edges of lengths a, b, and c. All three strains are shown positive (elongation) in the figure. The strains can be expressed in terms of the stresses (Fig. 6-24) by superimposing the effects of the individual stresses.

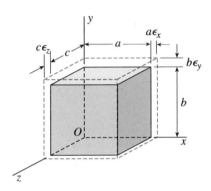

FIG. 6-25 Element of material subjected to normal strains ϵ_x, ϵ_y, and ϵ_z

For instance, the strain ϵ_x in the x direction due to the stress σ_x is equal to σ_x/E, where E is the modulus of elasticity. Also, the strain ϵ_x due to the stress σ_y is equal to $-\nu\sigma_y/E$, where ν is Poisson's ratio (see Section 1.5). Of course, the shear stress τ_{xy} produces no normal strains in the x, y, or z directions. Thus, the resultant strain in the x direction is

$$\epsilon_x = \frac{1}{E}(\sigma_x - \nu\sigma_y) \tag{6-34a}$$

In a similar manner, we obtain the strains in the y and z directions:

$$\epsilon_y = \frac{1}{E}(\sigma_y - \nu\sigma_x) \qquad \epsilon_z = -\frac{\nu}{E}(\sigma_x + \sigma_y) \tag{6-34b,c}$$

These equations may be used to find the normal strains (in plane stress) when the stresses are known.

The shear stress τ_{xy} (Fig. 6-24) causes a distortion of the element such that each z face becomes a rhombus (Fig. 6-26). The **shear strain** γ_{xy} is the decrease in angle between the x and y faces of the element and is related to the shear stress by Hooke's law in shear, as follows:

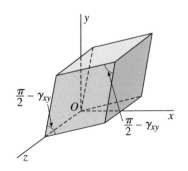

FIG. 6-26 Shear strain γ_{xy}

$$\gamma_{xy} = \frac{\tau_{xy}}{G} \tag{6-35}$$

where G is the shear modulus of elasticity. Note that the normal stresses σ_x and σ_y have no effect on the shear strain γ_{xy}. Consequently, Eqs. (6-34) and (6-35) give the strains (in plane stress) when all stresses (σ_x, σ_y, and τ_{xy}) act simultaneously.

The first two equations (Eqs. 6-34a and 6-34b) give the strains ϵ_x and ϵ_y in terms of the stresses. These equations can be solved simultaneously for the stresses in terms of the strains:

$$\sigma_x = \frac{E}{1 - \nu^2}(\epsilon_x + \nu\epsilon_y) \qquad \sigma_y = \frac{E}{1 - \nu^2}(\epsilon_y + \nu\epsilon_x) \qquad \text{(6-36a,b)}$$

In addition, we have the following equation for the shear stress in terms of the shear strain:

$$\tau_{xy} = G\gamma_{xy} \qquad \text{(6-37)}$$

Equations (6-36) and (6-37) may be used to find the stresses (in plane stress) when the strains are known. Of course, the normal stress σ_z in the z direction is equal to zero.

Equations (6-34) through (6-37) are known collectively as **Hooke's law for plane stress**. They contain three material constants (E, G, and ν), but only two are independent because of the relationship

$$G = \frac{E}{2(1 + \nu)} \qquad \text{(6-38)}$$

which was derived previously in Section 3.6.

Special Cases of Hooke's Law

In the special case of **biaxial stress** (Fig. 6-11b), we have $\tau_{xy} = 0$, and therefore Hooke's law for plane stress simplifies to

$$\epsilon_x = \frac{1}{E}(\sigma_x - \nu\sigma_y) \qquad \epsilon_y = \frac{1}{E}(\sigma_y - \nu\sigma_x)$$

$$\epsilon_z = -\frac{\nu}{E}(\sigma_x + \sigma_y) \qquad \text{(6-39a,b,c)}$$

$$\sigma_x = \frac{E}{1 - \nu^2}(\epsilon_x + \nu\epsilon_y) \qquad \sigma_y = \frac{E}{1 - \nu^2}(\epsilon_y + \nu\epsilon_x) \qquad \text{(6-40a,b)}$$

These equations are the same as Eqs. (6-34) and (6-36) because the effects of normal and shear stresses are independent of each other.

For **uniaxial stress**, with $\sigma_y = 0$ (Fig. 6-11a), the equations of Hooke's law simplify even further:

$$\epsilon_x = \frac{\sigma_x}{E} \qquad \epsilon_y = \epsilon_z = -\frac{\nu\sigma_x}{E} \qquad \sigma_x = E\epsilon_x \qquad \text{(6-41a,b,c)}$$

Finally, we consider **pure shear** (Fig. 6-12a), which means that $\sigma_x = \sigma_y = 0$. Then we obtain

$$\epsilon_x = \epsilon_y = \epsilon_z = 0 \qquad \gamma_{xy} = \frac{\tau_{xy}}{G} \qquad \text{(6-42a,b)}$$

In all three of these special cases, the normal stress σ_z is equal to zero.

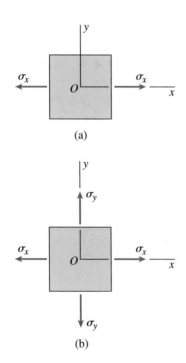

(a)

(b)

FIG. 6-11 (Repeated)

Volume Change

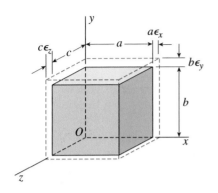

FIG. 6-25 (Repeated)

When a solid object undergoes strains, both its dimensions and its volume will change. The change in volume can be determined if the normal strains in three perpendicular directions are known. To show how this is accomplished, let us again consider the small element of material shown in Fig. 6-25. The original element is a rectangular parallelepiped having sides of lengths a, b, and c in the x, y, and z directions, respectively. The strains ϵ_x, ϵ_y, and ϵ_z produce the changes in dimensions shown by the dashed lines. Thus, the increases in the lengths of the sides are $a\epsilon_x$, $b\epsilon_y$, and $c\epsilon_z$.

The original volume of the element is

$$V_0 = abc \tag{a}$$

and its final volume is

$$
\begin{aligned}
V_1 &= (a + a\epsilon_x)(b + b\epsilon_y)(c + c\epsilon_z) \\
&= abc(1 + \epsilon_x)(1 + \epsilon_y)(1 + \epsilon_z)
\end{aligned}
\tag{b}
$$

By referring to Eq. (a), we can express the final volume of the element (Eq. b) in the form

$$V_1 = V_0(1 + \epsilon_x)(1 + \epsilon_y)(1 + \epsilon_z) \tag{6-43a}$$

Upon expanding the terms on the right-hand side, we obtain the following equivalent expression:

$$V_1 = V_0(1 + \epsilon_x + \epsilon_y + \epsilon_z + \epsilon_x\epsilon_y + \epsilon_x\epsilon_z + \epsilon_y\epsilon_z + \epsilon_x\epsilon_y\epsilon_z) \tag{6-43b}$$

The preceding equations for V_1 are valid for both large and small strains.

If we now limit our discussion to structures having only very small strains (as is usually the case), we can disregard the terms in Eq. (6-43b) that consist of products of small strains. Such products are themselves small in comparison to the individual strains ϵ_x, ϵ_y, and ϵ_z. Then the expression for the final volume simplifies to

$$V_1 = V_0(1 + \epsilon_x + \epsilon_y + \epsilon_z) \tag{6-44}$$

and the **volume change** is

$$\Delta V = V_1 - V_0 = V_0(\epsilon_x + \epsilon_y + \epsilon_z) \tag{6-45}$$

This expression can be used for any volume of material *provided the strains are small and remain constant throughout the volume.* Note also that the material does not have to follow Hooke's law. Furthermore, the expression is not limited to plane stress, but is valid for any stress conditions. (As a final note, we should mention that shear strains produce no change in volume.)

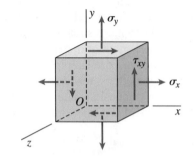

FIG. 6-24 (Repeated)

The **unit volume change** e, also known as the **dilatation**, is defined as the change in volume divided by the original volume; thus,

$$e = \frac{\Delta V}{V_0} = \epsilon_x + \epsilon_y + \epsilon_z \qquad (6\text{-}46)$$

By applying this equation to a differential element of volume and then integrating, we can obtain the change in volume of a body even when the normal strains vary throughout the body.

The preceding equations for volume changes apply to both tensile and compressive strains, inasmuch as the strains ϵ_x, ϵ_y, and ϵ_z are algebraic quantities (positive for elongation and negative for shortening). With this sign convention, positive values for ΔV and e represent increases in volume, and negative values represent decreases.

Let us now return to materials that follow **Hooke's law** and are subjected only to **plane stress** (Fig. 6-24). In this case the strains ϵ_x, ϵ_y, and ϵ_z are given by Eqs. (6-34a, b, and c). Substituting those relationships into Eq. (6-46), we obtain the following expression for the unit volume change in terms of stresses:

$$e = \frac{\Delta V}{V_0} = \frac{1 - 2\nu}{E}(\sigma_x + \sigma_y) \qquad (6\text{-}47)$$

Note that this equation also applies to **biaxial stress**.

In the case of a prismatic bar in tension, that is, **uniaxial stress**, Eq. (6-47) simplifies to

$$e = \frac{\Delta V}{V_0} = \frac{\sigma_x}{E}(1 - 2\nu) \qquad (6\text{-}48)$$

From this equation we see that the maximum possible value of Poisson's ratio for common materials is 0.5, because a larger value means that the volume decreases when the material is in tension, which is contrary to ordinary physical behavior.

6.6 TRIAXIAL STRESS

An element of material subjected to normal stresses σ_x, σ_y, and σ_z acting in three mutually perpendicular directions is said to be in a state of **triaxial stress** (Fig. 6-27a). Since there are no shear stresses on the x, y, and z faces, the stresses σ_x, σ_y, and σ_z are the *principal stresses* in the material.

If an inclined plane parallel to the z axis is cut through the element (Fig. 6-27b), the only stresses on the inclined face are the normal stress σ and shear stress τ, both of which act parallel to the xy plane. These stresses are analogous to the stresses σ_{x_1} and $\tau_{x_1 y_1}$ encountered in our earlier discussions of plane stress (see, for instance, Fig. 6-2a).

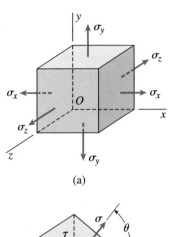

(a)

(b)

FIG. 6-27 Element in triaxial stress

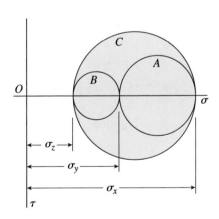

FIG. 6-28 Mohr's circles for an element in triaxial stress

Because the stresses σ and τ (Fig. 6-27b) are found from equations of force equilibrium in the xy plane, they are independent of the normal stress σ_z. Therefore, we can use the transformation equations of plane stress, as well as Mohr's circle for plane stress, when determining the stresses σ and τ in triaxial stress. The same general conclusion holds for the normal and shear stresses acting on inclined planes cut through the element parallel to the x and y axes.

Maximum Shear Stresses

From our previous discussions of plane stress, we know that the maximum shear stresses occur on planes oriented at 45° to the principal planes. Therefore, for a material in triaxial stress (Fig. 6-27a), the maximum shear stresses occur on elements oriented at angles of 45° to the x, y, and z axes. For example, consider an element obtained by a 45° rotation about the z axis. The maximum positive and negative shear stresses acting on this element are

$$(\tau_{\max})_z = \pm \frac{\sigma_x - \sigma_y}{2} \qquad (6\text{-}49a)$$

Similarly, by rotating about the x and y axes through angles of 45°, we obtain the following maximum shear stresses:

$$(\tau_{\max})_x = \pm \frac{\sigma_y - \sigma_z}{2} \qquad (\tau_{\max})_y = \pm \frac{\sigma_x - \sigma_z}{2} \qquad (6\text{-}49b,c)$$

The absolute maximum shear stress is the numerically largest of the stresses determined from Eqs. (6-49a, b, and c). It is equal to one-half the difference between the algebraically largest and algebraically smallest of the three principal stresses.

The stresses acting on elements oriented at various angles to the x, y, and z axes can be visualized with the aid of **Mohr's circles**. For elements oriented by rotations about the z axis, the corresponding circle is labeled A in Fig. 6-28. Note that this circle is drawn for the case in which $\sigma_x > \sigma_y$ and both σ_x and σ_y are tensile stresses.

In a similar manner, we can construct circles B and C for elements oriented by rotations about the x and y axes, respectively. The radii of the circles represent the maximum shear stresses given by Eqs. (6-49a, b, and c), and the absolute maximum shear stress is equal to the radius of the largest circle. The normal stresses acting on the planes of maximum shear stresses have magnitudes given by the abscissas of the centers of the respective circles.

In the preceding discussion of triaxial stress we only considered stresses acting on planes obtained by rotating about the x, y, and z axes. Thus, every plane we considered is parallel to one of the axes. For instance, the inclined plane of Fig. 6-27b is parallel to the z axis, and its normal is parallel to the xy plane. Of course, we can also cut through the element in **skew directions**, so that the resulting inclined planes are skew to all three coordinate axes. The normal and shear stresses acting on such planes can be obtained by a more complicated three-dimensional analysis. However, the normal stresses acting on skew planes are intermediate in value between the algebraically maximum and minimum principal stresses, and the shear stresses on those planes are smaller (in absolute value) than the absolute maximum shear stress obtained from Eqs. (6-49a, b, and c).

Hooke's Law for Triaxial Stress

If the material follows Hooke's law, we can obtain the relationships between the normal stresses and normal strains by using the same procedure as for plane stress (see Section 6.5). The strains produced by the stresses σ_x, σ_y, and σ_z acting independently are superimposed to obtain the resultant strains. Thus, we readily arrive at the following equations for the **strains in triaxial stress**:

$$\epsilon_x = \frac{\sigma_x}{E} - \frac{\nu}{E}(\sigma_y + \sigma_z) \tag{6-50a}$$

$$\epsilon_y = \frac{\sigma_y}{E} - \frac{\nu}{E}(\sigma_z + \sigma_x) \tag{6-50b}$$

$$\epsilon_z = \frac{\sigma_z}{E} - \frac{\nu}{E}(\sigma_x + \sigma_y) \tag{6-50c}$$

In these equations, the standard sign conventions are used; that is, tensile stress σ and extensional strain ϵ are positive.

The preceding equations can be solved simultaneously for the **stresses in terms of the strains**:

$$\sigma_x = \frac{E}{(1 + \nu)(1 - 2\nu)}\left[(1 - \nu)\epsilon_x + \nu(\epsilon_y + \epsilon_z)\right] \tag{6-51a}$$

$$\sigma_y = \frac{E}{(1 + \nu)(1 - 2\nu)}\left[(1 - \nu)\epsilon_y + \nu(\epsilon_z + \epsilon_x)\right] \tag{6-51b}$$

$$\sigma_z = \frac{E}{(1 + \nu)(1 - 2\nu)}\left[(1 - \nu)\epsilon_z + \nu(\epsilon_x + \epsilon_y)\right] \tag{6-51c}$$

Equations (6-50) and (6-51) represent **Hooke's law for triaxial stress**.

In the special case of **biaxial stress** (Fig. 6-11b), we can obtain the equations of Hooke's law by substituting $\sigma_z = 0$ into the preceding equations. The resulting equations reduce to Eqs. (6-39) and (6-40) of Section 6.5.

Unit Volume Change

The unit volume change (or *dilatation*) for an element in triaxial stress is obtained in the same manner as for plane stress (see Section 6.5). If the element is subjected to strains ϵ_x, ϵ_y, and ϵ_z, we may use Eq. (6-46) for the unit volume change:

$$e = \epsilon_x + \epsilon_y + \epsilon_z \tag{6-52}$$

This equation is valid for any material provided the strains are small.

If Hooke's law holds for the material, we can substitute for the strains ϵ_x, ϵ_y, and ϵ_z from Eqs. (6-50a, b, and c) and obtain

$$e = \frac{1 - 2\nu}{E}(\sigma_x + \sigma_y + \sigma_z) \tag{6-53}$$

Equations (6-52) and (6-53) give the unit volume change in triaxial stress in terms of the strains and stresses, respectively.

Spherical Stress

A special type of triaxial stress, called **spherical stress**, occurs whenever all three normal stresses are equal (Fig. 6-29):

$$\sigma_x = \sigma_y = \sigma_z = \sigma_0 \tag{6-54}$$

Under these stress conditions, *any* plane cut through the element will be subjected to the same normal stress σ_0 and will be free of shear stress. Thus, we have equal normal stresses in every direction and no shear stresses anywhere in the material. Every plane is a principal plane, and the three Mohr's circles shown in Fig. 6-28 reduce to a single point.

The normal strains in spherical stress are also the same in all directions, provided the material is homogeneous and isotropic. If Hooke's law applies, the normal strains are

$$\epsilon_0 = \frac{\sigma_0}{E}(1 - 2\nu) \tag{6-55}$$

as obtained from Eqs. (6-50a, b, and c).

Since there are no shear strains, an element in the shape of a cube changes in size but remains a cube. In general, any body subjected to spherical stress will maintain its relative proportions but will expand or contract in volume depending upon whether σ_0 is tensile or compressive.

The expression for the unit volume change can be obtained from Eq. (6-52) by substituting for the strains from Eq. (6-55). The result is

$$e = 3\epsilon_0 = \frac{3\sigma_0(1 - 2\nu)}{E} \tag{6-56}$$

Equation (6-56) is usually expressed in more compact form by introducing a new quantity K called the **volume modulus of elasticity**, or **bulk modulus of elasticity**, which is defined as follows:

$$K = \frac{E}{3(1 - 2\nu)} \tag{6-57}$$

With this notation, the expression for the unit volume change becomes

$$e = \frac{\sigma_0}{K} \tag{6-58}$$

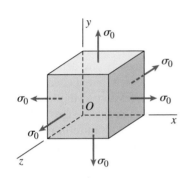

FIG. 6-29 Element in spherical stress

and the volume modulus is

$$K = \frac{\sigma_0}{e} \qquad (6\text{-}59)$$

Thus, the volume modulus can be defined as the ratio of the spherical stress to the volumetric strain, which is analogous to the definition of the modulus E in uniaxial stress. Note that the preceding formulas for e and K are based upon the assumptions that *the strains are small and Hooke's law holds for the material*.

From Eq. (6-57) for K, we see that if Poisson's ratio ν equals 1/3, the moduli K and E are numerically equal. If $\nu = 0$, then K has the value $E/3$, and if $\nu = 0.5$, K becomes infinite, which corresponds to a rigid material having no change in volume (that is, the material is incompressible).

The preceding formulas for spherical stress were derived for an element subjected to uniform tension in all directions, but of course the formulas also apply to an element in uniform compression. In the case of uniform compression, the stresses and strains have negative signs. Uniform compression occurs when the material is subjected to uniform pressure in all directions; for example, an object submerged in water or rock deep within the earth. This state of stress is often called **hydrostatic stress**.

Although uniform compression is relatively common, a state of uniform tension is difficult to achieve. It can be realized by suddenly and uniformly heating the outer surface of a solid metal sphere, so that the outer layers are at a higher temperature than the interior. The tendency of the outer layers to expand produces uniform tension in all directions at the center of the sphere.

CHAPTER SUMMARY & REVIEW

In Chapter 6, we investigated the **state of stress** at a point on a stressed body and then displayed it on a stress element. In two dimensions, **plane stress** was discussed and we derived transformation equations that gave different, but equivalent, expressions of the state of normal and shear stresses at that point. **Principal normal stresses** and **maximum shear stress**, and their orientations, were seen to be the most important information for design. A graphical representation of the transformation equations, **Mohr's circle**, was found to be a convenient way of exploring various representations of the state of stress at a point, including those orientations of the stress element at which principal stresses and maximum shear stress occur. Later, strains were introduced and **Hooke's law for plane stress** was derived (for homogeneous and isotropic materials) and then specialized to obtain stress-strain relationships for biaxial stress, uniaxial stress, and pure shear. The stress state in three dimensions, referred to as triaxial stress, was then introduced along with Hooke's law for triaxial stress. **Spherical stress** and **hydrostatic stress** were defined as special cases of triaxial stress. The major concepts presented in this chapter may be summarized as follows:

1. The stresses on inclined sections cut through a body, such as a beam, may be larger than the stresses acting on a stress element aligned with the cross section.

2. Stresses are tensors, not vectors, so we used equilibrium of a wedge element to transform the stress components from one set of axes to another. Since the transformation equations were derived solely from equilibrium of an element, they are applicable to stresses in any kind of material, whether linear, nonlinear, elastic, or inelastic. The **transformation equations for plane stress** are:

$$\sigma_{x_1} = \frac{\sigma_x + \sigma_y}{2} + \frac{\sigma_x - \sigma_y}{2}\cos 2\theta + \tau_{xy}\sin 2\theta$$

$$\tau_{x_1 y_1} = -\frac{\sigma_x - \sigma_y}{2}\sin 2\theta + \tau_{xy}\cos 2\theta$$

$$\sigma_{y_1} = \frac{\sigma_x + \sigma_y}{2} - \frac{\sigma_x - \sigma_y}{2}\cos 2\theta - \tau_{xy}\sin 2\theta$$

3. If we use two elements with different orientations to display the state of **plane stress** at the same point in a body, the stresses acting on the faces of the two elements are different, but they still represent the same intrinsic state of stress at that point.

4. From equilibrium, we showed that the shear stresses acting on all four side faces of a stress element in plane stress are known if we determine the shear stress acting on any one of those faces.

5. The sum of the normal stresses acting on perpendicular faces of plane-stress elements (at a given point in a stressed body) is constant and independent of the angle θ:

$$\sigma_{x_1} + \sigma_{y_1} = \sigma_x + \sigma_y$$

6. The maximum and minimum normal stresses (called the **principal stresses** σ_1, σ_2) can be found from the transformation equation for normal stress as follows:

$$\sigma_{1,2} = \frac{\sigma_x + \sigma_y}{2} \pm \sqrt{\left(\frac{\sigma_x - \sigma_y}{2}\right)^2 + \tau_{xy}^2}$$

We also can find the principal planes, at orientation θ_p, on which they act. The shear stresses are zero on the principal planes, the planes of maximum shear stress occur at 45° to the principal planes, and the maximum shear stress is equal to one-half the difference of the principal stresses. Maximum shear stress can be computed from the normal and shear stresses on the original element, or from the principal stresses as follows:

$$\tau_{max} = \sqrt{\left(\frac{\sigma_x - \sigma_y}{2}\right)^2 + \tau_{xy}^2}$$

$$\tau_{max} = \frac{\sigma_1 - \sigma_2}{2}$$

7. The transformation equations for plane stress can be represented in graphical form by a plot known as **Mohr's circle** which displays the relationship between normal and shear stresses acting on various inclined planes at a point in a stressed body. It also is used for calculating principal stresses, maximum shear stresses, and the orientations of the elements on which they act.

8. **Hooke's law for plane stress** provides the relationships between normal strains and stresses for homogeneous and isotropic materials which follow Hooke's law. These relationships contain three material constants *(E, G,* and *v)*. When the normal stresses in plane stress are known, the normal strains in the *x, y* and *z* directions are:

$$\epsilon_x = \frac{1}{E}(\sigma_x - v\sigma_y)$$

$$\epsilon_y = \frac{1}{E}(\sigma_y - v\sigma_x)$$

$$\epsilon_z = -\frac{v}{E}(\sigma_x + \sigma_y)$$

These equations can be solved simultaneously to give the *x* and *y* normal stresses in terms of the strains:

$$\sigma_x = \frac{E}{1 - v^2}(\epsilon_x + v\epsilon_y)$$

$$\sigma_y = \frac{E}{1 - v^2}(\epsilon_y + v\epsilon_x)$$

9. The **unit volume change e**, or the **dilatation** of a solid body, is defined as the change in volume divided by the original volume and is equal to the sum of the normal strains in three perpendicular directions:

$$e = \frac{\Delta V}{V_0} = \epsilon_x + \epsilon_y + \epsilon_z$$

10. A state of **triaxial stress** exists in an element if it is subjected to normal stresses in three mutually perpendicular directions and there are no shear stresses on the faces of the element; the stresses are seen to be the principal stresses in the material. A special type of triaxial stress (called **spherical stress**) occurs when all three normal stresses are equal and tensile. If all three stresses are equal and compressive, the triaxial stress state is referred to as **hydrostatic stress**.

PROBLEMS CHAPTER 6

Plane Stress

6.2-1 An element in *plane stress* is subjected to stresses $\sigma_x = 4750$ psi, $\sigma_y = 1200$ psi, and $\tau_{xy} = 950$ psi, as shown in the figure.

Determine the stresses acting on an element oriented at an angle $\theta = 60°$ from the x axis, where the angle θ is positive when counterclockwise. Show these stresses on a sketch of an element oriented at the angle θ.

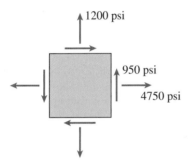

PROB. 6.2-1

6.2-2 Solve the preceding problem for an element in *plane stress* subjected to stresses $\sigma_x = 100$ MPa, $\sigma_y = 80$ MPa, and $\tau_{xy} = 28$ MPa, as shown in the figure.

Determine the stresses acting on an element oriented at an angle $\theta = 30°$ from the x axis, where the angle θ is positive when counterclockwise. Show these stresses on a sketch of an element oriented at the angle θ.

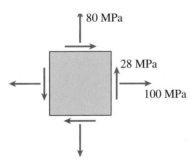

PROB. 6.2-2

6.2-3 Solve Problem 6.2-1 for an element in *plane stress* subjected to stresses $\sigma_x = -5700$ psi, $\sigma_y = -2300$ psi, and $\tau_{xy} = 2500$ psi, as shown in the figure.

Determine the stresses acting on an element oriented at an angle $\theta = 50°$ from the x axis, where the angle θ is positive when counterclockwise. Show these stresses on a sketch of an element oriented at the angle θ.

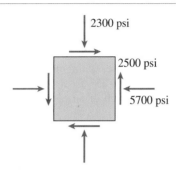

PROB. 6.2-3

6.2-4 The stresses acting on element A in the web of a train rail are found to be 40 MPa tension in the horizontal direction and 160 MPa compression in the vertical direction (see figure). Also, shear stresses of magnitude 54 MPa act in the directions shown.

Determine the stresses acting on an element oriented at a counterclockwise angle of 52° from the horizontal. Show these stresses on a sketch of an element oriented at this angle.

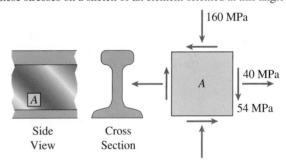

PROB. 6.2-4

6.2-5 Solve the preceding problem if the normal and shear stresses acting on element A are 6500 psi, 18,500 psi, and 3800 psi (in the directions shown in the figure).

Determine the stresses acting on an element oriented at a counterclockwise angle of 30° from the horizontal. Show these stresses on a sketch of an element oriented at this angle.

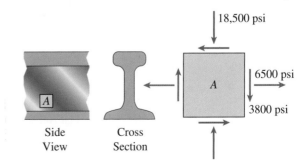

PROB. 6.2-5

6.2-6 An element in *plane stress* from the fuselage of an airplane is subjected to compressive stresses of magnitude 27 MPa in the horizontal direction and tensile stresses of magnitude 5.5 MPa in the vertical direction (see figure). Also, shear stresses of magnitude 10.5 MPa act in the directions shown.

Determine the stresses acting on an element oriented at a clockwise angle of 35° from the horizontal. Show these stresses on a sketch of an element oriented at this angle.

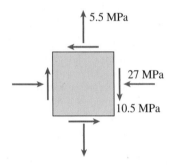

PROB. 6.2-6

6.2-7 The stresses acting on element B in the web of a wide-flange beam are found to be 14,000 psi compression in the horizontal direction and 2600 psi compression in the vertical direction (see figure). Also, shear stresses of magnitude 3800 psi act in the directions shown.

Determine the stresses acting on an element oriented at a counterclockwise angle of 40° from the horizontal. Show these stresses on a sketch of an element oriented at this angle.

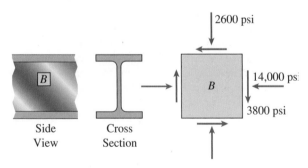

PROB. 6.2-7

6.2-8 Solve the preceding problem if the normal and shear stresses acting on element B are 46 MPa, 13 MPa, and 21 MPa (in the directions shown in the figure) and the angle is 42.5° (clockwise).

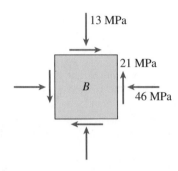

PROB. 6.2-8

6.2-9 The polyethylene liner of a settling pond is subjected to stresses $\sigma_x = 350$ psi, $\sigma_y = 112$ psi, and $\tau_{xy} = -120$ psi, as shown by the plane-stress element in the first part of the figure.

Determine the normal and shear stresses acting on a seam oriented at an angle of 30° to the element, as shown in the second part of the figure. Show these stresses on a sketch of an element having its sides parallel and perpendicular to the seam.

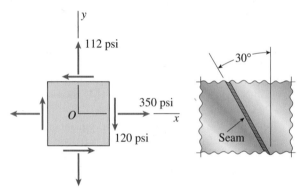

PROB. 6.2-9

6.2-10 Solve the preceding problem if the normal and shear stresses acting on the element are $\sigma_x = 2100$ kPa, $\sigma_y = 300$ kPa, and $\tau_{xy} = -560$ kPa, and the seam is oriented at an angle of 22.5° to the element.

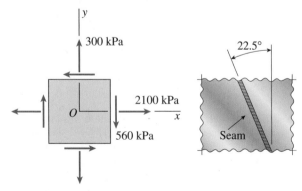

PROB. 6.2-10

6.2-11 A rectangular plate of dimensions 3.0 in. × 5.0 in. is formed by welding two triangular plates (see figure). The plate is subjected to a tensile stress of 500 psi in the long direction and a compressive stress of 350 psi in the short direction.

Determine the normal stress σ_w acting perpendicular to the line of the weld and the shear stress τ_w acting parallel to the weld. (Assume that the normal stress σ_w is positive when it acts in tension against the weld and the shear stress τ_w is positive when it acts counterclockwise against the weld.)

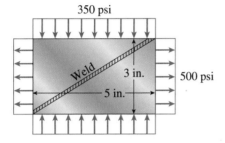

PROB. 6.2-11

6.2-12 Solve the preceding problem for a plate of dimensions 100 mm × 250 mm subjected to a compressive stress of 2.5 MPa in the long direction and a tensile stress of 12.0 MPa in the short direction (see figure).

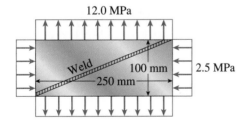

PROB. 6.2-12

6.2-13 At a point on the surface of a machine the material is in *biaxial stress* with $\sigma_x = 3600$ psi and $\sigma_y = -1600$ psi, as shown in the first part of the figure. The second part of the figure shows an inclined plane *aa* cut through the same point in the material but oriented at an angle θ.

Determine the value of the angle θ between zero and 90° such that no normal stress acts on plane *aa*. Sketch a stress element having plane *aa* as one of its sides and show all stresses acting on the element.

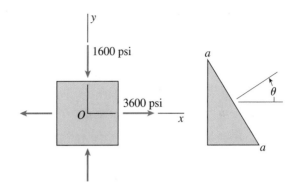

PROB. 6.2-13

6.2-14 Solve the preceding problem for $\sigma_x = 32$ MPa and $\sigma_y = -50$ MPa (see figure).

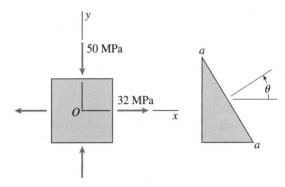

PROB. 6.2-14

6.2-15 An element in *plane stress* from the frame of a racing car is oriented at a known angle θ (see figure). On this inclined element, the normal and shear stresses have the magnitudes and directions shown in the figure.

Determine the normal and shear stresses acting on an element whose sides are parallel to the *xy* axes, that is, determine σ_x, σ_y, and τ_{xy}. Show the results on a sketch of an element oriented at $\theta = 0°$.

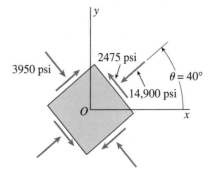

PROB. 6.2-15

6.2-16 Solve the preceding problem for the element shown in the figure.

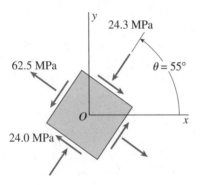

PROB. 6.2-16

6.2-17 A plate in *plane stress* is subjected to normal stresses σ_x and σ_y and *shear stress* τ_{xy}, as shown in the figure. At counterclockwise angles $\theta = 35°$ and $\theta = 75°$ from the x axis, the normal stress is 4800 psi tension.

If the stress σ_x equals 2200 psi tension, what are the stresses σ_y and τ_{xy}?

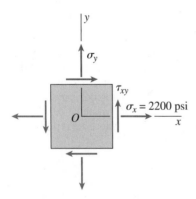

PROB. 6.2-17

6.2-18 The surface of an airplane wing is subjected to *plane stress* with normal stresses σ_x and σ_y and shear stress τ_{xy}, as shown in the figure. At a counterclockwise angle $\theta = 32°$ from the x axis, the normal stress is 37 MPa tension, and at an angle $\theta = 48°$, it is 12 MPa compression.

If the stress σ_x equals 110 MPa tension, what are the stresses σ_y and τ_{xy}?

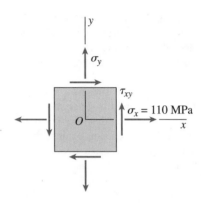

PROB. 6.2-18

6.2-19 At a point in a structure subjected to *plane stress*, the stresses are $\sigma_x = -4100$ psi, $\sigma_y = 2200$ psi, and $\tau_{xy} = 2900$ psi (the sign convention for these stresses is shown in Fig. 6-1). A stress element located at the same point in the structure (but oriented at a counterclockwise angle θ_1 with respect to the x axis) is subjected to the stresses shown in the figure (σ_b, τ_b, and 1800 psi).

Assuming that the angle θ_1 is between zero and 90°, calculate the normal stress σ_b, the shear stress τ_b, and the angle θ_1

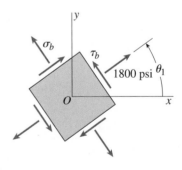

PROB. 6.2-19

Principal Stresses and Maximum Shear Stresses

When solving the problems for Section 6.3, consider only the in-plane stresses (the stresses in the xy plane).

6.3-1 An element in plane stress is subjected to stresses $\sigma_x = 4750$ psi, $\sigma_y = 1200$ psi, and $\tau_{xy} = 950$ psi (see the figure for Problem 6.2-1).

Determine the principal stresses and show them on a sketch of a properly oriented element.

6.3-2 An element in *plane stress* is subjected to stresses $\sigma_x = 100$ MPa, $\sigma_y = 80$ MPa, and $\tau_{xy} = 28$ MPa (see the figure for Problem 6.2-2).

Determine the principal stresses and show them on a sketch of a properly oriented element.

6.3-3 An element in *plane stress* is subjected to stresses $\sigma_x = -5700$ psi, $\sigma_y = -2300$ psi, and $\tau_{xy} = 2500$ psi (see the figure for Problem 6.2-3).

Determine the principal stresses and show them on a sketch of a properly oriented element.

6.3-4 The stresses acting on element A in the web of a train rail are found to be 40 MPa tension in the horizontal direction and 160 MPa compression in the vertical direction (see figure). Also, shear stresses of magnitude 54 MPa act in the directions shown (see the figure for Problem 6.2-4).

Determine the principal stresses and show them on a sketch of a properly oriented element.

6.3-5 The normal and shear stresses acting on element A are 6500 psi, 18,500 psi, and 3800 psi (in the directions shown in the figure) (see the figure for Problem 6.2-5).

Determine the maximum shear stresses and associated normal stresses and show them on a sketch of a properly oriented element.

6.3-6 An element in *plane stress* from the fuselage of an airplane is subjected to compressive stresses of magnitude 27 MPa in the horizontal direction and tensile stresses of magnitude 5.5 MPa in the vertical direction. Also, shear stresses of magnitude 10.5 MPa act in the directions shown (see the figure for Problem 6.2-6).

Determine the maximum shear stresses and associated normal stresses and show them on a sketch of a properly oriented element.

6.3-7 The stresses acting on element B in the web of a wide-flange beam are found to be 14,000 psi compression in the horizontal direction and 2600 psi compression in the vertical direction. Also, shear stresses of magnitude 3800 psi act in the directions shown (see the figure for Problem 6.2-7).

Determine the maximum shear stresses and associated normal stresses and show them on a sketch of a properly oriented element.

6.3-8 The normal and shear stresses acting on element B are $\sigma_x = -46$ MPa, $\sigma_y = -13$ MPa, and $\tau_{xy} = 21$ MPa (see figure for Problem 6.2-8).

Determine the maximum shear stresses and associated normal stresses and show them on a sketch of a properly oriented element.

6.3-9 A shear wall in a reinforced concrete building is subjected to a vertical uniform load of intensity q and a horizontal force H, as shown in the first part of the figure.

(The force H represents the effects of wind and earthquake loads.) As a consequence of these loads, the stresses at point A on the surface of the wall have the values shown in the second part of the figure (compressive stress equal to 1100 psi and shear stress equal to 480 psi).

(a) Determine the principal stresses and show them on a sketch of a properly oriented element.

(b) Determine the maximum shear stresses and associated normal stresses and show them on a sketch of a properly oriented element.

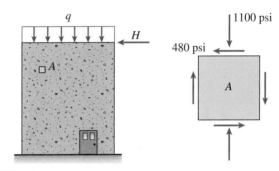

PROB. 6.3-9

6.3-10 A propeller shaft subjected to combined torsion and axial thrust is designed to resist a shear stress of 56 MPa and a compressive stress of 85 MPa (see figure).

(a) Determine the principal stresses and show them on a sketch of a properly oriented element.

(b) Determine the maximum shear stresses and associated normal stresses and show them on a sketch of a properly oriented element.

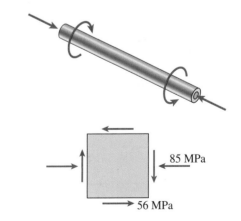

PROB. 6.3-10

6.3-11 through 6.3-16 An element in *plane stress* (see figure) is subjected to stresses σ_x, σ_y, and τ_{xy}.

(a) Determine the principal stresses and show them on a sketch of a properly oriented element.

(b) Determine the maximum shear stresses and associated normal stresses and show them on a sketch of a properly oriented element.

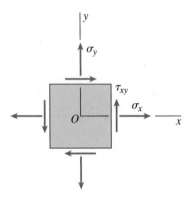

PROBS. 6.3-11 through 6.3-16

6.3-11 $\sigma_x = 2500$ psi, $\sigma_y = 1020$ psi, $\tau_{xy} = -900$ psi

6.3-12 $\sigma_x = 2150$ kPa, $\sigma_y = 375$ kPa, $\tau_{xy} = -460$ kPa

6.3-13 $\sigma_x = 14{,}500$ psi, $\sigma_y = 1070$ psi, $\tau_{xy} = 1900$ psi

6.3-14 $\sigma_x = 16.5$ MPa, $\sigma_y = -91$ MPa, $\tau_{xy} = -39$ MPa

6.3-15 $\sigma_x = -3300$ psi, $\sigma_y = -11{,}000$ psi, $\tau_{xy} = 4500$ psi

6.3-16 $\sigma_x = -108$ MPa, $\sigma_y = 58$ MPa, $\tau_{xy} = -58$ MPa

6.3-17 At a point on the surface of a machine component, the stresses acting on the x face of a stress element are $\sigma_x = 5900$ psi and $\tau_{xy} = 1950$ psi (see figure).

What is the allowable range of values for the stress σ_y if the maximum shear stress is limited to $\tau_0 = 2500$ psi?

6.3-18 At a point on the surface of a machine component the stresses acting on the x face of a stress element are $\sigma_x = 42$ MPa and $\tau_{xy} = 33$ MPa (see figure).

What is the allowable range of values for the stress σ_y if the maximum shear stress is limited to $\tau_0 = 35$ MPa?

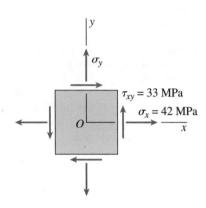

PROB. 6.3-18

6.3-19 An element in *plane stress* is subjected to stresses $\sigma_x = 5700$ psi and $\tau_{xy} = -2300$ psi (see figure). It is known that one of the principal stresses equals 6700 psi in tension.

(a) Determine the stress σ_y.

(b) Determine the other principal stress and the orientation of the principal planes, then show the principal stresses on a sketch of a properly oriented element.

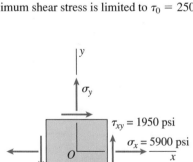

PROB. 6.3-17

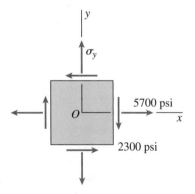

PROB. 6.3-19

6.3-20 An element in *plane stress* is subjected to stresses $\sigma_x = -50$ MPa and $\tau_{xy} = 42$ MPa (see figure). It is known that one of the principal stresses equals 33 MPa in tension.

(a) Determine the stress σ_y.

(b) Determine the other principal stress and the orientation of the principal planes, then show the principal stresses on a sketch of a properly oriented element.

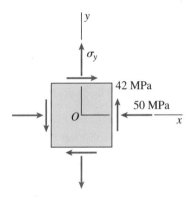

PROB. 6.3-20

Mohr's Circle

The problems for Section 6.4 are to be solved using Mohr's circle. Consider only the in-plane stresses (the stresses in the xy *plane).*

6.4-1 An element in *uniaxial stress* is subjected to tensile stresses $\sigma_x = 11{,}375$ psi, as shown in the figure. Using Mohr's circle, determine:

(a) The stresses acting on an element oriented at a counterclockwise angle $\theta = 24°$ from the x axis.

(b) The maximum shear stresses and associated normal stresses.

Show all results on sketches of properly oriented elements.

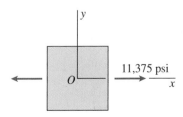

PROB. 6.4-1

6.4-2 An element in *uniaxial stress* is subjected to tensile stresses $\sigma_x = 49$ MPa, as shown in the figure Using Mohr's circle, determine:

(a) The stresses acting on an element oriented at an angle $\theta = -27°$ from the x axis (minus means clockwise).

(b) The maximum shear stresses and associated normal stresses.

Show all results on sketches of properly oriented elements.

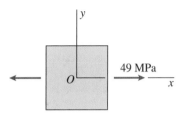

PROB. 6.4-2

6.4-3 An element in *uniaxial stress* is subjected to compressive stresses of magnitude 6100 psi, as shown in the figure. Using Mohr's circle, determine:

(a) The stresses acting on an element oriented at a slope of 1 on 2 (see figure).

(b) The maximum shear stresses and associated normal stresses.

Show all results on sketches of properly oriented elements.

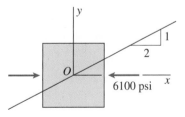

PROB. 6.4-3

6.4-4 An element in *biaxial stress* is subjected to stresses $\sigma_x = -48$ MPa and $\sigma_y = 19$ MPa, as shown in the figure. Using Mohr's circle, determine:

(a) The stresses acting on an element oriented at a counterclockwise angle $\theta = 25°$ from the x axis.

(b) The maximum shear stresses and associated normal stresses.

Show all results on sketches of properly oriented elements.

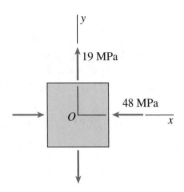

PROB. 6.4-4

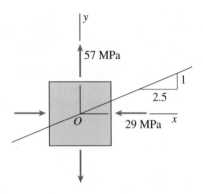

PROB. 6.4-6

6.4-7 An element in *pure shear* is subjected to stresses $\tau_{xy} = 2700$ psi, as shown in the figure. Using Mohr's circle, determine:

(a) The stresses acting on an element oriented at a counterclockwise angle $\theta = 52°$ from the *x* axis.

(b) The principal stresses.

Show all results on sketches of properly oriented elements.

6.4-5 An element in *biaxial stress* is subjected to stresses $\sigma_x = 6250$ psi and $\sigma_y = -1750$ psi, as shown in the figure. Using Mohr's circle, determine:

(a) The stresses acting on an element oriented at a counterclockwise angle $\theta = 55°$ from the *x* axis.

(b) The maximum shear stresses and associated normal stresses.

Show all results on sketches of properly oriented elements.

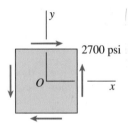

PROB. 6.4-7

6.4-8 An element in *pure shear* is subjected to stresses $\tau_{xy} = -14.5$ MPa, as shown in the figure. Using Mohr's circle, determine:

(a) The stresses acting on an element oriented at a counterclockwise angle $\theta = 22.5°$ from the x axis.

(b) The principal stresses.

Show all results on sketches of properly oriented elements.

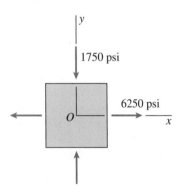

PROB. 6.4-5

6.4-6 An element in *biaxial stress* is subjected to stresses $\sigma_x = -29$ MPa and $\sigma_y = 57$ MPa, as shown in the figure. Using Mohr's circle, determine:

(a) The stresses acting on an element oriented at a slope of 1 on 2.5 (see figure).

(b) The maximum shear stresses and associated normal stresses.

Show all results on sketches of properly oriented elements.

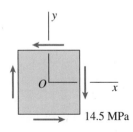

PROB. 6.4-8

6.4-9 An element in *pure shear* is subjected to stresses $\tau_{xy} = 3750$ psi, as shown in the figure. Using Mohr's circle, determine:

(a) The stresses acting on an element oriented at a slope of 3 on 4 (see figure).

(b) The principal stresses.

Show all results on sketches of properly oriented elements.

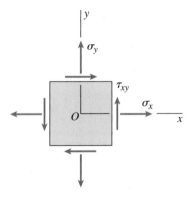

PROB. 6.4-9

6.4-10 through 6.4-15 An element in *plane stress* is subjected to stresses σ_x, σ_y, and τ_{xy} (see figure).

Using Mohr's circle, determine the stresses acting on an element oriented at an angle θ from the x axis. Show these stresses on a sketch of an element oriented at the angle θ. (*Note:* The angle θ is positive when counterclockwise and negative when clockwise.)

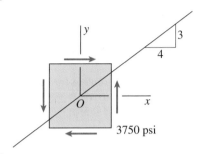

PROBS. 6.4-10 through 6.4-15

6.4-10 $\sigma_x = 27$ MPa, $\sigma_y = 14$ MPa, $\tau_{xy} = 6$ MPa, $\theta = 40°$

6.4-11 $\sigma_x = 3500$ psi, $\sigma_y = 12{,}200$ psi, $\tau_{xy} = -3300$ psi, $\theta = -51°$

6.4-.12 $\sigma_x = -47$ MPa, $\sigma_y = -186$ MPa, $\tau_{xy} = -29$ MPa, $\theta = -33°$

6.4-13 $\sigma_x = -1720$ psi, $\sigma_y = -680$ psi, $\tau_{xy} = 320$ psi, $\theta = 14°$

6.4-14 $\sigma_x = 33$ MPa, $\sigma_y = -9$ MPa, $\tau_{xy} = 29$ MPa, $\theta = 35°$

6.4-15 $\sigma_x = -5700$ psi, $\sigma_y = 950$ psi, $\tau_{xy} = -2100$ psi, $\theta = 65°$

6.4-16 through 6.4-23 An element in *plane stress* is subjected to stresses σ_x, σ_y, and τ_{xy} (see figure).

Using Mohr's circle, determine (a) the principal stresses and (b) the maximum shear stresses and associated normal stresses. Show all results on sketches of properly oriented elements.

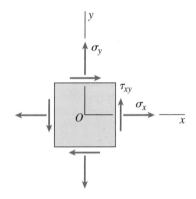

PROBS. 6.4-16 through 6.4-23

6.4-16 $\sigma_x = -29.5$ MPa, $\sigma_y = 29.5$ MPa, $\tau_{xy} = 27$ MPa

6.4-17 $\sigma_x = 7300$ psi, $\sigma_y = 0$ psi, $\tau_{xy} = 1300$ psi

6.4-18 $\sigma_x = 0$ MPa, $\sigma_y = -23.4$ MPa, $\tau_{xy} = -9.6$ MPa

6.4-19 $\sigma_x = 2050$ psi, $\sigma_y = 6100$ psi, $\tau_{xy} = 2750$ psi

6.4-20 $\sigma_x = 2900$ kPa, $\sigma_y = 9100$ kPa, $\tau_{xy} = -3750$ kPa

6.4-21 $\sigma_x = -11{,}500$ psi, $\sigma_y = -18{,}250$ psi, $\tau_{xy} = -7200$ psi

6.4-22 $\sigma_x = -3.3$ MPa, $\sigma_y = 8.9$ MPa, $\tau_{xy} = -14.1$ MPa

6.4-23 $\sigma_x = 800$ psi, $\sigma_y = -2200$ psi, $\tau_{xy} = 2900$ psi

Hooke's Law for Plane Stress

When solving the problems for Section 6.5, assume that the material is linearly elastic with modulus of elasticity E and Poisson's ratio ν.

6.5-1 A rectangular steel plate with thickness $t = 0.25$ in. is subjected to uniform normal stresses σ_x and σ_y, as shown in the figure. Strain gages A and B, oriented in the x and y directions, respectively, are attached to the plate. The gage readings give normal strains $\epsilon_x = 0.0010$ (elongation) and $\epsilon_y = -0.0007$ (shortening).

Knowing that $E = 30 \times 10^6$ psi and $\nu = 0.3$, determine the stresses σ_x and σ_y and the change Δt in the thickness of the plate.

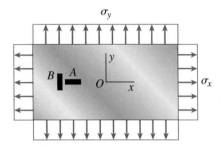

PROBS. 6.5-1 and 6.5-2

6.5-2 Solve the preceding problem if the thickness of the steel plate is $t = 10$ mm, the gage readings are $\epsilon_x = 480 \times 10^{-6}$ (elongation) and $\epsilon_y = 130 \times 10^{-6}$ (elongation), the modulus is $E = 200$ GPa, and Poisson's ratio is $\nu = 0.30$.

6.5-3 Assume that the normal strains ϵ_x and ϵ_y for an element in *plane stress* (see figure) are measured with strain gages.

(a) Obtain a formula for the normal strain ϵ_z in the z direction in terms of ϵ_x, ϵ_y, and Poisson's ratio ν.

(b) Obtain a formula for the dilatation e in terms of ϵ_x, ϵ_y, and Poisson's ratio ν.

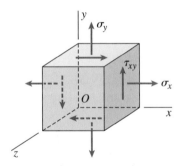

PROB. 6.5-3

6.5-4 A magnesium plate in *biaxial stress* is subjected to tensile stresses $\sigma_x = 24$ MPa and $\sigma_y = 12$ MPa (see figure). The corresponding strains in the plate are $\epsilon_x = 440 \times 10^{-6}$ and $\epsilon_y = 80 \times 10^{-6}$.

Determine Poisson's ratio ν and the modulus of elasticity E for the material.

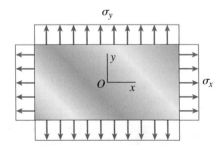

PROBS. 6.5-4 through 6.5-7

6.5-5 Solve the preceding problem for a steel plate with $\sigma_x = 10,800$ psi (tension), $\sigma_y = -5,400$ psi (compression), $\epsilon_x = 420 \times 10^{-6}$ (elongation), and $\epsilon_y = -300 \times 10^{-6}$ (shortening).

6.5-6 A rectangular plate in *biaxial stress* (see figure) is subjected to normal stresses $\sigma_x = 90$ MPa (tension) and $\sigma_y = -20$ MPa (compression). The plate has dimensions $400 \times 800 \times 20$ mm and is made of steel with $E = 200$ GPa and $\nu = 0.30$.

(a) Determine the maximum in-plane shear strain γ_{max} in the plate.

(b) Determine the change Δt in the thickness of the plate.

(c) Determine the change ΔV in the volume of the plate.

6.5-7 Solve the preceding problem for an aluminum plate with $\sigma_x = 12,000$ psi (tension), $\sigma_y = -3,000$ psi (compression), dimensions $20 \times 30 \times 0.5$ in., $E = 10.5 \times 10^6$ psi, and $\nu = 0.33$.

6.5-8 A brass cube 50 mm on each edge is compressed in two perpendicular directions by forces $P = 175$ kN (see figure).

Calculate the change ΔV in the volume of the cube assuming $E = 100$ GPa and $\nu = 0.34$.

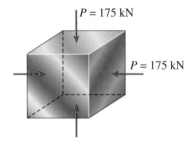

PROB. 6.5-8

6.5-9 A 4.0-inch cube of concrete ($E = 3.0 \times 10^6$ psi, $\nu = 0.1$) is compressed in *biaxial stress* by means of a framework that is loaded as shown in the figure.

Assuming that each load F equals 20 k, determine change ΔV in the volume of the cube.

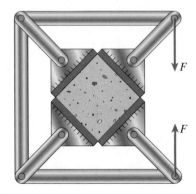

PROB. 6.5-9

6.5-10 A square plate of width b and thickness t is loaded by normal forces P_x and P_y, and by shear forces V, as shown in the figure. These forces produce uniformly distributed stresses acting on the side faces of the plate.

Calculate the change ΔV in the volume of the plate if the dimensions are $b = 600$ mm and $t = 40$ mm, the plate is made of magnesium with $E = 45$ GPa and $\nu = 0.35$, and the forces are $P_x = 480$ kN, $P_y = 180$ kN, and $V = 120$ kN.

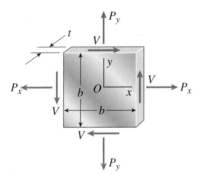

PROBS. 6.5-10 and 6.5-11

6.5-11 Solve the preceding problem for an aluminum plate with $b = 12$ in., $t = 1.0$ in., $E = 10,600$ ksi, $\nu = 0.33$, $P_x = 90$ k, $P_y = 20$ k, and $V = 15$ k.

6.5-12 A circle of diameter $d = 200$ mm is etched on a brass plate (see figure). The plate has dimensions $400 \times 400 \times 20$ mm. Forces are applied to the plate, producing uniformly distributed normal stresses $\sigma_x = 42$ MPa and $\sigma_y = 14$ MPa.

Calculate the following quantities: (a) the change in length Δac of diameter ac; (b) the change in length Δbd of diameter bd; (c) the change Δt in the thickness of the plate; (d) the change ΔV in the volume of the plate. (Assume $E = 100$ GPa and $\nu = 0.34$.)

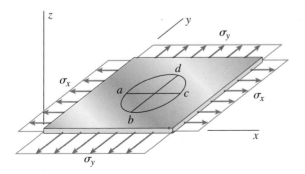

PROB. 6.5-12

Triaxial Stress

When solving the problems for Section 6.6, assume that the material is linearly elastic with modulus of elasticity E and Poisson's ratio ν.

6.6-1 An element of aluminum in the form of a rectangular parallelepiped (see figure) of dimensions $a = 6.0$ in., $b = 4.0$ in, and $c = 3.0$ in. is subjected to *triaxial stresses* $\sigma_x = 12,000$ psi, $\sigma_y = -4,000$ psi, and $\sigma_z = -1,000$ psi acting on the x, y, and z faces, respectively.

Determine the following quantities: (a) the maximum shear stress τ_{max} in the material; (b) the changes Δa, Δb, and Δc in the dimensions of the element; (c) the change ΔV in the volume; (Assume $E = 10,400$ ksi and $\nu = 0.33$.)

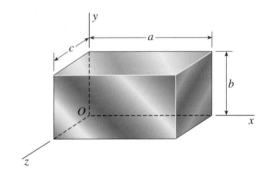

PROBS. 6.6-1 and 6.6-2

6.6-2 Solve the preceding problem if the element is steel ($E = 200$ GPa, $\nu = 0.30$) with dimensions $a = 300$ mm, $b = 150$ mm, and $c = 150$ mm and the stresses are $\sigma_x = -60$ MPa, $\sigma_y = -40$ MPa, and $\sigma_z = -40$ MPa.

6.6-3 A cube of cast iron with sides of length $a = 4.0$ in. (see figure) is tested in a laboratory under *triaxial stress*. Gages mounted on the testing machine show that the compressive strains in the material are $\epsilon_x = -225 \times 10^{-6}$ and $\epsilon_y = \epsilon_z = -37.5 \times 10^{-6}$.

Determine the following quantities: (a) the normal stresses σ_x, σ_y, and σ_z acting on the x, y, and z faces of the cube; (b) the maximum shear stress τ_{max} in the material; (c) the change ΔV in the volume of the cube; (Assume $E = 14,000$ ksi and $\nu = 0.25$.)

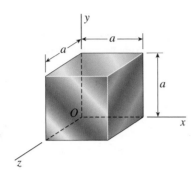

PROBS. 6.6-3 and 6.6-4

6.6-4 Solve the preceding problem if the cube is granite ($E = 60$ GPa, $\nu = 0.25$) with dimensions $a = 75$ mm and compressive strains $\epsilon_x = -720 \times 10^{-6}$ and $\epsilon_y = \epsilon_z = -270 \times 10^{-6}$.

6.6-5 An element of aluminum in *triaxial stress* (see figure) is subjected to stresses $\sigma_x = 5200$ psi (tension), $\sigma_y = -4750$ psi (compression), and $\sigma_z = -3090$ psi (compression). It is also known that the normal strains in the x and y directions are $\epsilon_x = 713.8 \times 10^{-6}$ (elongation) and $\epsilon_y = -502.3 \times 10^{-6}$ (shortening).

What is the bulk modulus K for the aluminum?

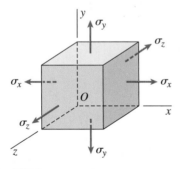

PROBS. 6.6-5 and 6.6-6

6.6-6 Solve the preceding problem if the material is nylon subjected to compressive stresses $\sigma_x = -4.5$ MPa, $\sigma_y = -3.6$ MPa, and $\sigma_z = -2.1$ MPa, and the normal strains are $\epsilon_x = -740 \times 10^{-6}$ and $\epsilon_y = -320 \times 10^{-6}$ (shortenings).

6.6-7 A rubber cylinder R of length L and cross-sectional area A is compressed inside a steel cylinder S by a force F that applies a uniformly distributed pressure to the rubber (see figure).

(a) Derive a formula for the lateral pressure p between the rubber and the steel. (Disregard friction between the rubber and the steel, and assume that the steel cylinder is rigid when compared to the rubber.)

(b) Derive a formula for the shortening δ of the rubber cylinder.

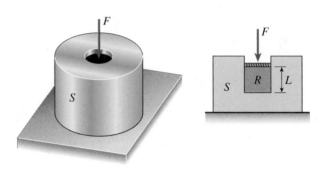

PROB. 6.6-7

6.6-8 A block R of rubber is confined between plane parallel walls of a steel block S (see figure). A uniformly distributed pressure p_0 is applied to the top of the rubber block by a force F.

(a) Derive a formula for the lateral pressure p between the rubber and the steel. (Disregard friction between the rubber and the steel, and assume that the steel block is rigid when compared to the rubber.)

(b) Derive a formula for the dilatation e of the rubber.

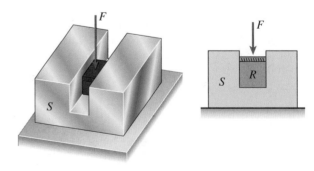

PROB. 6.6-8

6.6-9 A solid spherical ball of brass ($E = 15 \times 10^6$ psi, $\nu = 0.34$) is lowered into the ocean to a depth of 10,000 ft. The diameter of the ball is 11.0 in.

Determine the decrease Δd in diameter, and the decrease ΔV in volume of the ball.

6.6-10 A solid steel sphere ($E = 210$ GPa, $\nu = 0.3$) is subjected to hydrostatic pressure p such that its volume is reduced by 0.4%.

(a) Calculate the pressure p.

(b) Calculate the volume modulus of elasticity K for the steel.

6.6-11 A solid bronze sphere (volume modulus of elasticity $K = 14.5 \times 10^6$ psi) is suddenly heated around its outer surface. The tendency of the heated part of the sphere to expand produces uniform tension in all directions at the center of the sphere.

If the stress at the center is 12,000 psi, what is the strain? Also, calculate the unit volume change e at the center.

Airships such as this blimp rely on internal pressure to maintain their shape using a gas lighter than air for buoyant lift.
(Courtesy of Christian Michel, www.modernairships.info)

Applications of Plane Stress (Pressure Vessels and Combined Loadings)

CHAPTER OVERVIEW

Chapter 7 deals with a number of applications of plane stress, a topic discussed in detail in Sections 6.2 through 6.5 of the previous chapter. Plane stress is a common stress condition that exists in all ordinary structures, including buildings, machines, vehicles, and aircraft. First, thin-wall shell theory is presented describing the behavior of **spherical** (Section 7.2) and **cylindrical** (Section 7.3) **pressure vessels** under internal pressure and having walls whose thickness t is small compared with radius r of the cross section (i.e., $r/t > 10$). We will determine the stresses and strains in the walls of these structures due to the internal pressures from the compressed gases or liquids. Only **positive internal pressure** (not the effects of external loads, reactions, the weight of the contents, and the weight of the structure) is considered. Linear-elastic behavior is assumed, and the formulas for **membrane stresses** in spherical tanks and **hoop and axial stresses** in cylindrical tanks are only valid in regions of the tank away from stress concentrations caused by openings and support brackets or legs. Finally, stresses at points of interest in structures under combined loadings (axial, shear, torsion, bending, and possibly internal pressure) are assessed (Section 7.4). Our objective is to determine the maximum normal and shear stresses at various points in these structures. Linear-elastic behavior is assumed so that superposition can be used to combine normal and shear stresses due to various loadings, all of which contribute to the **state of plane stress** at that point.

Chapter 7 is organized as follows:

7.1 INTRODUCTION

We will now investigate some practical examples of structures and components in states of plane stress building upon the concepts presented in Chapter 6. First, stresses and strains in the walls of thin pressure vessels are examined. Then structures acted upon by combined loadings will be evaluated to find the maximum normal and shear stresses which govern their design.

7.2 SPHERICAL PRESSURE VESSELS

Pressure vessels are closed structures containing liquids or gases under pressure. Familiar examples include tanks, pipes, and pressurized cabins in aircraft and space vehicles. When pressure vessels have walls that are thin in comparison to their overall dimensions, they are included within a more general category known as **shell structures**. Other examples of shell structures are roof domes, airplane wings, and submarine hulls.

In this section we consider thin-walled pressure vessels of spherical shape, like the compressed-air tank shown in Fig. 7-1. The term **thin-walled** is not precise, but as a general rule, pressure vessels are considered to be thin-walled when the ratio of radius r to wall thickness t (Fig. 7-2) is greater than 10. When this condition is met, we can determine the stresses in the walls with reasonable accuracy using statics alone.

We assume in the following discussions that the internal pressure p (Fig. 7-2) exceeds the pressure acting on the outside of the shell. Otherwise, the vessel may collapse inward due to buckling.

A sphere is the theoretically ideal shape for a vessel that resists internal pressure. We only need to contemplate the familiar soap bubble to recognize that a sphere is the "natural" shape for this purpose. To determine the stresses in a spherical vessel, let us cut through the sphere on a vertical diametral plane (Fig. 7-3a) and isolate half of the shell *and its fluid contents* as a single free body (Fig. 7-3b). Acting on this free body are the tensile stresses σ in the wall of the vessel and the fluid pressure p. This pressure acts horizontally against the plane circular area of fluid remaining inside the hemisphere. Since the pressure is uniform, the resultant pressure force P (Fig. 7-3b) is

$$P = p(\pi r^2) \tag{a}$$

where r is the inner radius of the sphere.

Note that the pressure p is not the absolute pressure inside the vessel but is the net internal pressure, or the **gage pressure**. Gage pressure is the internal pressure *above* the pressure acting on the outside of the vessel. If the internal and external pressures are the same, no stresses are developed in the wall of the vessel—only the excess of internal pressure over external pressure has any effect on these stresses.

Thin-walled spherical pressure vessel used for storage of propane in this oil refinery (Wayne Eastep/Getty Images)

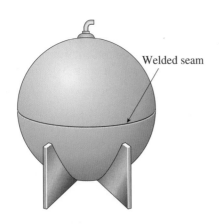

Welded seam

FIG. 7-1 Spherical pressure vessel

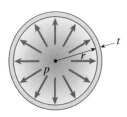

FIG. 7-2 Cross section of spherical pressure vessel showing inner radius r, wall thickness t, and internal pressure p

Because of the symmetry of the vessel and its loading (Fig. 7-3b), the tensile stress σ is uniform around the circumference. Furthermore, since the wall is thin, we can assume with good accuracy that the stress is uniformly distributed across the thickness t. The accuracy of this approximation increases as the shell becomes thinner and decreases as it becomes thicker.

The resultant of the tensile stresses σ in the wall is a horizontal force equal to the stress σ times the area over which it acts, or

$$\sigma(2\pi r_m t)$$

where t is the thickness of the wall and r_m is its mean radius:

$$r_m = r + \frac{t}{2} \tag{b}$$

Thus, equilibrium of forces in the horizontal direction (Fig. 7-3b) gives

$$\Sigma F_{\text{horiz}} = 0 \qquad \sigma(2\pi r_m t) - p(\pi r^2) = 0 \tag{c}$$

from which we obtain the *tensile stresses* in the wall of the vessel:

$$\sigma = \frac{pr^2}{2r_m t} \tag{d}$$

Since our analysis is valid only for thin shells, we can disregard the small difference between the two radii appearing in Eq. (d) and replace r by r_m or replace r_m by r. While either choice is satisfactory for this approximate analysis, it turns out that the stresses are closer to the theoretically exact stresses if we use the inner radius r instead of the mean radius r_m. Therefore, we will adopt the following formula for calculating the **tensile stresses in the wall of a spherical shell**:

$$\sigma = \frac{pr}{2t} \tag{7-1}$$

As is evident from the symmetry of a spherical shell, we obtain the same equation for the tensile stresses when we cut a plane through the center of the sphere in any direction whatsoever. Thus, we reach the following conclusion: *The wall of a pressurized spherical vessel is subjected to uniform tensile stresses σ in all directions.* This stress condition is represented in Fig. 7-3c by the small stress element with stresses σ acting in mutually perpendicular directions.

Stresses that act tangentially to the curved surface of a shell, such as the stresses σ shown in Fig. 7-3c, are known as **membrane stresses**. The name arises from the fact that these are the only stresses that exist in true membranes, such as soap films.

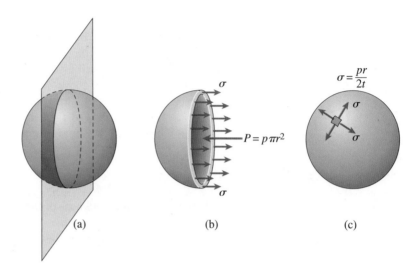

FIG. 7-3 Tensile stresses σ in the wall of a spherical pressure vessel

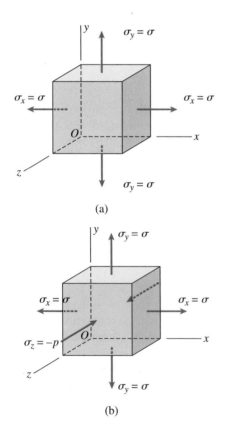

FIG. 7-4 Stresses in a spherical pressure vessel at (a) the outer surface and (b) the inner surface

Stresses at the Outer Surface

The outer surface of a spherical pressure vessel is usually free of any loads. Therefore, the element shown in Fig. 7-3c is in *biaxial stress*. To aid in analyzing the stresses acting on this element, we show it again in Fig. 7-4a, where a set of coordinate axes is oriented parallel to the sides of the element. The x and y axes are tangential to the surface of the sphere, and the z axis is perpendicular to the surface. Thus, the normal stresses σ_x and σ_y are the same as the membrane stresses σ, and the normal stress σ_z is zero. No shear stresses act on the sides of this element.

If we analyze the element of Fig. 7-4a by using the transformation equations for plane stress (see Fig. 6-1 and Eqs. 6-4a and 6-4b of Section 6.2), we find

$$\sigma_{x_1} = \sigma \quad \text{and} \quad \tau_{x_1y_1} = 0$$

as expected. In other words, when we consider elements obtained by rotating the axes about the z axis, the normal stresses remain constant and there are no shear stresses. *Every plane is a principal plane and every direction is a principal direction.* Thus, the **principal stresses** for the element are

$$\sigma_1 = \sigma_2 = \frac{pr}{2t} \qquad \sigma_3 = 0 \qquad \text{(7-2a,b)}$$

The stresses σ_1 and σ_2 lie in the xy plane and the stress σ_3 acts in the z direction.

To obtain the **maximum shear stresses**, we must consider out-of-plane rotations, that is, rotations about the x and y axes (because all in-plane shear stresses are zero). Elements oriented by making 45° rotations about the x and y axes have maximum shear stresses equal to $\sigma/2$ and normal stresses equal to $\sigma/2$. Therefore,

$$\tau_{max} = \frac{\sigma}{2} = \frac{pr}{4t} \qquad (7\text{-}3)$$

These stresses are the largest shear stresses in the element.

Stresses at the Inner Surface

At the inner surface of the wall of a spherical vessel, a stress element (Fig. 7-4b) has the same membrane stresses σ_x and σ_y as does an element at the outer surface (Fig. 7-4a). In addition, a compressive stress σ_z equal to the pressure p acts in the z direction (Fig. 7-4b). This compressive stress decreases from p at the inner surface of the sphere to zero at the outer surface.

The element shown in Fig. 7-4b is in triaxial stress with principal stresses

$$\sigma_1 = \sigma_2 = \frac{pr}{2t} \qquad \sigma_3 = -p \qquad (e,f)$$

The in-plane shear stresses are zero, but the maximum out-of-plane shear stress (obtained by a 45° rotation about either the x or y axis) is

$$\tau_{max} = \frac{\sigma + p}{2} = \frac{pr}{4t} + \frac{p}{2} = \frac{p}{2}\left(\frac{r}{2t} + 1\right) \qquad (g)$$

When the vessel is thin-walled and the ratio r/t is large, we can disregard the number 1 in comparison with the term $r/2t$. In other words, the principal stress σ_3 in the z direction is small when compared with the principal stresses σ_1 and σ_2. Consequently, we can consider the stress state at the inner surface to be the same as at the outer surface (biaxial stress). This approximation is consistent with the approximate nature of thin-shell theory, and therefore we will use Eqs. (7-1), (7-2), and (7-3) to obtain the stresses in the wall of a spherical pressure vessel.

General Comments

Pressure vessels usually have openings in their walls (to serve as inlets and outlets for the fluid contents) as well as fittings and supports that exert forces on the shell (Fig. 7-1). These features result in nonuniformities in the stress distribution, or *stress concentrations*, that cannot be analyzed by the elementary formulas given here. Instead, more advanced methods of analysis are needed. Other factors that affect the design of pressure vessels include corrosion, accidental impacts, and temperature changes.

Some of the limitations of thin-shell theory as applied to pressure vessels are listed here:

1. The wall thickness must be small in comparison to the other dimensions (the ratio r/t should be 10 or more).
2. The internal pressure must exceed the external pressure (to avoid inward buckling).

3. The analysis presented in this section is based only on the effects of internal pressure (the effects of external loads, reactions, the weight of the contents, and the weight of the structure are not considered).
4. The formulas derived in this section are valid throughout the wall of the vessel *except* near points of stress concentrations.

The following example illustrates how the principal stresses and maximum shear stresses are used in the analysis of a spherical shell.

Example 7-1

FIG. 7-5 Example 7-1. Spherical pressure vessel. (Attachments and supports are not shown.)

A compressed-air tank having an inner diameter of 18 inches and a wall thickness of 1/4 inch is formed by welding two steel hemispheres (Fig. 7-5).

(a) If the allowable tensile stress in the steel is 14,000 psi, what is the maximum permissible air pressure p_a in the tank?

(b) If the allowable shear stress in the steel is 5,700 psi, what is the maximum permissible pressure p_b?

(c) If the normal strain at the outer surface of the tank is not to exceed 0.0003, what is the maximum permissible pressure p_c? (Assume that Hooke's law is valid and that the modulus of elasticity for the steel is 29×10^6 psi and Poisson's ratio is 0.28.)

(d) Tests on the welded seam show that failure occurs when the tensile load on the welds exceeds 8.1 kips per inch of weld. If the required factor of safety against failure of the weld is 2.5, what is the maximum permissible pressure p_d?

(e) Considering the four preceding factors, what is the allowable pressure p_{allow} in the tank?

Solution

(a) *Allowable pressure based upon the tensile stress in the steel.* The maximum tensile stress in the wall of the tank is given by the formula $\sigma = pr/2t$ (see Eq. 7-1). Solving this equation for the pressure in terms of the allowable stress, we get

$$p_a = \frac{2t\sigma_{\text{allow}}}{r} = \frac{2(0.25 \text{ in.})(14{,}000 \text{ psi})}{9.0 \text{ in.}} = 777.8 \text{ psi} \quad \Longleftarrow$$

Thus, the maximum allowable pressure based upon tension in the wall of the tank is $p_a = 777$ psi. (Note that in a calculation of this kind, we round downward, not upward.)

(b) *Allowable pressure based upon the shear stress in the steel.* The maximum shear stress in the wall of the tank is given by Eq. (7-3), from which we get the following equation for the pressure:

$$p_b = \frac{4t\tau_{\text{allow}}}{r} = \frac{4(0.25 \text{ in.})(5{,}700 \text{ psi})}{9.0 \text{ in.}} = 633.3 \text{ psi} \quad \Longleftarrow$$

Therefore, the allowable pressure based upon shear is $p_b = 633$ psi.

(c) *Allowable pressure based upon the normal strain in the steel.* The normal strain is obtained from Hooke's law for biaxial stress (Eq. 6-39a):

$$\epsilon_x = \frac{1}{E}(\sigma_x - \nu\sigma_y) \qquad (h)$$

Substituting $\sigma_x = \sigma_y = \sigma = pr/2t$ (see Fig. 7-4a), we obtain

$$\epsilon_x = \frac{\sigma}{E}(1 - \nu) = \frac{pr}{2tE}(1 - \nu) \qquad (7\text{-}4)$$

This equation can be solved for the pressure p_c:

$$p_c = \frac{2tE\epsilon_{\text{allow}}}{r(1 - \nu)} = \frac{2(0.25 \text{ in.})(29 \times 10^6 \text{ psi})(0.0003)}{(9.0 \text{ in.})(1 - 0.28)} = 671.3 \text{ psi} \quad \Longleftarrow$$

Thus, the allowable pressure based upon the normal strain in the wall is $p_c = 671$ psi.

(d) *Allowable pressure based upon the tension in the welded seam.* The allowable tensile load on the welded seam is equal to the failure load divided by the factor of safety:

$$T_{\text{allow}} = \frac{T_{\text{failure}}}{n} = \frac{8.1 \text{ k/in.}}{2.5} = 3.24 \text{ k/in.} = 3240 \text{ lb/in.}$$

The corresponding allowable tensile stress is equal to the allowable load on a one-inch length of weld divided by the cross-sectional area of a one-inch length of weld:

$$\sigma_{\text{allow}} = \frac{T_{\text{allow}}(1.0 \text{ in.})}{(1.0 \text{ in.})(t)} = \frac{(3240 \text{ lb/in.})(1.0 \text{ in})}{(1.0 \text{ in.})(0.25 \text{ in.})} = 12{,}960 \text{ psi}$$

Finally, we solve for the internal pressure by using Eq. (7-1):

$$p_d = \frac{2t\sigma_{\text{allow}}}{r} = \frac{2(0.25 \text{ in.})(12{,}960 \text{ psi})}{9.0 \text{ in.}} = 720.0 \text{ psi} \quad \Longleftarrow$$

This result gives the allowable pressure based upon tension in the welded seam.

(e) *Allowable pressure.* Comparing the preceding results for p_a, p_b, p_c, and p_d, we see that shear stress in the wall governs and the allowable pressure in the tank is

$$p_{\text{allow}} = 633 \text{ psi} \quad \Longleftarrow$$

This example illustrates how various stresses and strains enter into the design of a spherical pressure vessel.

Note: When the internal pressure is at its maximum allowable value (633 psi), the tensile stresses in the shell are

$$\sigma = \frac{pr}{2t} = \frac{(633 \text{ psi})(9.0 \text{ in.})}{2(0.25 \text{ in.})} = 11{,}400 \text{ psi}$$

Thus, at the inner surface of the shell (Fig. 7-4b), the ratio of the principal stress in the z direction (633 psi) to the in-plane principal stresses (12,000 psi) is only 0.056. Therefore, our earlier assumption that we can disregard the principal stress σ_3 in the z direction and consider the entire shell to be in biaxial stress is justified.

7.3 CYLINDRICAL PRESSURE VESSELS

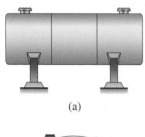

(a)

(b)

FIG. 7-6 Cylindrical pressure vessels with circular cross sections

Cylindrical storage tanks in a petrochemical plant
(William H. Edwards/Getty Images)

Cylindrical pressure vessels with a circular cross section (Fig. 7-6) are found in industrial settings (compressed air tanks and rocket motors), in homes (fire extinguishers and spray cans), and in the countryside (propane tanks and grain silos). Pressurized pipes, such as water-supply pipes and penstocks, are also classified as cylindrical pressure vessels.

We begin our analysis of cylindrical vessels by determining the normal stresses in a *thin-walled circular tank AB* subjected to internal pressure (Fig. 7-7a). A *stress element* with its faces parallel and perpendicular to the axis of the tank is shown on the wall of the tank. The normal stresses σ_1 and σ_2 acting on the side faces of this element are the membrane stresses in the wall. No shear stresses act on these faces because of the symmetry of the vessel and its loading. Therefore, the stresses σ_1 and σ_2 are principal stresses.

Because of their directions, the stress σ_1 is called the **circumferential stress** or the **hoop stress**, and the stress σ_2 is called the **longitudinal stress** or the **axial stress**. Each of these stresses can be calculated from equilibrium by using appropriate free-body diagrams.

Circumferential Stress

To determine the circumferential stress σ_1, we make two cuts (*mn* and *pq*) perpendicular to the longitudinal axis and distance *b* apart (Fig. 7-7a). Then we make a third cut in a vertical plane through the longitudinal axis of the tank, resulting in the free body shown in Fig. 7-7b. This free body consists not only of the half-circular piece of the tank but also of the fluid contained within the cuts. Acting on the longitudinal cut (plane *mpqn*) are the circumferential stresses σ_1 and the internal pressure *p*.

Stresses and pressures also act on the left-hand and right-hand faces of the free body. However, these stresses and pressures are not shown in the figure because they do not enter the equation of equilibrium that we will use. As in our analysis of a spherical vessel, we will disregard the weight of the tank and its contents.

The circumferential stresses σ_1 acting in the wall of the vessel have a resultant equal to $\sigma_1(2bt)$, where *t* is the thickness of the wall. Also, the resultant force P_1 of the internal pressure is equal to $2pbr$, where *r* is the inner radius of the cylinder. Hence, we have the following equation of equilibrium:

$$\sigma_1(2bt) - 2pbr = 0$$

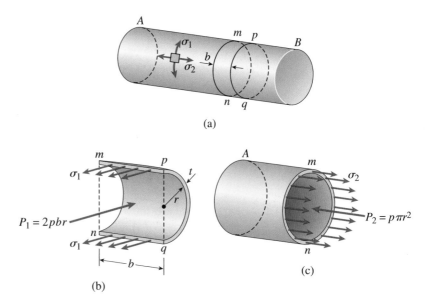

FIG. 7-7 Stresses in a circular cylindrical pressure vessel

From this equation we obtain the following formula for the *circumferential stress in a pressurized cylinder*:

$$\sigma_1 = \frac{pr}{t} \tag{7-5}$$

This stress is uniformly distributed over the thickness of the wall, provided the thickness is small compared to the radius.

Longitudinal Stress

The longitudinal stress σ_2 is obtained from the equilibrium of a free body of the part of the vessel to the left of cross section *mn* (Fig. 7-7c). Again, the free body includes not only part of the tank but also its contents. The stresses σ_2 act longitudinally and have a resultant force equal to $\sigma_2(2\pi rt)$. Note that we are using the inner radius of the shell in place of the mean radius, as explained in Section 7.2.

The resultant force P_2 of the internal pressure is a force equal to $p\pi r^2$. Thus, the equation of equilibrium for the free body is

$$\sigma_2(2\pi rt) - p\pi r^2 = 0$$

Solving this equation for σ_2, we obtain the following formula for the *longitudinal stress* in a cylindrical pressure vessel:

$$\sigma_2 = \frac{pr}{2t} \tag{7-6}$$

This stress is equal to the membrane stress in a spherical vessel (Eq. 7-1).

Comparing Eqs. (7-5) and (7-6), we see that the circumferential stress in a cylindrical vessel is equal to twice the longitudinal stress:

$$\sigma_1 = 2\sigma_2 \tag{7-7}$$

From this result we note that a longitudinal welded seam in a pressurized tank must be twice as strong as a circumferential seam.

Stresses at the Outer Surface

The principal stresses σ_1 and σ_2 at the outer surface of a cylindrical vessel are shown on the stress element of Fig. 7-8a. Since the third principal stress (acting in the z direction) is zero, the element is in *biaxial stress*.

The maximum *in-plane shear stresses* occur on planes that are rotated 45° about the z axis; these stresses are

$$(\tau_{\max})_z = \frac{\sigma_1 - \sigma_2}{2} = \frac{\sigma_1}{4} = \frac{pr}{4t} \tag{7-8}$$

The maximum *out-of-plane shear stresses* are obtained by 45° rotations about the x and y axes, respectively; thus,

$$(\tau_{\max})_x = \frac{\sigma_1}{2} = \frac{pr}{2t} \qquad (\tau_{\max})_y = \frac{\sigma_2}{2} = \frac{pr}{4t} \tag{7-9a,b}$$

Comparing the preceding results, we see that the *absolute maximum shear stress* is

$$\tau_{\max} = \frac{\sigma_1}{2} = \frac{pr}{2t} \tag{7-10}$$

This stress occurs on a plane that has been rotated 45° about the x axis.

Stresses at the Inner Surface

The stress conditions at the inner surface of the wall of the vessel are shown in Fig. 7-8b. The principal stresses are

$$\sigma_1 = \frac{pr}{t} \qquad \sigma_2 = \frac{pr}{2t} \qquad \sigma_3 = -p \tag{a,b,c}$$

The three maximum shear stresses, obtained by 45° rotations about the x, y, and z axes, are

$$(\tau_{\max})_x = \frac{\sigma_1 - \sigma_3}{2} = \frac{pr}{2t} + \frac{p}{2} \qquad (\tau_{\max})_y = \frac{\sigma_2 - \sigma_3}{2} = \frac{pr}{4t} + \frac{p}{2} \tag{d,e}$$

$$(\tau_{\max})_z = \frac{\sigma_1 - \sigma_2}{2} = \frac{pr}{4t} \tag{f}$$

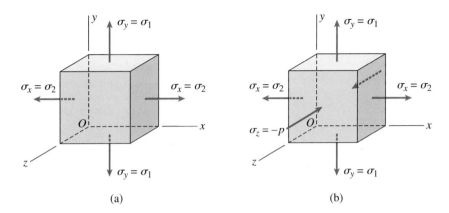

FIG. 7-8 Stresses in a circular cylindrical pressure vessel at (a) the outer surface and (b) the inner surface

The first of these three stresses is the largest. However, as explained in the discussion of shear stresses in a spherical shell, we may disregard the additional term $p/2$ in Eqs. (d) and (e) when the shell is thin-walled. Equations (d), (e), and (f) then become the same as Eqs. (7-9) and (7-8), respectively.

Therefore, in all of our examples and problems pertaining to cylindrical pressure vessels, *we will disregard the presence of the compressive stress in the z direction.* (This compressive stress varies from p at the inner surface to zero at the outer surface.) With this approximation, the stresses at the inner surface become the same as the stresses at the outer surface (biaxial stress). As explained in the discussion of spherical pressure vessels, this procedure is satisfactory when we consider the numerous other approximations in this theory.

General Comments

The preceding formulas for stresses in a circular cylinder are valid in parts of the cylinder away from any discontinuities that cause stress concentrations, as discussed previously for spherical shells. An obvious discontinuity exists at the ends of the cylinder where the heads are attached, because the geometry of the structure changes abruptly. Other stress concentrations occur at openings, at points of support, and wherever objects or fittings are attached to the cylinder. The stresses at such points cannot be determined solely from equilibrium equations; instead, more advanced methods of analysis (such as shell theory and finite-element analysis) must be used.

Some of the limitations of the elementary theory for thin-walled shells are listed in Section 7.2.

Example 7-2

A cylindrical pressure vessel is constructed from a long, narrow steel plate by wrapping the plate around a mandrel and then welding along the edges of the plate to make a helical joint (Fig. 7-9). The helical weld makes an angle $\alpha = 55°$ with the longitudinal axis. The vessel has inner radius $r = 1.8$ m and wall thickness $t = 20$ mm. The material is steel with modulus $E = 200$ GPa and Poisson's ratio $\nu = 0.30$. The internal pressure p is 800 kPa.

Calculate the following quantities for the cylindrical part of the vessel: (a) the circumferential and longitudinal stresses σ_1 and σ_2, respectively; (b) the maximum in-plane and out-of-plane shear stresses; (c) the circumferential and longitudinal strains ϵ_1 and ϵ_2, respectively; and (d) the normal stress σ_w and shear stress τ_w acting perpendicular and parallel, respectively, to the welded seam.

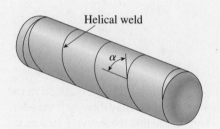

FIG. 7-9 Example 7-2. Cylindrical pressure vessel with a helical weld

Solution

(a) *Circumferential and longitudinal stresses.* The circumferential and longitudinal stresses σ_1 and σ_2, respectively, are pictured in Fig. 7-10a, where they are shown acting on a stress element at point A on the wall of the vessel. The magnitudes of the stresses can be calculated from Eqs. (7-5) and (7-6):

$$\sigma_1 = \frac{pr}{t} = \frac{(800 \text{ kPa})(1.8 \text{ m})}{20 \text{ mm}} = 72 \text{ MPa} \qquad \sigma_2 = \frac{pr}{2t} = \frac{\sigma_1}{2} = 36 \text{ MPa} \quad \Longleftarrow$$

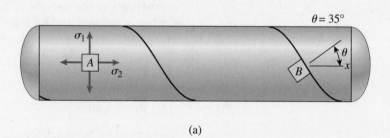

(a)

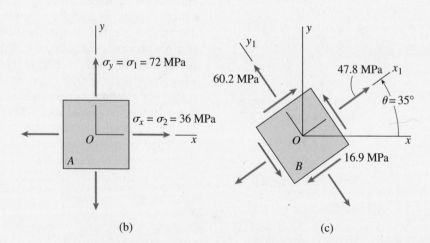

(b)   (c)

The stress element at point A is shown again in Fig. 7-10b, where the x axis is in the longitudinal direction of the cylinder and the y axis is in the circumferential direction. Since there is no stress in the z direction ($\sigma_3 = 0$), the element is in biaxial stress.

Note that the ratio of the internal pressure (800 kPa) to the smaller in-plane principal stress (36 MPa) is 0.022. Therefore, our assumption that we may disregard any stresses in the z direction and consider all elements in the cylindrical shell, even those at the inner surface, to be in biaxial stress is justified.

(b) *Maximum shear stresses.* The largest in-plane shear stress is obtained from Eq. (7-8):

$$(\tau_{\max})_z = \frac{\sigma_1 - \sigma_2}{2} = \frac{\sigma_1}{4} = \frac{pr}{4t} = 18 \text{ MPa}$$

Because we are disregarding the normal stress in the z direction, the largest out-of-plane shear stress is obtained from Eq. (7-9a):

$$\tau_{\max} = \frac{\sigma_1}{2} = \frac{pr}{2t} = 36 \text{ MPa}$$

This last stress is the absolute maximum shear stress in the wall of the vessel.

(c) *Circumferential and longitudinal strains.* Since the largest stresses are well below the yield stress of steel (see Table I-3, Appendix I available online), we may assume that Hooke's law applies to the wall of the vessel. Then we can obtain the strains in the x and y directions (Fig. 7-10b) from Eqs. (6-39a) and (6-39b) for biaxial stress:

$$\epsilon_x = \frac{1}{E}(\sigma_x - \nu\sigma_y) \qquad \epsilon_y = \frac{1}{E}(\sigma_y - \nu\sigma_x) \qquad \text{(g,h)}$$

We note that the strain ϵ_x is the same as the principal strain ϵ_2 in the longitudinal direction and that the strain ϵ_y is the same as the principal strain ϵ_1 in the circumferential direction. Also, the stress σ_x is the same as the stress σ_2, and the stress σ_y is the same as the stress σ_1. Therefore, the preceding two equations can be written in the following forms:

$$\epsilon_2 = \frac{\sigma_2}{E}(1 - 2\nu) = \frac{pr}{2tE}(1 - 2\nu) \qquad \text{(7-11a)}$$

$$\epsilon_1 = \frac{\sigma_1}{2E}(2 - \nu) = \frac{pr}{2tE}(2 - 2\nu) \qquad \text{(7-11b)}$$

Substituting numerical values, we find

$$\epsilon_2 = \frac{\sigma_2}{E}(1 - 2\nu) = \frac{(36 \text{ MPa})[1 - 2(0.30)]}{200 \text{ GPa}} = 72 \times 10^{-6}$$

$$\epsilon_1 = \frac{\sigma_1}{2E}(2 - \nu) = \frac{(72 \text{ MPa})(2 - 0.30)}{2(200 \text{ GPa})} = 306 \times 10^{-6}$$

These are the longitudinal and circumferential strains in the cylinder.

continued

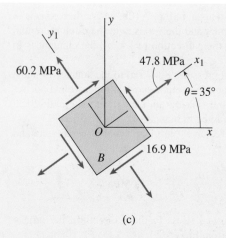

(c)

FIG. 7-10c (Repeated)

(d) *Normal and shear stresses acting on the welded seam.* The stress element at point B in the wall of the cylinder (Fig. 7-10a) is oriented so that its sides are parallel and perpendicular to the weld. The angle θ for the element is

$$\theta = 90° - \alpha = 35°$$

as shown in Fig. 7-10c. Either the stress-transformation equations or Mohr's circle may be used to obtain the normal and shear stresses acting on the side faces of this element.

Stress-transformation equations. The normal stress σ_{x_1} and the shear stress $\tau_{x_1 y_1}$ acting on the x_1 face of the element (Fig. 7-10c) are obtained from Eqs. (6-4a) and (6-4b), which are repeated here:

$$\sigma_{x_1} = \frac{\sigma_x + \sigma_y}{2} + \frac{\sigma_x - \sigma_y}{2} \cos 2\theta + \tau_{xy} \sin 2\theta \qquad (7\text{-}12a)$$

$$\tau_{x_1 y_1} = -\frac{\sigma_x - \sigma_y}{2} \sin 2\theta + \tau_{xy} \cos 2\theta \qquad (7\text{-}12b)$$

Substituting $\sigma_x = \sigma_2 = pr/2t$, $\sigma_y = \sigma_1 = pr/t$, and $\tau_{xy} = 0$, we obtain

$$\sigma_{x_1} = \frac{pr}{4t}(3 - \cos 2\theta) \qquad \tau_{x_1 y_1} = \frac{pr}{4t} \sin 2\theta \qquad (7\text{-}13a,b)$$

These equations give the normal and shear stresses acting on an inclined plane oriented at an angle θ with the longitudinal axis of the cylinder.

Substituting $pr/4t = 18$ MPa and $\theta = 35°$ into Eqs. (7-13a) and (7-13b), we obtain

$$\sigma_{x_1} = 47.8 \text{ MPa} \qquad \tau_{x_1 y_1} = 16.9 \text{ MPa}$$

These stresses are shown on the stress element of Fig. 7-10c.

To complete the stress element, we can calculate the normal stress σ_{y_1} acting on the y_1 face of the element from the sum of the normal stresses on perpendicular faces (Eq. 6-6):

$$\sigma_1 + \sigma_2 = \sigma_{x_1} + \sigma_{y_1} \qquad (7\text{-}14)$$

Substituting numerical values, we get

$$\sigma_{y_1} = \sigma_1 + \sigma_2 - \sigma_{x_1} = 72 \text{ MPa} + 36 \text{ MPa} - 47.8 \text{ MPa} = 60.2 \text{ MPa}$$

as shown in Fig. 7-10c.

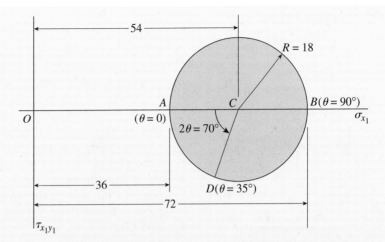

FIG. 7-11 Mohr's circle for the biaxial stress element of Fig. 7-10b. (*Note:* All stresses on the circle have units of MPa.)

From the figure, we see that the normal and shear stresses acting perpendicular and parallel, respectively, to the welded seam are

$$\sigma_w = 47.8 \text{ MPa} \qquad \tau_w = 16.9 \text{ MPa}$$

Mohr's circle. The Mohr's circle construction for the biaxial stress element of Fig. 7-10b is shown in Fig. 7-11. Point A represents the stress $\sigma_2 = 36$ MPa on the x face ($\theta = 0$) of the element, and point B represents the stress $\sigma_1 = 72$ MPa on the y face ($\theta = 90°$). The center C of the circle is at a stress of 54 MPa, and the radius of the circle is

$$R = \frac{72 \text{ MPa} - 36 \text{ MPa}}{2} = 18 \text{ MPa}$$

A counterclockwise angle $2\theta = 70°$ (measured on the circle from point A) locates point D, which corresponds to the stresses on the x_1 face ($\theta = 35°$) of the element. The coordinates of point D (from the geometry of the circle) are

$$\sigma_{x_1} = 54 \text{ MPa} - R\cos 70° = 54 \text{ MPa} - (18 \text{ MPa})(\cos 70°) = 47.8 \text{ MPa}$$

$$\tau_{x_1 y_1} = R \sin 70° = (18 \text{ MPa})(\sin 70°) = 16.9 \text{ MPa}$$

These results are the same as those found earlier from the stress-transformation equations.

Note: When seen in a side view, a **helix** follows the shape of a sine curve (Fig. 7-12). The pitch of the helix is

$$p = \pi d \tan \theta \tag{7-15}$$

where d is the diameter of the circular cylinder and θ is the angle between a normal to the helix and a longitudinal line. The width of the flat plate that wraps into the cylindrical shape is

$$w = \pi d \sin \theta \tag{7-16}$$

Thus, if the diameter of the cylinder and the angle θ are given, both the pitch and the plate width are established. For practical reasons, the angle θ is usually in the range from 20° to 35°.

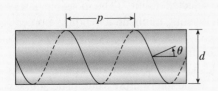

FIG. 7-12 Side view of a helix

7.4 COMBINED LOADINGS

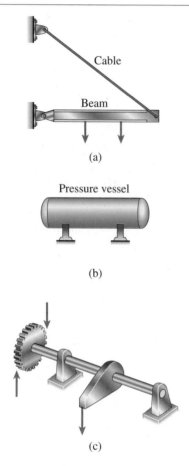

FIG. 7-13 Examples of structures subjected to combined loadings: (a) wide-flange beam supported by a cable (combined bending and axial load), (b) cylindrical pressure vessel supported as a beam, and (c) shaft in combined torsion and bending

In previous chapters we analyzed structural members subjected to a single type of loading. For instance, we analyzed axially loaded bars in Chapters 1 and 2, shafts in torsion in Chapter 3, and beams in bending in Chapters 4, and 5. We also analyzed pressure vessels earlier in this chapter. For each type of loading, we developed methods for finding stresses, strains, and deformations.

However, in many structures the members are required to resist more than one kind of loading. For example, a beam may be subjected to the simultaneous action of bending moments and axial forces (Fig. 7-13a), a pressure vessel may be supported so that it also functions as a beam (Fig. 7-13b), or a shaft in torsion may carry a bending load (Fig. 7-13c). Known as **combined loadings**, situations similar to those shown in Fig. 7-13 occur in a great variety of machines, buildings, vehicles, tools, equipment, and many other kinds of structures.

A structural member subjected to combined loadings can often be analyzed by superimposing the stresses and strains caused by each load acting separately. However, superposition of both stresses and strains is permissible only under certain conditions, as explained in earlier chapters. One requirement is that the stresses and strains must be linear functions of the applied loads, which in turn requires that the material follow Hooke's law and the displacements remain small.

A second requirement is that there must be no interaction between the various loads, that is, the stresses and strains due to one load must not be affected by the presence of the other loads. Most ordinary structures satisfy these two conditions, and therefore the use of superposition is very common in engineering work.

Method of Analysis

While there are many ways to analyze a structure subjected to more than one type of load, the procedure usually includes the following steps:

1. Select a point in the structure where the stresses and strains are to be determined. (The point is usually selected at a cross section where the stresses are large, such as at a cross section where the bending moment has its maximum value.)
2. For each load on the structure, determine the stress resultants at the cross section containing the selected point. (The possible stress resultants are an axial force, a twisting moment, a bending moment, and a shear force.)
3. Calculate the normal and shear stresses at the selected point due to each of the stress resultants. Also, if the structure is a pressure vessel, determine the stresses due to the internal pressure. (The stresses are found from the stress formulas derived previously; for instance, $\sigma = P/A$, $\tau = T\rho/I_P$, $\sigma = My/I$, $\tau = VQ/Ib$, and $\sigma = pr/t$.)
4. Combine the individual stresses to obtain the resultant stresses at the selected point. In other words, obtain the stresses σ_x, σ_y, and τ_{xy}

acting on a stress element at the point. (Note that in this chapter we are dealing only with elements in plane stress.)

5. Determine the principal stresses and maximum shear stresses at the selected point, using either the stress-transformation equations or Mohr's circle. If required, determine the stresses acting on other inclined planes.

6. Determine the strains at the point with the aid of Hooke's law for plane stress.

7. Select additional points and repeat the process. Continue until enough stress and strain information is available to satisfy the purposes of the analysis.

Illustration of the Method

To illustrate the procedure for analyzing a member subjected to combined loadings, we will discuss in general terms the stresses in the cantilever bar of circular cross section shown in Fig. 7-14a. This bar is subjected to two types of load—a torque T and a vertical load P, both acting at the free end of the bar.

Let us begin by arbitrarily selecting two points A and B for investigation (Fig. 7-14a). Point A is located at the top of the bar and point B is located on the side. Both points are located at the same cross section.

The stress resultants acting at the cross section (Fig. 7-14b) are a twisting moment equal to the torque T, a bending moment M equal to the load P times the distance b from the free end of the bar to the cross section, and a shear force V equal to the load P.

The stresses acting at points A and B are shown in Fig. 7-14c. The twisting moment T produces torsional shear stresses

$$\tau_1 = \frac{Tr}{I_P} = \frac{2T}{\pi r^3} \tag{a}$$

in which r is the radius of the bar and $I_P = \pi r^4/2$ is the polar moment of inertia of the cross-sectional area. The stress τ_1 acts horizontally to the left at point A and vertically downward at point B, as shown in the figure.

The bending moment M produces a tensile stress at point A:

$$\sigma_A = \frac{Mr}{I} = \frac{4M}{\pi r^3} \tag{b}$$

in which $I = \pi r^4/4$ is the moment of inertia about the neutral axis. However, the bending moment produces no stress at point B, because B is located on the neutral axis.

The shear force V produces no shear stress at the top of the bar (point A), but at point B the shear stress is as follows (see Eq. 5-39 in Chapter 5):

$$\tau_2 = \frac{4V}{3A} = \frac{4V}{3\pi r^2} \tag{c}$$

in which $A = \pi r^2$ is the cross-sectional area.

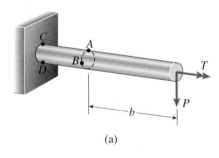

(a)

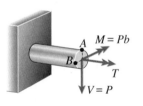

(b)

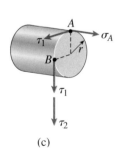

(c)

FIG. 7-14 Cantilever bar subjected to combined torsion and bending: (a) loads acting on the bar, (b) stress resultants at a cross section, and (c) stresses at points A and B

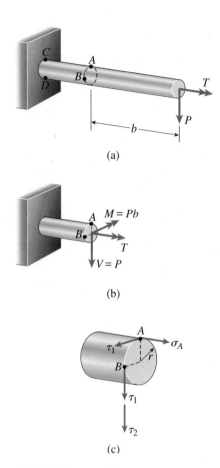

(a)

(b)

(c)

FIG. 7-14 (Repeated)

The stresses σ_A and τ_1 acting at point A (Fig. 7-14c) are shown acting on a stress element in Fig. 7-15a. This element is cut from the top of the bar at point A. A two-dimensional view of the element, obtained by looking vertically downward on the element, is shown in Fig. 7-15b. For the purpose of determining the principal stresses and maximum shear stresses, we construct x and y axes through the element. The x axis is parallel to the longitudinal axis of the circular bar (Fig. 7-14a) and the y axis is horizontal. Note that the element is in plane stress with $\sigma_x = \sigma_A$, $\sigma_y = 0$, and $\tau_{xy} = -\tau_1$.

A stress element at point B (also in plane stress) is shown in Fig. 7-16a. The only stresses acting on this element are the shear stresses, equal to $\tau_1 + \tau_2$ (see Fig. 7-14c). A two-dimensional view of the stress element is shown in Fig. 7-16b, with the x axis parallel to the longitudinal axis of the bar and the y axis in the vertical direction. The stresses acting on the element are $\sigma_x = \sigma_y = 0$ and $\tau_{xy} = -(\tau_1 + \tau_2)$.

Now that we have determined the stresses acting at points A and B and constructed the corresponding stress elements, we can use the transformation equations of plane stress (Sections 6.2 and 6.3) or Mohr's circle (Section 6.4) to determine principal stresses, maximum shear stresses, and stresses acting in inclined directions. We can also use Hooke's law (Section 6.5) to determine the strains at points A and B.

The procedure described previously for analyzing the stresses at points A and B (Fig. 7-14a) can be used at other points in the bar. Of particular interest are the points where the stresses calculated from the flexure and shear formulas have maximum or minimum values, called **critical points**. For instance, the normal stresses due to bending are largest at the cross section of maximum bending moment, which is at the support. Therefore, points C and D at the top and bottom of the beam at the fixed end (Fig. 7-14a) are critical points where the stresses should be calculated. Another critical point is point B itself, because the shear stresses are a maximum at this point. (Note that in this example the shear stresses do not change if point B is moved along the bar in the longitudinal direction.)

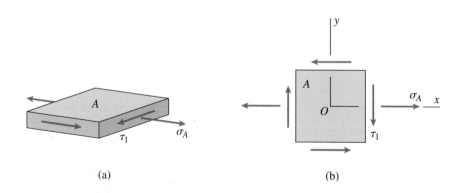

FIG. 7-15 Stress element at point A (a)   (b)

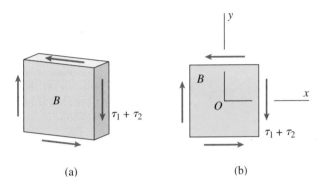

FIG. 7-16 Stress element at point B

(a)

(b)

As a final step, the principal stresses and maximum shear stresses at the critical points can be compared with one another in order to determine the absolute maximum normal and shear stresses in the bar.

This example illustrates the general procedure for determining the stresses produced by combined loadings. Note that no new theories are involved—only applications of previously derived formulas and concepts. Since the variety of practical situations seems to be endless, we will not derive general formulas for calculating the maximum stresses. Instead, we will treat each structure as a special case.

Selection of Critical Points

If the objective of the analysis is to determine the largest stresses *anywhere* in the structure, then the critical points should be selected at cross sections where the stress resultants have their largest values. Furthermore, within those cross sections, the points should be selected where either the normal stresses or the shear stresses have their largest values. By using good judgment in the selection of the points, we often can be reasonably certain of obtaining the absolute maximum stresses in the structure.

However, it is sometimes difficult to recognize in advance where the maximum stresses in the member are to be found. Then it may be necessary to investigate the stresses at a large number of points, perhaps even using trial-and-error in the selection of points. Other strategies may also prove fruitful—such as deriving equations specific to the problem at hand or making simplifying assumptions to facilitate an otherwise difficult analysis.

The following examples illustrate the methods used to calculate stresses in structures subjected to combined loadings.

Example 7-3

(a)

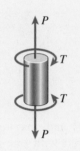

(b)

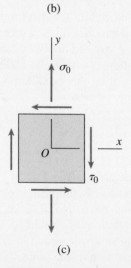

(c)

FIG. 7-17 Example 7-3. Rotor shaft of a helicopter (combined torsion and axial force)

The rotor shaft of a helicopter drives the rotor blades that provide the lifting force to support the helicopter in the air (Fig. 7-17a). As a consequence, the shaft is subjected to a combination of torsion and axial loading (Fig. 7-17b).

For a 50-mm diameter shaft transmitting a torque $T = 2.4$ kN·m and a tensile force $P = 125$ kN, determine the maximum tensile stress, maximum compressive stress, and maximum shear stress in the shaft.

Solution

The stresses in the rotor shaft are produced by the combined action of the axial force P and the torque T (Fig. 7-17b). Therefore, the stresses at any point on the surface of the shaft consist of a tensile stress σ_0 and shear stresses τ_0, as shown on the stress element of Fig. 7-17c. Note that the y axis is parallel to the longitudinal axis of the shaft.

The tensile stress σ_0 equals the axial force divided by the cross-sectional area:

$$\sigma_0 = \frac{P}{A} = \frac{4P}{\pi d^2} = \frac{4(125 \text{ kN})}{\pi (50 \text{ mm})^2} = 63.66 \text{ MPa}$$

The shear stress τ_0 is obtained from the torsion formula (see Eqs. 3-11 and 3-12 of Section 3.3):

$$\tau_0 = \frac{Tr}{I_P} = \frac{16T}{\pi d^3} = \frac{16(2.4 \text{ kN·m})}{\pi (50 \text{ mm})^3} = 97.78 \text{ MPa}$$

The stresses σ_0 and τ_0 act directly on cross sections of the shaft.

Knowing the stresses σ_0 and τ_0, we can now obtain the principal stresses and maximum shear stresses by the methods described in Section 6.3. The principal stresses are obtained from Eq. (6-17):

$$\sigma_{1,2} = \frac{\sigma_x + \sigma_y}{2} \pm \sqrt{\left(\frac{\sigma_x - \sigma_y}{2}\right)^2 + \tau_{xy}^2} \qquad \text{(d)}$$

Substituting $\sigma_x = 0$, $\sigma_y = \sigma_0 = 63.66$ MPa, and $\tau_{xy} = -\tau_0 = -97.78$ MPa, we get

$$\sigma_{1,2} = 32 \text{ MPa} \pm 103 \text{ MPa} \quad \text{or} \quad \sigma_1 = 135 \text{ MPa} \qquad \sigma_2 = -71 \text{ MPa} \quad \Longleftarrow$$

These are the maximum tensile and compressive stresses in the rotor shaft.

The maximum in-plane shear stresses (Eq. 6-25) are

$$\tau_{\max} = \sqrt{\left(\frac{\sigma_x - \sigma_y}{2}\right)^2 + \tau_{xy}^2} \qquad \text{(e)}$$

This term was evaluated previously, so we see immediately that

$$\tau_{\max} = 103 \text{ MPa} \qquad \Longleftarrow$$

Because the principal stresses σ_1 and σ_2 have opposite signs, the maximum in-plane shear stresses are larger than the maximum out-of-plane shear stresses (see Eqs. 6-28a, b, and c and the accompanying discussion). Therefore, the maximum shear stress in the shaft is 103 MPa.

Example 7-4

A thin-walled cylindrical pressure vessel with a circular cross section is subjected to internal gas pressure p and simultaneously compressed by an axial load $P = 12$ k (Fig. 7-18a). The cylinder has inner radius $r = 2.1$ in. and wall thickness $t = 0.15$ in.

Determine the maximum allowable internal pressure p_{allow} based upon an allowable shear stress of 6500 psi in the wall of the vessel.

Solution

The stresses in the wall of the pressure vessel are caused by the combined action of the internal pressure and the axial force. Since both actions produce uniform normal stresses throughout the wall, we can select any point on the surface for investigation. At a typical point, such as point A (Fig. 7-18a), we isolate a stress element as shown in Fig. 7-18b. The x axis is parallel to the longitudinal axis of the pressure vessel and the y axis is circumferential. Note that there are no shear stresses acting on the element.

Principal stresses. The longitudinal stress σ_x is equal to the tensile stress σ_2 produced by the internal pressure (see Fig. 7-7a and Eq. 7-6) minus the compressive stress produced by the axial force; thus,

$$\sigma_x = \frac{pr}{2t} - \frac{P}{A} = \frac{pr}{2t} - \frac{P}{2\pi rt} \tag{f}$$

in which $A = 2\pi rt$ is the cross-sectional area of the cylinder. (Note that for convenience we are using the inner radius r in all calculations.)

The circumferential stress σ_y is equal to the tensile stress σ_1 produced by the internal pressure (Fig. 7-7a and Eq. 7-5):

$$\sigma_y = \frac{pr}{t} \tag{g}$$

Note that σ_y is algebraically larger than σ_x.

Since no shear stresses act on the element (Fig. 7-18), the normal stresses σ_x and σ_y are also the principal stresses:

$$\sigma_1 = \sigma_y = \frac{pr}{t} \qquad \sigma_2 = \sigma_x = \frac{pr}{2t} - \frac{P}{2\pi rt} \tag{h,i}$$

Now substituting numerical values, we obtain

$$\sigma_1 = \frac{pr}{t} = \frac{p(2.1 \text{ in.})}{0.15 \text{ in.}} = 14.0p$$

$$\sigma_2 = \frac{pr}{2t} - \frac{P}{2\pi rt} = \frac{p(2.1 \text{ in.})}{2(0.15 \text{ in.})} - \frac{12 \text{ k}}{2\pi(2.1 \text{ in.})(0.15 \text{ in.})}$$

$$= 7.0p - 6063 \text{ psi}$$

in which σ_1, σ_2, and p have units of pounds per square inch (psi).

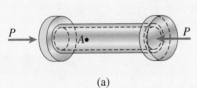

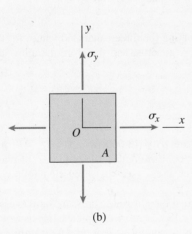

(a)

(b)

FIG. 7-18 Example 7-4. Pressure vessel subjected to combined internal pressure and axial force

continued

In-plane shear stresses. The maximum in-plane shear stress (Eq. 6-26) is

$$\tau_{max} = \frac{\sigma_1 - \sigma_2}{2} = \frac{1}{2}(14.0p - 7.0p + 6063 \text{ psi}) = 3.5p + 3032 \text{ psi}$$

Since τ_{max} is limited to 6500 psi, the preceding equation becomes

$$6500 \text{ psi} = 3.5p + 3032 \text{ psi}$$

from which we get

$$p = \frac{3468 \text{ psi}}{3.5} = 990.9 \text{ psi} \quad \text{or} \quad (p_{allow})_1 = 990 \text{ psi}$$

because we round downward.

Out-of-plane shear stresses. The maximum out-of-plane shear stress (see Eqs. 6-28a and 6-28b) is either

$$\tau_{max} = \frac{\sigma_2}{2} \quad \text{or} \quad \tau_{max} = \frac{\sigma_1}{2}$$

From the first of these two equations we get

$$6500 \text{ psi} = 3.5p - 3032 \text{ psi} \quad \text{or} \quad (p_{allow})_2 = 2720 \text{ psi}$$

From the second equation we get

$$6500 \text{ psi} = 7.0p \quad \text{or} \quad (p_{allow})_3 = 928 \text{ psi}$$

Allowable internal pressure. Comparing the three calculated values for the allowable pressure, we see that $(p_{allow})_3$ governs, and therefore the allowable internal pressure is

$$p_{allow} = 928 \text{ psi} \quad \longleftarrow$$

At this pressure the principal stresses are $\sigma_1 = 13,000$ psi and $\sigma_2 = 430$ psi. These stresses have the same signs, thus confirming that one of the out-of-plane shear stresses must be the largest shear stress (see the discussion following Eqs. 6-28a, b, and c).

Note: In this example, we determined the allowable pressure in the vessel assuming that the axial load was equal to 12 k. A more complete analysis would include the possibility that the axial force may not be present. (As it turns out, the allowable pressure does not change if the axial force is removed from this example.)

Example 7-5

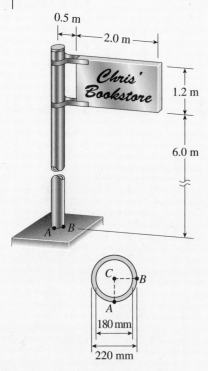

0.5 m

2.0 m

Chris' Bookstore

1.2 m

6.0 m

A B

C B

A

180 mm

220 mm

FIG. 7-19 Example 7-5. Wind pressure against a sign (combined bending, torsion, and shear of the pole)

A sign of dimensions 2.0 m × 1.2 m is supported by a hollow circular pole having outer diameter 220 mm and inner diameter 180 mm (Fig. 7-19). The sign is offset 0.5 m from the centerline of the pole and its lower edge is 6.0 m above the ground.

Determine the principal stresses and maximum shear stresses at points A and B at the base of the pole due to a wind pressure of 2.0 kPa against the sign.

Solution

Stress resultants. The wind pressure against the sign produces a resultant force W that acts at the midpoint of the sign (Fig. 7-20a) and is equal to the pressure p times the area A over which it acts:

$$W = pA = (2.0 \text{ kPa})(2.0 \text{ m} \times 1.2 \text{ m}) = 4.8 \text{ kN}$$

The line of action of this force is at height $h = 6.6$ m above the ground and at distance $b = 1.5$ m from the centerline of the pole.

The wind force acting on the sign is statically equivalent to a lateral force W and a torque T acting on the pole (Fig. 7-20b). The torque is equal to the force W times the distance b:

$$T = Wb = (4.8 \text{ kN})(1.5 \text{ m}) = 7.2 \text{ kN·m}$$

The stress resultants at the base of the pole (Fig. 7-20c) consist of a bending moment M, a torque T, and a shear force V. Their magnitudes are

$$M = Wh = (4.8 \text{ kN})(6.6 \text{ m}) = 31.68 \text{ kN·m}$$

$$T = 7.2 \text{ kN·m} \qquad V = W = 4.8 \text{ kN}$$

Examination of these stress resultants shows that maximum bending stresses occur at point A and maximum shear stresses at point B. Therefore, A and B are critical points where the stresses should be determined. (Another critical point is diametrically opposite point A, as explained in the *Note* at the end of this example.)

Stresses at points A and B. The bending moment M produces a tensile stress σ_A at point A (Fig. 7-20d) but no stress at point B (which is located on the neutral axis). The stress σ_A is obtained from the flexure formula:

$$\sigma_A = \frac{M(d_2/2)}{I}$$

in which d_2 is the outer diameter (220 mm) and I is the moment of inertia of the cross section. The moment of inertia is

$$I = \frac{\pi}{64}\left(d_2^4 - d_1^4\right) = \frac{\pi}{64}\left[(220 \text{ mm})^4 - (180 \text{ mm})^4\right] = 63.46 \times 10^{-6} \text{ m}^4$$

in which d_1 is the inner diameter. Therefore, the stress σ_A is

$$\sigma_A = \frac{Md_2}{2I} = \frac{(31.68 \text{ kN·m})(220 \text{ mm})}{2(63.46 \times 10^{-6} \text{ m}^4)} = 54.91 \text{ MPa}$$

continued

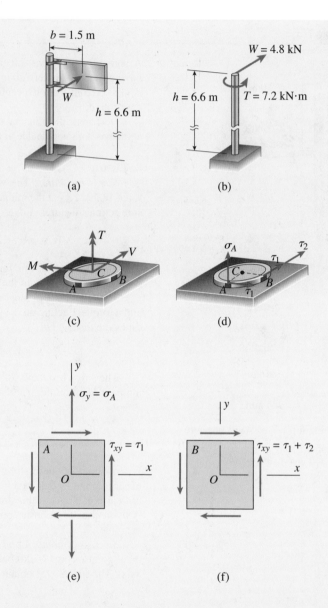

FIG. 7-20 Solution to Example 7-5

The torque T produces shear stresses τ_1 at points A and B (Fig. 7-20d). We can calculate these stresses from the torsion formula:

$$\tau_1 = \frac{T(d_2/2)}{I_P}$$

in which I_P is the polar moment of inertia:

$$I_P = \frac{\pi}{32}\left(d_2^4 - d_1^4\right) = 2I = 126.92 \times 10^{-6}\,\text{m}^4$$

Thus,

$$\tau_1 = \frac{Td_2}{2I_P} = \frac{(7.2 \text{ kN·m})(220 \text{ mm})}{2(126.92 \times 10^{-6} \text{ m}^4)} = 6.24 \text{ MPa}$$

Finally, we calculate the shear stresses at points A and B due to the shear force V. The shear stress at point A is zero, and the shear stress at point B (denoted τ_2 in Fig. 7-20d) is obtained from the shear formula for a circular tube (Eq. 5-41 of Section 5.8):

$$\tau_2 = \frac{4V}{3A}\left(\frac{r_2^2 + r_2 r_1 + r_1^2}{r_2^2 + r_1^2}\right) \tag{j}$$

in which r_2 and r_1 are the outer and inner radii, respectively, and A is the cross-sectional area:

$$r_2 = \frac{d_2}{2} = 110 \text{ mm} \qquad r_1 = \frac{d_1}{2} = 90 \text{ mm}$$

$$A = \pi(r_2^2 - r_1^2) = 12{,}570 \text{ mm}^2$$

Substituting numerical values into Eq. (j), we obtain

$$\tau_2 = 0.76 \text{ MPa}$$

The stresses acting on the cross section at points A and B have now been calculated.

Stress elements. The next step is to show these stresses on stress elements (Figs. 7-20e and f). For both elements, the y axis is parallel to the longitudinal axis of the pole and the x axis is horizontal. At point A the stresses acting on the element are

$$\sigma_x = 0 \qquad \sigma_y = \sigma_A = 54.91 \text{ MPa} \qquad \tau_{xy} = \tau_1 = 6.24 \text{ MPa}$$

At point B the stresses are

$$\sigma_x = \sigma_y = 0 \qquad \tau_{xy} = \tau_1 + \tau_2 = 6.24 \text{ MPa} + 0.76 \text{ MPa} = 7.00 \text{ MPa}$$

Since there are no normal stresses acting on the element, point B is in pure shear.

Now that all stresses acting on the stress elements (Figs. 7-20e and f) are known, we can use the equations given in Section 6.3 to determine the principal stresses and maximum shear stresses.

continued

Principal stresses and maximum shear stresses at point A. The principal stresses are obtained from Eq. (6-17), which is repeated here:

$$\sigma_{1,2} = \frac{\sigma_x + \sigma_y}{2} \pm \sqrt{\left(\frac{\sigma_x - \sigma_y}{2}\right)^2 + \tau_{xy}^2} \qquad \text{(k)}$$

Substituting $\sigma_x = 0$, $\sigma_y = 54.91$ MPa, and $\tau_{xy} = 6.24$ MPa, we get

$$\sigma_{1,2} = 27.5 \text{ MPa} \pm 28.2 \text{ MPa}$$

or

$$\sigma_1 = 55.7 \text{ MPa} \qquad \sigma_2 = -0.7 \text{ MPa} \qquad \Longleftarrow$$

The maximum in-plane shear stresses may be obtained from Eq. (6-25):

$$\tau_{max} = \sqrt{\left(\frac{\sigma_x - \sigma_y}{2}\right)^2 + \tau_{xy}^2} \qquad \text{(l)}$$

This term was evaluated previously, so we see immediately that

$$\tau_{max} = 28.2 \text{ MPa} \qquad \Longleftarrow$$

Because the principal stresses σ_1 and σ_2 have opposite signs, the maximum in-plane shear stresses are larger than the maximum out-of-plane shear stresses (see Eqs. 6-28a, b, and c and the accompanying discussion). Therefore, the maximum shear stress at point A is 28.2 MPa.

Principal stresses and maximum shear stresses at point B. The stresses at this point are $\sigma_x = 0$, $\sigma_y = 0$, and $\tau_{xy} = 7.0$ MPa. Since the element is in pure shear, the principal stresses are

$$\sigma_1 = 7.0 \text{ MPa} \qquad \sigma_2 = -7.0 \text{ MPa} \qquad \Longleftarrow$$

and the maximum in-plane shear stress is

$$\tau_{max} = 7.0 \text{ MPa} \qquad \Longleftarrow$$

The maximum out-of-plane shear stresses are half this value.

Note: If the largest stresses anywhere in the pole are needed, then we must also determine the stresses at the critical point diametrically opposite point A, because at that point the compressive stress due to bending has its largest value. The principal stresses at that point are

$$\sigma_1 = 0.7 \text{ MPa} \qquad \sigma_2 = -55.7 \text{ MPa}$$

and the maximum shear stress is 28.2 MPa. Therefore, the largest tensile stress in the pole is 55.7 MPa, the largest compressive stress is -55.7 MPa, and the largest shear stress is 28.2 MPa. (Keep in mind that only the effects of the wind pressure are considered in this analysis. Other loads, such as the weight of the structure, also produce stresses at the base of the pole.)

Example 7-6

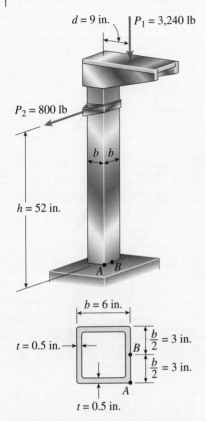

$d = 9$ in. $P_1 = 3,240$ lb

$P_2 = 800$ lb

b b

$h = 52$ in.

A B

$b = 6$ in.

$t = 0.5$ in.

B $\dfrac{b}{2} = 3$ in.

$\dfrac{b}{2} = 3$ in.

A

$t = 0.5$ in.

FIG. 7-21 Example 7-6. Loads on a post
(combined axial load, bending, and
shear)

A tubular post of square cross section supports a horizontal platform (Fig. 7-21). The tube has outer dimension $b = 6$ in. and wall thickness $t = 0.5$ in. The plat-form has dimensions 6.75 in. $\times$ 24.0 in. and supports a uniformly distributed load of 20 psi acting over its upper surface. The resultant of this distributed load is a vertical force P_1:

$$P_1 = (20 \text{ psi})(6.75 \text{ in.} \times 24.0 \text{ in.}) = 3240 \text{ lb}$$

This force acts at the midpoint of the platform, which is at distance $d = 9$ in. from the longitudinal axis of the post. A second load $P_2 = 800$ lb acts horizon-tally on the post at height $h = 52$ in. above the base.

Determine the principal stresses and maximum shear stresses at points A and B at the base of the post due to the loads P_1 and P_2.

Solution

Stress resultants. The force P_1 acting on the platform (Fig. 7-21) is statically equivalent to a force P_1 and a moment $M_1 = P_1 d$ acting at the centroid of the cross section of the post (Fig. 7-22a). The load P_2 is also shown in this figure.

The stress resultants at the base of the post due to the loads P_1 and P_2 and the moment M_1 are shown in Fig. 7-22b. These stress resultants are the following:

1. An axial compressive force $P_1 = 3240$ lb
2. A bending moment M_1 produced by the force P_1:

$$M_1 = P_1 d = (3240 \text{ lb})(9 \text{ in.}) = 29{,}160 \text{ lb-in.}$$

3. A shear force $P_2 = 800$ lb
4. A bending moment M_2 produced by the force P_2:

$$M_2 = P_2 h = (800 \text{ lb})(52 \text{ in.}) = 41{,}600 \text{ lb-in.}$$

Examination of these stress resultants (Fig. 7-22b) shows that both M_1 and M_2 produce maximum compressive stresses at point A and the shear force produces maximum shear stresses at point B. Therefore, A and B are critical points where the stresses should be determined. (Another critical point is diagonally opposite point A, as explained in the *Note* at the end of this example.)

Stresses at points A and B.

(1) The axial force P_1 (Fig. 7-22b) produces uniform compressive stresses throughout the post. These stresses are

$$\sigma_{P_1} = \frac{P_1}{A}$$

in which A is the cross-sectional area of the post:

$$A = b^2 - (b - 2t)^2 = 4t(b - t)$$

$$= 4(0.5 \text{ in.})(6 \text{ in.} - 0.5 \text{ in.}) = 11.00 \text{ in.}^2$$

continued

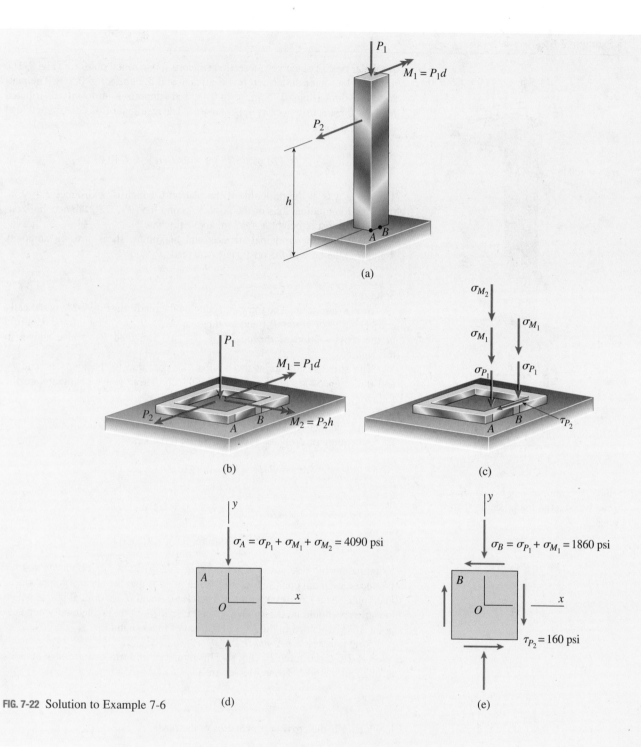

FIG. 7-22 Solution to Example 7-6

Therefore, the axial compressive stress is

$$\sigma_{P_1} = \frac{P_1}{A} = \frac{3240 \text{ lb}}{11.00 \text{ in.}^2} = 295 \text{ psi}$$

The stress σ_{P_1} is shown acting at points A and B in Fig. 7-22c.

(2) The bending moment M_1 (Fig. 7-22b) produces compressive stresses σ_{M_1} at points A and B (Fig. 7-22c). These stresses are obtained from the flexure formula:

$$\sigma_{M_1} = \frac{M_1(b/2)}{I} = \frac{M_1 b}{2I}$$

in which I is the moment of inertia of the cross-sectional area:

$$I = \frac{b^4}{12} - \frac{(b - 2t)^4}{12} = \frac{1}{12}\left[(6 \text{ in.})^4 - (5 \text{ in.})^4\right] = 55.92 \text{ in.}^4$$

Thus, the stress σ_{M_1} is

$$\sigma_{M_1} = \frac{M_1 b}{2I} = \frac{(29{,}160 \text{ lb-in.})(6 \text{ in.})}{2(55.92 \text{ in.}^4)} = 1564 \text{ psi}$$

(3) The shear force P_2 (Fig. 7-22b) produces a shear stress at point B but not at point A. From the discussion of shear stresses in the webs of beams with flanges (Section 5.9), we know that an approximate value of the shear stress can be obtained by dividing the shear force by the web area (see Eq. 5-47 in Section 5.9). Thus, the shear stress produced at point B by the force P_2 is

$$\tau_{P_2} = \frac{P_2}{A_{\text{web}}} = \frac{P_2}{2t(b - 2t)} = \frac{800 \text{ lb}}{2(0.5 \text{ in.})(6 \text{ in.} - 1 \text{ in.})} = 160 \text{ psi}$$

The stress τ_{P_2} acts at point B in the direction shown in Fig. 7-22c.

If desired, we can calculate the shear stress τ_{P_2} from the more accurate formula of Eq. (5-45a) in Section 5.9. The result of that calculation is $\tau_{P_2} = 163$ psi, which shows that the shear stress obtained from the approximate formula is satisfactory.

(4) The bending moment M_2 (Fig. 7-22b) produces a compressive stress at point A but no stress at point B. The stress at A is

$$\sigma_{M_2} = \frac{M_2(b/2)}{I} = \frac{M_2 b}{2I} = \frac{(41{,}600 \text{ lb-in.})(6 \text{ in.})}{2(55.92 \text{ in.}^4)} = 2232 \text{ psi}$$

This stress is also shown in Fig. 7-22c.

Stress elements. The next step is to show the stresses acting on stress elements at points A and B (Figs. 7-22d and e). Each element is oriented so that the y axis is vertical (that is, parallel to the longitudinal axis of the post) and the x axis is horizontal. At point A the only stress is a compressive stress σ_A in the y direction (Fig. 7-22d):

$$\sigma_A = \sigma_{P_1} + \sigma_{M_1} + \sigma_{M_2}$$

$$= 295 \text{ psi} + 1564 \text{ psi} + 2232 \text{ psi} = 4090 \text{ psi (compression)}$$

Thus, this element is in uniaxial stress.

continued

At point B the compressive stress in the y direction (Fig. 7-22e) is

$$\sigma_B = \sigma_{P_1} + \sigma_{M_1} = 295 \text{ psi} + 1564 \text{ psi} = 1860 \text{ psi (compression)}$$

and the shear stress is

$$\tau_{P_2} = 160 \text{ psi}$$

The shear stress acts leftward on the top face of the element and downward on the x face of the element.

Principal stresses and maximum shear stresses at point A. Using the standard notation for an element in plane stress (Fig. 7-23), we write the stresses for element A (Fig. 7-22d) as follows:

$$\sigma_x = 0 \qquad \sigma_y = -\sigma_A = -4090 \text{ psi} \qquad \tau_{xy} = 0$$

Since the element is in uniaxial stress, the principal stresses are

$$\sigma_1 = 0 \qquad \sigma_2 = -4090 \text{ psi} \qquad \longleftarrow$$

and the maximum in-plane shear stress (Eq. 7-26) is

$$\tau_{\max} = \frac{\sigma_1 - \sigma_2}{2} = \frac{4090 \text{ psi}}{2} = 2050 \text{ psi} \qquad \longleftarrow$$

The maximum out-of-plane shear stress (Eq. 6-28a) has the same magnitude.

Principal stresses and maximum shear stresses at point B. Again using the standard notation for plane stress (Fig. 7-23), we see that the stresses at point B (Fig. 7-22e) are

$$\sigma_x = 0 \qquad \sigma_y = -\sigma_B = -1860 \text{ psi} \qquad \tau_{xy} = -\tau_{P_2} = -160 \text{ psi}$$

To obtain the principal stresses, we use Eq. (6-17), which is repeated here:

$$\sigma_{1,2} = \frac{\sigma_x + \sigma_y}{2} \pm \sqrt{\left(\frac{\sigma_x - \sigma_y}{2}\right)^2 + \tau_{xy}^2} \qquad \text{(m)}$$

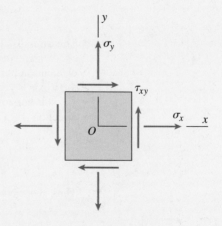

FIG. 7-23 Notation for an element in plane stress

Substituting for σ_x, σ_y, and τ_{xy}, we get

$$\sigma_{1,2} = -930 \text{ psi} \pm 944 \text{ psi}$$

or

$$\sigma_1 = 14 \text{ psi} \qquad \sigma_2 = -1870 \text{ psi} \quad \Longleftarrow$$

The maximum in-plane shear stresses may be obtained from Eq. (6-25):

$$\tau_{max} = \sqrt{\left(\frac{\sigma_x - \sigma_y}{2}\right)^2 + \tau_{xy}^2} \qquad (n)$$

This term was evaluated previously, so we see immediately that

$$\tau_{max} = 944 \text{ psi} \quad \Longleftarrow$$

Because the principal stresses σ_1 and σ_2 have opposite signs, the maximum in-plane shear stresses are larger than the maximum out-of-plane shear stresses (see Eqs. 6-28a, b, and c and the accompanying discussion). Therefore, the maximum shear stress at point B is 944 psi.

Note: If the largest stresses anywhere at the base of the post are needed, then we must also determine the stresses at the critical point diagonally opposite point A (Fig. 7-22c), because at that point each bending moment produces the maximum tensile stress. Thus, the tensile stress acting at that point is

$$\sigma_y = -\sigma_{P_1} + \sigma_{M_1} + \sigma_{M_2} = -295 \text{ psi} + 1564 \text{ psi} + 2232 \text{ psi} = 3500 \text{ psi}$$

The stresses acting on a stress element at that point (see Fig. 7-23) are

$$\sigma_x = 0 \qquad \sigma_y = 3500 \text{ psi} \qquad \tau_{xy} = 0$$

and therefore the principal stresses and maximum shear stress are

$$\sigma_1 = 3500 \text{ psi} \qquad \sigma_2 = 0 \qquad \tau_{max} = 1750 \text{ psi}$$

Thus, the largest tensile stress anywhere at the base of the post is 3500 psi, the largest compressive stress is 4090 psi, and the largest shear stress is 2050 psi. (Keep in mind that only the effects of the loads P_1 and P_2 are considered in this analysis. Other loads, such as the weight of the structure, also produce stresses at the base of the post.)

CHAPTER SUMMARY & REVIEW

In Chapter 7, we investigated some practical examples of structures in states of plane stress, building upon the material presented in Sections 6.2 through 6.5 in the previous chapter. First, we considered the stresses in thin-walled spherical and cylindrical vessels, such as storage tanks containing compressed gases or liquids. Then we evaluated the maximum normal and shear stresses at various points in structures or components acted upon by combined loadings. The major concepts and findings presented in this chapter are as follows:

1. Plane stress is a common stress condition that exists in all ordinary structures, such as in the walls of pressure vessels, in the webs and/or flanges of beams of various shapes, and in a wide variety of structures subject to the combined effects of axial, shear, and bending loads, as well as internal pressure.

2. The wall of a pressurized thin-walled **spherical vessel** is in a state of plane stress—specifically, biaxial stress—with uniform tensile stresses known as membrane stresses σ acting in all directions. The tensile stresses σ in the wall of a spherical shell may be calculated as:

$$\sigma = \frac{pr}{2t}$$

Only the excess of internal pressure over external pressure, or gage pressure, has any effect on these stresses. Additional important considerations for more detailed analysis or design of spherical vessels include: stress concentrations around openings, effects of external loads and self weight (including contents), and influence of corrosion, impacts, and temperature changes.

3. The walls of thin-walled **cylindrical pressure vessels** with circular cross sections are also in a state of biaxial stress. The circumferential stress σ_1 is referred to as the hoop stress, and the stress parallel to the axis of the tank is called the longitudinal stress or the axial stress σ_2. The circumferential stress is equal to twice the longitudinal stress. Both are principal stresses. The formulas for σ_1 and σ_2 are

$$\sigma_1 = \frac{pr}{t} \qquad \sigma_2 = \frac{pr}{2t}$$

These formulas were derived using elementary theory for thin-walled shells and are only valid in parts of the cylinder away from any discontinuities that cause stress concentrations.

4. A structural member subjected to **combined loadings** often can be analyzed by superimposing the stresses and strains caused by each load acting separately. However, the stresses and strains must be linear functions of the applied loads, which in turn requires that the material follow Hooke's law and the displacements remain small. There must be no interaction between the various loads, that is, the stresses and strains due to one load must not be affected by the presence of the other loads.

5. A detailed approach for analysis of critical points in a structure or component subjected to more than one type of load is presented in Section 7.4.

PROBLEMS CHAPTER 7

Spherical Pressure Vessels

When solving the problems for Section 7.2, assume that the given radius or diameter is an inside dimension and that all internal pressures are gage pressures.

7.2-1 A large spherical tank (see figure) contains gas at a pressure of 450 psi. The tank is 42 ft in diameter and is constructed of high-strength steel having a yield stress in tension of 80 ksi.

Determine the required thickness (to the nearest 1/4 inch) of the wall of the tank if a factor of safety of 3.5 with respect to yielding is required.

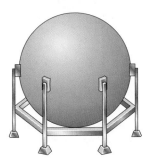

PROBS. 7.2-1 and 7.2-2

7.2-2 Solve the preceding problem if the internal pressure is 3.75 MPa, the diameter is 19 m, the yield stress is 570 MPa, and the factor of safety is 3.0.

Determine the required thickness to the nearest millimeter.

7.2-3 A hemispherical window (or *viewport*) in a decompression chamber (see figure) is subjected to an internal air pressure of 80 psi. The port is attached to the wall of the chamber by 18 bolts.

Find the tensile force F in each bolt and the tensile stress σ in the viewport if the radius of the hemisphere is 7.0 in. and its thickness is 1.0 in.

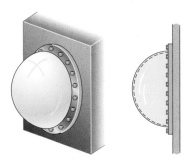

PROB. 7.2-3

7.2-4 A rubber ball (see figure) is inflated to a pressure of 60 kPa. At that pressure the diameter of the ball is 230 mm and the wall thickness is 1.2 mm. The rubber has modulus of elasticity $E = 3.5$ MPa and Poisson's ratio $\nu = 0.45$.

Determine the maximum stress and strain in the ball.

PROB. 7.2-4

7.2-5 Solve the preceding problem if the pressure is 9.0 psi, the diameter is 9.0 in., the wall thickness is 0.05 in., the modulus of elasticity is 500 psi, and Poisson's ratio is 0.45.

PROB. 7.2-5

7.2-6 A spherical steel pressure vessel (diameter 480 mm, thickness 8.0 mm) is coated with brittle lacquer that cracks when the strain reaches 150×10^{-6} (see figure).

What internal pressure p will cause the lacquer to develop cracks? (Assume $E = 205$ GPa and $\nu = 0.30$.)

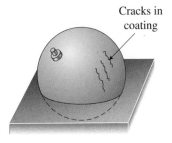

Cracks in coating

PROB. 7.2-6

7.2-7 A spherical tank of diameter 48 in. and wall thickness 1.75 in. contains compressed air at a pressure of 2200 psi. The tank is constructed of two hemispheres joined by a welded seam (see figure).

(a) What is the tensile load f (lb per in. of length of weld) carried by the weld?

(b) What is the maximum shear stress τ_{max} in the wall of the tank?

(c) What is the maximum normal strain ε in the wall? (For steel, assume $E = 30 \times 10^6$ psi and $\nu = 0.29$.)

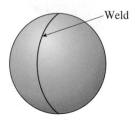

PROBS. 7.2-7 and 7.2-8

7.2-8 Solve the preceding problem for the following data: diameter 1.0 m, thickness 48 mm, pressure 22 MPa, modulus 210 GPa, and Poisson's ratio 0.29.

7.2-9 A spherical stainless-steel tank having a diameter of 22 in. is used to store propane gas at a pressure of 2450 psi. The properties of the steel are as follows: yield stress in tension, 140,000 psi; yield stress in shear, 65,000 psi; modulus of elasticity, 30×10^6 psi; and Poisson's ratio, 0.28. The desired factor of safety with respect to yielding is 2.8. Also, the normal strain must not exceed 1100×10^{-6}.

Determine the minimum permissible thickness t_{min} of the tank.

7.2-10 Solve the preceding problem if the diameter is 500 mm, the pressure is 18 MPa, the yield stress in tension is 975 MPa, the yield stress in shear is 460 MPa, the factor of safety is 2.5, the modulus of elasticity is 200 GPa, Poisson's ratio is 0.28, and the normal strain must not exceed 1210×10^{-6}.

7.2-11 A hollow pressurized sphere having radius $r = 4.8$ in. and wall thickness $t = 0.4$ in. is lowered into a lake (see figure). The compressed air in the tank is at a pressure of 24 psi (gage pressure when the tank is out of the water).

At what depth D_0 will the wall of the tank be subjected to a compressive stress of 90 psi?

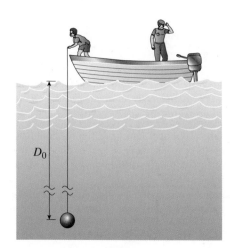

PROB. 7.2-11

Cylindrical Pressure Vessels

When solving the problems for Section 7.3, assume that the given radius or diameter is an inside dimension and that all internal pressures are gage pressures.

7.3-1 A scuba tank (see figure) is being designed for an internal pressure of 1600 psi with a factor of safety of 2.0 with respect to yielding. The yield stress of the steel is 35,000 psi in tension and 16,000 psi in shear.

If the diameter of the tank is 7.0 in., what is the minimum required wall thickness?

PROB. 7.3-1

7.3-2 A tall standpipe with an open top (see figure) has diameter $d = 2.2$ m and wall thickness $t = 20$ mm.

(a) What height h of water will produce a circumferential stress of 12 MPa in the wall of the standpipe?

(b) What is the axial stress in the wall of the tank due to the water pressure?

PROB. 7.3-2

7.3-3 An inflatable structure used by a traveling circus has the shape of a half-circular cylinder with closed ends (see figure). The fabric and plastic structure is inflated by a small blower and has a radius of 40 ft when fully inflated. A longitudinal seam runs the entire length of the "ridge" of the structure.

If the longitudinal seam along the ridge tears open when it is subjected to a tensile load of 540 pounds per inch of seam, what is the factor of safety n against tearing when the internal pressure is 0.5 psi and the structure is fully inflated?

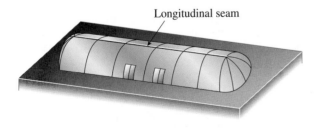

Longitudinal seam

PROB. 7.3-3

7.3-4 A thin-walled cylindrical pressure vessel of radius r is subjected simultaneously to internal gas pressure p and a compressive force F acting at the ends (see figure).

What should be the magnitude of the force F in order to produce pure shear in the wall of the cylinder?

PROB. 7.3-4

7.3-5 A strain gage is installed in the longitudinal direction on the surface of an aluminum beverage can (see figure). The radius-to-thickness ratio of the can is 200. When the lid of the can is popped open, the strain changes by $\epsilon_0 = 170 \times 10^{-6}$.

What was the internal pressure p in the can? (Assume $E = 10 \times 10^6$ psi and $\nu = 0.33$.)

PROB. 7.3-5

7.3-6 A circular cylindrical steel tank (see figure) contains a volatile fuel under pressure. A strain gage at point A records the longitudinal strain in the tank and transmits this information to a control room. The ultimate shear stress in the wall of the tank is 84 MPa, and a factor of safety of 2.5 is required.

At what value of the strain should the operators take action to reduce the pressure in the tank? (Data for the steel are as follows: modulus of elasticity $E = 205$ GPa and Poisson's ratio $\nu = 0.30$.)

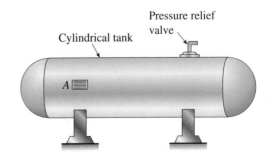

Pressure relief valve

Cylindrical tank

A

PROB. 7.3-6

7.3-7 A cylinder filled with oil is under pressure from a piston, as shown in the figure. The diameter d of the piston is 1.80 in. and the compressive force F is 3500 lb. The maximum allowable shear stress τ_{allow} in the wall of the cylinder is 5500 psi.

What is the minimum permissible thickness $t_{\min}$ of the cylinder wall? (See figure.)

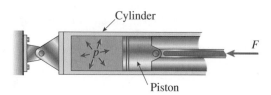

Cylinder

F

Piston

PROBS. 7.3-7 and 7.3-8

7.3-8 Solve the preceding problem if $d = 90$ mm, $F = 42$ kN, and $\tau_{\text{allow}} = 40$ MPa.

7.3-9 A standpipe in a water-supply system (see figure) is 12 ft in diameter and 6 inches thick. Two horizontal pipes carry water out of the standpipe; each is 2 ft in diameter and 1 inch thick. When the system is shut down and water fills the pipes but is not moving, the hoop stress at the bottom of the standpipe is 130 psi.

(a) What is the height h of the water in the standpipe?

(b) If the bottoms of the pipes are at the same elevation as the bottom of the standpipe, what is the hoop stress in the pipes?

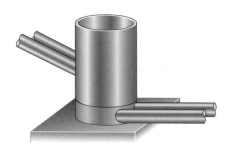

PROB. 7.3-9

7.3-10 A cylindrical tank with hemispherical heads is constructed of steel sections that are welded circumferentially (see figure). The tank diameter is 1.25 m, the wall thickness is 22 mm, and the internal pressure is 1750 kPa.

(a) Determine the maximum tensile stress σ_h in the heads of the tank.

(b) Determine the maximum tensile stress σ_c in the cylindrical part of the tank.

(c) Determine the tensile stress σ_w acting perpendicular to the welded joints.

(d) Determine the maximum shear stress τ_h in the heads of the tank.

(e) Determine the maximum shear stress τ_c in the cylindrical part of the tank.

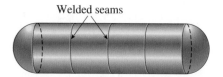

Welded seams

PROBS. 7.3-10 and 7.3-11

7.3-11 A cylindrical tank with diameter $d = 18$ in. is subjected to internal gas pressure $p = 450$ psi. The tank is constructed of steel sections that are welded circumferentially (see figure). The heads of the tank are hemispherical. The allowable tensile and shear stresses are 8200 psi and 3000 psi, respectively. Also, the allowable tensile stress perpendicular to a weld is 6250 psi.

Determine the minimum required thickness $t_{\min}$ of (a) the cylindrical part of the tank and (b) the hemispherical heads.

7.3-12 A pressurized steel tank is constructed with a helical weld that makes an angle $\alpha = 55°$ with the longitudinal axis (see figure). The tank has radius $r = 0.6$ m, wall thickness $t = 18$ mm, and internal pressure $p = 2.8$ MPa. Also, the steel has modulus of elasticity $E = 200$ GPa and Poisson's ratio $\nu = 0.30$.

Determine the following quantities for the cylindrical part of the tank.

(a) The circumferential and longitudinal stresses.

(b) The maximum in-plane and out-of-plane shear stresses.

(c) The circumferential and longitudinal strains.

(d) The normal and shear stresses acting on planes parallel and perpendicular to the weld (show these stresses on a properly oriented stress element).

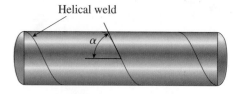

Helical weld

α

PROBS. 7.3-12 and 7.3-13

7.3-13 Solve the preceding problem for a welded tank with $\alpha = 62°$, $r = 19$ in., $t = 0.65$ in., $p = 240$ psi, $E = 30 \times 10^6$ psi, and $\nu = 0.30$.

Combined Loadings

The problems for Section 7.4 are to be solved assuming that the structures behave linearly elastically and that the stresses caused by two or more loads may be superimposed to obtain the resultant stresses acting at a point. Consider both in-plane and out-of-plane shear stresses unless otherwise specified.

7.4-1 A bracket $ABCD$ having a hollow circular cross section consists of a vertical arm AB, a horizontal arm BC parallel to the x_0 axis, and a horizontal arm CD parallel to the z_0 axis (see figure). The arms BC and CD have lengths $b_1 = 3.6$ ft and $b_2 = 2.2$ ft, respectively. The outer and inner diameters of the bracket are $d_2 = 7.5$ in. and $d_1 = 6.8$ in. A vertical load $P = 1400$ lb acts at point D. Determine the maximum tensile, compressive, and shear stresses in the vertical arm.

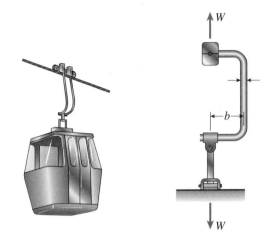

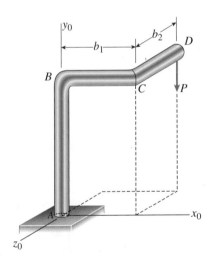

PROB. 7.4-1

7.4-2 A gondola on a ski lift is supported by two bent arms, as shown in the figure. Each arm is offset by the distance $b = 180$ mm from the line of action of the weight force W. The allowable stresses in the arms are 100 MPa in tension and 50 MPa in shear.

If the loaded gondola weighs 12 kN, what is the minimum diameter d of the arms?

PROB. 7.4-2

7.4-3 The hollow drill pipe for an oil well (see figure) is 6.2 in. in outer diameter and 0.75 in. in thickness. Just above the bit, the compressive force in the pipe (due to the weight of the pipe) is 62 k and the torque (due to drilling) is 185 k-in.

Determine the maximum tensile, compressive, and shear stresses in the drill pipe.

PROB. 7.4-3

7.4-4 A segment of a generator shaft is subjected to a torque T and an axial force P, as shown in the figure. The shaft is hollow (outer diameter $d_2 = 300$ mm and inner diameter $d_1 = 250$ mm) and delivers 1800 kW at 4.0 Hz.

If the compressive force $P = 540$ kN, what are the maximum tensile, compressive, and shear stresses in the shaft?

PROBS. 7.4-4 and 7.4-5

7.4-5 A segment of a generator shaft of hollow circular cross section is subjected to a torque $T = 240$ k-in. (see figure). The outer and inner diameters of the shaft are 8.0 in. and 6.25 in., respectively.

What is the maximum permissible compressive load P that can be applied to the shaft if the allowable in-plane shear stress is $\tau_{\text{allow}} = 6250$ psi?

7.4-6 A cylindrical tank subjected to internal pressure p is simultaneously compressed by an axial force $F = 72$ kN (see figure). The cylinder has diameter $d = 100$ mm and wall thickness $t = 4$ mm.

Calculate the maximum allowable internal pressure p_{max} based upon an allowable shear stress in the wall of the tank of 60 MPa.

PROB. 7.4-6

7.4-7 A cylindrical tank having diameter $d = 2.5$ in. is subjected to internal gas pressure $p = 600$ psi and an external tensile load $T = 1000$ lb (see figure).

Determine the minimum thickness t of the wall of the tank based on an allowable shear stress of 3000 psi.

7.4-8 The torsional pendulum shown in the figure consists of a horizontal circular disk of mass $M = 60$ kg suspended by a vertical steel wire ($G = 80$ GPa) of length $L = 2$ m and diameter $d = 4$ mm.

Calculate the maximum permissible angle of rotation ϕ_{max} of the disk (that is, the maximum amplitude of torsional vibrations) so that the stresses in the wire do not exceed 100 MPa in tension or 50 MPa in shear.

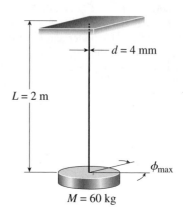

PROB. 7.4-8

7.4-9 Determine the maximum tensile, compressive, and shear stresses at points A and B on the bicycle pedal crank shown in the figure.

The pedal and crank are in a horizontal plane and points A and B are located on the top of the crank. The load $P = 160$ lb acts in the vertical direction and the distances (in the horizontal plane) between the line of action of the load and points A and B are $b_1 = 5.0$ in., $b_2 = 2.5$ in. and $b_3 = 1.0$ in. Assume that the crank has a solid circular cross section with diameter $d = 0.6$ in.

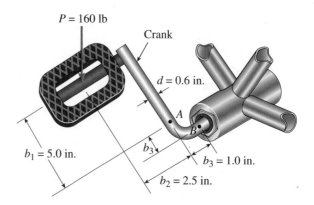

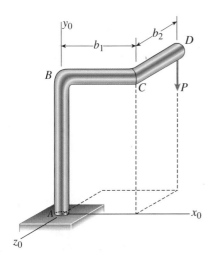

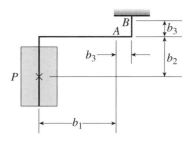

Top view

PROB. 7.4-9

7.4-10 A cylindrical pressure vessel having radius $r = 300$ mm and wall thickness $t = 15$ mm is subjected to internal pressure $p = 2.5$ MPa. In addition, a torque $T = 120$ kN·m acts at each end of the cylinder (see figure).

(a) Determine the maximum tensile stress σ_{max} and the maximum in-plane shear stress τ_{max} in the wall of the cylinder.

(b) If the allowable in-plane shear stress is 30 MPa, what is the maximum allowable torque T?

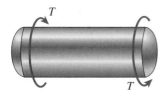

PROB. 7.4-10

7.4-11 An L-shaped bracket lying in a horizontal plane supports a load $P = 150$ lb (see figure). The bracket has a hollow rectangular cross section with thickness $t = 0.125$ in. and outer dimensions $b = 2.0$ in. and $h = 3.5$ in. The centerline lengths of the arms are $b_1 = 20$ in. and $b_2 = 30$ in.

Considering only the load P, calculate the maximum tensile stress σ_t, maximum compressive stress σ_c, and maximum shear stress τ_{max} at point A, which is located on the top of the bracket at the support.

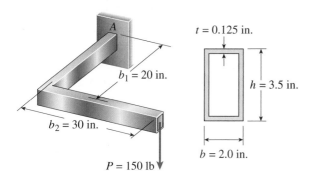

PROB. 7.4-11

7.4-12 A semicircular bar AB lying in a horizontal plane is supported at B (see figure). The bar has centerline radius R and weight q per unit of length (total weight of the bar equals πqR). The cross section of the bar is circular with diameter d.

Obtain formulas for the maximum tensile stress σ_t, maximum compressive stress σ_c, and maximum in-plane shear stress τ_{max} at the top of the bar at the support due to the weight of the bar.

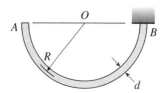

PROB. 7.4-12

7.4-13 An arm ABC lying in a horizontal plane and supported at A (see figure) is made of two identical solid steel bars AB and BC welded together at a right angle. Each bar is 20 in. long.

Knowing that the maximum tensile stress (principal stress) at the top of the bar at support A due solely to the weights of the bars is 932 psi, determine the diameter d of the bars.

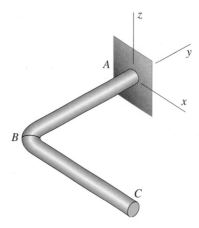

PROB. 7.4-13

7.5-14 A pressurized cylindrical tank with flat ends is loaded by torques T and tensile forces P (see figure on the next page). The tank has radius $r = 50$ mm and wall thickness $t = 3$ mm. The internal pressure $p = 3.5$ MPa and the torque $T = 450$ N·m.

What is the maximum permissible value of the forces P if the allowable tensile stress in the wall of the cylinder is 72 MPa?

PROB. 7.4-14

7.4-15 A post having a hollow circular cross section supports a horizontal load $P = 240$ lb acting at the end of an arm that is 5 ft long (see figure). The height of the post is 27 ft, and its section modulus is $S = 15$ in.[3] Assume that outer radius of the post, $r_2 = 4.5$ inches, and inner radius $r_1 = 4.243$ inches.

(a) Calculate the maximum tensile stress σ_{max} and maximum in-plane shear stress τ_{max} at point A on the outer surface of the post along the x-axis due to the load P. Load P acts in a horizontal plane at an angle of 30° from a line which is parallel to the $(-x)$ axis.

(b) If the maximum tensile stress and maximum in-plane shear stress at point A are limited to 16,000 psi and 6000 psi, respectively, what is the largest permissible value of the load P?

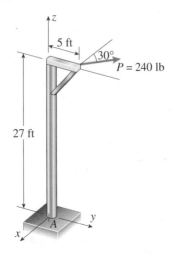

PROB. 7.4-15

7.4-16 A sign is supported by a pipe (see figure) having outer diameter 110 mm and inner diameter 90 mm. The dimensions of the sign are 2.0 m × 1.0 m, and its lower edge is 3.0 m above the base. Note that the center of gravity of the sign is 1.05 m from the axis of the pipe. The wind pressure against the sign is 1.5 kPa.

Determine the maximum in-plane shear stresses due to the wind pressure on the sign at points A, B, and C, located on the outer surface at the base of the pipe.

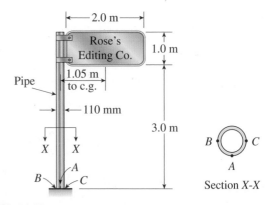

PROB. 7.4-16

7.4-17 A sign is supported by a pole of hollow circular cross section, as shown in the figure. The outer and inner diameters of the pole are 10.5 in. and 8.5 in., respectively. The pole is 42 ft high and weighs 4.0 k. The sign has dimensions 8 ft × 3 ft and weighs 500 lb. Note that its center of gravity is 53.25 in. from the axis of the pole. The wind pressure against the sign is 35 lb/ft².

(a) Determine the stresses acting on a stress element at point A, which is on the outer surface of the pole at the "front" of the pole, that is, the part of the pole nearest to the viewer.

(b) Determine the maximum tensile, compressive, and shear stresses at point A.

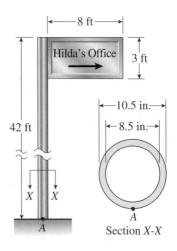

PROB. 7.4-17

7.4-18 A horizontal bracket *ABC* consists of two perpendicular arms *AB* of length 0.75 m, and *BC* of length of 0.5 m. The bracket has a solid circular cross section with diameter equal to 65 mm. The bracket is inserted in a frictionless sleeve at *A* (which is slightly larger in diameter) so is free to rotate about the z_0 axis at *A*, and is supported by a pin at *C*. Moments are applied at point *C* as follows: $M_1 = 1.5$ kN · m in the *x*-direction and $M_2 = 1.0$ kN · m acts in the $(-z)$ direction.

Considering only the moments M_1 and M_2, calculate the maximum tensile stress σ_t, the maximum compressive stress σ_c, and the maximum in-plane shear stress τ_{max} at point *p*, which is located at support *A* on the side of the bracket at midheight.

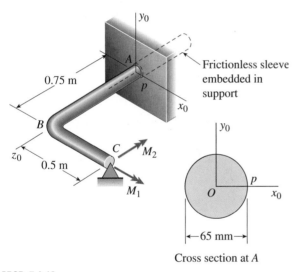

PROB. 7.4-18

7.4-19 A cylindrical pressure vessel with flat ends is subjected to a torque *T* and a bending moment *M* (see figure). The outer radius is 12.0 in. and the wall thickness is 1.0 in. The loads are as follows: $T = 800$ k-in., $M = 1000$ k-in., and the internal pressure $p = 900$ psi.

Determine the maximum tensile stress σ_t, maximum compressive stress σ_c, and maximum shear stress τ_{max} in the wall of the cylinder.

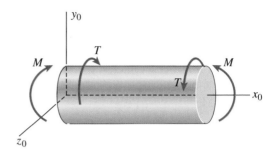

PROB. 7.4-19

7.4-20 For purposes of analysis, a segment of the crankshaft in a vehicle is represented as shown in the figure. Two loads *P* act as shown, one parallel to $(-x_0)$ and another parallel to z_0; each load *P* equals 1.0 kN. The crankshaft dimensions are $b_1 = 80$ mm, $b_2 = 120$ mm, and $b_3 = 40$ mm. The diameter of the upper shaft is $d = 20$ mm.

(a) Determine the maximum tensile, compressive, and shear stresses at point *A*, which is located on the surface of the upper shaft at the z_0 axis.

(b) Determine the maximum tensile, compressive, and shear stresses at point *B*, which is located on the surface of the shaft at the y_0 axis.

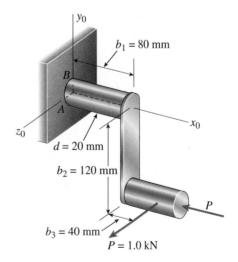

PROB. 7.4-20

7.4-21 A moveable steel stand supports an automobile engine weighing $W = 750$ lb as shown in figure part (a). The stand is constructed of 2.5 in. $\times$ 2.5 in. $\times$ 1/8in. thick steel tubing. Once in position the stand is restrained by pin supports at B and C. Of interest are stresses at point A at the base of the vertical post; point A has coordinates ($x = 1.25$, $y = 0$, $z = 1.25$) (inches). Neglect the weight of the stand.

(a) Initially, the engine weight acts in the (-z) direction through point Q which has coordinates (24,0,1.25); find the maximum tensile, compressive, and shear stresses at point A.

(b) Repeat (a) assuming now that, during repair, the engine is rotated about its own longitudinal axis (which is parallel to the x axis) so that W acts through $\mathbf{Q'}$ (with coordinates (24,6,1.25)) and force $F_y = 200$ lb is applied parallel to the y axis at distance $d = 30$ in.

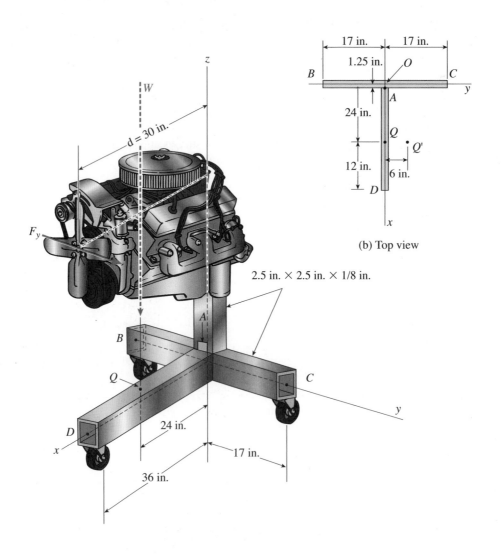

(b) Top view

PROB. 7.4-21

7.4-22 A mountain bike rider going uphill applies force $P = 65$ N to each end of the handlebars $ABCD$, made of aluminum alloy 7075-T6, by pulling on the handlebar extenders (DF on right handlebar segment). Consider the right half of the handlebar assembly only (assume the bars are fixed at the fork at A). Segments AB and CD are prismatic with lengths L_1 and L_3 and with outer diameters and thicknesses d_{01}, t_{01} and d_{03}, t_{03}, respectively, as shown. Segment BC of length L_2, however, is tapered and, outer diameter and thickness vary linearly between dimensions at B and C. Consider shear, torsion, and bending effects only for segment AD; assume DF is rigid.

Find maximum tensile, compressive, and shear stresses adjacent to support A. Show where each maximum stress value occurs.

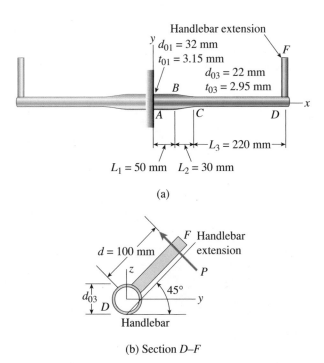

(a)

(b) Section D–F

PROB. 7.4-22

7.4-23 Determine the maximum tensile, compressive, and shear stresses acting on the cross section of the tube at point A of the hitch bicycle rack shown in the figure.

The rack is made up of 2 in. $\times$ 2 in. steel tubing which is 1/8 in. thick. Assume that the weight of each of four bicycles is distributed evenly between the two support arms so that the rack can be represented as a cantilever beam ($ABCDEF$) in the x-y plane. The overall weight of the rack alone is $W = 60$ lb. directed through C, and the weight of each bicycle is $B = 30$ lb.

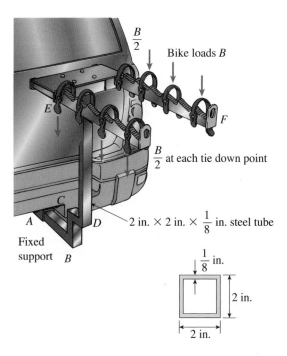

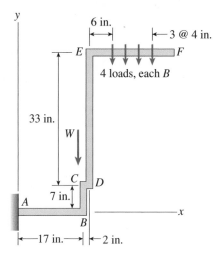

PROB. 7.4-23

Deflection of beams is an important consideration in their initial design; deflections also must be monitored during construction. (Courtesy of the National Information Service for Earthquake Engineering EERC, University of California, Berkeley)

8

Deflections of Beams

CHAPTER OVERVIEW

In Chapter 8, methods for calculation of beam deflections are presented. Beam deflections, in addition to beam stresses and strains discussed in Chapter 5, are an essential consideration in their analysis and design. A beam may be strong enough to carry a range of static or dynamic loadings (see the discussion in Sections 1.7 and 5.6), but if it deflects too much or vibrates under applied loadings, it fails to meet the "serviceability" requirements which are an important element of its overall design. Chapter 8 covers several different methods that can be used to compute either deflections (both translations and rotations) **at specific points** along the beam or **the deflected shape of the entire beam**. In general, the beam is assumed to behave in a linearly elastic manner and is restricted to small displacements (i.e., small compared to its own length). Methods based on **integration of the differential equation of the elastic curve** are discussed (Sections 8.2 through 8.4). Beam deflection results for a wide range of loadings acting on either cantilever or simple beams are summarized in Appendix H (available online) and are available for use in the **method of superposition** (Section 8.5).

Chapter 8 is organized as follows:

8.1 INTRODUCTION

When a beam with a straight longitudinal axis is loaded by lateral forces, the axis is deformed into a curve, called the **deflection curve** of the beam. In Chapter 5, we used the curvature of the bent beam to determine the normal strains and stresses in the beam. However, we did not develop a method for finding the deflection curve itself. In this chapter, we will determine the equation of the deflection curve and also find deflections at specific points along the axis of the beam.

The calculation of deflections is an important part of structural analysis and design. For example, finding deflections is an essential ingredient in the analysis of statically indeterminate structures. Deflections are also important in dynamic analyses, as when investigating the vibrations of aircraft or the response of buildings to earthquakes.

Deflections are sometimes calculated in order to verify that they are within tolerable limits. For instance, specifications for the design of buildings usually place upper limits on the deflections. Large deflections in buildings are unsightly (and even unnerving) and can cause cracks in ceilings and walls. In the design of machines and aircraft, specifications may limit deflections in order to prevent undesirable vibrations.

8.2 DIFFERENTIAL EQUATIONS OF THE DEFLECTION CURVE

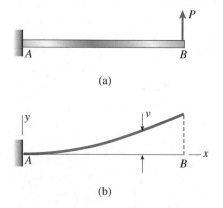

FIG. 8-1 Deflection curve of a cantilever beam

Most procedures for finding beam deflections are based on the differential equations of the deflection curve and their associated relationships. Consequently, we will begin by deriving the basic equation for the deflection curve of a beam.

For discussion purposes, consider a cantilever beam with a concentrated load acting upward at the free end (Fig. 8-1a). Under the action of this load, the axis of the beam deforms into a curve, as shown in Fig. 8-1b. The reference axes have their origin at the fixed end of the beam, with the x axis directed to the right and the y axis directed upward. The z axis is directed outward from the figure (toward the viewer).

As in our previous discussions of beam bending in Chapter 5, we assume that the xy plane is a plane of symmetry of the beam, and we assume that all loads act in this plane (the *plane of bending*).

The **deflection** v is the displacement in the y direction of any point on the axis of the beam (Fig. 8-1b). Because the y axis is positive upward, the deflections are also positive when upward.[*]

To obtain the equation of the deflection curve, we must express the deflection v as a function of the coordinate x. Therefore, let us now consider the deflection curve in more detail. The deflection v at any point

---

[*]As mentioned in Section 5.1, the traditional symbols for displacements in the x, y, and z directions are u, v, and w, respectively. The advantage of this notation is that it emphasizes the distinction between a *coordinate* and a *displacement*.

m_1 on the deflection curve is shown in Fig. 8-2a. Point m_1 is located at distance x from the origin (measured along the x axis). A second point m_2, located at distance $x + dx$ from the origin, is also shown. The deflection at this second point is $v + dv$, where dv is the increment in deflection as we move along the curve from m_1 to m_2.

When the beam is bent, there is not only a deflection at each point along the axis but also a rotation. The **angle of rotation** θ of the axis of the beam is the angle between the x axis and the tangent to the deflection curve, as shown for point m_1 in the enlarged view of Fig. 8-2b. For our choice of axes (x positive to the right and y positive upward), the angle of rotation is positive when counterclockwise. (Other names for the angle of rotation are *angle of inclination* and *angle of slope*.)

The angle of rotation at point m_2 is $\theta + d\theta$, where $d\theta$ is the increase in angle as we move from point m_1 to point m_2. It follows that if we construct lines normal to the tangents (Figs. 8-2a and b), the angle between these normals is $d\theta$. Also, as discussed earlier in Section 5.3, the point of intersection of these normals is the **center of curvature** O' (Fig. 8-2a) and the distance from O' to the curve is the **radius of curvature** ρ. From Fig. 8-2a we see that

$$\rho \, d\theta = ds \qquad\qquad \text{(a)}$$

in which $d\theta$ is in radians and ds is the distance along the deflection curve between points m_1 and m_2. Therefore, the **curvature** κ (equal to the reciprocal of the radius of curvature) is given by the equation

$$\kappa = \frac{1}{\rho} = \frac{d\theta}{ds} \qquad\qquad \text{(8-1)}$$

The **sign convention** for curvature is pictured in Fig. 8-3, which is repeated from Fig. 5-6 of Section 5.3. Note that curvature is positive

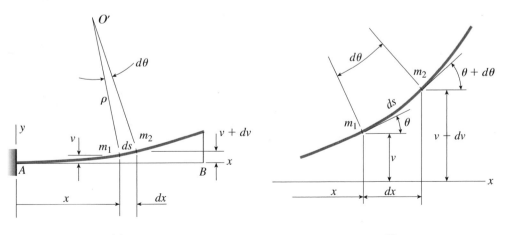

FIG. 8-2 Deflection curve of a beam   (a)   (b)

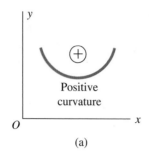

Positive curvature

(a)

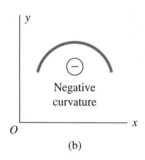

Negative curvature

(b)

FIG. 8-3 Sign convention for curvature

when the angle of rotation increases as we move along the beam in the positive x direction.

The **slope of the deflection curve** is the first derivative dv/dx of the expression for the deflection v. In geometric terms, the slope is the increment dv in the deflection (as we go from point m_1 to point m_2 in Fig. 8-2) divided by the increment dx in the distance along the x axis. Since dv and dx are infinitesimally small, the slope dv/dx is equal to the tangent of the angle of rotation θ (Fig. 8-2b). Thus,

$$\frac{dv}{dx} = \tan \theta \qquad \theta = \arctan \frac{dv}{dx} \qquad \text{(8-2a,b)}$$

In a similar manner, we also obtain the following relationships:

$$\cos \theta = \frac{dx}{ds} \qquad \sin \theta = \frac{dv}{ds} \qquad \text{(8-3a,b)}$$

Note that when the x and y axes have the directions shown in Fig. 8-2a, the slope dv/dx is positive when the tangent to the curve slopes upward to the right.

Equations (8-1) through (8-3) are based only upon geometric considerations, and therefore they are valid for beams of any material. Furthermore, there are no restrictions on the magnitudes of the slopes and deflections.

Beams with Small Angles of Rotation

The structures encountered in everyday life, such as buildings, automobiles, aircraft, and ships, undergo relatively small changes in shape while in service. The changes are so small as to be unnoticed by a casual observer. Consequently, the deflection curves of most beams and columns have very small angles of rotation, very small deflections, and very small curvatures. Under these conditions we can make some mathematical approximations that greatly simplify beam analysis.

Consider, for instance, the deflection curve shown in Fig. 8-2. If the angle of rotation θ is a very small quantity (and hence the deflection curve is nearly horizontal), we see immediately that the distance ds along the deflection curve is practically the same as the increment dx along the x axis. This same conclusion can be obtained directly from Eq. (8-3a). Since $\cos \approx 1$ when the angle θ is small, Eq. (8-3a) gives

$$ds \approx dx \qquad \text{(b)}$$

With this approximation, the curvature becomes (see Eq. 8-1)

$$\kappa = \frac{1}{\rho} = \frac{d\theta}{dx} \qquad \text{(8-4)}$$

Also, since $\tan \theta \approx \theta$ when θ is small, we can make the following approximation to Eq. (8-2a):

$$\theta \approx \tan \theta = \frac{dv}{dx} \tag{c}$$

Thus, if the rotations of a beam are small, we can assume that the angle of rotation θ and the slope dv/dx are equal. (Note that the angle of rotation must be measured in radians.)

Taking the derivative of θ with respect to x in Eq. (c), we get

$$\frac{d\theta}{dx} = \frac{d^2v}{dx^2} \tag{d}$$

Combining this equation with Eq. (8-4), we obtain a relation between the **curvature** of a beam and its deflection:

$$\kappa = \frac{1}{\rho} = \frac{d^2v}{dx^2} \tag{8-5}$$

This equation is valid for a beam of any material, provided the rotations are small quantities.

If the material of a beam is **linearly elastic** and follows Hooke's law, the curvature (from Eq. 5-12, Chapter 5) is

$$\kappa = \frac{1}{\rho} = \frac{M}{EI} \tag{8-6}$$

in which M is the bending moment and EI is the flexural rigidity of the beam. Equation (8-6) shows that a positive bending moment produces positive curvature and a negative bending moment produces negative curvature, as shown earlier in Fig. 5-10.

Combining Eq. (8-5) with Eq. (8-6) yields the basic **differential equation of the deflection curve** of a beam:

$$\frac{d^2v}{dx^2} = \frac{M}{EI} \tag{8-7}$$

This equation can be integrated in each particular case to find the deflection v, provided the bending moment M and flexural rigidity EI are known as functions of x.

As a reminder, the **sign conventions** to be used with the preceding equations are repeated here: (1) The x and y axes are positive to the right and upward, respectively; (2) the deflection v is positive upward; (3) the slope dv/dx and angle of rotation θ are positive when counterclockwise with respect to the positive x axis; (4) the curvature κ is positive when the beam is bent concave upward; and (5) the bending moment M is positive when it produces compression in the upper part of the beam.

Additional equations can be obtained from the relations between bending moment M, shear force V, and intensity q of distributed load. In

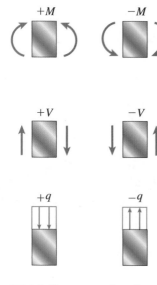

+M   −M

+V   −V

+q   −q

FIG. 8-4 Sign conventions for bending moment M, shear force V, and intensity q of distributed load

Chapter 4 we derived the following equations between M, V, and q (see Eqs. 4-4 and 4-6):

$$\frac{dV}{dx} = -q \qquad \frac{dM}{dx} = V \qquad (8\text{-}8a,b)$$

The sign conventions for these quantities are shown in Fig. 8-4. By differentiating Eq. (8-7) with respect to x and then substituting the preceding equations for shear force and load, we can obtain the additional equations. In so doing, we will consider two cases, nonprismatic beams and prismatic beams.

Nonprismatic Beams

In the case of a nonprismatic beam, the flexural rigidity EI is variable, and therefore we write Eq. (8-7) in the form

$$EI_x \frac{d^2v}{dx^2} = M \qquad (8\text{-}9a)$$

where the subscript x is inserted as a reminder that the flexural rigidity may vary with x. Differentiating both sides of this equation and using Eqs. (8-8a) and (8-8b), we obtain

$$\frac{d}{dx}\left(EI_x \frac{d^2v}{dx^2}\right) = \frac{dM}{dx} = V \qquad (8\text{-}9b)$$

$$\frac{d^2}{dx^2}\left(EI_x \frac{d^2v}{dx^2}\right) = \frac{dV}{dx} = -q \qquad (8\text{-}9c)$$

The deflection of a nonprismatic beam can be found by solving (either analytically or numerically) any one of the three preceding differential equations. The choice usually depends upon which equation provides the most efficient solution.

Prismatic Beams

In the case of a prismatic beam (constant EI), the differential equations become

$$EI\frac{d^2v}{dx^2} = M \qquad EI\frac{d^3v}{dx^3} = V \qquad EI\frac{d^4v}{dx^4} = -q \qquad (8\text{-}10a,b,c)$$

To simplify the writing of these and other equations, **primes** are often used to denote differentiation:

$$v' \equiv \frac{dv}{dx} \qquad v'' \equiv \frac{d^2v}{dx^2} \qquad v''' \equiv \frac{d^3v}{dx^3} \qquad v'''' \equiv \frac{d^4v}{dx^4} \qquad (8\text{-}11)$$

Using this notation, we can express the differential equations for a prismatic beam in the following forms:

$$EIv'' = M \qquad EIv''' = V \qquad EIv'''' = -q \qquad (8\text{-}12a,b,c)$$

We will refer to these equations as the **bending-moment equation**, the **shear-force equation**, and the **load equation**, respectively.

In the next two sections we will use the preceding equations to find deflections of beams. The general procedure consists of integrating the equations and then evaluating the constants of integration from boundary and other conditions pertaining to the beam.

When deriving the differential equations (Eqs. 8-9, 8-10, and 8-12), we assumed that the material followed Hooke's law and that the slopes of the deflection curve were very small. We also assumed that any shear deformations were negligible; consequently, we considered only the deformations due to pure bending. All of these assumptions are satisfied by most beams in common use.

Exact Expression for Curvature

If the deflection curve of a beam has large slopes, we cannot use the approximations given by Eqs. (b) and (c). Instead, we must resort to the exact expressions for curvature and angle of rotation (see Eqs. 8-1 and 8-2b). Combining those expressions, we get

$$\kappa = \frac{1}{\rho} = \frac{d\theta}{ds} = \frac{d(\arctan v')}{dx} \frac{dx}{ds} \tag{e}$$

From Fig. 8-2 we see that

$$ds^2 = dx^2 + dv^2 \quad \text{or} \quad ds = [dx^2 + dv^2]^{1/2} \tag{f,g}$$

Dividing both sides of Eq. (g) by dx gives

$$\frac{ds}{dx} = \left[1 + \left(\frac{dv}{dx} \right)^2 \right]^{1/2} = [1 + (v')^2]^{1/2} \quad \text{or} \quad \frac{dx}{ds} = \frac{1}{[1 + (v')^2]^{1/2}} \tag{h,i}$$

Also, differentiation of the arctangent function (see Appendix D available online) gives

$$\frac{d}{dx} (\arctan v') = \frac{v''}{1 + (v')^2} \tag{j}$$

Substitution of expressions (i) and (j) into the equation for curvature (Eq. e) yields

$$\kappa = \frac{1}{\rho} = \frac{v''}{[1 + (v')^2]^{3/2}} \tag{8-13}$$

Comparing this equation with Eq. (8-5), we see that the assumption of small rotations is equivalent to disregarding $(v')^2$ in comparison to one. Equation (8-13) should be used for the curvature whenever the slopes are large.[*]

---

[*]The basic relationship stating that the curvature of a beam is proportional to the bending moment (Eq. 8-6) was first obtained by Jacob Bernoulli, although he obtained an incorrect value for the constant of proportionality. The relationship was used later by Euler, who solved the differential equation of the deflection curve for both large deflections (using Eq. 8-13) and small deflections (using Eq. 8-7). For the history of deflection curves, see Ref. 8-1. A list of references is available online.

8.3 DEFLECTIONS BY INTEGRATION OF THE BENDING-MOMENT EQUATION

$v_A = 0$ $v_B = 0$

FIG. 8-5 Boundary conditions at simple supports

We are now ready to solve the differential equations of the deflection curve and obtain deflections of beams. The first equation we will use is the bending-moment equation (Eq. 8-12a). Since this equation is of second order, two integrations are required. The first integration produces the slope $v' = dv/dx$, and the second produces the deflection v.

We begin the analysis by writing the equation (or equations) for the bending moments in the beam. Since only statically determinate beams are considered in this chapter, we can obtain the bending moments from free-body diagrams and equations of equilibrium, using the procedures described in Chapter 4. In some cases a single bending-moment expression holds for the entire length of the beam, as illustrated in Examples 8-1 and 8-2. In other cases the bending moment changes abruptly at one or more points along the axis of the beam. Then we must write separate bending-moment expressions for each region of the beam between points where changes occur, as illustrated in Example 8-3.

Regardless of the number of bending-moment expressions, the general procedure for solving the differential equations is as follows. For each region of the beam, we substitute the expression for M into the differential equation and integrate to obtain the slope v'. Each such integration produces one constant of integration. Next, we integrate each slope equation to obtain the corresponding deflection v. Again, each integration produces a new constant. Thus, there are two constants of integration for each region of the beam. These constants are evaluated from known conditions pertaining to the slopes and deflections. The conditions fall into three categories: (1) boundary conditions, (2) continuity conditions, and (3) symmetry conditions.

Boundary conditions pertain to the deflections and slopes at the supports of a beam. For example, at a simple support (either a pin or a roller) the deflection is zero (Fig. 8-5), and at a fixed support both the deflection and the slope are zero (Fig. 8-6). Each such boundary condition supplies one equation that can be used to evaluate the constants of integration.

$v_A = 0$

$v_A = 0$

FIG. 8-6 Boundary conditions at a fixed support

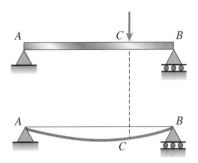

At point C: $(v)_{AC} = (v)_{CB}$
$(v')_{AC} = (v')_{CB}$

FIG. 8-7 Continuity conditions at point C

Continuity conditions occur at points where the regions of integration meet, such as at point C in the beam of Fig. 8-7. The deflection curve of this beam is physically continuous at point C, and therefore the deflection at point C as determined for the left-hand part of the beam must be equal to the deflection at point C as determined for the right-hand part. Similarly, the slopes found for each part of the beam must be equal at point C. Each of these continuity conditions supplies an equation for evaluating the constants of integration.

Symmetry conditions may also be available. For instance, if a simple beam supports a uniform load throughout its length, we know in advance that the slope of the deflection curve at the midpoint must be zero. This condition supplies an additional equation, as illustrated in Example 8-1.

Each boundary, continuity, and symmetry condition leads to an equation containing one or more of the constants of integration. Since the number of *independent* conditions always matches the number of constants of integration, we can always solve these equations for the constants. (The boundary and continuity conditions alone are always sufficient to determine the constants. Any symmetry conditions provide additional equations, but they are not independent of the other equations. The choice of which conditions to use is a matter of convenience.)

Once the constants are evaluated, they can be substituted back into the expressions for slopes and deflections, thus yielding the final equations of the deflection curve. These equations can then be used to obtain the deflections and angles of rotation at particular points along the axis of the beam.

The preceding method for finding deflections is sometimes called the **method of successive integrations.** The following examples illustrate the method in detail.

Note: When sketching deflection curves, such as those shown in the following examples and in Figs. 8-5, 8-6, and 8-7, we greatly exaggerate the deflections for clarity. However, it should always be kept in mind that the actual deflections are very small quantities.

Example 8-1

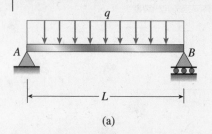

(a)

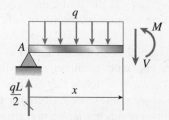

(b)

FIG. 8-8 Example 8-1. Deflections of a simple beam with a uniform load

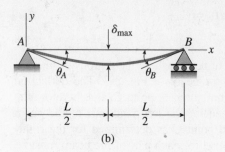

FIG. 8-9 Free-body diagram used in determining the bending moment M (Example 8-1)

Determine the equation of the deflection curve for a simple beam AB supporting a uniform load of intensity q acting throughout the span of the beam (Fig. 8-8a).

Also, determine the maximum deflection δ_{max} at the midpoint of the beam and the angles of rotation θ_A and θ_B at the supports (Fig. 8-8b). (*Note:* The beam has length L and constant flexural rigidity EI.)

Solution

Bending moment in the beam. The bending moment at a cross section distance x from the left-hand support is obtained from the free-body diagram of Fig. 8-9. Since the reaction at the support is $qL/2$, the equation for the bending moment is

$$M = \frac{qL}{2}(x) - qx\left(\frac{x}{2}\right) = \frac{qLx}{2} - \frac{qx^2}{2} \tag{8-14}$$

Differential equation of the deflection curve. By substituting the expression for the bending moment (Eq. 8-14) into the differential equation (Eq. 8-12a), we obtain

$$EIv'' = \frac{qLx}{2} - \frac{qx^2}{2} \tag{8-15}$$

This equation can now be integrated to obtain the slope and deflection of the beam.

Slope of the beam. Multiplying both sides of the differential equation by dx, we get the following equation:

$$EIv''\,dx = \frac{qLx}{2}\,dx - \frac{qx^2}{2}\,dx$$

Integrating each term, we obtain

$$EI\int v''\,dx = \int \frac{qLx}{2}\,dx - \int \frac{qx^2}{2}\,dx$$

or

$$EIv' = \frac{qLx^2}{4} - \frac{qx^3}{6} + C_1 \tag{a}$$

in which C_1 is a constant of integration.

To evaluate the constant C_1, we observe from the symmetry of the beam and its load that the slope of the deflection curve at midspan is equal to zero. Thus, we have the following symmetry condition:

$$v' = 0 \quad \text{when} \quad x = \frac{L}{2}$$

This condition may be expressed more succinctly as

$$v'\left(\frac{L}{2}\right) = 0$$

Applying this condition to Eq. (a) gives

$$0 = \frac{qL}{4}\left(\frac{L}{2}\right)^2 - \frac{q}{6}\left(\frac{L}{2}\right)^3 + C_1 \quad \text{or} \quad C_1 = -\frac{qL^3}{24}$$

The equation for the slope of the beam (Eq. a) then becomes

$$EIv' = \frac{qLx^2}{4} - \frac{qx^3}{6} - \frac{qL^3}{24} \tag{b}$$

or $$v' = -\frac{q}{24EI}(L^3 - 6Lx^2 + 4x^3) \tag{8-16} \Longleftarrow$$

As expected, the slope is negative (i.e., clockwise) at the left-hand end of the beam ($x = 0$), positive at the right-hand end ($x = L$), and equal to zero at the midpoint ($x = L/2$).

Deflection of the beam. The deflection is obtained by integrating the equation for the slope. Thus, upon multiplying both sides of Eq. (b) by dx and integrating, we obtain

$$EIv = \frac{qLx^3}{12} - \frac{qx^4}{24} - \frac{qL^3x}{24} + C_2 \tag{c}$$

The constant of integration C_2 may be evaluated from the condition that the deflection of the beam at the left-hand support is equal to zero; that is, $v = 0$ when $x = 0$, or

$$v(0) = 0$$

Applying this condition to Eq. (c) yields $C_2 = 0$; hence the equation for the deflection curve is

$$EIv = \frac{qLx^3}{12} - \frac{qx^4}{24} - \frac{qL^3x}{24} \tag{d}$$

or $$v = -\frac{qx}{24EI}(L^3 - 2Lx^2 + x^3) \tag{8-17} \Longleftarrow$$

This equation gives the deflection at any point along the axis of the beam. Note that the deflection is zero at both ends of the beam ($x = 0$ and $x = L$) and negative elsewhere (recall that downward deflections are negative).

continued

Maximum deflection. From symmetry we know that the maximum deflection occurs at the midpoint of the span (Fig. 8-8b). Thus, setting x equal to $L/2$ in Eq. (8-17), we obtain

$$v\left(\frac{L}{2}\right) = -\frac{5qL^4}{384EI}$$

in which the negative sign means that the deflection is downward (as expected). Since δ_{max} represents the magnitude of this deflection, we obtain

$$\delta_{max} = \left|v\left(\frac{L}{2}\right)\right| = \frac{5qL^4}{384EI} \qquad (8\text{-}18) \quad \Longleftarrow$$

Angles of rotation. The maximum angles of rotation occur at the supports of the beam. At the left-hand end of the beam, the angle θ_A, which is a clockwise angle (Fig. 8-8b), is equal to the negative of the slope v'. Thus, by substituting $x = 0$ into Eq. (8-16), we find

$$\theta_A = -v'(0) = \frac{qL^3}{24EI} \qquad (8\text{-}19) \quad \Longleftarrow$$

In a similar manner, we can obtain the angle of rotation θ_B at the right-hand end of the beam. Since θ_B is a counterclockwise angle, it is equal to the slope at the end:

$$\theta_B = v'(L) = \frac{qL^3}{24EI} \qquad (8\text{-}20) \quad \Longleftarrow$$

Because the beam and loading are symmetric about the midpoint, the angles of rotation at the ends are equal.

This example illustrates the process of setting up and solving the differential equation of the deflection curve. It also illustrates the process of finding slopes and deflections at selected points along the axis of a beam.

Note: Now that we have derived formulas for the maximum deflection and maximum angles of rotation (see Eqs. 8-18, 8-19, and 8-20), we can evaluate those quantities numerically and observe that the deflections and angles are indeed small, as the theory requires.

Consider a steel beam on simple supports with a span length $L = 6$ ft. The cross section is rectangular with width $b = 3$ in. and height $h = 6$ in. The intensity of uniform load is $q = 8000$ lb/ft, which is relatively large because it produces a stress in the beam of 24,000 psi. (Thus, the deflections and slopes are larger than would normally be expected.)

Substituting into Eq. (8-18), and using $E = 30 \times 10^6$ psi, we find that the maximum deflection is $\delta_{max} = 0.144$ in., which is only 1/500 of the span length. Also, from Eq. (8-19), we find that the maximum angle of rotation is $\theta_A = 0.0064$ radians, or 0.37°, which is a very small angle.

Thus, our assumption that the slopes and deflections are small is validated.

Example 8-2

Determine the equation of the deflection curve for a cantilever beam AB subjected to a uniform load of intensity q (Fig. 8-10a).

Also, determine the angle of rotation θ_B and the deflection δ_B at the free end (Fig. 8-10b). (*Note:* The beam has length L and constant flexural rigidity EI.)

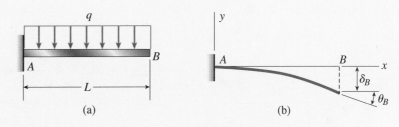

FIG. 8-10 Example 8-2. Deflections of a cantilever beam with a uniform load

(a)   (b)

Solution

Bending moment in the beam. The bending moment at distance x from the fixed support is obtained from the free-body diagram of Fig. 8-11. Note that the vertical reaction at the support is equal to qL and the moment reaction is equal to $qL^2/2$. Consequently, the expression for the bending moment M is

$$M = -\frac{qL^2}{2} + qLx - \frac{qx^2}{2} \tag{8-21}$$

Differential equation of the deflection curve. When the preceding expression for the bending moment is substituted into the differential equation (Eq. 8-12a), we obtain

$$EIv'' = -\frac{qL^2}{2} + qLx - \frac{qx^2}{2} \tag{8-22}$$

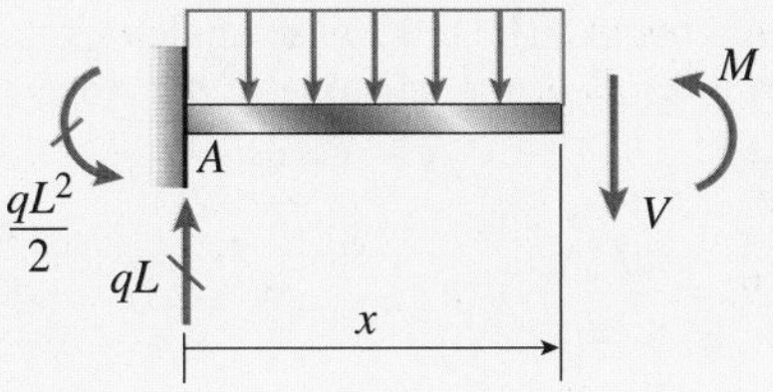

We now integrate both sides of this equation to obtain the slopes and deflections.

Slope of the beam. The first integration of Eq. (8-22) gives the following equation for the slope:

$$EIv' = -\frac{qL^2x}{2} + \frac{qLx^2}{2} - \frac{qx^3}{6} + C_1 \tag{e}$$

The constant of integration C_1 can be found from the boundary condition that the slope of the beam is zero at the support; thus, we have the following condition:

$$v'(0) = 0$$

continued

FIG. 8-11 Free-body diagram used in determining the bending moment M (Example 8-2)

When this condition is applied to Eq. (e) we get $C_1 = 0$. Therefore, Eq. (e) becomes

$$EIv' = -\frac{qL^2x}{2} + \frac{qLx^2}{2} - \frac{qx^3}{6} \tag{f}$$

and the slope is

$$v' = -\frac{qx}{6EI}(3L^2 - 3Lx + x^2) \tag{8-23}$$

As expected, the slope obtained from this equation is zero at the support ($x = 0$) and negative (i.e., clockwise) throughout the length of the beam.

Deflection of the beam. Integration of the slope equation (Eq. f) yields

$$EIv = -\frac{qL^2x^2}{4} + \frac{qLx^3}{6} - \frac{qx^4}{24} + C_2 \tag{g}$$

The constant C_2 is found from the boundary condition that the deflection of the beam is zero at the support:

$$v(0) = 0$$

When this condition is applied to Eq. (g), we see immediately that $C_2 = 0$. Therefore, the equation for the deflection v is

$$v = -\frac{qx^2}{24EI}(6L^2 - 4Lx + x^2) \tag{8-24}$$

As expected, the deflection obtained from this equation is zero at the support ($x = 0$) and negative (that is, downward) elsewhere.

Angle of rotation at the free end of the beam. The clockwise angle of rotation θ_B at end B of the beam (Fig. 8-10b) is equal to the negative of the slope at that point. Thus, using Eq. (8-23), we get

$$\theta_B = -v'(L) = \frac{qL^3}{6EI} \tag{8-25}$$

This angle is the maximum angle of rotation for the beam.

Deflection at the free end of the beam. Since the deflection δ_B is downward (Fig. 8-10b), it is equal to the negative of the deflection obtained from Eq. (8-24):

$$\delta_B = -v(L) = \frac{qL^4}{8EI} \tag{8-26}$$

This deflection is the maximum deflection of the beam.

Example 8-3

A simple beam AB supports a concentrated load P acting at distances a and b from the left-hand and right-hand supports, respectively (Fig. 8-12a).

Determine the equations of the deflection curve, the angles of rotation θ_A and θ_B at the supports, the maximum deflection δ_{max}, and the deflection δ_C at the midpoint C of the beam (Fig. 8-12b). (*Note:* The beam has length L and constant flexural rigidity EI.)

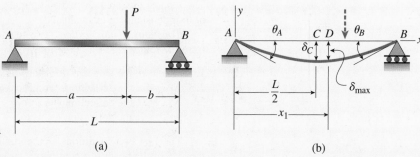

FIG. 8-12 Example 8-3. Deflections of a simple beam with a concentrated load

(a)   (b)

Solution

Bending moments in the beam. In this example the bending moments are expressed by two equations, one for each part of the beam. Using the free-body diagrams of Fig. 8-13, we arrive at the following equations:

$$M = \frac{Pbx}{L} \qquad (0 \le x \le a) \tag{8-27a}$$

$$M = \frac{Pbx}{L} - P(x - a) \qquad (a \le x \le L) \tag{8-27b}$$

Differential equations of the deflection curve. The differential equations for the two parts of the beam are obtained by substituting the bending-moment expressions (Eqs. 8-27a and b) into Eq. (8-12a). The results are

$$EIv'' = \frac{Pbx}{L} \qquad (0 \le x \le a) \tag{8-28a}$$

$$EIv'' = \frac{Pbx}{L} - P(x - a) \qquad (a \le x \le L) \tag{8-28b}$$

Slopes and deflections of the beam. The first integrations of the two differential equations yield the following expressions for the slopes:

$$EIv' = \frac{Pbx^2}{2L} + C_1 \qquad (0 \le x \le a) \tag{h}$$

$$EIv' = \frac{Pbx^2}{2L} - \frac{P(x - a)^2}{2} + C_2 \qquad (a \le x \le L) \tag{i}$$

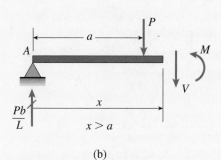

FIG. 8-13 Free-body diagrams used in determining the bending moments (Example 8-3)

(a)

(b)

continued

in which C_1 and C_2 are constants of integration. A second pair of integrations gives the deflections:

$$EIv = \frac{Pbx^3}{6L} + C_1x + C_3 \qquad (0 \leq x \leq a) \tag{j}$$

$$EIv = \frac{Pbx^3}{6L} - \frac{P(x-a)^3}{6} + C_2x + C_4 \qquad (a \leq x \leq L) \tag{k}$$

These equations contain two additional constants of integration, making a total of four constants to be evaluated.

Constants of integration. The four constants of integration can be found from the following four conditions:

1. At $x = a$, the slopes v' for the two parts of the beam are the same.
2. At $x = a$, the deflections v for the two parts of the beam are the same.
3. At $x = 0$, the deflection v is zero.
4. At $x = L$, the deflection v is zero.

The first two conditions are continuity conditions based upon the fact that the axis of the beam is a continuous curve. Conditions (3) and (4) are boundary conditions that must be satisfied at the supports.

Condition (1) means that the slopes determined from Eqs. (h) and (i) must be equal when $x = a$; therefore,

$$\frac{Pba^2}{2L} + C_1 = \frac{Pba^2}{2L} + C_2 \quad \text{or} \quad C_1 = C_2$$

Condition (2) means that the deflections found from Eqs. (j) and (k) must be equal when $x = a$; therefore,

$$\frac{Pba^3}{6L} + C_1a + C_3 = \frac{Pba^3}{6L} + C_2a + C_4$$

In as much as $C_1 = C_2$, this equation gives $C_3 = C_4$.

Next, we apply condition (3) to Eq. (j) and obtain $C_3 = 0$; therefore,

$$C_3 = C_4 = 0 \tag{l}$$

Finally, we apply condition (4) to Eq. (k) and obtain

$$\frac{PbL^2}{6} - \frac{Pb^3}{6} + C_2L = 0$$

Therefore,

$$C_1 = C_2 = -\frac{Pb(L^2 - b^2)}{6L} \tag{m}$$

Equations of the deflection curve. We now substitute the constants of integration (Eqs. l and m) into the equations for the deflections (Eqs. j and k) and obtain the deflection equations for the two parts of the beam. The resulting equations, after a slight rearrangement, are

$$v = -\frac{Pbx}{6LEI}(L^2 - b^2 - x^2) \qquad (0 \le x \le a) \qquad \text{(8-29a)}$$

$$v = -\frac{Pbx}{6LEI}(L^2 - b^2 - x^2) - \frac{P(x-a)^3}{6EI} \qquad (a \le x \le L) \quad \text{(8-29b)}$$

The first of these equations gives the deflection curve for the part of the beam to the left of the load P, and the second gives the deflection curve for the part of the beam to the right of the load.

The slopes for the two parts of the beam can be found either by substituting the values of C_1 and C_2 into Eqs. (h) and (i) or by taking the first derivatives of the deflection equations (Eqs. 8-29a and b). The resulting equations are

$$v' = -\frac{Pb}{6LEI}(L^2 - b^2 - 3x^2) \qquad (0 \le x \le a) \qquad \text{(8-30a)}$$

$$v' = -\frac{Pb}{6LEI}(L^2 - b^2 - 3x^2) - \frac{P(x-a)^2}{2EI} \qquad (a \le x \le L) \quad \text{(8-30b)}$$

The deflection and slope at any point along the axis of the beam can be calculated from Eqs. (8-29) and (8-30).

Angles of rotation at the supports. To obtain the angles of rotation θ_A and θ_B at the ends of the beam (Fig. 8-12b), we substitute $x = 0$ into Eq. (8-30a) and $x = L$ into Eq. (8-30b):

$$\theta_A = -v'(0) = \frac{Pb(L^2 - b^2)}{6LEI} = \frac{Pab(L+b)}{6LEI} \qquad \text{(8-31a)}$$

$$\theta_B = v'(L) = \frac{Pb(2L^2 - 3bL + b^2)}{6LEI} = \frac{Pab(L+a)}{6LEI} \qquad \text{(8-31b)}$$

Note that the angle θ_A is clockwise and the angle θ_B is counterclockwise, as shown in Fig. 8-12b.

The angles of rotation are functions of the position of the load and reach their largest values when the load is located near the midpoint of the beam. In the case of the angle of rotation θ_A, the maximum value of the angle is

$$(\theta_A)_{max} = \frac{PL^2\sqrt{3}}{27EI} \qquad \text{(8-32)}$$

and occurs when $b = L/\sqrt{3} = 0.577L$ (or $a = 0.423L$). This value of b is obtained by taking the derivative of θ_A with respect to b (using the first of the two expressions for θ_A in Eq. 8-31a) and then setting it equal to zero.

Maximum deflection of the beam. The maximum deflection δ_{max} occurs at point D (Fig. 8-12b) where the deflection curve has a horizontal tangent. If the load is to the right of the midpoint, that is, if $a > b$, point D is in the part of the beam to the left of the load. We can locate this point by equating the slope v' from Eq. (8-30a) to zero and solving for the distance x, which we now denote as x_1. In this manner we obtain the following formula for x_1:

$$x_1 = \sqrt{\frac{L^2 - b^2}{3}} \qquad (a \ge b) \qquad \text{(8-33)}$$

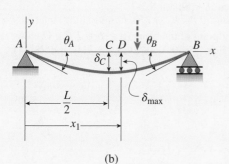

FIG. 8-12b (Repeated)

continued

From this equation we see that as the load P moves from the middle of the beam ($b = L/2$) to the right-hand end ($b = 0$), the distance x_1 varies from $L/2$ to $L/\sqrt{3} = 0.577L$. Thus, the maximum deflection occurs at a point very close to the midpoint of the beam, and this point is always between the midpoint of the beam and the load.

The maximum deflection δ_{max} is found by substituting x_1 (from Eq. 8-33) into the deflection equation (Eq. 8-29a) and then inserting a minus sign:

$$\delta_{max} = -(v)_{x=x_1} = \frac{Pb(L^2 - b^2)^{3/2}}{9\sqrt{3}\,LEI} \qquad (a \geq b) \quad (8\text{-}34) \quad \longleftarrow$$

The minus sign is needed because the maximum deflection is downward (Fig. 8-12b) whereas the deflection v is positive upward.

The maximum deflection of the beam depends on the position of the load P, that is, on the distance b. The maximum value of the maximum deflection (the "max-max" deflection) occurs when $b = L/2$ and the load is at the midpoint of the beam. This maximum deflection is equal to $PL^3/48EI$.

Deflection at the midpoint of the beam. The deflection δ_C at the midpoint C when the load is acting to the right of the midpoint (Fig. 8-12b) is obtained by substituting $x = L/2$ into Eq. (8-29a), as follows:

$$\delta_C = -v\left(\frac{L}{2}\right) = \frac{Pb(3L^2 - 4b^2)}{48EI} \qquad (a \geq b) \qquad (8\text{-}35) \quad \longleftarrow$$

Because the maximum deflection always occurs near the midpoint of the beam, Eq. (8-35) yields a close approximation to the maximum deflection. In the most unfavorable case (when b approaches zero), the difference between the maximum deflection and the deflection at the midpoint is less than 3% of the maximum deflection, as demonstrated in Problem 8.3-7.

Special case (load at the midpoint of the beam). An important special case occurs when the load P acts at the midpoint of the beam ($a = b = L/2$). Then we obtain the following results from Eqs. (8-30a), (8-29a), (8-31), and (8-34), respectively:

$$v' = -\frac{P}{16EI}(L^2 - 4x^2) \qquad \left(0 \leq x \leq \frac{L}{2}\right) \qquad (8\text{-}36)$$

$$v = -\frac{Px}{48EI}(3L^2 - 4x^2) \qquad \left(0 \leq x \leq \frac{L}{2}\right) \qquad (8\text{-}37)$$

$$\theta_A = \theta_B = \frac{PL^2}{16EI} \qquad (8\text{-}38)$$

$$\delta_{max} = \delta_C = \frac{PL^3}{48EI} \qquad (8\text{-}39)$$

Since the deflection curve is symmetric about the midpoint of the beam, the equations for v' and v are given only for the left-hand half of the beam (Eqs. 8-36 and 8-37). If needed, the equations for the right-hand half can be obtained from Eqs. (8-30b) and (8-29b) by substituting $a = b = L/2$.

8.4 DEFLECTIONS BY INTEGRATION OF THE SHEAR-FORCE AND LOAD EQUATIONS

The equations of the deflection curve in terms of the shear force V and the load q (Eqs. 8-12b and c, respectively) may also be integrated to obtain slopes and deflections. Since the loads are usually known quantities, whereas the bending moments must be determined from free-body diagrams and equations of equilibrium, many analysts prefer to start with the load equation. For this same reason, most computer programs for finding deflections begin with the load equation and then perform numerical integrations to obtain the shear forces, bending moments, slopes, and deflections.

The procedure for solving either the load equation or the shear-force equation is similar to that for solving the bending-moment equation, except that more integrations are required. For instance, if we begin with the load equation, four integrations are needed in order to arrive at the deflections. Thus, four constants of integration are introduced for each load equation that is integrated. As before, these constants are found from boundary, continuity, and symmetry conditions. However, these conditions now include conditions on the shear forces and bending moments as well as conditions on the slopes and deflections.

Conditions on the shear forces are equivalent to conditions on the third derivative (because $EIv''' = V$). In a similar manner, conditions on the bending moments are equivalent to conditions on the second derivative (because $EIv'' = M$). When the shear-force and bending-moment conditions are added to those for the slopes and deflections, we always have enough independent conditions to solve for the constants of integration.

The following examples illustrate the techniques of analysis in detail. The first example begins with the load equation and the second begins with the shear-force equation.

Example 8-4

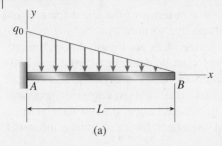

(a)

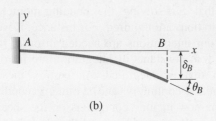

(b)

FIG. 8-14 Example 8-4. Deflections of a cantilever beam with a triangular load

Determine the equation of the deflection curve for a cantilever beam AB supporting a triangularly distributed load of maximum intensity q_0 (Fig. 8-14a).

Also, determine the deflection δ_B and angle of rotation θ_B at the free end (Fig. 8-14b). Use the fourth-order differential equation of the deflection curve (the load equation). (*Note:* The beam has length L and constant flexural rigidity EI.)

Solution

Differential equation of the deflection curve. The intensity of the distributed load is given by the following equation (see Fig. 8-14a):

$$q = \frac{q_0(L-x)}{L} \tag{8-40}$$

Consequently, the fourth-order differential equation (Eq. 8-12c) becomes

$$EIv'''' = -q = -\frac{q_0(L-x)}{L} \tag{a}$$

Shear force in the beam. The first integration of Eq. (a) gives

$$EIv''' = \frac{q_0}{2L}(L-x)^2 + C_1 \tag{b}$$

The right-hand side of this equation represents the shear force V (see Eq. 8-12b). Because the shear force is zero at $x = L$, we have the following boundary condition:

$$v'''(L) = 0$$

Using this condition with Eq. (b), we get $C_1 = 0$. Therefore, Eq. (b) simplifies to

$$EIv''' = \frac{q_0}{2L}(L-x)^2 \tag{c}$$

and the shear force in the beam is

$$V = EIv''' = \frac{q_0}{2L}(L-x)^2 \tag{8-41}$$

Bending moment in the beam. Integrating a second time, we obtain the following equation from Eq. (c):

$$EIv'' = -\frac{q_0}{6L}(L-x)^3 + C_2 \tag{d}$$

This equation is equal to the bending moment M (see Eq. 8-12a). Since the bending moment is zero at the free end of the beam, we have the following boundary condition:

$$v''(L) = 0$$

Cantilever portion of roof structure
(Courtesy of the National Information Service
for Earthquake Engineering EERC, University
of California, Berkeley)

Applying this condition to Eq. (d), we obtain $C_2 = 0$, and therefore the bending moment is

$$M = EIv'' = -\frac{q_0}{6L}(L - x)^3 \tag{8-42}$$

Slope and deflection of the beam. The third and fourth integrations yield

$$EIv' = \frac{q_0}{24L}(L - x)^4 + C_3 \tag{e}$$

$$EIv = -\frac{q_0}{120L}(L - x)^5 + C_3 x + C_4 \tag{f}$$

The boundary conditions at the fixed support, where both the slope and deflection equal zero, are

$$v'(0) = 0 \qquad v(0) = 0$$

Applying these conditions to Eqs. (e) and (f), respectively, we find

$$C_3 = -\frac{q_0 L^3}{24} \qquad C_4 = \frac{q_0 L^4}{120}$$

Substituting these expressions for the constants into Eqs. (e) and (f), we obtain the following equations for the slope and deflection of the beam:

$$v' = -\frac{q_0 x}{24LEI}(4L^3 - 6L^2 x + 4Lx^2 - x^3) \tag{8-43}$$

$$v = -\frac{q_0 x^2}{120LEI}(10L^3 - 10L^2 x + 5Lx^2 - x^3) \tag{8-44}$$

Angle of rotation and deflection at the free end of the beam. The angle of rotation θ_B and deflection δ_B at the free end of the beam (Fig. 8-14b) are obtained from Eqs. (8-43) and (8-44), respectively, by substituting $x = L$. The results are

$$\theta_B = -v'(L) = \frac{q_0 L^3}{24EI} \qquad \delta_B = -v(L) = \frac{q_0 L^4}{30EI} \tag{8-45a,b}$$

Thus, we have determined the required slopes and deflections of the beam by solving the fourth-order differential equation of the deflection curve.

Example 8-5

A simple beam AB with an overhang BC supports a concentrated load P at the end of the overhang (Fig. 8-15a). The main span of the beam has length L and the overhang has length $L/2$.

Determine the equations of the deflection curve and the deflection δ_C at the end of the overhang (Fig. 8-15b). Use the third-order differential equation of the deflection curve (the shear-force equation). (*Note:* The beam has constant flexural rigidity EI.)

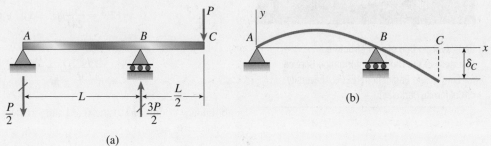

FIG. 8-15 Example 8-5. Deflections of a beam with an overhang

(a)

Solution

Differential equations of the deflection curve. Because reactive forces act at supports A and B, we must write separate differential equations for parts AB and BC of the beam. Therefore, we begin by finding the shear forces in each part of the beam.

The downward reaction at support A is equal to $P/2$, and the upward reaction at support B is equal to $3P/2$ (see Fig. 8-15a). It follows that the shear forces in parts AB and BC are

$$V = -\frac{P}{2} \qquad (0 < x < L) \tag{8-46a}$$

$$V = P \qquad \left(L < x < \frac{3L}{2}\right) \tag{8-46b}$$

in which x is measured from end A of the beam (Fig. 8-12b).

The third-order differential equations for the beam now become (see Eq. 8-12b):

$$EIv''' = -\frac{P}{2} \qquad (0 < x < L) \tag{g}$$

$$EIv''' = P \qquad \left(L < x < \frac{3L}{2}\right) \tag{h}$$

Bending moments in the beam. Integration of the preceding two equations yields the bending-moment equations:

$$M = EIv'' = -\frac{Px}{2} + C_1 \qquad (0 \le x \le L) \tag{i}$$

$$M = EIv'' = Px + C_2 \qquad \left(L \le x \le \frac{3L}{2}\right) \tag{j}$$

Bridge girder with overhang during transport to the construction site
(Tom Brakefield/Getty Images)

The bending moments at points A and C are zero; hence we have the following boundary conditions:

$$v''(0) = 0 \qquad v''\left(\frac{3L}{2}\right) = 0$$

Using these conditions with Eqs. (i) and (j), we get

$$C_1 = 0 \qquad C_2 = -\frac{3PL}{2}$$

Therefore, the bending moments are

$$M = EIv'' = -\frac{Px}{2} \qquad (0 \le x \le L) \tag{8-47a}$$

$$M = EIv'' = -\frac{P(3L - 2x)}{2} \qquad \left(L \le x \le \frac{3L}{2}\right) \tag{8-47b}$$

These equations can be verified by determining the bending moments from free-body diagrams and equations of equilibrium.

Slopes and deflections of the beam. The next integrations yield the slopes:

$$EIv' = -\frac{Px^2}{4} + C_3 \qquad (0 \le x \le L)$$

$$EIv' = -\frac{Px(3L - x)}{2} + C_4 \qquad \left(L \le x \le \frac{3L}{2}\right)$$

The only condition on the slopes is the continuity condition at support B. According to this condition, the slope at point B as found for part AB of the beam is equal to the slope at the same point as found for part BC of the beam. Therefore, we substitute $x = L$ into each of the two preceding equations for the slopes and obtain

$$-\frac{PL^2}{4} + C_3 = -PL^2 + C_4$$

This equation eliminates one constant of integration because we can express C_4 in terms of C_3:

$$C_4 = C_3 + \frac{3PL^2}{4} \tag{k}$$

The third and last integrations give

$$EIv = -\frac{Px^3}{12} + C_3x + C_5 \qquad (0 \le x \le L) \tag{l}$$

$$EIv = -\frac{Px^2(9L - 2x)}{12} + C_4x + C_6 \qquad \left(L \le x \le \frac{3L}{2}\right) \tag{m}$$

For part AB of the beam (Fig. 8-15a), we have two boundary conditions on the deflections, namely, the deflection is zero at points A and B:

$$v(0) = 0 \quad \text{and} \quad v(L) = 0$$

continued

Applying these conditions to Eq. (l), we obtain

$$C_5 = 0 \qquad C_3 = \frac{PL^2}{12} \tag{n,o}$$

Substituting the preceding expression for C_3 in Eq. (k), we get

$$C_4 = \frac{5PL^2}{6} \tag{p}$$

For part BC of the beam, the deflection is zero at point B. Therefore, the boundary condition is

$$v(L) = 0$$

Applying this condition to Eq. (m), and also substituting Eq. (p) for C_4, we get

$$C_6 = -\frac{PL^3}{4} \tag{q}$$

All constants of integration have now been evaluated.

The deflection equations are obtained by substituting the constants of integration (Eqs. n, o, p, and q) into Eqs. (l) and (m). The results are

$$v = \frac{Px}{12EI}(L^2 - x^2) \qquad (0 \le x \le L) \tag{8-48a}$$

$$v = -\frac{P}{12EI}(3L^3 - 10L^2x + 9Lx^2 - 2x^3) \qquad \left(L \le x \le \frac{3L}{2}\right) \tag{8-48b}$$

Note that the deflection is always positive (upward) in part AB of the beam (Eq. 8-48a) and always negative (downward) in the overhang BC (Eq. 8-48b).

Deflection at the end of the overhang. We can find the deflection δ_C at the end of the overhang (Fig. 8-15b) by substituting $x = 3L/2$ in Eq. (8-48b):

$$\delta_C = -v\left(\frac{3L}{2}\right) = \frac{PL^3}{8EI} \tag{8-49}$$

Thus, we have determined the required deflections of the overhanging beam (Eqs. 8-48 and 8-49) by solving the third-order differential equation of the deflection curve.

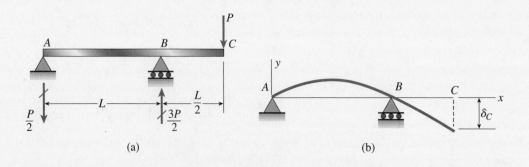

FIG. 8-15 (Repeated)   (a)   (b)

8.5 METHOD OF SUPERPOSITION

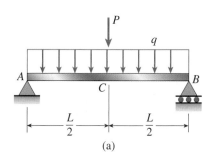

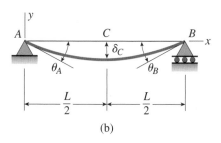

FIG. 8-16 Simple beam with two loads

The **method of superposition** is a practical and commonly used technique for obtaining deflections and angles of rotation of beams. The underlying concept is quite simple and may be stated as follows:

Under suitable conditions, the deflection of a beam produced by several different loads acting simultaneously can be found by superposing the deflections produced by the same loads acting separately.

For instance, if v_1 represents the deflection at a particular point on the axis of a beam due to a load q_1, and if v_2 represents the deflection at that same point due to a different load q_2, then the deflection at that point due to loads q_1 and q_2 acting simultaneously is $v_1 + v_2$. (The loads q_1 and q_2 are independent loads and each may act anywhere along the axis of the beam.)

The justification for superposing deflections lies in the nature of the differential equations of the deflection curve (Eqs. 8-12a, b, and c). These equations are *linear* differential equations, because all terms containing the deflection v and its derivatives are raised to the first power. Therefore, the solutions of these equations for several loading conditions may be added algebraically, or *superposed*.(The conditions for superposition to be valid are described later in the subsection "Principle of Superposition.")

As an **illustration** of the superposition method, consider the simple beam ACB shown in Fig. 8-16a. This beam supports two loads: (1) a uniform load of intensity q acting throughout the span, and (2) a concentrated load P acting at the midpoint. Suppose we wish to find the deflection δ_C at the midpoint and the angles of rotation θ_A and θ_B at the ends (Fig. 8-16b). Using the method of superposition, we obtain the effects of each load acting separately and then combine the results.

For the uniform load acting alone, the deflection at the midpoint and the angles of rotation are obtained from the formulas of Example 8-1 (see Eqs. 8-18, 8-19, and 8-20):

$$(\delta_C)_1 = \frac{5qL^4}{384EI} \qquad (\theta_A)_1 = (\theta_B)_1 = \frac{qL^3}{24EI}$$

in which EI is the flexural rigidity of the beam and L is its length.

For the load P acting alone, the corresponding quantities are obtained from the formulas of Example 8-3 (see Eqs. 8-38 and 8-39):

$$(\delta_C)_2 = \frac{PL^3}{48EI} \qquad (\theta_A)_2 = (\theta_B)_2 = \frac{PL^2}{16EI}$$

The deflection and angles of rotation due to the combined loading (Fig. 8-16a) are obtained by summation:

$$\delta_C = (\delta_C)_1 + (\delta_C)_2 = \frac{5qL^4}{384EI} + \frac{PL^3}{48EI} \qquad \text{(a)}$$

$$\theta_A = \theta_B = (\theta_A)_1 + (\theta_A)_2 = \frac{qL^3}{24EI} + \frac{PL^2}{16EI} \qquad \text{(b)}$$

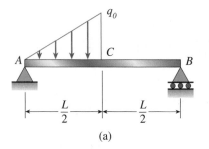

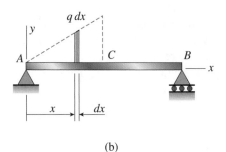

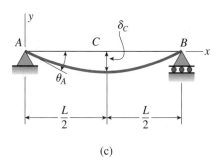

FIG. 8-17 Simple beam with a triangular load

The deflections and angles of rotation at other points on the beam axis can be found by this same procedure. However, the method of superposition is not limited to finding deflections and angles of rotation at single points. The method may also be used to obtain general equations for the slopes and deflections of beams subjected to more than one load.

Tables of Beam Deflections

The method of superposition is useful only when formulas for deflections and slopes are readily available. To provide convenient access to such formulas, tables for both cantilever and simple beams are given in Appendix H (available online). Similar tables can be found in engineering handbooks. Using these tables and the method of superposition, we can find deflections and angles of rotation for many different loading conditions, as illustrated in the examples at the end of this section.

Distributed Loads

Sometimes we encounter a distributed load that is not included in a table of beam deflections. In such cases, superposition may still be useful. We can consider an element of the distributed load as though it were a concentrated load, and then we can find the required deflection by integrating throughout the region of the beam where the load is applied.

To illustrate this process of integration, consider a simple beam ACB with a triangular load acting on the left-hand half (Fig. 8-17a). We wish to obtain the deflection δ_C at the midpoint C and the angle of rotation θ_A at the left-hand support (Fig. 8-17c).

We begin by visualizing an element $q\,dx$ of the distributed load as a concentrated load (Fig. 8-17b). Note that the load acts to the left of the midpoint of the beam. The deflection at the midpoint due to this concentrated load is obtained from Case 5 of Table H-2, Appendix H (available online). The formula given there for the midpoint deflection (for the case in which $a \leq b$) is

$$\frac{Pa}{48EI}(3L^2 - 4a^2)$$

In our example (Fig. 8-17b), we substitute $q\,dx$ for P and x for a:

$$\frac{(q\,dx)(x)}{48EI}(3L^2 - 4x^2) \tag{c}$$

This expression gives the deflection at point C due to the element $q\,dx$ of the load.

Next, we note that the intensity of the uniform load (Figs. 8-17a and b) is

$$q = \frac{2q_0 x}{L} \tag{d}$$

where q_0 is the maximum intensity of the load. With this substitution for q, the formula for the deflection (Eq. c) becomes

$$\frac{q_0 x^2}{24LEI}(3L^2 - 4x^2)dx$$

Finally, we integrate throughout the region of the load to obtain the deflection δ_C at the midpoint of the beam due to the entire triangular load:

$$\delta_C = \int_0^{L/2} \frac{q_0 x^2}{24LEI}(3L^2 - 4x^2)dx$$

$$= \frac{q_0}{24LEI} \int_0^{L/2} (3L^2 - 4x^2)x^2 dx = \frac{q_0 L^4}{240EI} \qquad (8\text{-}50)$$

By a similar procedure, we can calculate the angle of rotation θ_A at the left-hand end of the beam (Fig. 8-17c). The expression for this angle due to a concentrated load P (see Case 5 of Table H-2 available online) is

$$\frac{Pab(L + b)}{6LEI}$$

Replacing P with $2q_0 x\, dx/L$, a with x, and b with $L - x$, we obtain

$$\frac{2q_0 x^2(L - x)(L + L - x)}{6L^2 EI} dx \quad \text{or} \quad \frac{q_0}{3L^2 EI}(L - x)(2L - x)x^2 dx$$

Finally, we integrate throughout the region of the load:

$$\theta_A = \int_0^{L/2} \frac{q_0}{3L^2 EI}(L - x)(2L - x)x^2 dx = \frac{41q_0 L^3}{2880EI} \qquad (8\text{-}51)$$

This is the angle of rotation produced by the triangular load.

This example illustrates how we can use superposition and integration to find deflections and angles of rotation produced by distributed loads of almost any kind. If the integration cannot be performed easily by analytical means, numerical methods can be used.

Principle of Superposition

The method of superposition for finding beam deflections is an example of a more general concept known in mechanics as the **principle of superposition**. This principle is valid whenever the quantity to be determined is a linear function of the applied loads. When that is the case, the desired quantity may be found due to each load acting separately, and then these results may be superposed to obtain the desired quantity due to all loads acting simultaneously. In ordinary structures, the principle is usually valid for stresses, strains, bending moments, and many other quantities besides deflections.

In the particular case of **beam deflections**, the principle of superposition is valid under the following conditions: (1) Hooke's law holds for the material, (2) the deflections and rotations are small, and (3) the presence of the deflections does not alter the actions of the applied loads. These requirements ensure that the differential equations of the deflection curve are linear.

The following examples provide additional illustrations in which the principle of superposition is used to calculate deflections and angles of rotation of beams.

Example 8-6

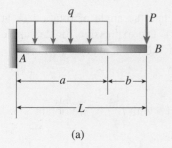

(a)

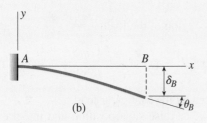

(b)

FIG. 8-18 Example 8-6. Cantilever beam with a uniform load and a concentrated load

A cantilever beam AB supports a uniform load of intensity q acting over part of the span and a concentrated load P acting at the free end (Fig. 8-18a).

Determine the deflection δ_B and angle of rotation θ_B at end B of the beam (Fig. 8-18b). (*Note:* The beam has length L and constant flexural rigidity EI.)

Solution

We can obtain the deflection and angle of rotation at end B of the beam by combining the effects of the loads acting separately. If the uniform load acts alone, the deflection and angle of rotation (obtained from Case 2 of Table H-1, Appendix H available online) are

$$(\delta_B)_1 = \frac{qa^3}{24EI}(4L - a) \qquad (\theta_B)_1 = \frac{qa^3}{6EI}$$

If the load P acts alone, the corresponding quantities (from Case 4, Table H-1) are

$$(\delta_B)_2 = \frac{PL^3}{3EI} \qquad (\theta_B)_2 = \frac{PL^2}{2EI}$$

Therefore, the deflection and angle of rotation due to the combined loading (Fig. 8-18a) are

$$\delta_B = (\delta_B)_1 + (\delta_B)_2 = \frac{qa^3}{24EI}(4L - a) + \frac{PL^3}{3EI} \qquad (8\text{-}52)$$

$$\theta_B = (\theta_B)_1 + (\theta_B)_2 = \frac{qa^3}{6EI} + \frac{PL^2}{2EI} \qquad (8\text{-}53)$$

Thus, we have found the required quantities by using tabulated formulas and the method of superposition.

Example 8-7

A cantilever beam AB with a uniform load of intensity q acting on the right-hand half of the beam is shown in Fig. 8-19a.

Obtain formulas for the deflection δ_B and angle of rotation θ_B at the free end (Fig. 8-19c). (*Note:* The beam has length L and constant flexural rigidity EI.)

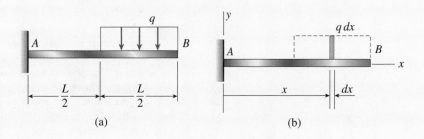

(a)　　　　　　　　　　　(b)

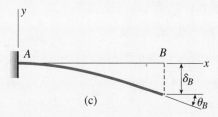

(c)

FIG. 8-19 Example 8-7. Cantilever beam with a uniform load acting on the right-hand half of the beam

Solution

In this example we will determine the deflection and angle of rotation by treating an element of the uniform load as a concentrated load and then integrating (see Fig. 8-19b). The element of load has magnitude $q\,dx$ and is located at distance x from the support. The resulting differential deflection $d\delta_B$ and differential angle of rotation $d\theta_B$ at the free end are found from the corresponding formulas in Case 5 of Table H-1, Appendix H available online, by replacing P with $q\,dx$ and a with x; thus,

$$d\delta_B = \frac{(qdx)(x^2)(3L - x)}{6EI} \qquad d\theta_B = \frac{(q\,dx)(x^2)}{2EI}$$

By integrating over the loaded region, we get

$$\delta_B = \int d\delta_B = \frac{q}{6EI} \int_{L/2}^{L} x^2(3L - x)\,dx = \frac{41qL^4}{384EI} \qquad (8\text{-}54)$$

$$\theta_B = \int d\theta_B = \frac{q}{2EI} \int_{L/2}^{L} x^2\,dx = \frac{7qL^3}{48EI} \qquad (8\text{-}55)$$

Note: These same results can be obtained by using the formulas in Case 3 of Table H-1 and substituting $a = b = L/2$.

Example 8-8

A compound beam ABC has a roller support at A, an internal hinge at B, and a fixed support at C (Fig. 8-20a). Segment AB has length a and segment BC has length b. A concentrated load P acts at distance $2a/3$ from support A and a uniform load of intensity q acts between points B and C.

Determine the deflection δ_B at the hinge and the angle of rotation θ_A at support A (Fig. 8-20d). (*Note:* The beam has constant flexural rigidity EI.)

Solution

For purposes of analysis, we will consider the compound beam to consist of two individual beams: (1) a simple beam AB of length a, and (2) a cantilever beam BC of length b. The two beams are linked together by a pin connection at B.

If we separate beam AB from the rest of the structure (Fig. 8-20b), we see that there is a vertical force F at end B equal to $2P/3$. This same force acts downward at end B of the cantilever (Fig. 8-20c). Consequently, the cantilever beam BC is subjected to two loads: a uniform load and a concentrated load. The deflection at the end of this cantilever (which is the same as the deflection δ_B of the hinge) is readily found from Cases 1 and 4 of Table H-1, Appendix H, available online:

$$\delta_B = \frac{qb^4}{8EI} + \frac{Fb^3}{3EI}$$

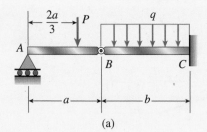

(a)

or, since $F = 2P/3$,

$$\delta_B = \frac{qb^4}{8EI} + \frac{2Pb^3}{9EI} \qquad (8\text{-}56)$$

The angle of rotation θ_A at support A (Fig. 8-20d) consists of two parts: (1) an angle BAB' produced by the downward displacement of the hinge, and (2) an additional angle of rotation produced by the bending of beam AB (or beam AB') as a simple beam. The angle BAB' is

$$(\theta_A)_1 = \frac{\delta_B}{a} = \frac{qb^4}{8aEI} + \frac{2Pb^3}{9aEI}$$

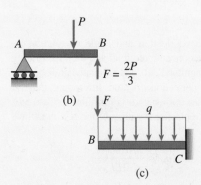

(b)

The angle of rotation at the end of a simple beam with a concentrated load is obtained from Case 5 of Table H-2. The formula given there is

$$\frac{Pab(L + b)}{6LEI}$$

in which L is the length of the simple beam, a is the distance from the left-hand support to the load, and b is the distance from the right-hand support to the load. Thus, in the notation of our example (Fig. 8-20a), the angle of rotation is

$$(\theta_A)_2 = \frac{P\left(\dfrac{2a}{3}\right)\left(\dfrac{a}{3}\right)\left(a + \dfrac{a}{3}\right)}{6aEI} = \frac{4Pa^2}{81EI}$$

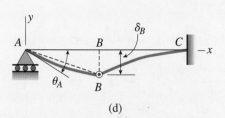

(d)

Combining the two angles, we obtain the total angle of rotation at support A:

$$\theta_A = (\theta_A)_1 + (\theta_A)_2 = \frac{qb^4}{8aEI} + \frac{2Pb^3}{9aEI} + \frac{4Pa^2}{81EI} \qquad (8\text{-}57)$$

FIG. 8-20 Example 8-8. Compound beam with a hinge

This example illustrates how the method of superposition can be adapted to handle a seemingly complex situation in a relatively simple manner.

Example 8-9

A simple beam AB of span length L has an overhang BC of length a (Fig. 8-21a). The beam supports a uniform load of intensity q throughout its length.

Obtain a formula for the deflection δ_C at the end of the overhang (Fig. 8-21c). (*Note:* The beam has constant flexural rigidity EI.)

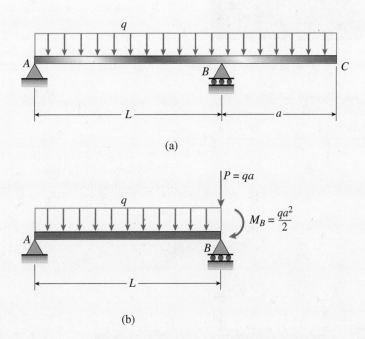

(a)

(b)

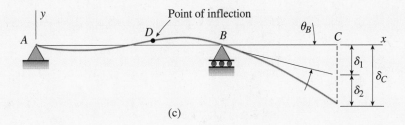

FIG. 8-21 Example 8-9. Simple beam with an overhang

(c)

Solution

We can find the deflection of point C by imagining the overhang BC (Fig. 8-21a) to be a cantilever beam subjected to two actions. The first action is the rotation of the support of the cantilever through an angle θ_B, which is the angle of rotation of beam ABC at support B (Fig. 8-21c). (We assume that a clockwise angle θ_B is positive.) This angle of rotation causes a rigid-body rotation of the overhang BC, resulting in a downward displacement δ_1 of point C.

continued

Beam with overhang loaded by gravity
uniform load
(Courtesy of the National Information Service
for Earthquake Engineering EERC, University
of California, Berkeley)

The second action is the bending of BC as a cantilever beam supporting a uniform load. This bending produces an additional downward displacement δ_2 (Fig. 8-21c). The superposition of these two displacements gives the total displacement δ_C at point C.

Deflection δ_1. Let us begin by finding the deflection δ_1 caused by the angle of rotation θ_B at point B. To find this angle, we observe that part AB of the beam is in the same condition as a simple beam (Fig. 8-21b) subjected to the following loads: (1) a uniform load of intensity q, (2) a couple M_B (equal to $qa^2/2$), and (3) a vertical load P (equal to qa). Only the loads q and M_B produce angles of rotation at end B of this simple beam. These angles are found from Cases 1 and 7 of Table H-2, Appendix H (available online). Thus, the angle θ_B is

$$\theta_B = -\frac{qL^3}{24EI} + \frac{M_B L}{3EI} = -\frac{qL^3}{24EI} + \frac{qa^2 L}{6EI} = \frac{qL(4a^2 - L^2)}{24EI} \tag{8-58}$$

in which a clockwise angle is positive, as shown in Fig. 8-21c.

The downward deflection δ_1 of point C, due solely to the angle of rotation θ_B, is equal to the length of the overhang times the angle (Fig. 8-21c):

$$\delta_1 = a\theta_B = \frac{qaL(4a^2 - L^2)}{24EI} \tag{e}$$

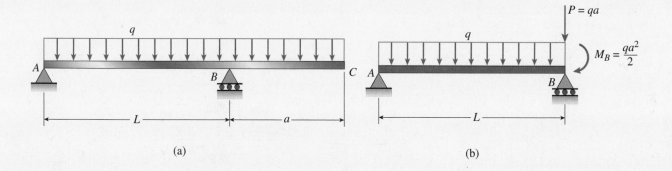

(a)   (b)

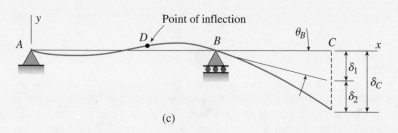

(c)

FIG. 8-21 (Repeated)

Deflection δ_2. Bending of the overhang *BC* produces an additional downward deflection δ_2 at point *C*. This deflection is equal to the deflection of a cantilever beam of length *a* subjected to a uniform load of intensity *q* (see Case 1 of Table H-1, available online):

$$\delta_2 = \frac{qa^4}{8EI} \tag{f}$$

Deflection δ_C. The total downward deflection of point *C* is the algebraic sum of δ_1 and δ_2:

$$\delta_C = \delta_1 + \delta_2 = \frac{qaL(4a^2 - L^2)}{24EI} + \frac{qa^4}{8EI} = \frac{qa}{24EI}\left[L(4a^2 - L^2) + 3a^3\right]$$

or

$$\delta_C = \frac{qa}{24EI}(a + L)(3a^2 + aL - L^2) \tag{8-59}$$

From the preceding equation we see that the deflection δ_C may be upward or downward, depending upon the relative magnitudes of the lengths *L* and *a*. If *a* is relatively large, the last term in the equation (the three-term expression in parentheses) is positive and the deflection δ_C is downward. If *a* is relatively small, the last term is negative and the deflection is upward. The deflection is zero when the last term is equal to zero:

$$3a^2 + aL - L^2 = 0$$

or

$$a = \frac{L(\sqrt{13} - 1)}{6} = 0.4343L \tag{g}$$

From this result, we see that if *a* is greater than 0.4343*L*, the deflection of point *C* is downward; if *a* is less than 0.4343*L*, the deflection is upward.

Deflection curve. The shape of the deflection curve for the beam in this example is shown in Fig. 8-21c for the case where *a* is large enough ($a > 0.4343L$) to produce a downward deflection at *C* and small enough ($a < L$) to ensure that the reaction at *A* is upward. Under these conditions the beam has a positive bending moment between support *A* and a point such as *D*. The deflection curve in region *AD* is concave upward (positive curvature). From *D* to *C*, the bending moment is negative, and therefore the deflection curve is concave downward (negative curvature).

Point of inflection. At point *D* the curvature of the deflection curve is zero because the bending moment is zero. A point such as *D* where the curvature and bending moment *change signs* is called a **point of inflection** (or *point of contraflexure*). The bending moment *M* and the second derivative d^2v/dx^2 always vanish at an inflection point.

However, a point where *M* and d^2v/dx^2 equal zero is not necessarily an inflection point because it is possible for those quantities to be zero without changing signs at that point; for example, they could have maximum or minimum values.

CHAPTER SUMMARY & REVIEW

In Chapter 8, we investigated the linear elastic, small displacement behavior of beams of different types, with different support conditions, acted upon by a wide variety of loadings. We studied methods based on integration of the second-, third- or fourth-order differential equation of the deflection curve. We computed displacements (both translations and rotations) at specific points along the beam and also found the equation describing the deflected shape of the entire beam. Using solutions for a number of standard cases (see Appendix H available online), we used the powerful principle of superposition to solve more complicated beams and loadings by combining the simpler standard solutions. The major concepts presented in this chapter may be summarized as follows:

1. By combining expressions for linear curvature ($\kappa = d^2v/dx^2$) and the moment curvature relation ($\kappa = M/EI$), we obtained the **ordinary differential equation of the deflection curve** for a beam, which is valid only for linear elastic behavior.

$$EI\frac{d^2v}{dx^2} = M$$

2. The differential equation of the deflection curve may be differentiated once to obtain a third-order equation relating shear force V and first derivative of moment, dM/dx, or twice to obtain a fourth-order equation relating intensity of distributed load q and first derivative of shear, dV/dx.

$$EI\frac{d^3v}{dx^3} = V$$

$$EI\frac{d^4v}{dx^4} = -q$$

The choice of second-, third- or fourth-order differential equations depends on which is most efficient for a particular beam support case and applied loading.

3. We must write expressions for either moment (M), shear (V), or load intensity (q) for each separate region of the beam (e.g., whenever q, V, M, or EI vary) and then apply **boundary**, **continuity**, or **symmetry conditions**, as appropriate, to solve for unknown constants of integration which arise as we apply the method of successive integrations; the beam deflection equation, $v(x)$, may be evaluated at a particular value of x to find the translational displacement at that point; evaluation of dv/dx at that same point provides the slope of the deflection equation.

4. The **method of superposition** may be used to solve for displacements and rotations for more complicated beams and loadings; the actual beam first must be broken down into the sum of a number of simpler cases whose solutions already are known (see Appendix H, available online); superposition is only applicable to beams undergoing small displacements and behaving in a linear elastic manner.

PROBLEMS CHAPTER 8

Differential Equations of the Deflection Curve

The beams described in the problems for Section 8.2 have constant flexural rigidity EI.

8.2-1 The deflection curve for a simple beam AB (see figure) is given by the following equation:

$$v = -\frac{q_0 x}{360 LEI}(7L^4 - 10L^2 x^2 + 3x^4)$$

Describe the load acting on the beam.

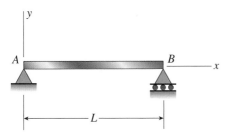

PROBS. 8.2-1 and 8.2-2

8.2-2 The deflection curve for a simple beam AB (see figure) is given by the following equation:

$$v = -\frac{q_0 L^4}{\pi^4 EI}\sin\frac{\pi x}{L}$$

(a) Describe the load acting on the beam.
(b) Determine the reactions R_A and R_B at the supports.
(c) Determine the maximum bending moment M_{max}.

8.2-3 The deflection curve for a cantilever beam AB (see figure) is given by the following equation:

$$v = -\frac{q_0 x^2}{120 LEI}(10L^3 - 10L^2 x + 5Lx^2 - x^3)$$

Describe the load acting on the beam.

PROBS. 8.2-3 and 8.2-4

8.2-4 The deflection curve for a cantilever beam AB (see figure) is given by the following equation:

$$v = -\frac{q_0 x^2}{360 L^2 EI}(45L^4 - 40L^3 x + 15L^2 x^2 - x^4)$$

(a) Describe the load acting on the beam.
(b) Determine the reactions R_A and M_A at the support.

Deflection Formulas

Problems 8.3-1 through 8.3-7 require the calculation of deflections using the formulas derived in Examples 8-1, 8-2, and 8-3. All beams have constant flexural rigidity EI.

8.3-1 A wide-flange beam (W 12×35) supports a uniform load on a simple span of length $L = 14$ ft (see figure).
Calculate the maximum deflection δ_{max} at the midpoint and the angles of rotation θ at the supports if $q = 1.8$ k/ft and $E = 30 \times 10^6$ psi. Use the formulas of Example 8-1.

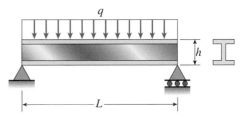

PROBS. 8.3-1, 8.3-2, and 8.3-3

8.3-2 A uniformly loaded steel wide-flange beam with simple supports (see figure) has a downward deflection of 10 mm at the midpoint and angles of rotation equal to 0.01 radians at the ends.
Calculate the height h of the beam if the maximum bending stress is 90 MPa and the modulus of elasticity is 200 GPa. (*Hint:* Use the formulas of Example 8-1.)

8.3-3 What is the span length L of a uniformly loaded simple beam of wide-flange cross section (see figure) if the maximum bending stress is 12,000 psi, the maximum deflection is 0.1 in., the height of the beam is 12 in., and the modulus of elasticity is 30×10^6 psi? (Use the formulas of Example 8-1.)

8.3-4 Calculate the maximum deflection δ_{max} of a uniformly loaded simple beam (see figure) if the span length $L = 2.0$ m, the intensity of the uniform load $q = 2.0$ kN/m, and the maximum bending stress $\sigma = 60$ MPa.
The cross section of the beam is square, and the material is aluminum having modulus of elasticity $E = 70$ GPa. (Use the formulas of Example 8-1.)

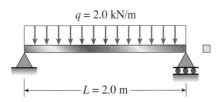

PROB. 8.3-4

8.3-5 A cantilever beam with a uniform load (see figure) has a height h equal to 1/8 of the length L. The beam is a steel wide-flange section with $E = 28 \times 10^6$ psi and an allowable bending stress of 17,500 psi in both tension and compression. Calculate the ratio δ/L of the deflection at the free end to the length, assuming that the beam carries the maximum allowable load. (Use the formulas of Example 8-2.)

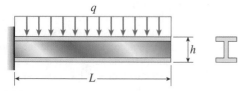

PROB. 8.3-5

8.3-6 A gold-alloy microbeam attached to a silicon wafer behaves like a cantilever beam subjected to a uniform load (see figure). The beam has length $L = 27.5$ μm and rectangular cross section of width $b = 4.0$ μm and thickness $t = 0.88$ μm. The total load on the beam is 17.2 μN. If the deflection at the end of the beam is 2.46 μm, what is the modulus of elasticity E_g of the gold alloy? (Use the formulas of Example 8-2.)

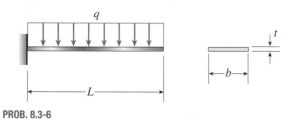

PROB. 8.3-6

8.3-7 Obtain a formula for the ratio δ_C/δ_{max} of the deflection at the midpoint to the maximum deflection for a simple beam supporting a concentrated load P (see figure).

From the formula, plot a graph of δ_C/δ_{max} versus the ratio a/L that defines the position of the load $(0.5 < a/L < 1)$. What conclusion do you draw from the graph? (Use the formulas of Example 8-3.)

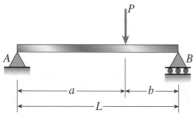

PROB. 8.3-7

Deflections by Integration of the Bending-Moment Equation

Problems 8.3-8 through 8.3-16 are to be solved by integrating the second-order differential equation of the deflection curve (the bending-moment equation). The origin of coordinates is at the left-hand end of each beam, and all beams have constant flexural rigidity EI.

8.3-8 Derive the equation of the deflection curve for a cantilever beam AB supporting a load P at the free end (see figure). Also, determine the deflection δ_B and angle of rotation θ_B at the free end. (*Note:* Use the second-order differential equation of the deflection curve.)

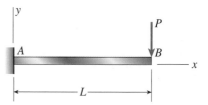

PROB. 8.3-8

8.3-9 Derive the equation of the deflection curve for a simple beam AB loaded by a couple M_0 at the left-hand support (see figure). Also, determine the maximum deflection δ_{max}. (*Note:* Use the second-order differential equation of the deflection curve.)

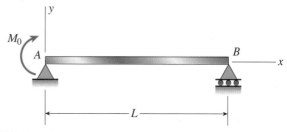

PROB. 8.3-9

8.3-10 A cantilever beam AB supporting a triangularly distributed load of maximum intensity q_0 is shown in the figure.

Derive the equation of the deflection curve and then obtain formulas for the deflection δ_B and angle of rotation θ_B at the free end. (*Note:* Use the second-order differential equation of the deflection curve.)

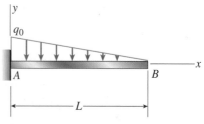

PROB. 8.3-10

8.3-11 A cantilever beam AB is acted upon by a uniformly distributed moment (bending moment, not torque) of intensity m per unit distance along the axis of the beam (see figure).

Derive the equation of the deflection curve and then obtain formulas for the deflection δ_B and angle of rotation θ_B at the free end. (*Note:* Use the second-order differential equation of the deflection curve.)

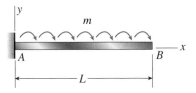

PROB. 8.3-11

8.3-12 The beam shown in the figure has a guided support at A and a spring support at B. The guided support permits vertical movement but no rotation. Derive the equation of the deflection curve and determine the deflection δ_B at end B due to the uniform load of intensity q. (*Note*: Use the second-order differential equation of the deflection curve.)

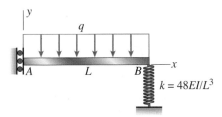

PROB. 8.3-12

8.3-13 Derive the equations of the deflection curve for a simple beam AB loaded by a couple M_0 acting at distance a from the left-hand support (see figure). Also, determine the deflection δ_0 at the point where the load is applied. (*Note:* Use the second-order differential equation of the deflection curve.)

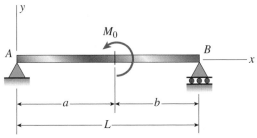

PROB. 8.3-13

8.3-14 Derive the equations of the deflection curve for a cantilever beam AB carrying a uniform load of intensity q over part of the span (see figure). Also, determine the deflection δ_B at the end of the beam. (*Note:* Use the second-order differential equation of the deflection curve.)

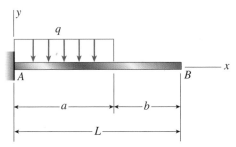

PROB. 8.3-14

8.3-15 Derive the equations of the deflection curve for a cantilever beam AB supporting a distributed load of peak intensity q_0 acting over one-half of the length (see figure). Also, obtain formulas for the deflections δ_B and δ_C at points B and C, respectively. (*Note:* Use the second-order differential equation of the deflection curve.)

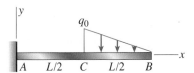

PROB. 8.3-15

8.3-16 Derive the equations of the deflection curve for a simple beam AB with a distributed load of peak intensity q_0 acting over the left-hand half of the span (see figure). Also, determine the deflection δ_C at the midpoint of the beam. (*Note:* Use the second-order differential equation of the deflection curve.)

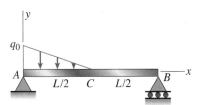

PROB. 8.3-16

8.3-17 The beam shown in the figure has a guided support at A and a roller support at B. The guided support permits vertical movement but no rotation. Derive the equation of the deflection curve and determine the deflection δ_A at end A and also δ_C at point C due to the uniform load of intensity $q = P/L$ applied over segment CB and load P at $x = L/3$. (*Note*: Use the second-order differential equation of the deflection curve.)

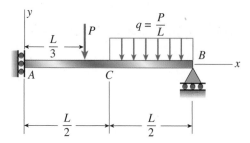

PROB. 8.3-17

Deflections by Integration of the Shear-Force and Load Equations

The beams described in the problems for Section 8.4 have constant flexural rigidity EI. Also, the origin of coordinates is at the left-hand end of each beam.

8.4-1 Derive the equation of the deflection curve for a cantilever beam AB when a couple M_0 acts counterclockwise at the free end (see figure). Also, determine the deflection δ_B and slope θ_B at the free end. Use the third-order differential equation of the deflection curve (the shear-force equation).

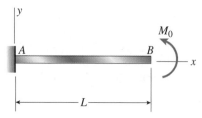

PROB. 8.4-1

8.4-2 A simple beam AB is subjected to a distributed load of intensity $q = q_0 \sin \pi x/L$, where q_0 is the maximum intensity of the load (see figure).

Derive the equation of the deflection curve, and then determine the deflection δ_{max} at the midpoint of the beam. Use the fourth-order differential equation of the deflection curve (the load equation).

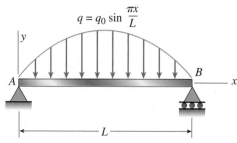

PROB. 8.4-2

8.4-3 The simple beam AB shown in the figure has moments $2M_0$ and M_0 acting at the ends.

Derive the equation of the deflection curve, and then determine the maximum deflection δ_{max}. Use the third-order differential equation of the deflection curve (the shear-force equation).

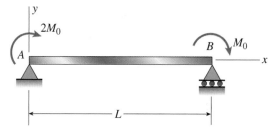

PROB. 8.4-3

8.4-4 A beam with a uniform load has a guided support at one end and spring support at the other. The spring has stiffness $k = 48EI/L^3$. Derive the equation of the deflection curve by starting with the third-order differential equation (the shear-force equation). Also, determine the angle of rotation θ_B at support B.

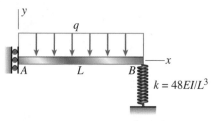

PROB. 8.4-4

8.4-5 The distributed load acting on a cantilever beam AB has an intensity q given by the expression $q_0 \cos \pi x /2L$, where q_0 is the maximum intensity of the load (see figure).

Derive the equation of the deflection curve, and then determine the deflection δ_B at the free end. Use the fourth-order differential equation of the deflection curve (the load equation).

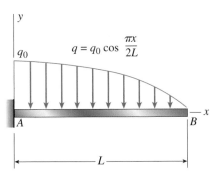

PROB. 8.4-5

8.4-6 A cantilever beam AB is subjected to a parabolically varying load of intensity $q = q_0(L^2 - x^2)/L^2$, where q_0 is the maximum intensity of the load (see figure).

Derive the equation of the deflection curve, and then determine the deflection δ_B and angle of rotation θ_B at the free end. Use the fourth-order differential equation of the deflection curve (the load equation).

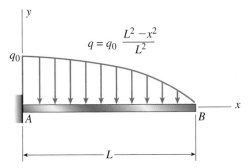

PROB. 8.4-6

8.4-7 A beam on simple supports is subjected to a parabolically distributed load of intensity $q = 4q_0x$ $(L - x)/L^2$, where q_0 is the maximum intensity of the load (see figure).

Derive the equation of the deflection curve, and then determine the maximum deflection δ_{max}. Use the fourth-order differential equation of the deflection curve (the load equation).

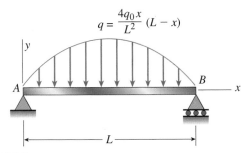

PROB. 8.4-7

8.4-8 Derive the equation of the deflection curve for beam AB, with guided support at A and roller at B, carrying a triangularly distributed load of maximum intensity q_0 (see figure). Also, determine the maximum deflection δ_{max} of the beam. Use the fourth-order differential equation of the deflection curve (the load equation).

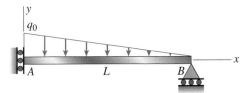

PROB. 8.4-8

8.4-9 Derive the equations of the deflection curve for beam ABC, with guided support at A and roller support at B, supporting a uniform load of intensity q acting on the overhang portion of the beam (see figure). Also, determine deflection δ_C and angle of rotation θ_C. Use the fourth-order differential equation of the deflection curve (the load equation).

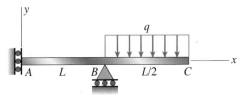

PROB. 8.4-9

8.4-10 Derive the equations of the deflection curve for beam AB, with guided support at A and roller support at B, supporting a distributed load of maximum intensity q_0 acting on the right-hand half of the beam (see figure). Also, determine deflection δ_A, angle of rotation θ_B, and deflection δ_C at the midpoint. Use the fourth-order differential equation of the deflection curve (the load equation).

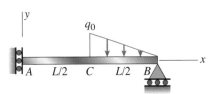

PROB. 8.4-10

Method of Superposition

The problems for Section 8.5 are to be solved by the method of superposition. All beams have constant flexural rigidity EI.

8.5-1 A cantilever beam *AB* carries three equally spaced concentrated loads, as shown in the figure. Obtain formulas for the angle of rotation θ_B and deflection δ_B at the free end of the beam.

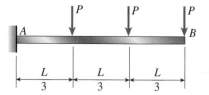

PROB. 8.5-1

8.5-2 A simple beam *AB* supports five equally spaced loads *P* (see figure).

(a) Determine the deflection δ_1 at the midpoint of the beam.

(b) If the same total load (5*P*) is distributed as a uniform load on the beam, what is the deflection δ_2 at the midpoint?

(c) Calculate the ratio of δ_1 to δ_2.

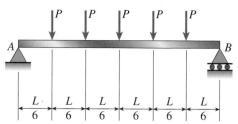

PROB. 8.5-2

8.5-3 The cantilever beam *AB* shown in the figure has an extension *BCD* attached to its free end. A force *P* acts at the end of the extension.

(a) Find the ratio a/L so that the vertical deflection of point *B* will be zero.

(b) Find the ratio a/L so that the angle of rotation at point *B* will be zero.

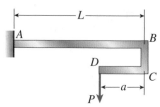

PROB. 8.5-3

8.5-4 Beam *ACB* hangs from two springs, as shown in the figure. The springs have stiffnesses k_1 and k_2 and the beam has flexural rigidity *EI*.

(a) What is the downward displacement of point *C*, which is at the midpoint of the beam, when the moment M_0 is

applied? Data for the structure are as follows: $M_0 = 10.0$ kN·m, $L = 1.8$ m, $EI = 216$ kN·m^2, $k_1 = 250$ kN/m, and $k_2 = 160$ kN/m.

(b) Repeat (a) but remove M_0 and apply uniform load $q = 3.5$ kN/m to the entire beam.

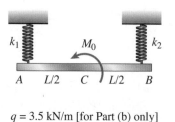

$q = 3.5$ kN/m [for Part (b) only]

PROB. 8.5-4

8.5-5 What must be the equation $y = f(x)$ of the axis of the slightly curved beam *AB* (see figure) *before* the load is applied in order that the load *P*, moving along the bar, always stays at the same level?

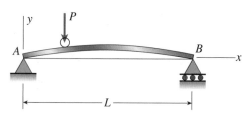

PROB. 8.5-5

8.5-6 Determine the angle of rotation θ_B and deflection δ_B at the free end of a cantilever beam *AB* having a uniform load of intensity *q* acting over the middle third of its length (see figure).

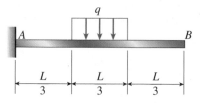

PROB. 8.5-6

8.5-7 The cantilever beam *ACB* shown in the figure has flexural rigidity $EI = 2.1 \times 10^6$ k-in.^{2} Calculate the downward deflections δ_C and δ_B at points *C* and *B*, respectively, due to the simultaneous action of the moment of 35 k-in. applied at point *C* and the concentrated load of 2.5 k applied at the free end *B*.

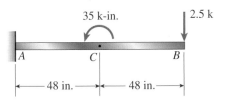

PROB. 8.5-7

8.5-8 A beam *ABCD* consisting of a simple span *BD* and an overhang *AB* is loaded by a force *P* acting at the end of the bracket *CEF* (see figure).

(a) Determine the deflection δ_A at the end of the overhang.

(b) Under what conditions is this deflection upward? Under what conditions is it downward?

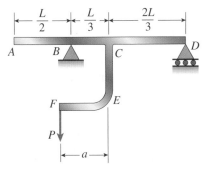

PROB. 8.5-8

8.5-9 A horizontal load *P* acts at end *C* of the bracket *ABC* shown in the figure.

(a) Determine the deflection δ_C of point *C*.

(b) Determine the maximum upward deflection δ_{max} of member *AB*.

Note: Assume that the flexural rigidity *EI* is constant throughout the frame. Also, disregard the effects of axial deformations and consider only the effects of bending due to the load *P*.

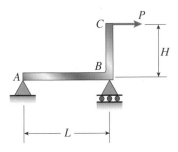

PROB. 8.5-9

8.5-10 A beam *ABC* having flexural rigidity $EI = 75$ kN·m² is loaded by a force $P = 800$ N at end *C* and tied down at end *A* by a wire having axial rigidity $EA = 900$ kN (see figure).

What is the deflection at point *C* when the load *P* is applied?

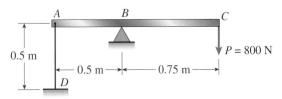

PROB. 8.5-10

8.5-11 Determine the angle of rotation θ_B and deflection δ_B at the free end of a cantilever beam *AB* supporting a parabolic load defined by the equation $q = q_0 x^2/L^2$ (see figure).

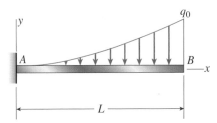

PROB. 8.5-11

8.5-12 A simple beam *AB* supports a uniform load of intensity *q* acting over the middle region of the span (see figure).

Determine the angle of rotation θ_A at the left-hand support and the deflection δ_{max} at the midpoint.

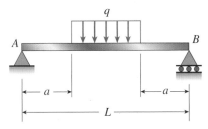

PROB. 8.5-12

8.5-13 The overhanging beam *ABCD* supports two concentrated loads *P* and *Q* (see figure).

(a) For what ratio *P/Q* will the deflection at point *B* be zero?

(b) For what ratio will the deflection at point *D* be zero?

(c) If *Q* is replaced by uniform load with intensity *q* (on the overhang), repeat (a) and (b) but find ratio $P/(qa)$.

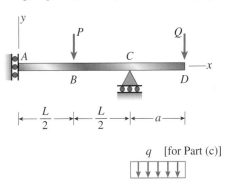

PROB. 8.5-13

8.5-14 A thin metal strip of total weight W and length L is placed across the top of a flat table of width $L/3$ as shown in the figure.

What is the clearance δ between the strip and the middle of the table? (The strip of metal has flexural rigidity EI.)

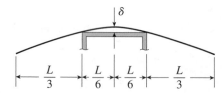

PROB. 8.5-14

8.5-15 An overhanging beam ABC with flexural rigidity $EI = 15$ k-in.^{2} is supported by a guided support at A and by a spring of stiffness k at point B (see figure). Span AB has length $L = 30$ in. and carries a uniform load. The overhang BC has length $b = 15$ in. For what stiffness k of the spring will the uniform load produce no deflection at the free end C?

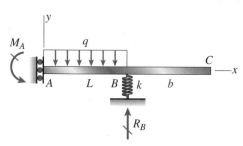

PROB. 8.5-15

8.5-16 A beam $ABCD$ rests on simple supports at B and C (see figure). The beam has a slight initial curvature so that end A is 18 mm above the elevation of the supports and end D is 12 mm above. What moments M_1 and M_2, acting at points A and D, respectively, will move points A and D downward to the level of the supports? (The flexural rigidity EI of the beam is 2.5×10^6 N·m^2 and $L = 2.5$m).

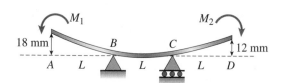

PROB. 8.5-16

8.5-17 The compound beam ABC shown in the figure has a guided support at A and a fixed support at C. The beam consists of two members joined by a pin connection (i.e., moment release) at B. Find the deflection δ under the load P.

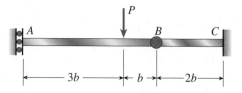

PROB. 8.5-17

8.5-18 A compound beam $ABCDE$ (see figure) consists of two parts (ABC and CDE) connected by a hinge (i.e., moment release) at C. The elastic support at B has stiffness $k = EI/b^3$ Determine the deflection δ_E at the free end E due to the load P acting at that point.

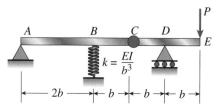

PROB. 8.5-18

8.5-19 A steel beam ABC is simply supported at A and held by a high-strength steel wire at B (see figure). A load $P = 240$ lb acts at the free end C. The wire has axial rigidity $EA = 1500 \times 10^3$ lb, and the beam has flexural rigidity $EI = 36 \times 10^6$ lb-in.^{2}

What is the deflection δ_C of point C due to the load P?

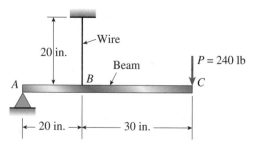

PROB. 8.5-19

8.5-20 The compound beam shown in the figure consists of a cantilever beam AB (length L) that is pin-connected to a simple beam BD (length $2L$). After the beam is constructed, a clearance c exists between the beam and a support at C, midway between points B and D. Subsequently, a uniform load is placed along the entire length of the beam.

What intensity q of the load is needed to close the gap at C and bring the beam into contact with the support?

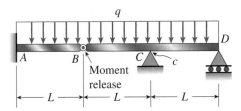

8.5-21 Find the horizontal deflection δ_h and vertical deflection δ_v at the free end C of the frame ABC shown in the figure. (The flexural rigidity EI is constant throughout the frame.)

Note: Disregard the effects of axial deformations and consider only the effects of bending due to the load P.

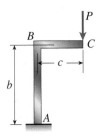

PROB. 8.5-21

8.5-22 The frame $ABCD$ shown in the figure is squeezed by two collinear forces P acting at points A and D. What is the decrease δ in the distance between points A and D when the loads P are applied? (The flexural rigidity EI is constant throughout the frame.)

Note: Disregard the effects of axial deformations and consider only the effects of bending due to the loads P.

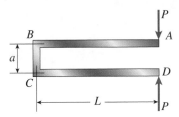

PROB. 8.5-22

8.5-23 A beam $ABCDE$ has simple supports at B and D and symmetrical overhangs at each end (see figure). The center span has length L and each overhang has length b. A uniform load of intensity q acts on the beam.

(a) Determine the ratio b/L so that the deflection δ_C at the midpoint of the beam is equal to the deflections δ_A and δ_E at the ends.

(b) For this value of b/L, what is the deflection δ_C at the midpoint?

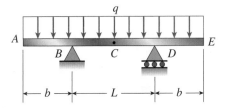

PROB. 8.5-23

8.5-24 A frame ABC is loaded at point C by a force P acting at an angle α to the horizontal (see figure). Both members of the frame have the same length and the same flexural rigidity.

Determine the angle α so that the deflection of point C is in the same direction as the load. (Disregard the effects of axial deformations and consider only the effects of bending due to the load P.)

Note: A direction of loading such that the resulting deflection is in the same direction as the load is called a *principal direction*. For a given load on a planar structure, there are two principal directions, perpendicular to each other.

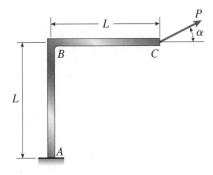

PROB. 8.5-24

Critical load carrying elements in structures such as columns and other slender compression members are susceptible to buckling failure.
(LUSHPIX/UNLISTED IMAGES, INC.)

9

Columns

CHAPTER OVERVIEW

Chapter 9 is primarily concerned with the **buckling** of slender columns which support compressive loads in structures. First, the **critical axial load** which indicates the onset of buckling is defined and computed for a number of simple models composed of rigid bars and elastic springs (Section 9.2). **Stable, neutral, and unstable equilibrium conditions** are described for these idealized rigid structures. Then, linear elastic buckling of slender columns with pinned-end conditions is considered (Section 9.3). The differential equation of the deflection curve is derived and solved to obtain expressions for the **Euler buckling load** (P_{cr}) and associated buckled shape for the fundamental mode. **Critical stress** (σ_{cr}) and **slenderness ratio** (L/r) are defined, and the effects of large deflections, column imperfections, inelastic behavior, and optimum shapes of columns are explained. Finally, critical loads and buckled mode shapes are computed for three additional column support cases (fixed-free, fixed-fixed, and fixed-pinned) (Section 9.4), and the concept of effective length (L_e) is introduced.

Chapter 9 is organized as follows:

523

9.1 INTRODUCTION

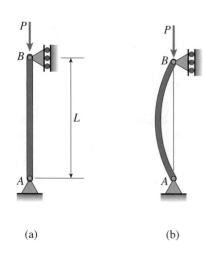

(a)   (b)

FIG. 9-1 Buckling of a slender column due to an axial compressive load P

Load-carrying structures may fail in a variety of ways, depending upon the type of structure, the conditions of support, the kinds of loads, and the materials used. For instance, an axle in a vehicle may fracture suddenly from repeated cycles of loading, or a beam may deflect excessively, so that the structure is unable to perform its intended functions. These kinds of failures are prevented by designing structures so that the maximum stresses and maximum displacements remain within tolerable limits. Thus, **strength** and **stiffness** are important factors in design, as discussed throughout the preceding chapters.

Another type of failure is **buckling**, which is the subject matter of this chapter. We will consider specifically the buckling of **columns**, which are long, slender structural members loaded axially in compression (Fig. 9-1a). If a compression member is relatively slender, it may deflect laterally and fail by bending (Fig. 9-1b) rather than failing by direct compression of the material. You can demonstrate this behavior by compressing a plastic ruler or other slender object. When lateral bending occurs, we say that the column has *buckled*. Under an increasing axial load, the lateral deflections will increase too, and eventually the column will collapse completely.

The phenomenon of buckling is not limited to columns. Buckling can occur in many kinds of structures and can take many forms. When you step on the top of an empty aluminum can, the thin cylindrical walls buckle under your weight and the can collapses. When a large bridge collapsed a few years ago, investigators found that failure was caused by the buckling of a thin steel plate that wrinkled under compressive stresses. Buckling is one of the major causes of failures in structures, and therefore the possibility of buckling should always be considered in design.

9.2 BUCKLING AND STABILITY

To illustrate the fundamental concepts of buckling and stability, we will analyze the **idealized structure**, or **buckling model**, shown in Fig. 9-2a. This hypothetical structure consists of two rigid bars AB and BC, each of length $L/2$. They are joined at B by a pin connection and held in a vertical position by a rotational spring having stiffness β_R.[*]

This idealized structure is analogous to the column of Fig. 9-1a, because both structures have simple supports at the ends and are compressed by an axial load P. However, the elasticity of the idealized

---

[*]The general relationship for a rotational spring is $M = \beta_R \theta$, where M is the moment acting on the spring, β_R is the rotational stiffness of the spring, and θ is the angle through which the spring rotates. Thus, rotational stiffness has units of moment divided by angle, such as lb-in./rad or N·m/rad. The analogous relationship for a translational spring is $F = \beta\delta$, where F is the force acting on the spring, β is the translational stiffness of the spring (or spring constant), and δ is the change in length of the spring. Thus, translational stiffness has units of force divided by length, such as lb/in. or N/m.

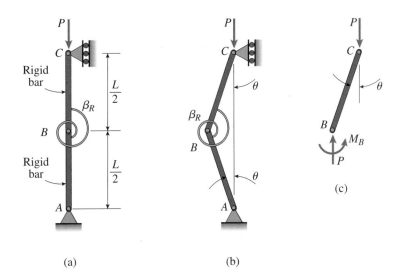

FIG. 9-2 Buckling of an idealized structure consisting of two rigid bars and a rotational spring

(a)   (b)

structure is "concentrated" in the rotational spring, whereas a real column can bend throughout its length (Fig. 9-1b).

In the idealized structure, the two bars are perfectly aligned and the axial load P has its line of action along the longitudinal axis (Fig. 9-2a). Consequently, the spring is initially unstressed and the bars are in direct compression.

Now suppose that the structure is disturbed by some external force that causes point B to move a small distance laterally (Fig. 9-2b). The rigid bars rotate through small angles θ and a moment develops in the spring. The direction of this moment is such that it tends to return the structure to its original straight position, and therefore it is called a **restoring moment**. At the same time, however, the tendency of the axial compressive force is to increase the lateral displacement. Thus, these two actions have opposite effects—the restoring moment tends to *decrease* the displacement and the axial force tends to *increase* it.

Now consider what happens when the disturbing force is removed. If the axial force P is relatively small, the action of the restoring moment will predominate over the action of the axial force and the structure will return to its initial straight position. Under these conditions, the structure is said to be **stable**. However, if the axial force P is large, the lateral displacement of point B will increase and the bars will rotate through larger and larger angles until the structure collapses. Under these conditions, the structure is **unstable** and fails by lateral buckling.

Critical Load

The transition between the stable and unstable conditions occurs at a special value of the axial force known as the **critical load** (denoted by the symbol P_{cr}). We can determine the critical load of our buckling model by considering the structure in the disturbed position (Fig. 9-2b) and investigating its equilibrium.

First, we consider the entire structure as a free body and sum moments about support A. This step leads to the conclusion that there is no horizontal reaction at support C. Second, we consider bar BC as a free body (Fig. 9-2c) and note that it is subjected to the action of the axial forces P and the moment M_B in the spring. The moment M_B is equal to the rotational stiffness β_R times the angle of rotation 2θ of the spring; thus,

$$M_B = 2\beta_R\theta \tag{a}$$

Since the angle θ is a small quantity, the lateral displacement of point B is $\theta L/2$. Therefore, we obtain the following equation of equilibrium by summing moments about point B for bar BC (Fig. 9-2c):

$$M_B - P\left(\frac{\theta L}{2}\right) = 0 \tag{b}$$

or, upon substituting from Eq. (a),

$$\left(2\beta_R - \frac{PL}{2}\right)\theta = 0 \tag{9-1}$$

One solution of this equation is $\theta = 0$, which is a trivial solution and merely means that the structure is in equilibrium when it is perfectly straight, regardless of the magnitude of the force P.

A second solution is obtained by setting the term in parentheses equal to zero and solving for the load P, which is the *critical load*:

$$P_{cr} = \frac{4\beta_R}{L} \tag{9-2}$$

At the critical value of the load the structure is in equilibrium regardless of the magnitude of the angle θ (provided the angle remains small, because we made that assumption when deriving Eq. b).

From the preceding analysis we see that the critical load is the *only* load for which the structure will be in equilibrium in the disturbed position. At this value of the load, the restoring effect of the moment in the spring just matches the buckling effect of the axial load. Therefore, the critical load represents the boundary between the stable and unstable conditions.

If the axial load is less than P_{cr}, the effect of the moment in the spring predominates and the structure returns to the vertical position after a slight disturbance; if the axial load is larger than P_{cr}, the effect of the axial force predominates and the structure buckles:

If $P < P_{cr}$, the structure is *stable*
If $P > P_{cr}$, the structure is *unstable*

From Eq. (9-2) we see that the stability of the structure is increased either by *increasing its stiffness* or by *decreasing its length*. Later in this chapter, when we determine critical loads for various types of columns, we will see that these same observations apply.

Summary

Let us now summarize the behavior of the idealized structure (Fig. 9-2a) as the axial load P increases from zero to a large value.

When the axial load is less than the critical load ($0 < P < P_{cr}$), the structure is in equilibrium when it is perfectly straight. Because the equilibrium is **stable**, the structure returns to its initial position after being disturbed. Thus, the structure is in equilibrium *only* when it is perfectly straight ($\theta = 0$).

When the axial load is greater than the critical load ($P > P_{cr}$), the structure is still in equilibrium when $\theta = 0$ (because it is in direct compression and there is no moment in the spring), but the equilibrium is **unstable** and cannot be maintained. The slightest disturbance will cause the structure to buckle.

At the critical load ($P = P_{cr}$), the structure is in equilibrium even when point B is displaced laterally by a small amount. In other words, the structure is in equilibrium for *any* small angle θ, including $\theta = 0$. However, the structure is neither stable nor unstable—it is at the boundary between stability and instability. This condition is referred to as **neutral equilibrium**.

The three equilibrium conditions for the idealized structure are shown in the graph of axial load P versus angle of rotation θ (Fig. 9-3). The two heavy lines, one vertical and one horizontal, represent the equilibrium conditions. Point B, where the equilibrium diagram branches, is called a *bifurcation point*.

The horizontal line for neutral equilibrium extends to the left and right of the vertical axis because the angle θ may be clockwise or counterclockwise. The line extends only a short distance, however, because our analysis is based upon the assumption that θ is a small angle. (This assumption is quite valid, because θ is indeed small when the structure first departs from its vertical position. If buckling continues and θ becomes large, the line labeled "Neutral equilibrium" curves upward, as shown later in Fig. 9-11.)

The three equilibrium conditions represented by the diagram of Fig. 9-3 are analogous to those of a ball placed upon a smooth surface (Fig. 9-4). If the surface is concave upward, like the inside of a dish, the equilibrium is stable and the ball always returns to the low point when disturbed. If the surface is convex upward, like a dome, the ball can theoretically be in equilibrium on top of the surface, but the equilibrium is unstable and in reality the ball rolls away. If the surface is perfectly flat, the ball is in neutral equilibrium and remains wherever it is placed.

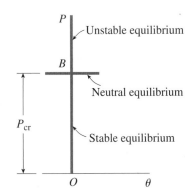

FIG. 9-3 Equilibrium diagram for buckling of an idealized structure

FIG. 9-4 Ball in stable, unstable, and neutral equilibrium

As we will see in the next section, the behavior of an ideal elastic column is analogous to that of the buckling model shown in Fig. 9-2. Furthermore, many other kinds of structural and mechanical systems fit this model.

9.3 COLUMNS WITH PINNED ENDS

We begin our investigation of the stability behavior of columns by analyzing a slender column with pinned ends (Fig. 9-5a). The column is loaded by a vertical force P that is applied through the centroid of the end cross section. The column itself is perfectly straight and is made of a linearly elastic material that follows Hooke's law. Since the column is assumed to have no imperfections, it is referred to as an **ideal column**.

For purposes of analysis, we construct a coordinate system with its origin at support A and with the x axis along the longitudinal axis of the column. The y axis is directed to the left in the figure, and the z axis (not shown) comes out of the plane of the figure toward the viewer. We assume that the xy plane is a plane of symmetry of the column and that any bending takes place in that plane (Fig. 9-5b). The coordinate system is identical to the one used in our previous discussions of beams, as can be seen by rotating the column clockwise through an angle of 90°.

When the axial load P has a small value, the column remains perfectly straight and undergoes direct axial compression. The only stresses are the uniform compressive stresses obtained from the equation $\sigma = P/A$. The column is in **stable equilibrium**, which means that it returns to the straight position after a disturbance. For instance, if we apply a small lateral load and cause the column to bend, the deflection will disappear and the column will return to its original position when the lateral load is removed.

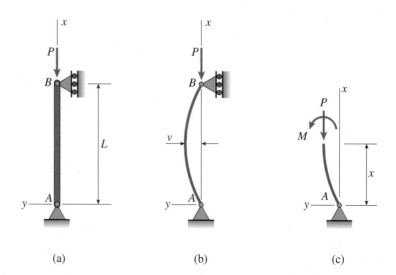

FIG. 9-5 Column with pinned ends: (a) ideal column, (b) buckled shape, and (c) axial force P and bending moment M acting at a cross section

(a)   (b)   (c)

As the axial load P is gradually increased, we reach a condition of **neutral equilibrium** in which the column may have a bent shape. The corresponding value of the load is the **critical load** P_{cr}. At this load the column may undergo small lateral deflections with no change in the axial force. For instance, a small lateral load will produce a bent shape that does not disappear when the lateral load is removed. Thus, the critical load can maintain the column in equilibrium *either* in the straight position or in a slightly bent position.

At higher values of the load, the column is **unstable** and may collapse by buckling, that is, by excessive bending. For the ideal case that we are discussing, the column will be in equilibrium in the straight position even when the axial force P is greater than the critical load. However, since the equilibrium is unstable, the smallest imaginable disturbance will cause the column to deflect sideways. Once that happens, the deflections will immediately increase and the column will fail by buckling. The behavior is similar to that described in the preceding section for the idealized buckling model (Fig. 9-2).

The behavior of an ideal column compressed by an axial load P (Figs. 9-5a and b) may be summarized as follows:

If $P < P_{cr}$, the column is in stable equilibrium in the straight position.

If $P = P_{cr}$, the column is in neutral equilibrium in either the straight or a slightly bent position.

If $P > P_{cr}$, the column is in unstable equilibrium in the straight position and will buckle under the slightest disturbance.

Of course, a real column does not behave in this idealized manner because imperfections are always present. For instance, the column is not *perfectly* straight, and the load is not *exactly* at the centroid. Nevertheless, we begin by studying ideal columns because they provide insight into the behavior of real columns.

Differential Equation for Column Buckling

To determine the critical loads and corresponding deflected shapes for an ideal pin-ended column (Fig. 9-5a), we use one of the differential equations of the deflection curve of a beam (see Eqs. 8-12a, b, and c in Section 8.2). These equations are applicable to a buckled column because the column bends as though it were a beam (Fig. 9-5b).

Although both the fourth-order differential equation (the load equation) and the third-order differential equation (the shear-force equation) are suitable for analyzing columns, we will elect to use the second-order equation (the bending-moment equation) because its general solution is usually the simplest. The **bending-moment equation** (Eq. 8-12a) is

$$EIv'' = M \tag{9-3}$$

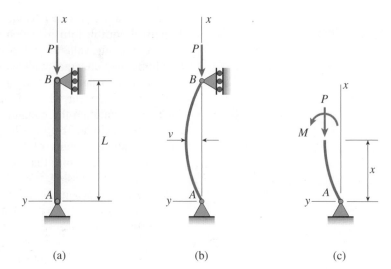

FIG. 9-5 (Repeated)

(a) (b) (c)

in which M is the bending moment at any cross section, v is the lateral deflection in the y direction, and EI is the flexural rigidity for bending in the xy plane.

The bending moment M at distance x from end A of the buckled column is shown acting in its positive direction in Fig. 9-5c. Note that the bending moment sign convention is the same as that used in earlier chapters, namely, positive bending moment produces positive curvature (see Figs. 8-3 and 8-4).

The axial force P acting at the cross section is also shown in Fig. 9-5c. Since there are no horizontal forces acting at the supports, there are no shear forces in the column. Therefore, from equilibrium of moments about point A, we obtain

$$M + Pv = 0 \quad \text{or} \quad M = -Pv \tag{9-4}$$

where v is the deflection at the cross section.

This same expression for the bending moment is obtained if we assume that the column buckles to the right instead of to the left (Fig. 9-6a). When the column deflects to the right, the deflection itself is $-v$ but the moment of the axial force about point A also changes sign. Thus, the equilibrium equation for moments about point A (see Fig. 9-6b) is

$$M - P(-v) = 0$$

which gives the same expression for the bending moment M as before.

The **differential equation of the deflection curve** (Eq. 9-3) now becomes

$$EIv'' + Pv = 0 \tag{9-5}$$

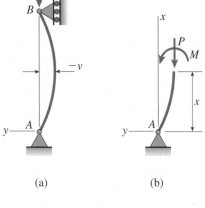

(a) (b)

FIG. 9-6 Column with pinned ends (alternative direction of buckling)

By solving this equation, which is a *homogeneous, linear, differential equation of second order with constant coefficients*, we can determine the magnitude of the critical load and the deflected shape of the buckled column.

Note that we are analyzing the buckling of columns by solving the same basic differential equation as the one we solved in Chapter 8 when finding beam deflections. However, there is a fundamental difference in the two types of analysis. In the case of beam deflections, the bending moment M appearing in Eq. (9-3) is a function of the loads only—it does not depend upon the deflections of the beam. In the case of buckling, the bending moment is a function of the deflections themselves (Eq. 9-4).

Thus, we now encounter a new aspect of bending analysis. In our previous work, the deflected shape of the structure was not considered, and the equations of equilibrium were based upon the geometry of the *undeformed* structure. Now, however, the geometry of the *deformed* structure is taken into account when writing equations of equilibrium.

Solution of the Differential Equation

For convenience in writing the solution of the differential equation (Eq. 9-5), we introduce the notation

$$k^2 = \frac{P}{EI} \quad \text{or} \quad k = \sqrt{\frac{P}{EI}} \tag{9-6a,b}$$

in which k is always taken as a positive quantity. Note that k has units of the reciprocal of length, and therefore quantities such as kx and kL are nondimensional.

Using this notation, we can rewrite Eq. (9-5) in the form

$$v'' + k^2 v = 0 \tag{9-7}$$

From mathematics we know that the **general solution** of this equation is

$$v = C_1 \sin kx + C_2 \cos kx \tag{9-8}$$

in which C_1 and C_2 are constants of integration (to be evaluated from the boundary conditions, or end conditions, of the column). Note that the number of arbitrary constants in the solution (two in this case) agrees with the order of the differential equation. Also, note that we can verify the solution by substituting the expression for v (Eq. 9-8) into the differential equation (Eq. 9-7) and reducing it to an identity.

To evaluate the **constants of integration** appearing in the solution (Eq. 9-8), we use the boundary conditions at the ends of the

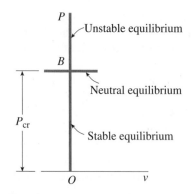

FIG. 9-7 Load-deflection diagram for an ideal, linearly elastic column

column; namely, the deflection is zero when $x = 0$ and $x = L$ (see Fig. 9-5b):

$$v(0) = 0 \quad \text{and} \quad v(L) = 0 \qquad \text{(a,b)}$$

The first condition gives $C_2 = 0$, and therefore

$$v = C_1 \sin kx \qquad \text{(c)}$$

The second condition gives

$$C_1 \sin kL = 0 \qquad \text{(d)}$$

From this equation we conclude that either $C_1 = 0$ or $\sin kL = 0$. We will consider both of these possibilities.

Case 1. If the constant C_1 equals zero, the deflection v is also zero (see Eq. c), and therefore the column remains straight. In addition, we note that when C_1 equals zero, Eq. (d) is satisfied for *any* value of the quantity kL. Consequently, the axial load P may also have any value (see Eq. 9-6b). This solution of the differential equation (known in mathematics as the *trivial solution*) is represented by the vertical axis of the load-deflection diagram (Fig. 9-7). It gives the behavior of an ideal column that is in equilibrium (either stable or unstable) in the straight position (no deflection) under the action of the compressive load P.

Case 2. The second possibility for satisfying Eq. (d) is given by the following equation, known as the **buckling equation**:

$$\sin kL = 0 \qquad \text{(9-9)}$$

This equation is satisfied when $kL = 0, \pi, 2\pi, \ldots$. However, since $kL = 0$ means that $P = 0$, this solution is not of interest. Therefore, the solutions we will consider are

$$kL = n\pi \qquad n = 1, 2, 3, \ldots \qquad \text{(e)}$$

or (see Eq. 9-6a):

$$P = \frac{n^2 \pi^2 EI}{L^2} \qquad n = 1, 2, 3, \ldots \qquad \text{(9-10)}$$

This formula gives the values of P that satisfy the buckling equation and provide solutions (other than the trivial solution) to the differential equation.

The equation of the **deflection curve** (from Eqs. c and e) is

$$v = C_1 \sin kx = C_1 \sin \frac{n \pi x}{L} \qquad n = 1, 2, 3, \ldots \qquad \text{(9-11)}$$

Only when P has one of the values given by Eq. (9-10) is it theoretically possible for the column to have a bent shape (given by Eq. 9-11). For all other values of P, the column is in equilibrium only if it remains straight. Therefore, the values of P given by Eq. (9-10) are the **critical loads** for this column.

Critical Loads

The lowest critical load for a column with pinned ends (Fig. 9-8a) is obtained when $n = 1$:

$$P_{cr} = \frac{\pi^2 EI}{L^2} \tag{9-12}$$

The corresponding buckled shape (sometimes called a *mode shape*) is

$$v = C_1 \sin \frac{\pi x}{L} \tag{9-13}$$

as shown in Fig. 9-8b. The constant C_1 represents the deflection at the midpoint of the column and may have any small value, either positive or negative. Therefore, the part of the load-deflection diagram corresponding to P_{cr} is a horizontal straight line (Fig. 9-7). Thus, the deflection at the critical load is *undefined*, although it must remain small for our equations to be valid. Above the bifurcation point B the equilibrium is unstable, and below point B it is stable.

Buckling of a pinned-end column in the first mode is called the **fundamental case** of column buckling.

The type of buckling described in this section is called **Euler buckling**, and the critical load for an ideal elastic column is often called the **Euler load**. The famous mathematician Leonhard Euler (1707–1783), generally recognized as the greatest mathematician of all time, was the first person to investigate the buckling of a slender column and determine its critical load (Euler published his results in 1744); see Ref. 9-1; a list of references is available online.

By taking higher values of the index n in Eqs. (9-10) and (9-11), we obtain an infinite number of critical loads and corresponding mode shapes. The mode shape for $n = 2$ has two half-waves, as pictured in Fig. 9-8c. The corresponding critical load is four times larger than the critical load for the fundamental case. The magnitudes of the critical loads are proportional to the square of n, and the number of half-waves in the buckled shape is equal to n.

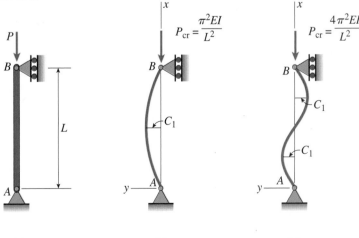

FIG. 9-8 Buckled shapes for an ideal column with pinned ends: (a) initially straight column, (b) buckled shape for $n = 1$, and (c) buckled shape for $n = 2$

Buckled shapes for the **higher modes** are often of no practical interest because the column buckles when the axial load P reaches its lowest critical value. The only way to obtain modes of buckling higher than the first is to provide lateral support of the column at intermediate points, such as at the midpoint of the column shown in Fig. 9-8 (see Example 9-1 at the end of this section).

General Comments

From Eq. (9-12), we see that the critical load of a column is proportional to the flexural rigidity EI and inversely proportional to the square of the length. Of particular interest is the fact that the *strength* of the material itself, as represented by a quantity such as the proportional limit or the yield stress, does not appear in the equation for the critical load. Therefore, increasing a strength property does not raise the critical load of a slender column. It can only be raised by increasing the flexural rigidity, reducing the length, or providing additional lateral support.

The *flexural rigidity* can be increased by using a "stiffer" material (that is, a material with larger modulus of elasticity E) or by distributing the material in such a way as to increase the moment of inertia I of the cross section, just as a beam can be made stiffer by increasing the moment of inertia. The moment of inertia is increased by distributing the material farther from the centroid of the cross section. Hence, a hollow tubular member is generally more economical for use as a column than a solid member having the same cross-sectional area.

Reducing the *wall thickness* of a tubular member and increasing its lateral dimensions (while keeping the cross-sectional area constant) also increases the critical load because the moment of inertia is increased. This process has a practical limit, however, because eventually the wall itself will become unstable. When that happens, localized buckling occurs in the form of small corrugations or wrinkles in the walls of the column. Thus, we must distinguish between *overall buckling* of a column, which is discussed in this chapter, and *local buckling* of its parts. The latter requires more detailed investigations and is beyond the scope of this book.

In the preceding analysis (see Fig. 9-8), we assumed that the xy plane was a plane of symmetry of the column and that buckling took place in that plane. The latter assumption will be met if the column has lateral supports perpendicular to the plane of the figure, so that the column is constrained to buckle in the xy plane. If the column is supported only at its ends and is free to buckle in *any* direction, then bending will occur about the principal centroidal axis having the smaller moment of inertia.

For instance, consider the rectangular and wide-flange cross sections shown in Fig. 9-9. In each case, the moment of inertia I_1 is greater than the moment of inertia I_2; hence the column will buckle in the 1–1 plane, and the smaller moment of inertia I_2 should be used in the formula for the critical load. If the cross section is square or circular, all centroidal axes have the same moment of inertia and buckling may occur in any longitudinal plane.

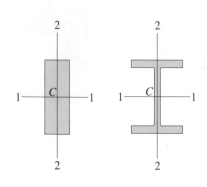

FIG. 9-9 Cross sections of columns showing principal centroidal axes with $I_1 > I_2$

Critical Stress

After finding the critical load for a column, we can calculate the corresponding **critical stress** by dividing the load by the cross-sectional area. For the fundamental case of buckling (Fig. 9-8b), the critical stress is

$$\sigma_{cr} = \frac{P_{cr}}{A} = \frac{\pi^2 EI}{AL^2} \tag{9-14}$$

in which I is the moment of inertia for the principal axis about which buckling occurs. This equation can be written in a more useful form by introducing the notation

$$r = \sqrt{\frac{I}{A}} \tag{9-15}$$

in which r is the **radius of gyration** of the cross section in the plane of bending.[*] Then the equation for the critical stress becomes

$$\sigma_{cr} = \frac{\pi^2 E}{(L/r)^2} \tag{9-16}$$

in which L/r is a nondimensional ratio called the **slenderness ratio**:

$$\text{Slenderness ratio} = \frac{L}{r} \tag{9-17}$$

Note that the slenderness ratio depends only on the dimensions of the column. A column that is long and slender will have a high slenderness ratio and therefore a low critical stress. A column that is short and stubby will have a low slenderness ratio and will buckle at a high stress. Typical values of the slenderness ratio for actual columns are between 30 and 150.

The critical stress is the average compressive stress on the cross section at the instant the load reaches its critical value. We can a plot a graph of this stress as a function of the slenderness ratio and obtain a curve known as **Euler's curve** (Fig. 9-10). The curve shown in the figure is plotted for a structural steel with $E = 30 \times 10^3$ ksi. The curve is valid only when the critical stress is less than the proportional limit of the steel, because the equations were derived using Hooke's law. Therefore, we draw a horizontal line on the graph at the proportional limit of the steel (assumed to be 36 ksi) and terminate Euler's curve at that level of stress.[**]

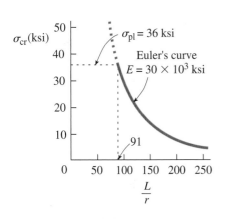

FIG. 9-10 Graph of Euler's curve (from Eq. 9-16) for structural steel with $E = 30 \times 10^3$ ksi and $\sigma_{pl} = 36$ ksi

---

[*]Radius of gyration is described in Section 10.4 (available online).

[**]Euler's curve is not a common geometric shape. It is sometimes mistakenly called a hyperbola, but hyperbolas are plots of polynomial equations of the second degree in two variables, whereas Euler's curve is a plot of an equation of the third degree in two variables.

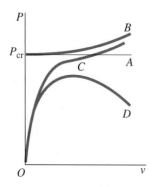

FIG. 9-11 Load-deflection diagram for columns: Line *A*, ideal elastic column with small deflections; Curve *B*, ideal elastic column with large deflections; Curve *C*, elastic column with imperfections; and Curve *D*, inelastic column with imperfections

Effects of Large Deflections, Imperfections, and Inelastic Behavior

The equations for critical loads were derived for ideal columns, that is, columns for which the loads are precisely applied, the construction is perfect, and the material follows Hooke's law. As a consequence, we found that the magnitudes of the small deflections at buckling were undefined.[*] Thus, when $P = P_{cr}$, the column may have any small deflection, a condition represented by the horizontal line labeled *A* in the load-deflection diagram of Fig. 9-11. (In this figure, we show only the right-hand half of the diagram, but the two halves are symmetric about the vertical axis.)

The theory for ideal columns is limited to small deflections because we used the second derivative v'' for the curvature. A more exact analysis, based upon the exact expression for curvature (Eq. 8-13 in Section 8.2), shows that there is no indefiniteness in the magnitudes of the deflections at buckling. Instead, for an ideal, linearly elastic column, the load-deflection diagram goes upward in accord with curve *B* of Fig. 9-11. Thus, after a linearly elastic column begins to buckle, an increasing load is required to cause an increase in the deflections.

Now suppose that the column is not constructed perfectly; for instance, the column might have an imperfection in the form of a small initial curvature, so that the unloaded column is not perfectly straight. Such imperfections produce deflections from the onset of loading, as shown by curve *C* in Fig. 9-11. For small deflections, curve *C* approaches line *A* as an asymptote. However, as the deflections become large, it approaches curve *B*. The larger the imperfections, the further curve *C* moves to the right, away from the vertical line. Conversely, if the column is constructed with considerable accuracy, curve *C* approaches the vertical axis and the horizontal line labeled *A*. By comparing lines *A*, *B*, and *C*, we see that for practical purposes the critical load represents the maximum load-carrying capacity of an elastic column, because large deflections are not acceptable in most applications.

Finally, consider what happens when the stresses exceed the proportional limit and the material no longer follows Hooke's law. Of course, the load-deflection diagram is unchanged up to the level of load at which the proportional limit is reached. Then the curve for inelastic behavior (curve *D*) departs from the elastic curve, continues upward, reaches a maximum, and turns downward.

---

[*]In mathematical terminology, we solved a *linear eigenvalue problem*. The critical load is an *eigenvalue* and the corresponding buckled mode shape is an *eigenfunction*.

The precise shapes of the curves in Fig. 9-11 depend upon the material properties and column dimensions, but the general nature of the behavior is typified by the curves shown.

Only extremely slender columns remain elastic up to the critical load. Stockier columns behave inelastically and follow a curve such as *D*. Thus, the maximum load that can be supported by an inelastic column may be considerably less than the Euler load for that same column. Furthermore, the descending part of curve *D* represents sudden and catastrophic collapse, because it takes smaller and smaller loads to maintain larger and larger deflections. By contrast, the curves for elastic columns are quite stable, because they continue upward as the deflections increase, and therefore it takes larger and larger loads to cause an increase in deflection.

Optimum Shapes of Columns

Compression members usually have the same cross sections throughout their lengths, and therefore only prismatic columns are analyzed in this chapter. However, prismatic columns are not the optimum shape if minimum weight is desired. The critical load of a column consisting of a given amount of material may be increased by varying the shape so that the column has larger cross sections in those regions where the bending moments are larger. Consider, for instance, a column of solid circular cross section with pinned ends. A column shaped as shown in Fig. 9-12a will have a larger critical load than a prismatic column made from the same volume of material. As a means of approximating this optimum shape, prismatic columns are sometimes reinforced over part of their lengths (Fig. 9-12b).

Now consider a prismatic column with pinned ends that is free to buckle in *any* lateral direction (Fig. 9-13a). Also, assume that the column has a solid cross section, such as a circle, square, triangle, rectangle, or hexagon (Fig. 9-13b). An interesting question arises: For a given cross-sectional area, which of these shapes makes the most efficient column? Or, in more precise terms, which cross section gives the largest critical load? Of course, we are assuming that the critical load is calculated from the Euler formula $P_{cr} = \pi^2 EI/L^2$ using the smallest moment of inertia for the cross section.

While a common answer to this question is "the circular shape," you can readily demonstrate that a cross section in the shape of an equilateral triangle gives a 21% higher critical load than does a circular cross section of the same area (see Problem 9.3-11). The critical load for an equilateral triangle is also higher than the loads obtained for the other shapes; hence, an equilateral triangle is the optimum cross section (based only upon theoretical considerations). For a mathematical analysis of optimum column shapes, including columns with varying cross sections, see Ref. 9-4 (a list of references is available online).

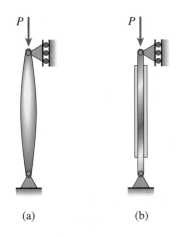

(a)   (b)

FIG. 9-12 Nonprismatic columns

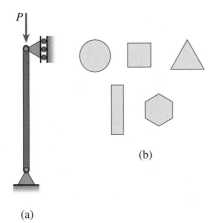

(b)

(a)

FIG. 9-13 Which cross-sectional shape is the optimum shape for a prismatic column?

Example 9-1

Slender steel column with lateral support near mid-height
(Lester Lefkowitz/Getty Images)

FIG. 9-14 Example 9-1. Euler buckling of a slender column

A long, slender column *ABC* is pin-supported at the ends and compressed by an axial load *P* (Fig. 9-14). Lateral support is provided at the midpoint *B* in the plane of the figure. However, lateral support perpendicular to the plane of the figure is provided only at the ends.

The column is constructed of a steel wide-flange section (W 8 × 28) having modulus of elasticity $E = 29 \times 10^3$ ksi and proportional limit $\sigma_{pl} = 42$ ksi. The total length of the column is $L = 25$ ft.

Determine the allowable load P_{allow} using a factor of safety $n = 2.5$ with respect to Euler buckling of the column.

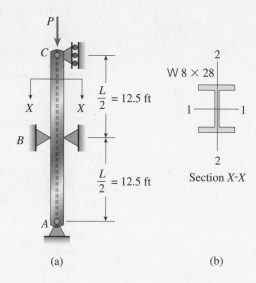

(a)   (b)

Solution

Because of the manner in which it is supported, this column may buckle in either of the two principal planes of bending. As one possibility, it may buckle in the plane of the figure, in which case the distance between lateral supports is $L/2 = 12.5$ ft and bending occurs about axis 2–2 (see Fig. 9-8c for the mode shape of buckling).

As a second possibility, the column may buckle perpendicular to the plane of the figure with bending about axis 1–1. Because the only lateral support in this direction is at the ends, the distance between lateral supports is $L = 25$ ft (see Fig. 9-8b for the mode shape of buckling).

Column properties. From Table F-1, Appendix F (available online), we obtain the following moments of inertia and cross-sectional area for a W 8 × 28 column:

$$I_1 = 98.0 \text{ in.}^4 \qquad I_2 = 21.7 \text{ in.}^4 \qquad A = 8.25 \text{ in.}^2$$

Critical loads. If the column buckles in the plane of the figure, the critical load is

$$P_{\text{cr}} = \frac{\pi^2 E I_2}{(L/2)^2} = \frac{4\pi^2 E I_2}{L^2}$$

Substituting numerical values, we obtain

$$P_{cr} = \frac{4\pi^2 EI_2}{L^2} = \frac{4\pi^2 (29 \times 10^3 \text{ ksi})(21.7 \text{ in.}^4)}{[(25 \text{ ft})(12 \text{ in./ft})]^2} = 276 \text{ k}$$

If the column buckles perpendicular to the plane of the figure, the critical load is

$$P_{cr} = \frac{\pi^2 EI_1}{L^2} = \frac{\pi^2 (29 \times 10^3 \text{ ksi})(98.0 \text{ in.}^4)}{[(25 \text{ ft})(12 \text{ in./ft})]^2} = 312 \text{ k}$$

Therefore, the critical load for the column (the smaller of the two preceding values) is

$$P_{cr} = 276 \text{ k}$$

and buckling occurs in the plane of the figure.

Critical stresses. Since the calculations for the critical loads are valid only if the material follows Hooke's law, we need to verify that the critical stresses do not exceed the proportional limit of the material. In the case of the larger critical load, we obtain the following critical stress:

$$\sigma_{cr} = \frac{P_{cr}}{A} = \frac{312 \text{ k}}{8.25 \text{ in.}^2} = 37.8 \text{ ksi}$$

Since this stress is less than the proportional limit ($\sigma_{pl} = 42$ ksi), both critical-load calculations are satisfactory.

Allowable load. The allowable axial load for the column, based on Euler buckling, is

$$P_{allow} = \frac{P_{cr}}{n} = \frac{276 \text{ k}}{2.5} = 110 \text{ k}$$

in which $n = 2.5$ is the desired factor of safety.

9.4 COLUMNS WITH OTHER SUPPORT CONDITIONS

Slender concrete columns fixed at the base and free at the top during construction (Digital Vision/Getty Images)

Buckling of a column with pinned ends (described in the preceding section) is usually considered as the most basic case of buckling. However, in practice we encounter many other end conditions, such as fixed ends, free ends, and elastic supports. The critical loads for columns with various kinds of support conditions can be determined from the differential equation of the deflection curve by following the same procedure that we used when analyzing a pinned-end column.

The **procedure** is as follows. First, with the column assumed to be in the buckled state, we obtain an expression for the bending moment in the column. Second, we set up the differential equation of the deflection curve, using the bending-moment equation ($EIv'' = M$). Third, we solve the equation and obtain its general solution, which contains two constants of integration plus any other unknown quantities. Fourth, we apply boundary conditions pertaining to the deflection v and the slope v' and obtain a set of simultaneous equations. Finally, we solve those equations to obtain the critical load and the deflected shape of the buckled column.

This straightforward mathematical procedure is illustrated in the following discussion of three types of columns.

Column Fixed at the Base and Free at the Top

The first case we will consider is an ideal column that is fixed at the base, free at the top, and subjected to an axial load P (Fig. 9-15a).[*] The deflected shape of the buckled column is shown in Fig. 9-15b. From this figure we see that the bending moment at distance x from the base is

$$M = P(\delta - v) \tag{9-18}$$

where δ is the deflection at the free end of the column. The **differential equation** of the deflection curve then becomes

$$EIv'' = M = P(\delta - v) \tag{9-19}$$

in which I is the moment of inertia for buckling in the xy plane.

Using the notation $k^2 = P/EI$ (Eq. 9-6a), we can rearrange Eq. (9-19) into the form

$$v'' + k^2 v = k^2 \delta \tag{9-20}$$

which is a linear differential equation of second order with constant coefficients. However, it is a more complicated equation than the equation for a column with pinned ends (see Eq. 9-7) because it has a nonzero term on the right-hand side.

The **general solution** of Eq. (9-20) consists of two parts: (1) the *homogeneous solution*, which is the solution of the homogeneous equation obtained by replacing the right-hand side with zero, and (2) the *particular solution*, which is the solution of Eq. (9-20) that produces the term on the right-hand side.

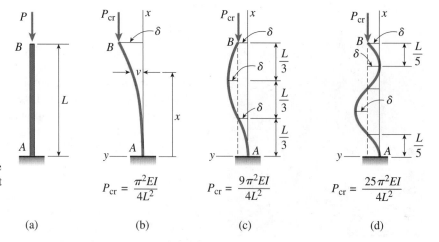

FIG. 9-15 Ideal column fixed at the base and free at the top: (a) initially straight column, (b) buckled shape for $n = 1$, (c) buckled shape for $n = 3$, and (d) buckled shape for $n = 5$

$$P_{cr} = \frac{\pi^2 EI}{4L^2} \qquad P_{cr} = \frac{9\pi^2 EI}{4L^2} \qquad P_{cr} = \frac{25\pi^2 EI}{4L^2}$$

(a)   (b)   (c)   (d)

---

[*]This column is of special interest because it is the one first analyzed by Euler in 1744.

The homogeneous solution (also called the *complementary solution*) is the same as the solution of Eq. (9-7); hence

$$v_H = C_1 \sin kx + C_2 \cos kx \tag{a}$$

where C_1 and C_2 are constants of integration. Note that when v_H is substituted into the left-hand side of the differential equation (Eq. 9-20), it produces zero.

The particular solution of the differential equation is

$$v_P = \delta \tag{b}$$

When v_P is substituted into the left-hand side of the differential equation, it produces the right-hand side, that is, it produces the term $k^2\delta$. Consequently, the *general solution* of the equation, equal to the sum of v_H and v_P, is

$$v = C_1 \sin kx + C_2 \cos kx + \delta \tag{9-21}$$

This equation contains three unknown quantities (C_1, C_2, and δ), and therefore three **boundary conditions** are needed to complete the solution.

At the base of the column, the deflection and slope are each equal to zero. Therefore, we obtain the following boundary conditions:

$$v(0) = 0 \qquad v'(0) = 0$$

Applying the first condition to Eq. (9-21), we find

$$C_2 = -\delta \tag{c}$$

To apply the second condition, we first differentiate Eq. (9-21) to obtain the slope:

$$v' = C_1 k \cos kx - C_2 k \sin kx \tag{d}$$

Applying the second condition to this equation, we find $C_1 = 0$.

Now we can substitute the expressions for C_1 and C_2 into the general solution (Eq. 9-21) and obtain the **equation of the deflection curve** for the buckled column:

$$v = \delta(1 - \cos kx) \tag{9-22}$$

Note that this equation gives only the *shape* of the deflection curve—the amplitude δ remains undefined. Thus, when the column buckles, the deflection given by Eq. (9-22) may have any arbitrary magnitude, except that it must remain small (because the differential equation is based upon small deflections).

The third boundary condition applies to the upper end of the column, where the deflection v is equal to δ:

$$v(L) = \delta$$

Using this condition with Eq. (9-22), we get

$$\delta \cos kL = 0 \tag{9-23}$$

From this equation we conclude that either $\delta = 0$ or cos $kL = 0$. If $\delta = 0$, there is no deflection of the bar (see Eq. 9-22) and we have the *trivial solution*—the column remains straight and buckling does not occur. In that case, Eq. (9-23) will be satisfied for any value of the quantity kL, that is, for any value of the load P. This conclusion is represented by the vertical line in the load-deflection diagram of Fig. 9-7.

The other possibility for solving Eq. (9-23) is

$$\cos kL = 0 \tag{9-24}$$

which is the **buckling equation**. In this case, Eq. (9-23) is satisfied regardless of the value of the deflection δ. Thus, as already observed, δ is undefined and may have any small value.

The equation cos $kL = 0$ is satisfied when

$$kL = \frac{n\pi}{2} \qquad n = 1, 3, 5, \ldots \tag{9-25}$$

Using the expression $k^2 = P/EI$, we obtain the following formula for the **critical loads**:

$$P_{cr} = \frac{n^2\pi^2 EI}{4L^2} \qquad n = 1, 3, 5, \ldots \tag{9-26}$$

Also, the **buckled mode shapes** are obtained from Eq. (9-22):

$$v = \delta\left(1 - \cos\frac{n\pi x}{2L}\right) \qquad n = 1, 3, 5, \ldots \tag{9-27}$$

The lowest critical load is obtained by substituting $n = 1$ in Eq. (9-26):

$$P_{cr} = \frac{\pi^2 EI}{4L^2} \tag{9-28}$$

The corresponding buckled shape (from Eq. 9-27) is

$$v = \delta\left(1 - \cos\frac{\pi x}{2L}\right) \tag{9-29}$$

and is shown in Fig. 9-15b.

By taking higher values of the index n, we can theoretically obtain an infinite number of critical loads from Eq. (9-26). The corresponding buckled mode shapes have additional waves in them. For instance, when $n = 3$ the buckled column has the shape shown in Fig. 9-15c and P_{cr} is nine times larger than for $n = 1$. Similarly, the buckled shape for $n = 5$ has even more waves (Fig. 9-15d) and the critical load is twenty-five times larger.

Effective Lengths of Columns

The critical loads for columns with various support conditions can be related to the critical load of a pinned-end column through the concept of an **effective length**. To demonstrate this idea, consider the deflected

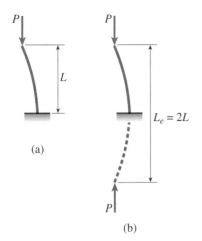

FIG. 9-16 Deflection curves showing the effective length L_e for a column fixed at the base and free at the top

shape of a column fixed at the base and free at the top (Fig. 9-16a). This column buckles in a curve that is one-quarter of a complete sine wave. If we extend the deflection curve (Fig. 9-16b), it becomes one-half of a complete sine wave, which is the deflection curve for a pinned-end column.

The effective length L_e for any column is the length of the equivalent pinned-end column, that is, it is the length of a pinned-end column having a deflection curve that exactly matches all or part of the deflection curve of the original column.

Another way of expressing this idea is to say that the effective length of a column is the distance between points of inflection (that is, points of zero moment) in its deflection curve, assuming that the curve is extended (if necessary) until points of inflection are reached. Thus, for a fixed-free column (Fig. 9-16), the effective length is

$$L_e = 2L \tag{9-30}$$

Because the effective length is the length of an equivalent pinned-end column, we can write a general formula for critical loads as follows:

$$P_{\mathrm{cr}} = \frac{\pi^2 EI}{L_e^2} \tag{9-31}$$

If we know the effective length of a column (no matter how complex the end conditions may be), we can substitute into the preceding equation and determine the critical load. For instance, in the case of a fixed-free column, we can substitute $L_e = 2L$ and obtain Eq. (9-28).

The effective length is often expressed in terms of an **effective-length factor** K:

$$L_e = KL \tag{9-32}$$

where L is the actual length of the column. Thus, the critical load is

$$P_{\mathrm{cr}} = \frac{\pi^2 EI}{(KL)^2} \tag{9-33}$$

The factor K equals 2 for a column fixed at the base and free at the top and equals 1 for a pinned-end column. The effective-length factor is often included in design formulas for columns.

Column with Both Ends Fixed Against Rotation

Next, let us consider a column with both ends fixed against rotation (Fig. 9-17a). Note that in this figure we use the standard symbol for the fixed support at the base of the column. However, since the column is free to shorten under an axial load, we must introduce a new symbol at the top of the column. This new symbol shows a rigid block that is constrained in such a manner that rotation and horizontal displacement are

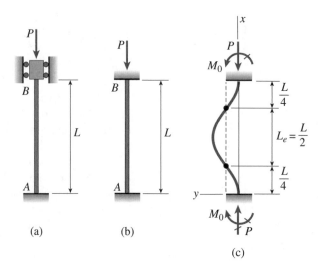

Buckling of a column with both ends fixed against rotation

prevented but vertical movement can occur. (As a convenience when drawing sketches, we often replace this more accurate symbol with the standard symbol for a fixed support—see Fig. 9-17b—with the understanding that the column is free to shorten.)

The buckled shape of the column in the first mode is shown in Fig. 9-17c. Note that the deflection curve is symmetrical (with zero slope at the midpoint) and has zero slope at the ends. Because rotation at the ends is prevented, reactive moments M_0 develop at the supports. These moments, as well as the reactive force at the base, are shown in the figure.

From our previous solutions of the differential equation, we know that the equation of the deflection curve involves sine and cosine functions. Also, we know that the curve is symmetric about the midpoint. Therefore, we see immediately that the curve must have inflection points at distances $L/4$ from the ends. It follows that the middle portion of the deflection curve has the same shape as the deflection curve for a pinned-end column. Thus, the effective length of a column with fixed ends, equal to the distance between inflection points, is

$$L_e = \frac{L}{2} \tag{9-34}$$

Substituting into Eq. (9-31) gives the critical load:

$$P_{\text{cr}} = \frac{4\pi^2 EI}{L^2} \tag{9-35}$$

This formula shows that the critical load for a column with fixed ends is four times that for a column with pinned ends. As a check, this result may be verified by solving the differential equation of the deflection curve (see Problem 9.4-9).

Column Fixed at the Base and Pinned at the Top

The critical load and buckled mode shape for a column that is fixed at the base and pinned at the top (Fig. 9-18a) can be determined by solving the differential equation of the deflection curve. When the column buckles (Fig. 9-18b), a reactive moment M_0 develops at the base because there can be no rotation at that point. Then, from the equilibrium of the entire column, we know that there must be horizontal reactions R at each end such that

$$M_0 = RL \tag{e}$$

The bending moment in the buckled column, at distance x from the base, is

$$M = M_0 - Pv - Rx = -Pv + R(L - x) \tag{9-36}$$

and therefore the **differential equation** is

$$EIv'' = M = -Pv + R(L - x) \tag{9-37}$$

Again substituting $k^2 = P/EI$ and rearranging, we get

$$v'' + k^2 v = \frac{R}{EI}(L - x) \tag{9-38}$$

The **general solution** of this equation is

$$v = C_1 \sin kx + C_2 \cos kx + \frac{R}{P}(L - x) \tag{9-39}$$

in which the first two terms on the right-hand side constitute the homogeneous solution and the last term is the particular solution. This solution can be verified by substitution into the differential equation (Eq. 9-37).

Since the solution contains three unknown quantities (C_1, C_2, and R), we need three **boundary conditions**. They are

$$v(0) = 0 \qquad v'(0) = 0 \qquad v(L) = 0$$

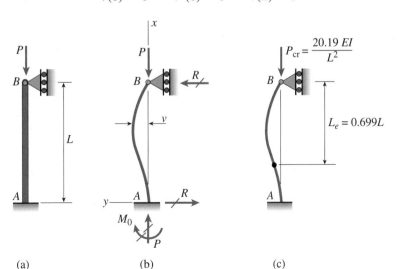

FIG. 9-18 Column fixed at the base and pinned at the top

(a)   (b)   (c)

Applying these conditions to Eq. (9-39) yields

$$C_2 + \frac{RL}{P} = 0 \qquad C_1 k - \frac{R}{P} = 0 \qquad C_1 \tan kL + C_2 = 0 \qquad \text{(f,g,h)}$$

All three equations are satisfied if $C_1 = C_2 = R = 0$, in which case we have the trivial solution and the deflection is zero.

To obtain the solution for buckling, we must solve Eqs. (f), (g), and (h) in a more general manner. One method of solution is to eliminate R from the first two equations, which yields

$$C_1 kL + C_2 = 0 \quad \text{or} \quad C_2 = -C_1 kL \tag{i}$$

Next, we substitute this expression for C_2 into Eq. (h) and obtain the **buckling equation**:

$$kL = \tan kL \tag{9-40}$$

The solution of this equation gives the critical load.

Since the buckling equation is a transcendental equation, it cannot be solved explicitly.[*] Nevertheless, the values of kL that satisfy the equation can be determined numerically by using a computer program for finding roots of equations. The smallest nonzero value of kL that satisfies Eq. (9-40) is

$$kL = 4.4934 \tag{9-41}$$

The corresponding **critical load** is

$$P_{\text{cr}} = \frac{20.19 EI}{L^2} = \frac{2.046 \pi^2 EI}{L^2} \tag{9-42}$$

which (as expected) is higher than the critical load for a column with pinned ends and lower than the critical load for a column with fixed ends (see Eqs. 9-12 and 9-35).

The **effective length** of the column may be obtained by comparing Eqs. (9-42) and (9-31); thus,

$$L_e = 0.699L \approx 0.7L \tag{9-43}$$

This length is the distance from the pinned end of the column to the point of inflection in the buckled shape (Fig. 9-18c).

The equation of the **buckled mode shape** is obtained by substituting $C_2 = -C_1 kL$ (Eq. i) and $R/P = C_1 k$ (Eq. g) into the general solution (Eq. 9-39):

$$v = C_1[\sin kx - kL \cos kx + k(L - x)] \tag{9-44}$$

---

[*]In a transcendental equation, the variables are contained within transcendental functions. A transcendental function cannot be expressed by a finite number of algebraic operations; hence trigonometric, logarithmic, exponential, and other such functions are transcendental.

in which $k = 4.4934/L$. The term in brackets gives the mode shape for the deflection of the buckled column. However, the amplitude of the deflection curve is undefined because C_1 may have any value (within the usual limitation that the deflections must remain small).

Limitations

In addition to the requirement of small deflections, the Euler buckling theory used in this section is valid only if the column is perfectly straight before the load is applied, the column and its supports have no imperfections, and the column is made of a linearly elastic material that follows Hooke's law. These limitations were explained previously in Section 9.3.

Summary of Results

The lowest critical loads and corresponding effective lengths for the four columns we have analyzed are summarized in Fig. 9-19.

(a) Pinned-pinned column	(b) Fixed-free column	(c) Fixed-fixed column	(d) Fixed-pinned column
$P_{cr} = \dfrac{\pi^2 EI}{L^2}$	$P_{cr} = \dfrac{\pi^2 EI}{4L^2}$	$P_{cr} = \dfrac{4\pi^2 EI}{L^2}$	$P_{cr} = \dfrac{2.046\,\pi^2 EI}{L^2}$
$L_e = L$	$L_e = 2L$	$L_e = 0.5L$	$L_e = 0.699L$
$K = 1$	$K = 2$	$K = 0.5$	$K = 0.699$

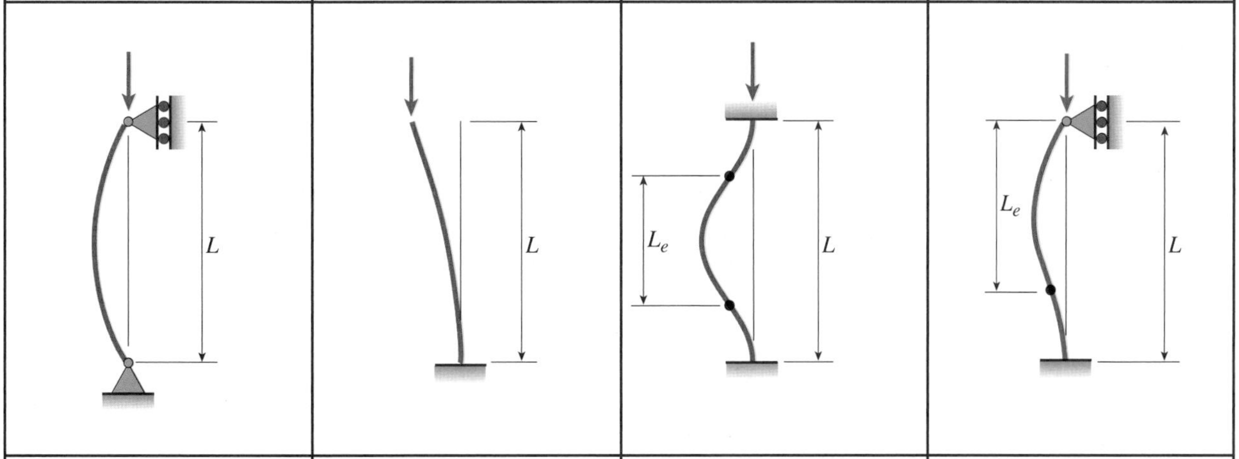

FIG. 9-19 Critical loads, effective lengths, and effective-length factors for ideal columns

Example 9-2

A viewing platform in a wild-animal park (Fig. 9-20a) is supported by a row of aluminum pipe columns having length $L = 3.25$ m and outer diameter $d = 100$ mm. The bases of the columns are set in concrete footings and the tops of the columns are supported laterally by the platform. The columns are being designed to support compressive loads $P = 100$ kN.

Determine the minimum required thickness t of the columns (Fig. 9-20b) if a factor of safety $n = 3$ is required with respect to Euler buckling. (For the aluminum, use 72 GPa for the modulus of elasticity and use 480 MPa for the proportional limit.)

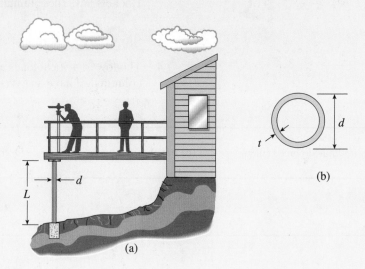

(b)

FIG. 9-20 Example 9-2. Aluminum pipe column

(a)

Solution

Critical load. Because of the manner in which the columns are constructed, we will model each column as a fixed-pinned column (see Fig. 9-19d). Therefore, the critical load is

$$P_{\text{cr}} = \frac{2.046\pi^2 EI}{L^2} \tag{j}$$

in which I is the moment of inertia of the tubular cross section:

$$I = \frac{\pi}{64}\Big[d^4 - (d - 2t)^4\Big] \tag{k}$$

Substituting $d = 100$ mm (or 0.1 m), we get

$$I = \frac{\pi}{64}\Big[(0.1 \text{ m})^4 - (0.1 \text{ m} - 2t)^4\Big] \tag{l}$$

in which t is expressed in meters.

Required thickness of the columns. Since the load per column is 100 kN and the factor of safety is 3, each column must be designed for the following critical load:

$$P_{cr} = nP = 3(100 \text{ kN}) = 300 \text{ kN}$$

Substituting this value for P_{cr} in Eq. (j), and also replacing I with its expression from Eq. (l), we obtain

$$300,000 \text{ N} = \frac{2.046\pi^2(72 \times 10^9 \text{ Pa})}{(3.25 \text{ m})^2}\left(\frac{\pi}{64}\right)\left[(0.1 \text{ m})^4 - (0.1 \text{ m} - 2t)^4\right]$$

Note that all terms in this equation are expressed in units of newtons and meters. After multiplying and dividing, the preceding equation simplifies to

$$44.40 \times 10^{-6} \text{ m}^4 = (0.1 \text{ m})^4 - (0.1 \text{ m} - 2t)^4$$

or

$$(0.1 \text{ m} - 2t)^4 = (0.1 \text{ m})^4 - 44.40 \times 10^{-6} \text{ m}^4 = 55.60 \times 10^{-6} \text{ m}^4$$

from which we obtain

$$0.1 \text{ m} - 2t = 0.08635 \text{ m} \quad \text{and} \quad t = 0.006825 \text{ m}$$

Therefore, the minimum required thickness of the column to meet the specified conditions is

$$t_{min} = 6.83 \text{ mm}$$

Supplementary calculations. Knowing the diameter and thickness of the column, we can now calculate its moment of inertia, cross-sectional area, and radius of gyration. Using the minimum thickness of 6.83 mm, we obtain

$$I = \frac{\pi}{64}\left[d^4 - (d - 2t)^4\right] = 2.18 \times 10^6 \text{ mm}^4$$

$$A = \frac{\pi}{4}\left[d^2 - (d - 2t)^2\right] = 1999 \text{ mm}^2 \qquad r = \sqrt{\frac{I}{A}} = 33.0 \text{ mm}$$

The slenderness ratio L/r of the column is approximately 98, which is in the customary range for slender columns, and the diameter-to-thickness ratio d/t is approximately 15, which should be adequate to prevent local buckling of the walls of the column.

The critical stress in the column must be less than the proportional limit of the aluminum if the formula for the critical load (Eq. j) is to be valid. The critical stress is

$$\sigma_{cr} = \frac{P_{cr}}{A} = \frac{300 \text{ kN}}{1999 \text{ mm}^2} = 150 \text{ MPa}$$

which is less than the proportional limit (480 MPa). Therefore, our calculation for the critical load using Euler buckling theory is satisfactory.

CHAPTER SUMMARY & REVIEW

In Chapter 9, we investigated the elastic behavior of axially loaded members known as columns. First, the concepts of buckling and stability of these slender compression elements were discussed using equilibrium of simple column models made up of rigid bars and elastic springs. Then, elastic columns with pinned ends, acted on by centroidal compressive loads, were considered and the differential equation of the deflection curve was solved to obtain the buckling load (P_{cr}) and buckled mode shape; linear elastic behavior was assumed. Three additional support cases were investigated, and the buckling load for each case was expressed in terms of the column's effective length, that is, the length of an equivalent pinned-end column.

The major concepts presented in this chapter are as follows:

1. Buckling instability of slender columns is an important mode of failure which must be considered in their design (in addition to strength and stiffness).

2. A slender column with pinned ends and length L, acted on by a compressive load at the centroid of the cross section, and restricted to linear elastic behavior, will buckle at the **Euler buckling load**

$$P_{cr} = \frac{\pi^2 EI}{L^2}$$

in the fundamental mode; hence, the buckling load depends on the flexural rigidity (EI) and length (L) but not the strength of the material.

3. Changing the support conditions, or providing additional lateral supports, changes the critical buckling load. However, P_{cr} for these other support cases may be obtained by replacing the actual column length (L) by the **effective length** (L_e) in the formula for P_{cr} above. **Three additional support cases are shown in Fig. 9-19**.

4. **Long columns** (i.e., large slenderness ratios L/r) buckle at low values of compressive stress; **short columns** (i.e., low L/r) fail by yielding and crushing of the material; and **intermediate columns** (with values of L/r which lie between those for long and short columns) fail by inelastic buckling. The critical buckling load for inelastic buckling is always less than the Euler buckling load. Only linear elastic behavior associated with Euler buckling was discussed in this chapter.

PROBLEMS CHAPTER 9

Idealized Buckling Models

9.2-1 The figure shows an idealized structure consisting of one or more **rigid bars** with pinned connections and linearly elastic springs. Rotational stiffness is denoted β_R, and translational stiffness is denoted β.

Determine the critical load P_{cr} for the structure.

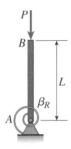

PROB. 9.2-1

9.2-2 The figure shows an idealized structure consisting of one or more rigid bars with pinned connections and linearly elastic springs. Rotational stiffness is denoted β_R, and translational stiffness is denoted β.

(a) Determine the critical load P_{cr} for the structure from figure part (a).

(b) Find P_{cr} if another rotational spring is added at B from figure part (b).

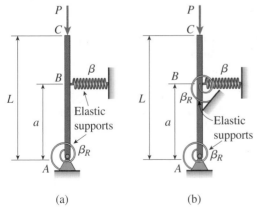

(a)   (b)

PROB. 9.2-2

9.2-3 The figure shows an idealized structure consisting of one or more **rigid bars** with pinned connections and linearly

elastic springs. Rotational stiffness is denoted β_R, and translational stiffness is denoted β.

Determine the critical load P_{cr} for the structure.

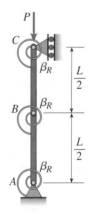

PROB. 9.2-3

9.2-4 The figure shows an idealized structure consisting of bars AB and BC which are connected using a hinge at B and linearly elastic springs at A and B. Rotational stiffness is denoted β_R and translational stiffness is denoted β.

(a) Determine the critical load P_{cr} for the structure from figure part (a).

(b) Find P_{cr} if an elastic connection is now used to connect bar segments AB and BC from figure part (b).

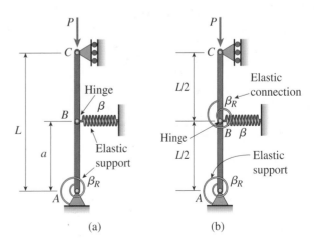

(a)   (b)

PROB. 9.2-4

9.2-5 The figure shows an idealized structure consisting of two rigid bars joined by an elastic connection with rotational stiffness β_R. Determine the critical load P_{cr} for the structure.

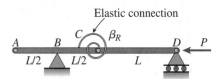

Elastic connection

PROB. 9.2-5

9.2-6 The figure shows an idealized structure consisting of rigid bars ABC and DEF joined by linearly elastic spring β between C and D. The structure is also supported by translational elastic support β at B and rotational elastic support at β_R at E.
Determine the critical load P_{cr} for the structure.

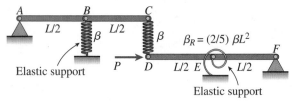

Elastic support

Elastic support

PROB. 9.2-6

9.2-7 The figure shows an idealized structure consisting of an L-shaped rigid bar structure supported by linearly elastic springs at A and C. Rotational stiffness in denoted β_R and translational stiffness is denoted β.
  Determine the critical load P_{cr} for the structure.

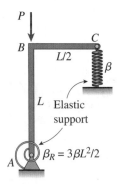

PROB. 9.2-7

Critical Loads of Columns with Pinned Supports

The problems for Section 9.3 are to be solved using the assumptions of ideal, slender, prismatic, linearly elastic columns (Euler buckling). Buckling occurs in the plane of the figure unless stated otherwise.

9.3-1 Calculate the critical load P_{cr} for a W 8 × 35 steel column (see figure) having length $L = 24$ ft and $E = 30 \times 10^6$ psi under the following conditions:
  (a) The column buckles by bending about its strong axis (axis 1–1), and (b) the column buckles by bending about its weak axis (axis 2–2). In both cases, assume that the column has pinned ends.

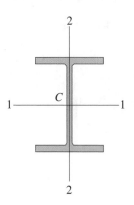

PROBS. 9.3-1 through 9.3-3

9.3-2 Solve the preceding problem for a W250 × 89 steel column having length $L = 10$ m. Let $E = 200$ GPa.

9.3-3 Solve Problem 9.3-1 for a W 10 × 45 steel column having length $L = 28$ ft.

9.3-4 A horizontal beam AB is pin-supported at end A and carries a CW moment M at joint B, as shown in the figure. The beam is also supported at C by a pinned-end column of length L; the column is restrained laterally at $0.6L$ from the base at D. Assume the column can only buckle in the plane of the frame. The column is a solid steel bar ($E = 200$ GPa) of square cross section having length $L = 2.4$ m side dimensions $b = 70$ mm. Let dimensions $d = L/2$. Based upon the critical load of the column, determine the allowable moment M if the factor of safety with respect to buckling is $n = 2.0$.

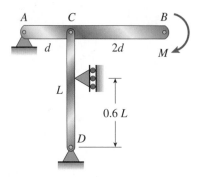

PROB. 9.3-4

9.3-5 A horizontal beam AB is pin-supported at end A and carries a load Q at joint B, as shown in the figure. The beam is also supported at C by a pinned-end column of length L; the column is restrained laterally at $0.6L$ from the base at D. Assume the column can only buckle in the plane of the frame. The column is a solid aluminum bar $(E = 10 \times 10^6 \text{ psi})$ of square cross section having length $L = 30$ in. and side dimensions $b = 1.5$ in. Let dimension $d = L/2$. Based upon the critical load of the column, determine the allowable force Q if the factor of safety with respect to buckling is $n = 1.8$.

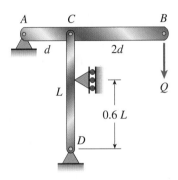

PROB. 9.3-5

9.3-6 A horizontal beam AB is supported at end A and carries a load Q at joint B, as shown in the figure part (a). The beam is also supported at C by a pinned-end column of length L. The column has flexural rigidity EI.

(a) For the case of a guided support at A [figure part (a)], what is the critical load Q_{cr}? (In other words, at what load Q_{cr} does the system collapse because of Euler buckling of the column DC?)

(b) Repeat (a) if the guided support at A is replaced by column AF with length $3L/2$ and flexural rigidity EI [see figure part (b)].

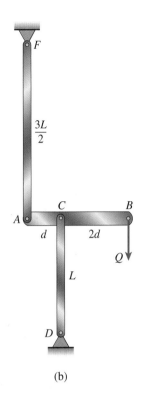

(b)

PROB. 9.3-6

9.3-7 A horizontal beam AB has a guided support at end A and carries a load Q at end B, as shown in the figure part (a). The beam is supported at C and D by two identical pinned-end columns of length L. Each column has flexural rigidity EI.

(a) Find an expression for the critical load Q_{cr}. (In other words, at what load Q_{cr} does the system collapses because of Euler buckling of the columns?)

(b) Repeats (a) but assume a pin support at A. Find an expression for the critical moment M_{cr} (i.e., find the moment M at B at which the system collapses because of Euler buckling of the columns).

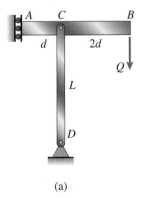

(a)

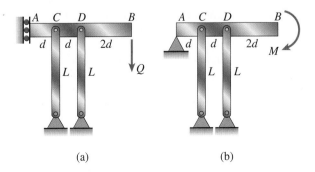

(a)   (b)

PROB. 9.3-7

9.3-8 A slender bar *AB* with pinned ends and length *L* is held between immovable supports (see figure).

What increase Δ*T* in the temperature of the bar will produce buckling at the Euler load?

PROB. 9.3-8

9.3-9 A rectangular column with cross-sectional dimensions *b* and *h* is pin-supported at ends *A* and *C* (see figure). At midheight, the column is restrained in the plane of the figure but is free to deflect perpendicular to the plane of the figure.

Determine the ratio *h/b* such that the critical load is the same for buckling in the two principal planes of the column.

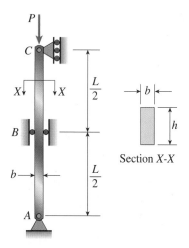

PROB. 9.3-9

9.3-10 Three identical, solid circular rods, each of radius *r* and length *L*, are placed together to form a compression member (see the cross section shown in the figure).

Assuming pinned-end conditions, determine the critical load *P*cr as follows: (a) The rods act independently as individual columns, and (b) the rods are bonded by epoxy throughout their lengths so that they function as a single member.

What is the effect on the critical load when the rods act as a single member?

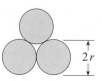

PROB. 9.3-10

9.3-11 Three pinned-end columns of the same material have the same length and the same cross-sectional area (see figure). The columns are free to buckle in any direction. The columns have cross sections as follows: (1) a circle, (2) a square, and (3) an equilateral triangle.

Determine the ratios $P_1 : P_2 : P_3$ of the critical loads for these columns.

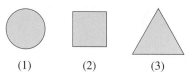

(1)   (2)   (3)

PROB. 9.3-11

9.3-12 A long slender column *ABC* is pinned at ends *A* and *C* and compressed by an axial force *P* (see figure). At the midpoint *B*, lateral support is provided to prevent deflection in the plane of the figure. The column is a steel wide-flange section (W 250 × 67) with *E* = 200 GPa. The distance between lateral supports is *L* = 5.5 m.

Calculate the allowable load *P* using a factor of safety *n* = 2.4, taking into account the possibility of Euler buckling about either principal centroidal axis (i.e., axis 1–1 or axis 2–2).

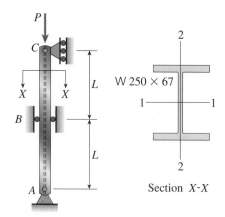

PROB. 9.3-12

9.3-13 The roof over a concourse at an airport is supported by the use of pretensioned cables. At a typical joint in the roof structure, a strut *AB* is compressed by the action of tensile forces *F* in a cable that makes an angle $\alpha = 75°$ with the strut (see figure and photo). The strut is a circular tube of steel ($E = 30,000$ ksi) with outer diameter $d_2 = 2.5$ in. and inner diameter $d_1 = 2.0$ in. The strut is 5.75 ft long and is assumed to be pin-connected at both ends.

Using a factor of safety $n = 2.5$ with respect to the critical load, determine the allowable force *F* in the cable.

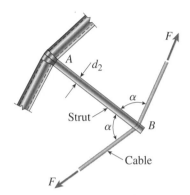

PROB. 9.3-13

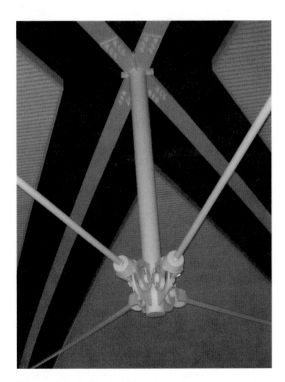

Cable and strut at typical joint of airport concourse roof (© Barry Goodno)

9.3-14 The hoisting arrangement for lifting a large pipe is shown in the figure. The spreader is a steel tubular section with outer diameter 70 mm and inner diameter 57 mm. Its length is 2.6 m and its modulus of elasticity is 200 GPa.

Based upon a factor of safety of 2.25 with respect to Euler buckling of the spreader, what is the maximum weight of pipe that can be lifted? (Assume pinned conditions at the ends of the spreader.)

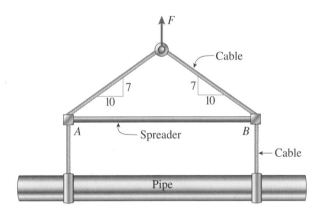

PROB. 9.3-14

9.3-15 A pinned-end strut of aluminum ($E = 10,400$ ksi) with length $L = 6$ ft is constructed of circular tubing with outside diameter $d = 2$ in. (see figure). The strut must resist an axial load $P = 4$ kips with a factor of safety $n = 2.0$ with respect to the critical load.

Determine the required thickness *t* of the tube.

$d = 2$ in.

PROB. 9.3-15

9.3-16 The cross section of a column built up of two steel I-beams (S 150 × 25.7 sections) is shown in the figure. The beams are connected by spacer bars, or *lacing*, to ensure that they act together as a single column. (The lacing is represented by dashed lines in the figure.)

The column is assumed to have pinned ends and may buckle in any direction. Assuming $E = 200$ GPa and $L = 8.5$ m, calculate the critical load P_{cr} for the column.

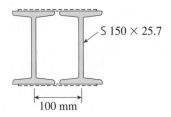

PROB. 9.3-16

9.3-17 The truss *ABC* shown in the figure supports a vertical load *W* at joint *B*. Each member is a slender circular steel pipe (E = 30,000 ksi) with outside diameter 4 in. And wall thickness 0.25 in. The distance between supports is 23 ft. Joint *B* is restrained against displacement perpendicular to the plane of the truss.

Determine the critical value W_{cr} of the load.

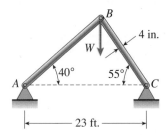

PROB. 9.3-17

9.3-18 A truss *ABC* supports a load *W* at joint *B*, as shown in the figure. The length L_1 of member *AB* is fixed, but the length of strut *BC* varies as the angle θ is changed. Strut *BC* has a solid circular cross section. Joint *B* is restrained against displacement perpendicular to the plane of the truss.

Assuming that collapse occurs by Euler buckling of the strut, determine the angle θ for minimum weight of the strut.

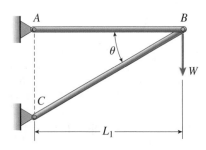

PROB. 9.3-18

9.3-19 An S 6 × 12.5 steel cantilever beam *AB* is supported by a steel tie rod at *B* as shown. The tie rod is just taut when a roller support is added at *C* at a distance *S* to the left of *B*, then the distributed load *q* is applied to beam segment *AC*. Assume $E = 30 \times 10^6$ psi and neglect the self weight of the beam and tie rod. See Table F-2(a) in Appendix F available online for the properties of the S-shape beam.

(a) What value of uniform load *q* will, if exceeded, result in buckling of the tie rod if L_1 = 6 ft, *S* = 2 ft, *H* = 3 ft, *d* = 0.25 in.?

(b) What minimum beam moment of inertia I_b is required to prevent buckling of the tie rod if *q* = 200 lb/ft, L_1 = 6 ft, *H* = 3 ft, *d* = 0.25 in., *S* = 2 ft?

(c) For what distance *S* will the tie rod be just on the verge of buckling if *q* = 200 lb/ft, L_1 = 6 ft, *H* = 3 ft, *d* = 0.25 in.?

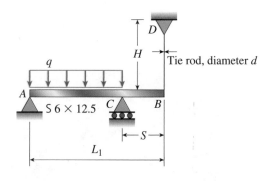

PROB. 9.3-19

Columns with Other Support Conditions

The problems for Section 9.4 are to be solved using the assumptions of ideal, slender, prismatic, linearly elastic columns (Euler buckling). Buckling occurs in the plane of the figure unless stated otherwise.

9.4-1 An aluminum pipe column (E = 10,400 ksi) with length *L* = 10.0 ft has inside and outside diameters d_1 = 5.0 in. and d_2 = 6.0 in., respectively (see figure). The column is supported only at the ends and may buckle in any direction.

Calculate the critical load P_{cr} for the following end conditions: (1) pinned-pinned, (2) fixed-free, (3) fixed-pinned, and (4) fixed-fixed.

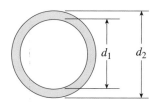

PROBS. 9.4-1 and 9.4-2

9.4-2 Solve the preceding problem for a steel pipe column ($E = 210$ GPa) with length $L = 1.2$ m, inner diameter $d_1 = 36$ mm, and outer diameter $d_2 = 40$ mm.

9.4-3 A wide-flange steel column ($E = 30 \times 10^6$ psi) of W 12×87 shape (see figure) has length $L = 28$ ft. It is supported only at the ends and may buckle in any direction.

Calculate the allowable load P_{allow} based upon the critical load with a factor of safety $n = 2.5$. Consider the following end conditions: (1) pinned-pinned, (2) fixed-free, (3) fixed-pinned, and (4) fixed-fixed.

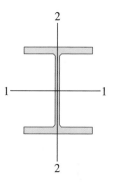

PROBS. 9.4-3 and 9.4-4

9.4-4 Solve the preceding problem for a W 250×89 shape with length $L = 7.5$m and $E = 200$ GPa.

9.4-5 The upper end of a W 8×21 wide-flange steel column ($E = 30 \times 10^3$ ksi) is supported laterally between two pipes (see figure). The pipes are not attached to the column, and friction between the pipes and the column is unreliable. The base of the column provides a fixed support, and the column is 13 ft long.

Determine the critical load for the column, considering Euler buckling in the plane of the web and also perpendicular to the plane of the web.

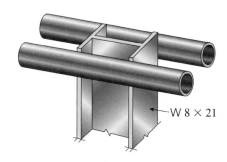

PROB. 9.4-5

9.4-6 A vertical post AB is embedded in a concrete foundation and held at the top by two cables (see figure). The post is a hollow steel tube with modulus of elasticity 200 GPa, outer diameter 40 mm, and thickness 5 mm. The cables are tightened equally by turnbuckles.

If a factor of safety of 3.0 against Euler buckling in the plane of the figure is desired, what is the maximum allowable tensile force T_{allow} in the cables?

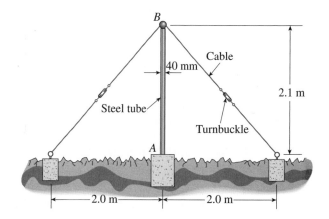

PROB. 9.4-6

9.4-7 The horizontal beam ABC shown in the figure is supported by columns BD and CE. The beam is prevented from moving horizontally by the pin support at end A. Each column is pinned at its upper end to the beam, but at the lower ends, support D is a guided support and support E is pinned. Both columns are solid steel bars ($E = 30 \times 10^6$ psi) of square cross section with width equal to 0.625 in. A load Q acts at distance a from column BD.

(a) If the distance $a = 12$ in., what is the critical value Q_{cr} of the load?

(b) If the distance a can be varied between 0 and 40 in., what is the maximum possible value of Q_{cr}? What is the corresponding value of the distance a?

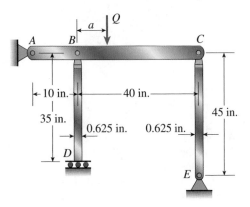

PROB. 9.4-7

9.4-8 The roof beams of a warehouse are supported by pipe columns (see figure) having outer diameter $d_2 = 100$ mm and inner diameter $d_1 = 90$ mm. The columns have length $L = 4.0$ m, modulus $E = 210$ GPa, and fixed supports at the base.

Calculate the critical load P_{cr} of one of the columns using the following assumptions: (1) the upper end is pinned and the beam prevents horizontal displacement; (2) the upper end is fixed against rotation and the beam prevents horizontal displacement; (3) the upper end is pinned but the beam is free to move horizontally; and (4) the upper end is fixed against rotation but the beam is free to move horizontally.

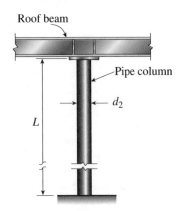

PROB. 9.4-8

9.4-9 Determine the critical load P_{cr} and the equation of the buckled shape for an ideal column with ends fixed against rotation (see figure) by solving the differential equation of the deflection curve. (See also Fig. 9-17.)

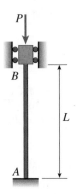

PROB. 9.4-9

9.4-10 An aluminum tube AB of circular cross section has a guided support at the base and is pinned at the top to a horizontal beam supporting a load $Q = 200$ kN (see figure).

Determine the required thickness t of the tube if its outside diameter d is 200 mm and the desired factor of safety with respect to Euler buckling is $n = 3.0$. (Assume $E = 72$ GPa.)

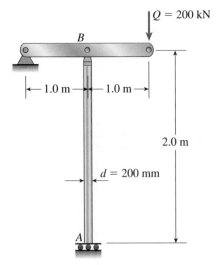

PROB. 9.4-10

9.4-11 The frame *ABC* consists of two members *AB* and *BC* that are rigidly connected at joint *B*, as shown in part (a) of the figure. The frame has pin supports at *A* and *C*. A concentrated load *P* acts at joint *B*, thereby placing member *AB* in direct compression.

To assist in determining the buckling load for member *AB*, we represent it as a pinned-end column, as shown in part (b) of the figure. At the top of the column, a rotational spring of stiffness β_R represents the restraining action of the horizontal beam *BC* on the column (note that the horizontal beam provides resistance to rotation of joint *B* when the column buckles). Also, consider only bending effects in the analysis (i.e., disregard the effects of axial deformations).

(a) By solving the differential equation of the deflection curve, derive the following buckling equation for this column:

$$\frac{\beta_R L}{EI}(kL \cot kL - 1) - k^2 L^2 = 0$$

in which *L* is the length of the column and *EI* is its flexural rigidity.

(b) For the particular case when member *BC* is identical to member *AB*, the rotational stiffness β_R equals $3EI/L$ (see Case 7, Table H-2, Appendix H available online). For this special case, determine the critical load P_{cr}.

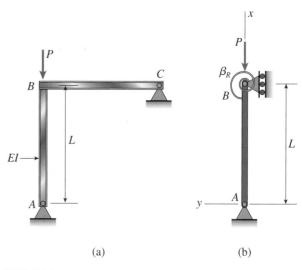

(a)　　　　(b)

PROB. 9.4-11

A

FE Exam Review Problems

The Fundamentals of Engineering (FE) examination [see *http://www.ncees. org/ Exams.php*] is the first step on the path to registration as a Professional Engineer (P.E.). In its current form, the FE exam is an 8-hour exam consisting of 120 multiple choice questions in the 4-hour morning session, followed by 60 multiple choice questions in the 4-hour afternoon session. The exam is usually taken by recent graduates of accredited college engineering programs and covers a broad range of topics presented in their undergraduate courses. The afternoon portion is usually focused on questions related to the student's specific engineering subdiscipline (chemical, civil, electrical, environmental, industrial, mechanical, and "other").

In the past, approximately 10–15% of the questions have been based on principles presented in undergraduate courses in engineering mechanics. This appendix presents 106 FE-type review problems in *Mechanics of Materials*, many of which are based upon modifications of problems presented at the end of each chapter throughout this text. The problems cover all of the major topics presented in the text and are thought to be representative of those likely to appear on an FE exam. Most of these problems are in SI units which is the system of units used on the FE Exam itself, and require use of an engineering calculator to carry out the solution. Each of the 106 problems is presented in the FE Exam format. The student must select from 4 available answers (A, B, C or D), only one of which is the correct answer. The correct answer choices are listed in the Answers section at the back of this text, and the detailed solution for each problem is available for download on the student website. It is expected that careful review of these problems will serve as a useful guide to the student in preparing for this important examination.

A-1.1: A hollow circular post ABC (see figure) supports a load $P_1 = 16$ kN acting at the top. A second load P_2 is uniformly distributed around the cap plate at B. The diameters and thicknesses of the upper and lower parts of the post are $d_{AB} = 30$ mm, $t_{AB} = 12$ mm, $d_{BC} = 60$ mm, and $t_{BC} = 9$ mm, respectively. The lower part of the post must have the same compressive stress as the upper part. The required magnitude of the load P_2 is approximately:

(A) 18 kN
(B) 22 kN
(C) 28 kN
(D) 46 kN

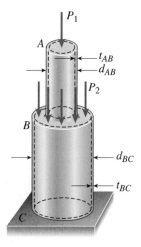

A-1.2: A circular aluminum tube of length $L = 650$ mm is loaded in compression by forces P. The outside and inside diameters are 80 mm and 68 mm, respectively. A strain gage on the outside of the bar records a normal strain in the longitudinal direction of 400×10^{-6}. The shortening of the bar is approximately:

(A) 0.12 mm
(B) 0.26 mm
(C) 0.36 mm
(D) 0.52 mm

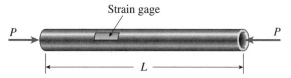

Strain gage

A-1.3: A steel plate weighing 27 kN is hoisted by a cable sling that has a clevis at each end. The pins through the clevises are 22 mm in diameter. Each half of the cable is at an angle of 35° to the vertical. The average shear stress in each pin is approximately:

(A) 22 MPa
(B) 28 MPa
(C) 40 MPa
(D) 48 MPa

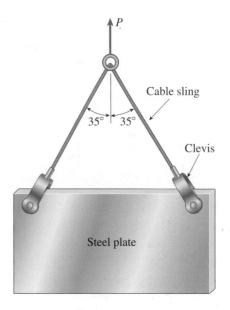

A-1.4: A steel wire hangs from a high-altitude balloon. The steel has unit weight 77kN/m³ and yield stress of 280 MPa. The required factor of safety against yield is 2.0. The maximum permissible length of the wire is approximately:

(A) 1800 m
(B) 2200 m
(C) 2600 m
(D) 3000 m

A-1.5: An aluminum bar (E = 72 GPa, v = 0.33) of diameter 50 mm cannot exceed a diameter of 50.1 mm when compressed by axial force P. The maximum acceptable compressive load P is approximately:

(A) 190 kN
(B) 200 kN
(C) 470 kN
(D) 860 kN

A-1.6: An aluminum bar (E = 70 GPa, v = 0.33) of diameter 20 mm is stretched by axial forces P, causing its diameter to decrease by 0.022 mm. The maximum acceptable compressive load P is approximately:

(A) 73 kN
(B) 100 kN
(C) 140 kN
(D) 339 kN

A-1.7: An polyethylene bar (E = 1.4 GPa, v = 0.4) of diameter 80 mm is inserted in a steel tube of inside diameter 80.2 mm and then compressed by axial force *P*. The gap between steel tube and polyethylene bar will close when compressive load P is approximately:

(A) 18 kN
(B) 25 kN
(C) 44 kN
(D) 60 kN

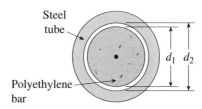

A-1.8: A pipe (E = 110 GPa) carries a load P_1 = 120 kN at A and a uniformly distributed load P_2 = 100 kN on the cap plate at B. Initial pipe diameters and thicknesses are: d_{AB} = 38 mm, t_{AB} = 12 mm, d_{BC} = 70 mm, t_{BC} = 10 mm. Under loads P_1 and P_2, wall thickness t_{BC} increases by 0.0036 mm. Poisson's ratio v for the pipe material is approximately:

(A) 0.27
(B) 0.30
(C) 0.31
(D) 0.34

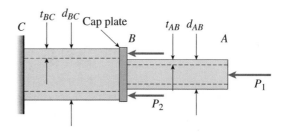

A-1.9: A titanium bar (E = 100 GPa, v = 0.33) with square cross section (b = 75 mm) and length L = 3.0 m is subjected to tensile load P = 900 kN. The increase in volume of the bar is approximately:

(A) 1400 mm^3
(B) 3500 mm^3
(C) 4800 mm^3
(D) 9200 mm^3

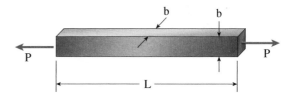

A-1.10: An elastomeric bearing pad is subjected to a shear force V during a static loading test. The pad has dimensions $a = 150$ mm and $b = 225$ mm, and thickness $t = 55$ mm. The lateral displacement of the top plate with respect to the bottom plate is 14 mm under a load $V = 16$ kN. The shear modulus of elasticity G of the elastomer is approximately:

(A) 1.0 MPa
(B) 1.5 MPa
(C) 1.7 MPa
(D) 1.9 MPa

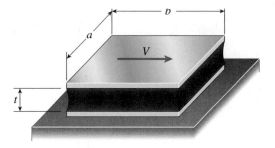

A-1.11: A bar of diameter d = 18 mm and length L = 0.75 m is loaded in tension by forces P. The bar has modulus E = 45 GPa and allowable normal stress of 180 MPa. The elongation of the bar must not exceed 2.7 mm. The allowable value of forces P is approximately:

(A) 41 kN
(B) 46 kN
(C) 56 kN
(D) 63 kN

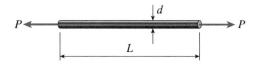

A-1.12: Two flanged shafts are connected by eight 18 mm bolts. The diameter of the bolt circle is 240 mm. The allowable shear stress in the bolts is 90 MPa. Ignore friction between the flange plates. The maximum value of torque T_0 is approximately:

(A) 19 kN m
(B) 22 kN m
(C) 29 kN m
(D) 37 kN m

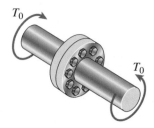

A-1.13: A copper tube with wall thickness of 8 mm must carry an axial tensile force of 175 kN. The allowable tensile stress is 90 MPa. The minimum required outer diameter is approximately:

(A) 60 mm
(B) 72 mm
(C) 85 mm
(D) 93 mm

A-2.1: Two wires, one copper and the other steel, of equal length stretch the same amount under an applied load P. The moduli of elasticity for each is: $E_s = 210$ GPa, $E_c = 120$ GPa. The ratio of the diameter of the copper wire to that of the steel wire is approximately:

(A) 1.00
(B) 1.08
(C) 1.19
(D) 1.32

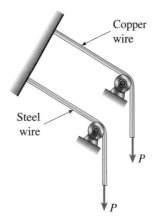

A-2.2: A plane truss with span length L = 4.5 m is constructed using cast iron pipes (E = 170 GPa) with cross sectional area of 4500 mm². The displacement of joint B cannot exceed 2.7 mm. The maximum value of loads P is approximately:

(A) 340 kN
(B) 460 kN
(C) 510 kN
(D) 600 kN

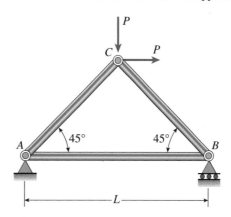

A-2.3: A brass rod (E = 110 GPa) with cross sectional area of 250 mm^2 is loaded by forces P_1 = 15 kN, P_2 = 10 kN, and P_3 = 8 kN. Segment lengths of the bar are a = 2.0 m, b = 0.75 m, and c = 1.2 m. The change in length of the bar is approximately:

(A) 0.9 mm
(B) 1.6 mm
(C) 2.1 mm
(D) 3.4 mm

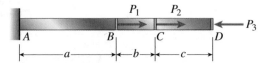

A-2.4: A brass bar (E = *110 MPa*) of length L = 2.5 m has diameter d_1 = 18 mm over one-half of its length and diameter d_2 = 12 mm over the other half. Compare this nonprismatic bar to a prismatic bar of the same volume of material with constant diameter d and length L. The elongation of the prismatic bar under the same load P = *25 kN* is approximately:

(A) 3 mm
(B) 4 mm
(C) 5 mm
(D) 6 mm

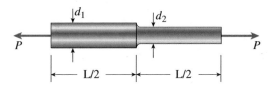

A-2.5: A nonprismatic cantilever bar has an internal cylindrical hole of diameter $d/2$ from 0 to x, so the net area of the cross section for Segment 1 is $(3/4)A$. Load P is applied at x, and load $-P/2$ is applied at $x = L$. Assume that E is constant. The length of the hollow segment, x, required to obtain axial displacement δ = PL/EA at the free end is:

(A) x = L/5
(B) x = L/4
(C) x = L/3
(D) x = 3L/5

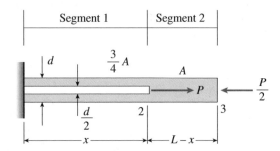

A-2.6: A nylon bar (E = 2.1 GPa) with diameter 12 mm, length 4.5 m, and weight 5.6 N hangs vertically under its own weight. The elongation of the bar at its free end is approximately:

(A) 0.05 mm
(B) 0.07 mm
(C) 0.11 mm
(D) 0.17 mm

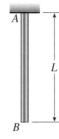

A-2.7: A monel shell (E_m = 170 GPa, d_3 = 12 mm, d_2 = 8 mm) encloses a brass core (E_b = 96 GPa, d_1 = 6 mm). Initially, both shell and core are of length 100 mm. A load P is applied to both shell and core through a cap plate. The load P required to compress both shell and core by 0.10 mm is approximately:

(A) 10.2 kN
(B) 13.4 kN
(C) 18.5 kN
(D) 21.0 kN

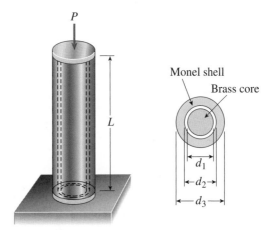

Monel shell
Brass core

A-2.8: A steel rod (E_s = 210 GPa, d_r = 12 mm, cte_s = 12 × 10^{-6}/degree Celsius) is held stress free between rigid walls by a clevis and pin (d_p = 15 mm) assembly at each end. If the allowable shear stress in the pin is 45 MPa and the allowable normal stress in the rod is 70 MPa, the maximum permissible temperature drop ΔT is approximately:

(A) 14 degrees Celsius
(B) 20 degrees Celsius
(C) 28 degrees Celsius
(D) 40 degrees Celsius

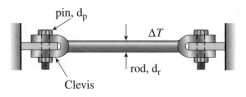

Clevis

A-2.9: A threaded steel rod (E_s = 210 GPa, d_r = 15 mm, cte_s = 12 × 10^{-6}/ degree Celsius) is held stress free between rigid walls by a nut and washer (d_w = 22 mm) assembly at each end. If the allowable bearing stress between the washer and wall is 55 MPa and the allowable normal stress in the rod is 90 MPa, the maximum permissible temperature drop ΔT is approximately:

(A) 25 degrees Celsius
(B) 30 degrees Celsius
(C) 38 degrees Celsius
(D) 46 degrees Celsius

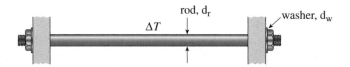

A-2.10: A steel bolt (area = 130 mm^2, E_s = 210 GPa) is enclosed by a copper tube (length = 0.5 m, area = 400 mm^2, E_c = 110 GPa) and the end nut is turned until it is just snug. The pitch of the bolt threads is 1.25 mm. The bolt is now tightened by a quarter turn of the nut. The resulting stress in the bolt is approximately:

(A) 56 MPa
(B) 62 MPa
(C) 74 MPa
(D) 81 MPa

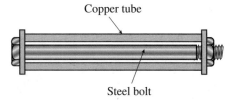

Copper tube

Steel bolt

A-2.11: A steel bar of rectangular cross section (a = 38 mm, b = 50 mm) carries a tensile load P. The allowable stresses in tension and shear are 100 MPa and 48 MPa respectively. The maximum permissible load P_{max} is approximately:

(A) 56 kN
(B) 62 kN
(C) 74 kN
(D) 91 kN

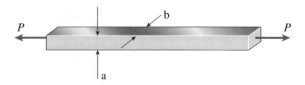

A-2.12: A brass wire ($d = 2.0$ mm, $E = 110$ GPa) is pretensioned to $T = 85$ N. The coefficient of thermal expansion for the wire is $19.5 \times 10^{-6}/°C$. The temperature change at which the wire goes slack is approximately:

(A) +5.7 degrees Celsius
(B) −12.6 degrees Celsius
(C) +12.6 degrees Celsius
(D) −18.2 degrees Celsius

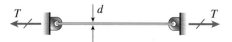

A-2.13: A copper bar ($d = 10$ mm, $E = 110$ GPa) is loaded by tensile load P = 11.5 kN. The maximum shear stress in the bar is approximately:

(A) 73 MPa
(B) 87 MPa
(C) 145 MPa
(D) 150 MPa

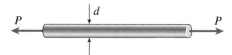

A-2.14: A steel plane truss is loaded at B and C by forces P = 200 kN. The cross sectional area of each member is A = 3970 mm². Truss dimensions are H = 3 m and L = 4 m. The maximum shear stress in bar AB is approximately:

(A) 27 MPa
(B) 33 MPa
(C) 50 MPa
(D) 69 MPa

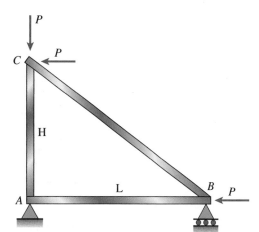

A-2.15: A plane stress element on a bar in uniaxial stress has tensile stress of $\sigma_\theta = 78$ MPa (see fig.). The maximum shear stress in the bar is approximately:

(A) 29 MPa
(B) 37 MPa
(C) 50 MPa
(D) 59 MPa

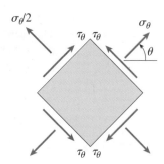

A-3.1: A brass rod of length $L = 0.75$ m is twisted by torques T until the angle of rotation between the ends of the rod is 3.5°. The allowable shear strain in the copper is 0.0005 rad. The maximum permissible diameter of the rod is approximately:

(A) 6.5 mm
(B) 8.6 mm
(C) 9.7 mm
(D) 12.3 mm

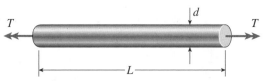

A-3.2: The angle of rotation between the ends of a nylon bar is 3.5°. The bar diameter is 70 mm and the allowable shear strain is 0.014 rad. The minimum permissible length of the bar is approximately:

(A) 0.15 m
(B) 0.27 m
(C) 0.40 m
(D) 0.55 m

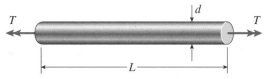

A-3.3: A brass bar twisted by torques T acting at the ends has the following properties: $L = 2.1$ m, $d = 38$ mm, and $G = 41$ GPa. The torsional stiffness of the bar is approximately:

(A) 1200 N·m
(B) 2600 N·m
(C) 4000 N·m
(D) 4800 N·m

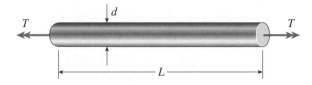

A-3.4: A brass pipe is twisted by torques $T = 800$ N·m acting at the ends causing an angle of twist of 3.5 degrees. The pipe has the following properties: $L = 2.1$ m, $d_1 = 38$ mm, and $d_2 = 56$ mm. The shear modulus of elasticity G of the pipe is approximately:

(A) 36.1 GPa
(B) 37.3 GPa
(C) 38.7 GPa
(D) 40.6 GPa

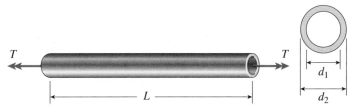

A-3.5: An aluminum bar of diameter d = 52 mm is twisted by torques T_1 at the ends. The allowable shear stress is 65 MPa. The maximum permissible torque T_1 is approximately:

(A) 1450 N·m
(B) 1675 N·m
(C) 1710 N·m
(D) 1800 N·m

A-3.6: A steel tube with diameters $d_2 = 86$ mm and $d_1 = 52$ mm is twisted by torques at the ends. The diameter of a solid steel shaft that resists the same torque at the same maximum shear stress is approximately:

(A) 56 mm
(B) 62 mm
(C) 75 mm
(D) 82 mm

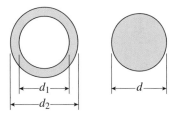

A-3.7: A stepped steel shaft with diameters $d_1 = 56$ mm and $d_2 = 52$ mm is twisted by torques $T_1 = 3.5$ kN·m and $T_2 = 1.5$ kN·m acting in opposite directions. The maximum shear stress is approximately:

(A) 54 MPa
(B) 58 MPa
(C) 62 MPa
(D) 79 MPa

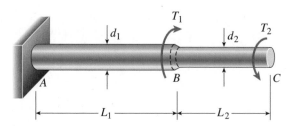

A-3.8: A stepped steel shaft (G = 75 GPa) with diameters d_1 = 36 mm and d_2 = 32 mm is twisted by torques T at each end. Segment lengths are L_1 = 0.9 m and L_2 = 0.75 m. If the allowable shear stress is 28 MPa and maximum allowable twist is 1.8 degrees, the maximum permissible torque is approximately:

(A) 142 N·m
(B) 180 N·m
(C) 185 N·m
(D) 257 N·m

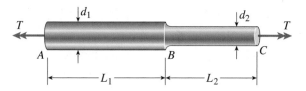

A-3.9: A gear shaft transmits torques T_A = 975 N·m, T_B = 1500 N·m, T_C = 650 N·m and T_D = 825 N·m. If the allowable shear stress is 50 MPa, the required shaft diameter is approximately:

(A) 38 mm
(B) 44 mm
(C) 46 mm
(D) 48 mm

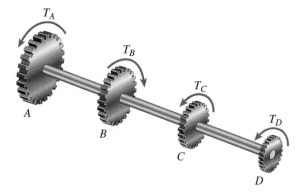

A-3.10: A hollow aluminum shaft (G = 27 GPa, d_2 = 96 mm, d_1 = 52 mm) has an angle of twist per unit length of 1.8°/*m* due to torques T. The resulting maximum tensile stress in the shaft is approximately:

(A) 38 MPa
(B) 41 MPa
(C) 49 MPa
(D) 58 MPa

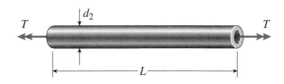

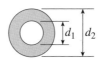

A-3.11: Torques T = 5.7 kN·m are applied to a hollow aluminum shaft (G = 27 GPa, d_1 = 52 mm). The allowable shear stress is 45 MPa and the allowable normal strain is 8.0×10^{-4}. The required outside diameter d_2 of the shaft is approximately:

(A) 38 mm
(B) 56 mm
(C) 87 mm
(D) 91 mm

A-3.12: A motor drives a shaft with diameter d = 46 mm at f = 5.25 Hz and delivers P = 25 kW of power. The maximum shear stress in the shaft is approximately:

(A) 32 MPa
(B) 40 MPa
(C) 83 MPa
(D) 91 MPa

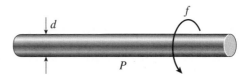

A-3.13: A motor drives a shaft at f = 10 Hz and delivers P = 35 kW of power. The allowable shear stress in the shaft is 45 MPa. The minimum diameter of the shaft is approximately:

(A) 35 mm
(B) 40 mm
(C) 47 mm
(D) 61 mm

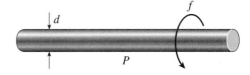

A-3.14: A drive shaft running at 2500 rpm has outer diameter 60 mm and inner diameter 40 mm. The allowable shear stress in the shaft is 35 MPa. The maximum power that can be transmitted is approximately:

(A) 220 kW
(B) 240 kW
(C) 288 kW
(D) 312 kW

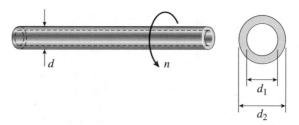

A-4.1: A simply-supported beam with proportional loading (P = 4.1 kN) has span length L = 5 m. Load P is 1.2 m from support A and load 2P is 1.5 m from support B. The bending moment just left of load 2P is approximately:

(A) 5.7 kN·m
(B) 6.2 kN·m
(C) 9.1 kN·m
(D) 10.1 kN·m

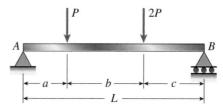

A-4.2: A simply-supported beam is loaded as shown in the figure. The bending moment at point C is approximately:

(A) 5.7 kN·m
(B) 6.1 kN·m
(C) 6.8 kN·m
(D) 9.7 kN·m

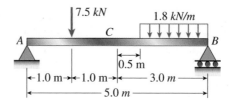

A-4.3: A cantilever beam is loaded as shown in the figure. The bending moment at 0.5 m from the support is approximately:

(A) 12.7 kN·m
(B) 14.2 kN·m
(C) 16.1 kN·m
(D) 18.5 kN·m

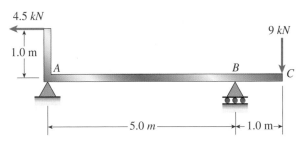

A-4.4: An L-shaped beam is loaded as shown in the figure. The bending moment at the midpoint of span AB is approximately:

(A) 6.8 kN·m
(B) 10.1 kN·m
(C) 12.3 kN·m
(D) 15.5 kN·m

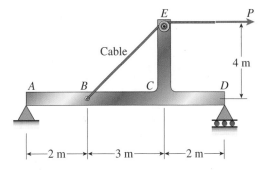

A-4.5: A T-shaped simple beam has a cable with force P anchored at B and passing over a pulley at E as shown in the figure. The bending moment just left of C is 1.25 kN·m. The cable force P is approximately:

(A) 2.7 kN
(B) 3.9 kN
(C) 4.5 kN
(D) 6.2 kN

A-4.6: A simple beam (L = 9 m) with attached bracket BDE has force P = 5 kN applied downward at E. The bending moment just right of B is approximately:

(A) 6 kN·m
(B) 10 kN·m
(C) 19 kN·m
(D) 22 kN·m

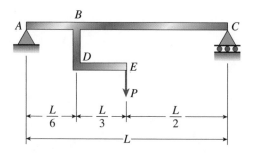

A-4.7: A simple beam AB with an overhang BC is loaded as shown in the figure. The bending moment at the midspan of AB is approximately:

(A) 8 kN·m
(B) 12 kN·m
(C) 17 kN·m
(D) 21 kN·m

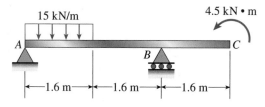

A-5.1: A copper wire (d = 1.5 mm) is bent around a tube of radius R = 0.6 m. The maximum normal strain in the wire is approximately:

(A) 1.25×10^{-3}
(B) 1.55×10^{-3}
(C) 1.76×10^{-3}
(D) 1.92×10^{-3}

A-5.2: A simply supported wood beam (L = 5 m) with rectangular cross section (b = 200 mm, h = 280 mm) carries uniform load q = 6.5 kN/m which includes the weight of the beam. The maximum flexural stress is approximately:

(A) 8.7 MPa
(B) 10.1 MPa
(C) 11.4 MPa
(D) 14.3 MPa

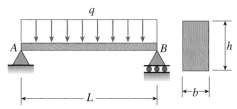

A-5.3: A cast iron pipe (L = 12 m, weight density = 72 kN/m³, d_2 = 100 mm, d_1 = 75 mm) is lifted by a hoist. The lift points are 6 m apart. The maximum bending stress in the pipe is approximately:

(A) 28 MPa
(B) 33 MPa
(C) 47 MPa
(D) 59 MPa

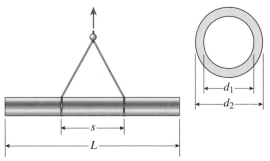

A-5.4: A beam with an overhang is loaded by a uniform load of 3 kN/m over its entire length. Moment of inertia I_z = 3.36 × 10⁶ mm⁴ and distances to top and bottom of the beam cross section are 20 mm and 66.4 mm, respectively. It is known that reactions at A and B are 4.5 kN and 13.5 kN, respectively. The maximum bending stress in the beam is approximately:

(A) 36 MPa
(B) 67 MPa
(C) 102 MPa
(D) 119 MPa

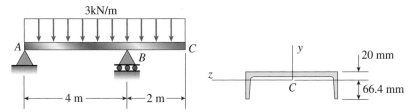

A-5.5: A steel hanger with solid cross section has horizontal force P = 5.5 kN applied at free end D. Dimension variable b = 175 mm and allowable normal stress is 150 MPa. Neglect self weight of the hanger. The required diameter of the hanger is approximately:

(A) 5 cm
(B) 7 cm
(C) 10 cm
(D) 13 cm

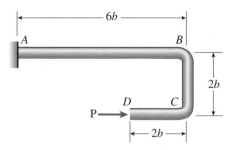

A-5.6: A cantilever wood pole carries force P = 300 N applied at its free end, as well as its own weight (weight density = 6 kN/m³). The length of the pole is L = 0.75 m and the allowable bending stress is 14 MPa. The required diameter of the pole is approximately:

(A) 4.2 cm
(B) 5.5 cm
(C) 6.1 cm
(D) 8.5 cm

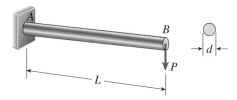

A-5.7: A simply supported steel beam of length L = 1.5 m and rectangular cross section (h = 75 mm, b = 20 mm) carries a uniform load of q = 48 kN/m, which includes its own weight. The maximum transverse shear stress on the cross section at 0.25 m from the left support is approximately:

(A) 20 MPa
(B) 24 MPa
(C) 30 MPa
(D) 36 MPa

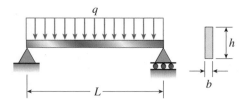

A-5.8: A simply supported laminated beam of length L = 0.5 m and square cross section weighs 4.8 N. Three strips are glued together to form the beam, with the allowable shear stress in the glued joint equal to 0.3 MPa. Considering also the weight of the beam, the maximum load P that can be applied at L/3 from the left support is approximately:

(A) 240 N
(B) 360 N
(C) 434 N
(D) 510 N

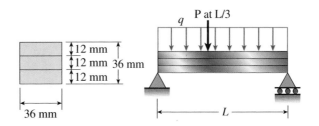

A-5.9: An aluminum cantilever beam of length L = 0.65 m carries a distributed load, which includes its own weight, of intensity q/2 at A and q at B. The beam cross section has width 50 mm and height 170 mm. Allowable bending stress is 95 MPa and allowable shear stress is 12 MPa. The permissible value of load intensity q is approximately:

(A) 110 kN/m
(B) 122 kN/m
(C) 130 kN/m
(D) 139 kN/m

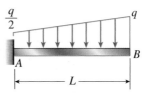

A-5.10: An aluminum light pole weighs 4300 N and supports an arm of weight 700 N, with arm center of gravity at 1.2 m left of the centroidal axis of the pole. A wind force of 1500 N acts to the right at 7.5 m above the base. The pole cross section at the base has outside diameter 235 mm and thickness 20 mm. The maximum compressive stress at the base is approximately:

(A) 16 MPa
(B) 18 MPa
(C) 21 MPa
(D) 24 MPa

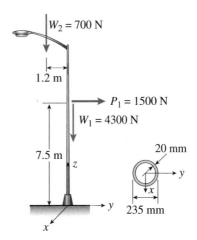

A-5.11: Two thin cables, each having diameter d = t/6 and carrying tensile loads P, are bolted to the top of a rectangular steel block with cross section dimensions b × t. The ratio of the maximum tensile to compressive stress in the block due to loads P is:

(A) 1.5
(B) 1.8
(C) 2.0
(D) 2.5

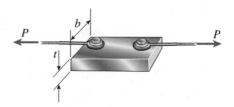

A-5.12: A composite beam is made up of a 200 mm × 300 mm core (E_c = 14 GPa) and an exterior cover sheet (300 mm × 12 mm, E_e = 100 GPa) on each side. Allowable stresses in core and exterior sheets are 9.5 MPa and 140 MPa, respectively. The ratio of the maximum permissible bending moment about the z-axis to that about the y-axis is most nearly:

(A) 0.5
(B) 0.7
(C) 1.2
(D) 1.5

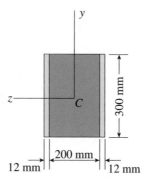

A-5.13: A composite beam is made up of a 90 mm × 160 mm wood beam (E_w = 11 GPa) and a steel bottom cover plate (90 mm × 8 mm, E_s = 190 GPa). Allowable stresses in wood and steel are 6.5 MPa and 110 MPa, respectively. The allowable bending moment about the z-axis of the composite beam is most nearly:

(A) 2.9 kN·m
(B) 3.5 kN·m
(C) 4.3 kN·m
(D) 9.9 kN·m

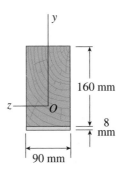

160 mm

8 mm

90 mm

A-5.14: A steel pipe (d_3 = 104 mm, d_2 = 96 mm) has a plastic liner with inner diameter d_1 = 82 mm. The modulus of elasticity of the steel is 75 times that of the modulus of the plastic. Allowable stresses in steel and plastic are 40 MPa and 550 kPa, respectively. The allowable bending moment for the composite pipe is approximately:

(A) *1100 N·m*
(B) *1230 N·m*
(C) *1370 N·m*
(D) *1460 N·m*

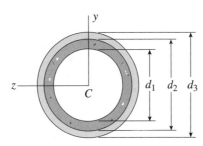

d_1 d_2 d_3

C

A-5.15: A bimetallic beam of aluminum (E_a = 70 GPa) and copper (E_c = 110 GPa) strips has width b = 25 mm; each strip has thickness t = 1.5 mm. A bending moment of 1.75 *N·m* is applied about the z axis. The ratio of the maximum stress in the aluminum to that in the copper is approximately:

(A) 0.6
(B) 0.8
(C) 1.0
(D) 1.5

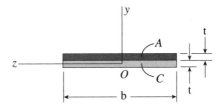

A

t

O C

b

t

A-5.16: A composite beam of aluminum (E_a = 72 GPa) and steel (E_s = 190 GPa) has width b = 25 mm and heights h_a = 42 mm, h_s = 68 mm. A bending moment is applied about the z axis resulting in a maximum stress in the aluminum of 55 MPa. The maximum stress in the steel is approximately:

(A) 86 MPa
(B) 90 MPa
(C) 94 MPa
(D) 98 MPa

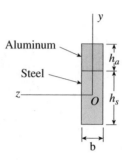

A-6.1: A rectangular plate (a = 120 mm, b = 160 mm) is subjected to compressive stress σ_x = − 4.5 MPa and tensile stress σ_y = 15 MPa. The ratio of the normal stress acting perpendicular to the weld to the shear stress acting along the weld is approximately:

(A) 0.27
(B) 0.54
(C) 0.85
(D) 1.22

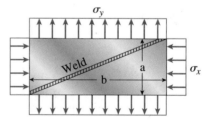

A-6.2: A rectangular plate in plane stress is subjected to normal stresses σ_x and σ_y and shear stress τ_{xy}. Stress σ_x is known to be 15 MPa but σ_y and τ_{xy} are unknown. However, the normal stress is known to be 33 MPa at counterclockwise angles of **35°** and **75°** from the x axis. Based on this, the normal stress σ_y on the element below is approximately:

(A) 14 MPa
(B) 21 MPa
(C) 26 MPa
(D) 43 MPa

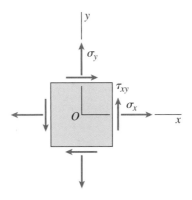

A-6.3: A rectangular plate in plane stress is subjected to normal stresses $\sigma_x =$ 35 MPa, $\sigma_y =$ 26 MPa, and shear stress $\tau_{xy} =$ 14 MPa. The ratio of the magnitudes of the principal stresses (σ_1/σ_2) is approximately:

(A) 0.8
(B) 1.5
(C) 2.1
(D) 2.9

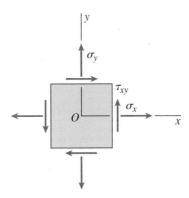

A-6.4: A drive shaft resists torsional shear stress of 45 MPa and axial compressive stress of 100 MPa. The ratio of the magnitudes of the principal stresses (σ_1/σ_2) is approximately:

(A) 0.15
(B) 0.55
(C) 1.2
(D) 1.9

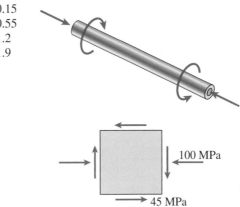

A-6.5: A drive shaft resists torsional shear stress of 45 MPa and axial compressive stress of 100 MPa. The maximum shear stress is approximately:

(A) 42 MPa
(B) 67 MPa
(C) 71 MPa
(D) 93 MPa

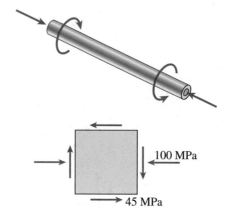

A-6.6: A drive shaft resists torsional shear stress of $\tau_{xy} = 40$ MPa and axial compressive stress $\sigma_x = -70$ MPa. One prinicipal normal stress is known to be 38 MPa (tensile). The stress σ_y is approximately:

(A) 23 MPa
(B) 35 MPa
(C) 62 MPa
(D) 75 MPa

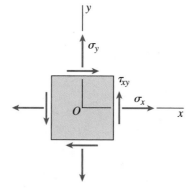

A-6.7: A cantilever beam with rectangular cross section (b = 95 mm, h = 300 mm) supports load P = 160 kN at its free end. The ratio of the magnitudes of the principal stresses (σ_1/σ_2) at point A (at distance c = 0.8 m from the free end and distance d = 200 mm up from the bottom) is approximately:

(A) 5
(B) 12
(C) 18
(D) 25

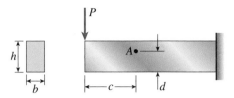

A-6.8: A simply supported beam (L = 4.5 m) with rectangular cross section (b = 95 mm, h = 280 mm) supports uniform load q = 25 kN/m. The ratio of the magnitudes of the principal stresses (σ_1/σ_2) at a point a = 1.0 m from the left support and distance d = 100 mm up from the bottom of the beam is approximately:

(A) 9
(B) 17
(C) 31
(D) 41

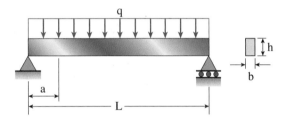

A-7.1: A thin wall spherical tank of diameter 1.5 m and wall thickness 65 mm has internal pressure of 20 MPa. The maximum shear stress in the wall of the tank is approximately:

(A) 58 MPa
(B) 67 MPa
(C) 115 MPa
(D) 127 MPa

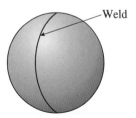

A-7.2: A thin wall spherical tank of diameter 0.75 m has internal pressure of 20 MPa. The yield stress in tension is 920 MPa, the yield stress in shear is 475 MPa, and the factor of safety is 2.5. The modulus of elasticity is 210 GPa, Poisson's ratio is 0.28, and maximum normal strain is 1220 × 10^{-6}. The minimum permissible thickness of the tank is approximately:

(A) 8.6 mm
(B) 9.9 mm
(C) 10.5 mm
(D) 11.1 mm

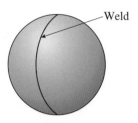

A-7.3: A thin wall cylindrical tank of diameter 200 mm has internal pressure of 11 MPa. The yield stress in tension is 250 MPa, the yield stress in shear is 140 MPa, and the factor of safety is 2.5. The minimum permissible thickness of the tank is approximately:

(A) 8.2 mm
(B) 9.1 mm
(C) 9.8 mm
(D) 11.0 mm

A-7.4: A thin wall cylindrical tank of diameter 2.0 m and wall thickness 18 mm is open at the top. The height h of water (weight density $= 9.81$ kN/m^3) in the tank at which the circumferential stress reaches 10 MPa in the tank wall is approximately:

(A) 14 m
(B) 18 m
(C) 20 m
(D) 24 m

A-7.5: The pressure relief valve is opened on a thin wall cylindrical tank, with radius to wall thickness ratio of 128, thereby decreasing the longitudinal strain by 150×10^{-6}. Assume $E = 73$ GPa and $v = 0.33$. The original internal pressure in the tank was approximately:

(A) 370 kPa
(B) 450 kPa
(C) 500 kPa
(D) 590 kPa

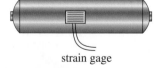

strain gage

A-7.6: A cylindrical tank is assembled by welding steel sections circumferentially. Tank diameter is 1.5 m, thickness is 20 mm, and internal pressure is 2.0 MPa. The maximum stress in the heads of the tank is approximately:

(A) 38 MPa
(B) 45 MPa
(C) 50 MPa
(D) 59 MPa

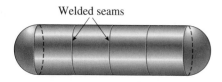

Welded seams

A-7.7: A cylindrical tank is assembled by welding steel sections circumferentially. Tank diameter is 1.5 m, thickness is 20 mm, and internal pressure is 2.0 MPa. The maximum tensile stress in the cylindrical part of the tank is approximately:

(A) 45 MPa
(B) 57 MPa
(C) 62 MPa
(D) 75 MPa

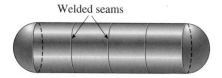

Welded seams

A-7.8: A cylindrical tank is assembled by welding steel sections circumferentially. Tank diameter is 1.5 m, thickness is 20 mm, and internal pressure is 2.0 MPa. The maximum tensile stress perpendicular to the welds is approximately:

(A) 22 MPa
(B) 29 MPa
(C) 33 MPa
(D) 37 MPa

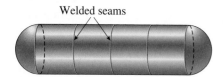

Welded seams

A-7.9: A cylindrical tank is assembled by welding steel sections circumferentially. Tank diameter is 1.5 m, thickness is 20 mm, and internal pressure is 2.0 MPa. The maximum shear stress in the heads is approximately:

(A) 19 MPa
(B) 23 MPa
(C) 33 MPa
(D) 35 MPa

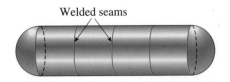

Welded seams

A-7.10: A cylindrical tank is assembled by welding steel sections circumferentially. Tank diameter is 1.5 m, thickness is 20 mm, and internal pressure is 2.0 MPa. The maximum shear stress in the cylindrical part of the tank is approximately:

(A) 17 MPa
(B) 26 MPa
(C) 34 MPa
(D) 38 MPa

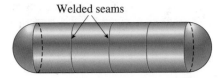

Welded seams

A-7.11: A cylindrical tank is assembled by welding steel sections in a helical pattern with angle $\alpha = 50$ degrees. Tank diameter is 1.6 m, thickness is 20 mm, and internal pressure is 2.75 MPa. Modulus E = 210 GPa and Poisson's ratio $v = 0.28$. The circumferential strain in the wall of the tank is approximately:

(A) 1.9×10^{-4}
(B) 3.2×10^{-4}
(C) 3.9×10^{-4}
(D) 4.5×10^{-4}

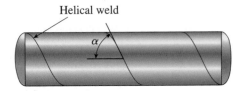

Helical weld

α

A-7.12: A cylindrical tank is assembled by welding steel sections in a helical pattern with angle $\alpha = 50$ degrees. Tank diameter is 1.6 m, thickness is 20 mm, and internal pressure is 2.75 MPa. Modulus E = 210 GPa and Poisson's ratio $v = 0.28$. The longitudinal strain in the the wall of the tank is approximately:

(A) 1.2×10^{-4}
(B) 2.4×10^{-4}
(C) 3.1×10^{-4}
(D) 4.3×10^{-4}

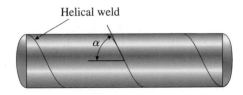

Helical weld

α

A-7.13: A cylindrical tank is assembled by welding steel sections in a helical pattern with angle $\alpha = 50$ degrees. Tank diameter is 1.6 m, thickness is 20 mm, and internal pressure is 2.75 MPa. Modulus E = 210 GPa and Poisson's ratio v = 0.28. The normal stress acting perpendicular to the weld is approximately:

(A) 39 MPa
(B) 48 MPa
(C) 78 MPa
(D) 84 MPa

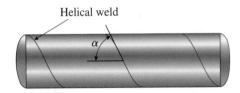

Helical weld

α

A-7.14: A segment of a drive shaft ($d_2 = 200$ mm, $d_1 = 160$ mm) is subjected to a torque T = 30 kN·m. The allowable shear stress in the shaft is 45 MPa. The maximum permissible compressive load P is approximately:

(A) 200 kN
(B) 286 kN
(C) 328 kN
(D) 442 kN

A-7.15: A thin walled cylindrical tank, under internal pressure p, is compressed by a force F = 75 kN. Cylinder diameter is d = 90 mm and wall thickness t = 5.5 mm. Allowable normal stress is 110 MPa and allowable shear stress is 60 MPa. The maximum allowable internal pressure p_{max} is approximately:

(A) 5 MPa
(B) 10 MPa
(C) 13 MPa
(D) 17 MPa

A-8.1: An aluminum beam (E = 72 GPa) with a square cross section and span length L = 2.5 m is subjected to uniform load q = 1.5 kN/m. The allowable bending stress is 60 MPa. The maximum deflection of the beam is approximately:

(A) 10 mm
(B) 16 mm
(C) 22 mm
(D) 26 mm

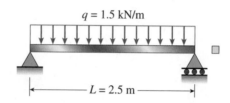

$q = 1.5$ kN/m

$L = 2.5$ m

A-8.2: An aluminum cantilever beam (E = 72 GPa) with a square cross section and span length L = 2.5 m is subjected to uniform load q = 1.5 kN/m. The allowable bending stress is 55 MPa. The maximum deflection of the beam is approximately:

(A) 10 mm
(B) 20 mm
(C) 30 mm
(D) 40 mm

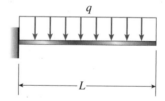

q

L

A-8.3: A steel beam (E = 210 GPa) with I = 119 × 10⁶ mm⁴ and span length L = 3.5 m is subjected to uniform load q = 9.5 kN/m. The maximum deflection of the beam is approximately:

(A) 10 mm
(B) 13 mm
(C) 17 mm
(D) 19 mm

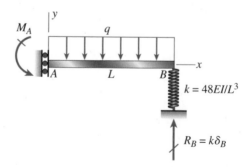

M_A

y

q

x

A   L   B

$k = 48EI/L^3$

$R_B = k\delta_B$

A-8.4: A steel bracket ABC ($EI = 4.2 \times 10^6$ N·m²) with span length L = 4.5 m and height H = 2 m is subjected to load P = 15 kN at C. The maximum rotation of joint B is approximately:

(A) 0.1 degrees
(B) 0.3 degrees
(C) 0.6 degrees
(D) 0.9 degrees

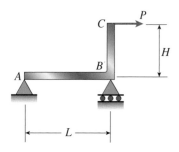

A-8.5: A steel bracket ABC ($EI = 4.2 \times 10^6$ N·m²) with span length L = 4.5 m and height H = 2 m is subjected to load P = 15 kN at C. The maximum horizontal displacement of joint C is approximately:

(A) 22 mm
(B) 31 mm
(C) 38 mm
(D) 40 mm

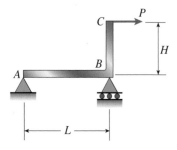

A-8.6: A nonprismatic cantilever beam of one material is subjected to load P at its free end. Moment of inertia $I_2 = 2 I_1$. The ratio r of the deflection δ_B to the deflection δ_1 at the free end of a prismatic cantilever with moment of inertia I_1 carrying the same load is approximately:

(A) 0.25
(B) 0.40
(C) 0.56
(D) 0.78

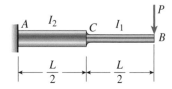

A-8.7: A steel bracket ABCD ($EI = 4.2 \times 10^6$ N·m^2), with span length L = 4.5 m and dimension a = 2 m, is subjected to load P = 10 kN at D. The maximum deflection at B is approximately:

(A) 10 mm
(B) 14 mm
(C) 19 mm
(D) 24 mm

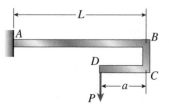

A-9.1: Beam ACB has a sliding support at A and is supported at C by a pinned end steel column with square cross section (E = 200 GPa, b = 40 mm) and height L = 3.75 m. The column must resist a load Q at B with a factor of safety 2.0 with respect to the critical load. The maximum permissible value of Q is approximately:

(A) 10.5 kN
(B) 11.8 kN
(C) 13.2 kN
(D) 15.0 kN

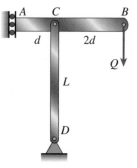

A-9.2: Beam ACB has a pin support at A and is supported at C by a steel column with square cross section (E = 190 GPa, b = 42 mm) and height L = 5.25 m. The column is pinned at C and fixed at D. The column must resist a load Q at B with a factor of safety 2.0 with respect to the critical load. The maximum permissible value of Q is approximately:

(A) 3.0 kN
(B) 6.0 kN
(C) 9.4 kN
(D) 10.1 kN

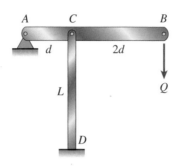

A-9.3: A steel pipe column (E = 190 GPa, $\alpha = 14 \times 10^{-6}$ per degree Celsius, d_2 = 82 mm, d_1 = 70 mm) of length L = 4.25 m is subjected to a temperature increase ΔT. The column is pinned at the top and fixed at the bottom. The temperature increase at which the column will buckle is approximately:

(A) 36 $°C$
(B) 42 $°C$
(C) 54 $°C$
(D) 58 $°C$

A-9.4: A steel pipe (E = 190 GPa, $\alpha = 14 \times 10^{-6}$ per degree Celsius, d_2 = 82 mm, d_1 = 70 mm) of length L = 4.25 m hangs from a rigid surface and is subjected to a temperature increase ΔT = 50 $°C$. The column is fixed at the top and has a small gap at the bottom. To avoid buckling, the minimum clearance at the bottom should be approximately:

(A) 2.55 mm
(B) 3.24 mm
(C) 4.17 mm
(D) 5.23 mm

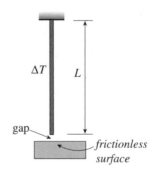

A-9.5: A pinned-end copper strut ($E = 110 \, GPa$) with length $L = 1.6 \, m$ is constructed of circular tubing with outside diameter $d = 38 \, mm$. The strut must resist an axial load $P = 14 \, kN$ with a factor of safety 2.0 with respect to the critical load. The required thickness t of the tube is:

(A) 2.75 mm
(B) 3.15 mm
(C) 3.89 mm
(D) 4.33 mm

A-9.6: A plane truss composed of two steel pipes ($E = 210 \, GPa$, $d = 100 \, mm$, wall thickness = 6.5 mm) is subjected to vertical load W at joint B. Joints A and C are $L = 7 \, m$ apart. The critical value of load W for buckling in the plane of the truss is nearly:

(A) 138 kN
(B) 146 kN
(C) 153 kN
(D) 164 kN

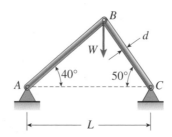

A-9.7: A beam is pin-connected to the tops of two identical pipe columns, each of height h, in a frame. The frame is restrained against sidesway at the top of column 1. Only buckling of columns 1 and 2 in the plane of the frame is of interest here. The ratio (a/L) defining the placement of load Q_{cr}, which causes both columns to buckle simultaneously, is approximately:

(A) 0.25
(B) 0.33
(C) 0.67
(D) 0.75

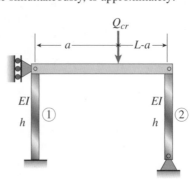

A-9.8: A steel pipe column ($E = 210\ GPa$) with length $L = 4.25\ m$ is constructed of circular tubing with outside diameter $d_2 = 90\ mm$ and inner diameter $d_1 = 64$ mm. The pipe column is fixed at the base and pinned at the top and may buckle in any direction. The Euler buckling load of the column is most nearly:

(A) 303 kN
(B) 560 kN
(C) 690 kN
(D) 720 kN

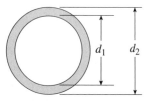

A-9.9: An aluminum tube (E = 72 GPa) *AB* of circular cross section has a pinned support at the base and is pin-connected at the top to a horizontal beam supporting a load $Q = 600$ kN. The outside diameter of the tube is 200 mm and the desired factor of safety with respect to Euler buckling is 3.0. The required thickness *t* of the tube is most nearly:

(A) 8 mm
(B) 10 mm
(C) 12 mm
(D) 14 mm

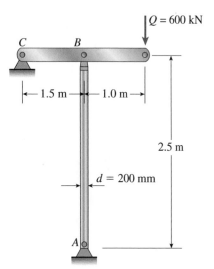

A-9.10: Two pipe columns are required to have the same Euler buckling load P_{cr}. Column 1 has flexural rigidity *E I* and height L_1; column 2 has flexural rigidity $(4/3)$*E I* and height L_2. The ratio (L_2/L_1) at which both columns will buckle under the same load is approximately:

(A) 0.55
(B) 0.72
(C) 0.81
(D) 1.10

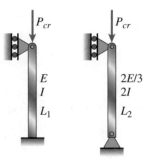

A-9.11: Two pipe columns are required to have the same Euler buckling load P_{cr}. Column 1 has flexural rigidity $E\,I_1$ and height L; column 2 has flexural rigidity $(2/3)E\,I_2$ and height L. The ratio (I_2/I_1) at which both columns will buckle under the same load is approximately:

(A) 0.8
(B) 1.0
(C) 2.2
(D) 3.1

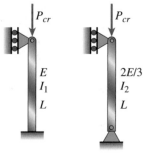

Answers to Problems

1.2-1 (a) σ_{AB} = 1443 psi; (b) P_2 = 1488 lbs;
(c) t_{BC} = 0.5 in.

1.2-2 (a) σ = 130.2 MPa; (b) ε = 4.65 $\times$ 10^{-4}

1.2-3 (a) R_B = 127.3 lb (cantilever), 191.3 lb (V-brakes);
σ_c = 204 psi (cantilever), 306 psi (V-brakes);
(b) (b) σ_{cable} = 26,946 psi (both)

1.2-4 (a) δ = 0.220 mm; (b) P = 34.6 kN

1.2-5 (a) σ_C = 2.13 ksi; x_C = 19.22 in., y_C = 19.22 in.

1.2-6 σ_t = 133 MPa

1.2-7 σ_1 = 25.5 ksi; σ_2 = 35.8 ksi;

1.2-8 σ_c = 5.21 MPa

1.2-9 (a) T = 184 lb, σ = 10.8 ksi; (b) ϵ_{cable} = 5 $\times$ 10^{-4}

1.2-10 (a) T = 819 N, σ = 74.5 MPa;
(b) ϵ_{cable} = 4.923 $\times$ 10^{-4}

1.2-11 (a) T_1 = 5877 lb, T_2 = 4679 lb, T_3 = 7159 lb;
(b) σ_1 = 49 ksi, σ_2 = 39 ksi, σ_3 = 60 ksi

1.2-12 (a) σ_x = $\gamma\omega^2(L^2 - x^2)/2g$; (b) σ_{max} = $\gamma\omega^2 L^2/2g$

1.2-13 (a) T_{AB} = 1620 lb, T_{BC} = 1536 lb, T_{CD} = 1640 lb
(b) σ_{AB} = 13,50 psi, σ_{BC} = 12,80 psi,
σ_{CD} = 13,67 psi

1.2-14 (a) T_{AQ} = T_{BQ} = 50.5 kN; (b) σ = 166.2 MPa

1.3-1 (a) L_{max} = 11,800 ft; (b) L_{max} = 13,500 ft

1.3-2 (a) L_{max} = 7900 m; (b) L_{max} = 8330 m

1.3-3 % elongation = 6.5, 24.0, 39.0;
% reduction = 8.1, 37.9, 74.9;
Brittle, ductile, ductile

1.3-4 11.9 $\times$ 10^3 m; 12.7 $\times$ 10^3 m; 6.1 $\times$ 10^3 m;
6.5 $\times$ 10^3 m; 23.9 $\times$ 10^3 m

1.3-5 $\sigma \approx$ 31 ksi

1.3-6 $\sigma_{pl} \approx$ 47 MPa, Slope $\approx$ 2.4 GPa, $\sigma_Y \approx$ 53 MPa;
Brittle

1.3-7 $\sigma_{pl} \approx$ 65,000 psi, Slope $\approx$ 30 $\times$ 10^6 psi,
$\sigma_Y \approx$ 69,000 psi, $\sigma_U \approx$ 113,000 psi;
Elongation = 6%, Reduction = 31%

1.4-1 0.13 in. longer

1.4-2 4.0 mm longer

1.4-3 (a) 2.81 in.; (b) 31.8 ksi

1.4-4 (a) 2.97 mm; (b) 180 MPa

1.4-5 (b) 0.71 in.; (c) 0.58 in.; (d) 49 ksi

1.5-1 P_{max} = 157 k

1.5-2 P = 27.4 kN (tension)

1.5-3 P = -15.71 kips

1.5-4 ΔL = 1.886 mm; % decrease in x-sec area = 0.072%

1.5-5 Δd = -1.56 $\times$ 10^{-4} in., P = 2.15 kips

1.5-6 (a) E = 104 GPa; (b) ν = 0.34

1.5-7 (a) $\Delta d_{BCinner}$ = 8 $\times$ 10^{-4} in.
(b) ν_{brass} = 0.34
(c) Δt_{AB} = 2.73 $\times$ 10^{-4} in.,
$\Delta d_{ABinner}$ = 1.366 $\times$ 10^{-4} in.

1.5-8 ΔV = 9789 mm^3

1.6-1 σ_b = 7.04 ksi, τ_{ave} = 10.76 ksi

1.6-2 σ_b = 139.9 MPa; P_{ult} = 144.4 kN

1.6-3 (a) τ = 12.73 ksi; (b) σ_{bf} = 20 ksi;
σ_{bg} = 26.7 ksi

1.6-4 (a) A_x = 255 N, A_y = 1072 N, B_x = -255 N
(b) $A_{resultant}$ = 1102 N
(c) τ = 5.48 MPa, σ_b = 6.89 MPa

1.6-5 (a) τ_{max} = 2979 psi; (b) σ_{bmax} = 936 psi

1.6-6 T_1 = 13.18 kN, T_2 = 10.77 kN,
τ_{1ave} = 25.9 MPa, τ_{2ave} = 21.2 MPa,
σ_{b1} = 9.15 MPa, σ_{b2} = 7.48 MPa

1.6-7 (a) Resultant = 1097 lb;
(b) σ_b = 4999 psi
(c) τ_{nut} = 2793 psi, τ_{pl} = 609 psi

1.6-8 G = 2.5 MPa

1.6-9 (a) γ_{aver} = 0.004; (b) V = 89.6 k

1.6-10 (a) γ_{aver} = 0.50; (b) δ = 4.50 mm

1.6-11 (a) τ_{aver} = 6050 psi; (b) σ_b = 9500 psi

1.6-12 τ_{aver} = 42.9 MPa

1.6-13 (a) A_x = 0, A_y = 170 lb, M_A = 4585 in.-lb
(b) B_x = 254 lb, B_y = 160 lb, B_{res} = 300 lb,
C_x = $-B_x$

(c) $\tau_B = 3054$ psi, $\tau_C = 1653$ psi

(d) $\sigma_{bB} = 4797$ psi, $\sigma_{bC} = C_x IA_{bC} = 3246$ psi

1.6-14 For a bicycle with $L/R = 1.8$:

(a) $T = 1440$ N; (b) $\tau_{aver} = 147$ MPa

1.6-15 (a) $\tau = \dfrac{P}{2\pi rh}$; (b) $\delta = \dfrac{P}{2\pi hG} \ln \dfrac{b}{d}$

1.6-16 (a) $A_x = 0$, $B_y = 0$, $A_y = 490$ kN; $F_{BC} = 0$,

$F_{AB} = 490$ kN, $F_{AC} = -693$ kN

(b) $\tau_p = 963$ MPa

(c) $\sigma_b = 1361$ MPa

1.6–17 (a) $O_x = 12.68$ lb, $O_y = 1.294$ lb, $O_{res} = 12.74$ lb

(b) $\tau_O = 519$ psi, $\sigma_{bO} = 816$ psi

(c) $\tau = 362$ psi

1.6–18 (a) $F_s = 153.9$ N, $\sigma = 3.06$ MPa

(b) $\tau_{ave} = 1.96$ MPa

(c) $\sigma_b = 1.924$ MPa

1.6–19 (a) $P = 395$ lb

(b) $C_x = 374$ lb, $C_y = -237$ lb, $C_{res} = 443$ lb

(c) $\tau = 18.04$ ksi, $\sigma_{bC} = 4.72$ psi

1.7-1 $P_{allow} = 3140$ lb

1.7-2 $T_{max} = 33.4$ kN·m

1.7-3 $P_{allow} = 607$ lb

1.7-4 (a) Tube BC (yield): $P_a = 7.69$ kN

(b) P_a (yield) $= 7.6$ kN

(c) Tube AB (yield): $P_a = 17.16$ kN

1.7-5 $P = 294$ k

1.7-6 (a) $F = 1.171$ kN

(b) Shear: $F_a = 2.86$ kN

1.7-7 $W_{max} = 5110$ lb

1.7-8 (a) $F_A = \sqrt{2}\,T$, $F_B = 2\,T$, $F_C = T$

(b) Shear at A: $W_{max} = 66.5$ kN

1.7-9 $P_a = 10.21$ kips

1.7-10 $C_{ult} = 5739$ N; $P_{max} = 445$ N

1.7-11 $W_{max} = 0.305$ kips

1.7-12 Shear in rivets in CG & CD controls:

$P_{allow} = 45.8$ kN

1.7-13 (a) $P_a = \sigma_a (0.587\ d^2)$; (b) $P_a = 21.6$ kips

1.7-14 $P_{allow} = 96.5$ kN

1.7-15 $p_{max} = 11.98$ psf

1.7-16 (a) $P_{allow} = \sigma_c(\pi d^2/4)\sqrt{1 - (R/L)^2}$;

(b) $P_{allow} = 9.77$ kN

1.8-1 (a) $d_{min} = 3.75$ in.; (b) $d_{min} = 4.01$ in.

1.8-2 (a) $d_{min} = 225$ mm; (b) $d_{min} = 242$ mm

1.8-3 (a) $d_{min} = 0.704$ in.; (b) $d_{min} = 0.711$ in.

1.8-4 $d_{min} = 63.3$ mm

1.8-5 $d_{min} = 0.64$ in.

1.8-6 (b) $A_{min} = 435$ mm^2

1.8-7 $d_{min} = 0.372$ in.

1.8-8 $d_{min} = 5.96$ mm

1.8-9 $n = 11.6$, or 12 bolts

1.8-10 $(d_2)_{min} = 131$ mm

1.8-11 $A_c = 1.189$ in^2

1.8-12 (a) $t_{min} = 18.8$ mm, use $t = 20$ mm;

(b) $D_{min} = 297$ mm

1.8-13 (a) $\sigma_{DF} = 10.38$ ksi $< \sigma_{allow}$; $\sigma_{bF} = 378$ psi $< \sigma_{ba}$

(b) required diameter of washer $= 1$-$5/16$ in. $=$
1.312 in.

1.8-14 (a) $d_m = 24.7$ mm; (b) $P_{max} = 49.4$ kN

1.8-15 $\theta = \arccos 1/\sqrt{3} = 54.7°$

CHAPTER 2

2.2-1 $\delta = 6W/(5k)$

2.2-2 (a) $\delta = 12.5$ mm; (b) $n = 5.8$

2.2-3 (a) $\delta_c/\delta_s = 1.67$; (b) $d_c/d_s = 1.29$

2.2-4 $h = 13.4$ mm

2.2-5 $h = L - \pi\rho_{max}d^2/4k$

2.2-6 $x = 118$ mm

2.2-7 $\delta_C = 16P/9k$

2.2-8 (a) $\delta_B = 2.5$ mm; (b) $P_{max} = 390$ kN

2.2-9 $P_{max} = 72.3$ lb

2.2-10 (a) $x = 134.7$ mm; (b) $k_1 = 0.204$ N/mm;

(c) $b = 74.1$ mm; (d) $k_3 = 0.638$ N/mm

2.2-11 (a) $t_{c,min} = 0.021$ in.; (b) $\delta_r = 0.031$ in.;

(c) $h_{min} = 0.051$ in.

2.2-12 $\delta_A = 0.200$ mm, $\delta_D = 0.880$ mm

2.2-13 $\theta = 35.1°$, $\delta = 1.78$ in.

2.2-14 $\theta = 35.1°$, $\delta = 44.5$ mm

2.3-1 $\delta = 0.0276$ in.

2.3-2 (a) $\delta = 0.675$ mm; (b) $P_{max} = 267$ kN

2.3-3 (a) $\delta = 0.0131$ in. (elongation); (b) $P = 1310$ lb

2.3-4 (a) $\delta = 7PL/6Ebt$; (b) $\delta = 0.500$ mm

2.3-5 (a) $\delta = 7PL/6Ebt$; (b) $\delta = 0.021$ in.

2.3-6 (a) $\delta_{AC} = 3.72$ mm; (b) $P_0 = 44.2$ kN

2.3-7 (a) $\delta = 0.0589$ in.; (b) $\delta = 0.0501$ in.

2.3-8 (a) $d_{max} = 23.9$ mm; (b) $b = 4.16$ mm;

(c) $x = 183.3$ mm

2.3-9 (a) $\delta = PL/2EA$; (b) $\sigma_c = Py/AL$

2.3-10 (a) $\delta_{2-4} = 0.024$ mm; (b) $P_{max} = 8.15$ kN;

(c) $L_2 = 9.16$ mm

2.3-11 (a) $R_1 = -3P/2$; (b) $N_1 = 3P/2$ (tension), $N_2 = P/2$
(tension); (c) $x = L/3$; (d) $\delta_2 = 2PL/3EA$;

(e) $\beta = 1/11$

2.3-12 (a) $\delta_C = W(L^2 - h^2)/2EAL$;

(b) $\delta_B = WL/2EA$; (c) $\beta = 3$

2.3-13 (b) $\delta = 0.010$ in.

2.3-14 $\delta = 2PH/3Eb^2$

2.3-15 $\delta = 2WL/\pi d^2 E$

2.3-16 (a) $\delta = 2.18$ mm; (b) $\delta = 6.74$ mm

2.3-17 (b) $\delta = 11.14$ ft

2.4-1 (a) $P = 1330$ lb; (b) $P_{allow} = 1300$ lb

2.4-2 (a) $P = 104$ kN; (b) $P_{max} = 116$ kN

2.4-3 (a) $P_B/P = 3/11$; (b) $\sigma_B/\sigma_A = 1/2$;
    (c) Ratio = 1

2.4-4 (a) If $x \le L/2$, $R_A = (-3PL)/(2(x + 3L))$,
        $R_B = -P(2x + 3L)/(2(x + 3L))$
        If $x \ge L/2$, $R_A = (-P(x + L))/(x + 3L)$,
        $R_B = (-2PL)/(x + 3L)$
    (b) If $x \le L/2$, $\delta = PL(2x + 3L)/[(x + 3L)E\pi d^2]$
        If $x \ge L/2$, $\delta = 8PL(x + L)/[3(x + 3L)E\pi d^2]$
    (c) $x = 3L/10$ or $x = 2L/3$
    (d) $R_B = -0.434\,P$, $R_A = -0.566\,P$
    (e) $R_B = \rho g \pi d^2 L/8$, $R_A = 3\,\rho g \pi d^2 L/32$

2.4-5 (a) 41.7%; (b) $\sigma_M = 32.7$ ksi, $\sigma_O = 51.4$ ksi

2.4-6 (a) $\delta = 1.91$ mm; (b) $\delta = 1.36$ mm;
    (c) $\delta = 2.74$ mm

2.4-7 (a) $R_A = 2R_D = 2P/3$; (b) $\delta_B = 2\delta_C = PL/6EA_1$

2.4-8 (a) $R_A = 10.5$ kN to the left;
        $R_D = 2.0$ kN to the right;
    (b) $F_{BC} = 15.0$ kN (compression)

2.4-9 (b) $\sigma_a = 1610$ psi (compression),
        $\sigma_s = 9350$ psi (tension)

2.4-10 (a) $R_A = (37/70)\,\rho gAL$, $R_C = (19/70)\,\rho gAL$
    (b) $\delta_B = (-24/175)\,\rho g L^2/E$
    (c) $\sigma_B = -\rho gL/14$, $\sigma_C = -19\,\rho gL/35$

2.4-11 (a) $P_1 = PE_1/(E_1 + E_2)$;
    (b) $e = b(E_2 - E_1)/[2(E_2 + E_1)]$;
    (c) $\sigma_1/\sigma_2 = E_1/E_2$

2.4-12 (a) $P_{allow} = 1504$ N; (b) $P_{allow} = 820$ N;
    (c) $P_{allow} = 703$ N

2.4-13 $d_2 = 0.338$ in., $L_2 = 48.0$ in.

2.4-14 $\delta_{AC} = 0.176$ mm

2.4-15 (a) $\sigma_C = 10{,}000$ psi, $\sigma_D = 12{,}500$ psi;
    (b) $\delta_B = 0.0198$ in.

2.4-16 $P_{max} = 1800$ N

2.4-17 $\sigma_s = 3.22$ ksi, $\sigma_b = 1.716$ ksi, $\sigma_c = 1.93$ ksi

2.5-1 $\sigma = 11{,}700$ psi

2.5-2 $T = 40.3°C$

2.5-3 $\Delta T = 185°F$

2.5-4 (a) $\Delta T = 24°C$; (b) clevis: $\sigma_{bc} = 42.4$ MPa;
    washer: $\sigma_{bw} = 74.1$ MPa

2.5-5 (a) $\sigma_c = E\alpha(\Delta T_B)/4$
    (b) $\sigma_c = E\alpha(\Delta T_B)/[4(EA/kL) + 1]$

2.5-6 (a) $N = 51.8$ kN, max. $\sigma_c = 26.4$ MPa,
        $\delta_C = -0.314$ mm
    (b) $N = 31.2$ kN, max. $\sigma_c = 15.91$ MPa,
        $\delta_C = -0.546$ mm

2.5-7 $\delta = 0.123$ in.

2.5-8 $\Delta T = 34°C$

2.5-9 $\tau = 15.0$ ksi

2.5-10 $P_{allow} = 39.5$ kN

2.5-11 (a) $T_A = 400$ lb, $T_B = 200$ lb;
    (b) $T_A = 454$ lb, $T_B = 92$ lb; (c) $\Delta T = 153°F$

2.5-12 (a) $\sigma = 98$ MPa; (b) $T = 35°C$

2.5-13 (a) $\sigma = -957$ psi; (b) $F_k = 3006$ lbs (C);
    (c) $\sigma = -2560$ psi

2.5-14 $s = PL/6EA$

2.5-15 (a) $P_1 = 231$ k; $R_A = -55.2$ k, $R_B = 55.2$ k
    (b) $P_2 = 145.1$ k; $R_A = -55.2$ k, $R_B = 55.2$ k
    (c) For P_1, $\tau_{max} = 13.39$ ksi; for P_2,
        $\tau_{max} = 19.44$ ksi
    (d) $\Delta T = 65.8°F$; $R_A = 0$, $R_B = 0$
    (e) $R_A = -55.2$ k, $R_B = 55.2$ k

2.5-16 (a) $R_A = [-s + \alpha\,\Delta T\,(L_1 + L_2)]/[(L_1/EA_1)$
        $+ (L_2/EA_2) + (1/k_3)]$, $R_D = -R_A$
    (b) $\delta_B = \alpha\,\Delta T\,(L_1) - R_A\,(L_1/EA_1)$, $\delta_C = \alpha\,\Delta T$
        $(L_1 + L_2) - R_A\,[(L_1/EA_1) + (L_2/EA_2)]$

2.5-17 $T_B = 660$ lb, $T_C = 780$ lb

2.5-18 $P_{allow} = 1.8$ MN

2.5-19 (a) $\sigma_p = -0.196$ ksi, $\sigma_r = 3.42$ ksi
    (b) $\sigma_b = 2.74$ ksi, $\tau_c = 0.285$ ksi

2.5-20 $\sigma_p = 25.0$ MPa

2.5-21 $\sigma_p = 2400$ psi

2.5-22 (a) $P_B = 25.4$ kN, $P_s = -P_B$
    (b) $S_{reqd} = 25.7$ mm
    (c) $\delta_{final} = 0.35$ mm

2.5-23 (a) $F_k = -0.174$ k; (b) $F_t = 0.174$ k;
    (c) $L_f = 12.01$ in.; (d) $\Delta T = 141.9°F$

2.5-24 $\sigma_s = 500$ MPa (tension),
    $\sigma_c = 10$ MPa (compression)

2.5-25 (a) $F_k = 0.174$ k; (b) $F_t = -0.174$ k;
    (c) $L_f = 11.99$ in.; (d) $\Delta T = -141.6\ °F$

2.6-1 $P_{max} = 42{,}600$ lb

2.6-2 $d_{min} = 6.81$ mm

2.6-3 $P_{max} = 24{,}000$ lb

2.6-4 (a) $\Delta T_{max} = -46°C$; (b) $\Delta T = +9.93°C$

2.6-5 (a) $\tau_{max} = 10{,}800$ psi; (b) $\Delta T_{max} = -49.9°F$;
    (c) $\Delta T = +75.9°F$

2.6-6 (a) $\sigma_{max} = 84.0$ MPa; (b) $\tau_{max} = 42.0$ MPa

2.6-7 (a) $\sigma_{max} = 18{,}000$ psi; (b) $\tau_{max} = 9{,}000$ psi

2.6-8 Element A: $\sigma_x = 105$ MPa (compression);
    Element B: $\tau_{max} = 52.5$ MPa

2.6-9 $N_{AB} = 90$ kips (C); (a) $\sigma_x = -10.91$ ksi;
    (b) $\sigma_\theta = -8.18$ ksi, $\tau_\theta = 4.72$ ksi;
    (c) $\sigma_\theta = -5.45$ ksi, $\tau_\theta = 5.45$ ksi;

2.6-10 (a) (1) $\sigma_x = -945$ kPa; (2) $\sigma_\theta = -807$ kPa,
$\tau_\theta = 334$ kPa; (3) $\sigma_\theta = -472$ kPa, $\tau_\theta = 472$ kPa;
$\sigma_{max} = -945$ kPa, $\tau_{max} = -472$ kPa
(b) $\sigma_{max} = -378$ kPa, $\tau_{max} = -189$ kPa

2.6-11 (a) $\tau_{pq} = 1154$ psi; (b) $\sigma_{pq} = -1700$ psi,
$\sigma(pq + \pi/2) = -784$ psi;
(c) $P_{max} = 14688$ lb

2.6-12 (a) $\Delta T_{max} = 31.3°C$; (b) $\sigma_{pq} = -21.0$ MPa (compression), $\tau_{pq} = 30$ MPa (CCW); (c) $\beta = 0.62$

2.6-13 $N_{AC} = 5.77$ kips; $d_{min} = 1.08$ in.

2.6-14 (a) $\sigma_\theta = 0.57$ MPa, $\tau_\theta = -1.58$ MPa;
(b) $\alpha = 33.3°$; (c) $\alpha = 26.6°$

2.6-15 (a) $\theta = 35.26°$, $\tau_\theta = -7070$ psi;
(b) $\sigma_{max} = 15,000$ psi, $\tau_{max} = 7,500$ psi

2.6-16 $\sigma_{\theta1} = 54.9$ MPa, $\sigma_{\theta2} = 18.3$ MPa,
$\tau_\theta = -31.7$ MPa

2.6-17 $\sigma_{max} = 10,000$ psi, $\tau_{max} = 5,000$ psi

2.6-18 (a) $\theta = 30.96°$; (b) $P_{max} = 1.53$ kN

2.6-19 (a) $\Delta T_{max} = 21.7°F$; (b) $\Delta T_{max} = 25.3°F$

CHAPTER 3

3.2-1 $d_{max} = 0.413$ in.

3.2-2 $L_{min} = 162.9$ mm

3.2-3 (a) $\gamma_1 = 267 \times 10^{-6}$ radians; (b) $r_{2,min} = 2.2$ inches

3.2-4 (a) $\gamma_1 = 393 \times 10^{-6}$ radians; (b) $r_{2,max} = 50.9$ mm

3.2-5 (a) $\gamma_1 = 195 \times 10^{-6}$ radians; (b) $r_{2,max} = 2.57$ inches

3.3-1 $\tau_{max} = 8340$ psi

3.3-2 (a) $\tau_{max} = 23.8$ MPa;
(b) $\theta = 9.12°/m$

3.3-3 (a) $\tau_{max} = 18,300$ psi;
(b) $\phi = 3.32°$

3.3-4 (a) $k_T = 2059$ N·m;
(b) $\tau_{max} = 27.9$ MPa, $\gamma_{max} = 997 \times 10^{-6}$ radians

3.3-5 $L_{min} = 38.0$ in.

3.3-6 $T_{max} = 6.03$ N·m, $\phi = 2.20°$

3.3-7 $\tau_{max} = 7965$ psi; $\gamma_{max} = 0.00255$ radians;
$G = 3.13 \times 10^6$ psi

3.3-8 $T_{max} = 9164$ N·m

3.3-9 $\tau_{max} = 4840$ psi

3.3-10 $d_{min} = 63.3$ mm

3.3-11 (a) $\tau_2 = 5170$ psi;
(b) $\tau_1 = 3880$ psi;
(c) $\theta = 0.00898°/in.$

3.3-12 (a) $\tau_2 = 30.1$ MPa;
(b) $\tau_1 = 20.1$ MPa;
(c) $\theta = 0.306°/m$

3.3-13 $d_{min} = 2.50$ in.

3.3-14 $d_{min} = 64.4$ mm

3.3-15 (a) $T_{1,max} = 4.60$ in.-k;
(b) $T_{1,max} = 4.31$ in.-k;
(c) torque: 6.25%, weight: 25%

3.3-16 (a) $\phi = 5.19°$;
(b) $d = 88.4$ mm;
(c) ratio $= 0.524$

3.3-17 $r_2 = 1.40$ in.

3.4-1 (a) $\tau_{max} = 7600$ psi;
(b) $\phi_C = 0.16°$

3.4-2 (a) $\tau_{bar} = 79.6$ MPa, $\tau_{tube} = 32.3$ MPa;
(b) $\phi_A = 9.43°$

3.4-3 (a) $\tau_{max} = 4.65$ ksi;
(b) $\phi_D = 0.978°$

3.4-4 $T_{allow} = 459$ N·m

3.4-5 $d_1 = 0.818$ in.

3.4-6 $d = 77.5$ mm

3.4-7 (a) $d = 1.78$ in.;
(b) $d = 1.83$ in.

3.4-8 $d_B/d_A = 1.45$

3.4-9 Minimum $d_A = 2.52$ in.

3.4-10 Minimum $d_B = 48.6$ mm

3.4-11 (a) $R_1 = -3T/2$;
(b) $T_1 = 1.5T$, $T_2 = 0.5T$;
(c) $x = 7L/17$;
(d) $\phi_2 = (12/17)(TL/GI_P)$

3.4-12 $\phi = 3TL/2\pi Gtd_A^3$

3.4-13 (a) $\phi = 2.79°$;
(b) $\phi = 2.21°$

3.4-14 (a) $R_1 = \dfrac{-T}{2}$ (b) $\phi_3 = \dfrac{19}{8} \cdot \dfrac{TL}{\pi Gtd^3}$

3.4-15 $\phi_D = \dfrac{4Fd}{\pi G}\left[\dfrac{L_1}{t_{01}d_{01}^{\,3}} \right.$

$+ \int_0^{L_2} \dfrac{L_2^{\,4}}{\left(d_{01}L_2 - d_{01}x + d_{03}x\right)^3 \left(t_{01}L_2 - t_{01}x + t_{03}x\right)} dx$

$\left. + \dfrac{L_3}{t_{03}d_{03}^{\,3}} \right]$

$\phi_D = 0.142°$

3.4-16 $\tau_{max} = 16tL/\pi d^3$;
(b) $\phi = 16tL^2/\pi Gd^4$

3.4-17 $\tau_{max} = 8t_AL/\pi d^3$;
(b) $\phi = 16t_AL^2/3\pi Gd^4$

3.4-18 (a) $R_A = -\dfrac{T_0}{6}$

(b) $T_{AB}(x) = \left(\dfrac{T_0}{6} - \dfrac{x^2}{L^2}T_0\right)\ 0 \le x \le \dfrac{L}{2}$

$T_{BC}(x) = -\left[\left(\dfrac{x-L}{L}\right)^2 \cdot \dfrac{T_0}{3}\right]\ \dfrac{L}{2} \le x \le L$

(c) $\phi_C = \dfrac{T_0 L}{144 G I_P}$

(d) $\tau_{max} = \dfrac{8}{3\pi} \cdot \dfrac{T_0}{d_{AB}^{\ 3}}$

3.4-19 $L_{max} = 4.42$ m; (b) $\phi = 170°$

3.4-20 (a) $T_{0,max} = \tau_{p,allow}\left(\dfrac{\pi d_2 d_p^{\ 2}}{4}\right)$

(b) $T_{0,max} = \tau_{t,allow}\left[\dfrac{\pi\left(d_3^{\ 4} - d_2^{\ 4}\right)}{16 d_3}\right]$

$T_{0,max} = \tau_{t,allow}\left[\dfrac{\pi\left(d_2^{\ 4} - d_1^{\ 4}\right)}{16 d_2}\right]$

(c) $\phi_{C,max} = \tau_{p,allow}\left(\dfrac{8 d_2 d_p^{\ 2}}{G}\right)$

$\dfrac{\left[\dfrac{L_A}{\left(d_3^{\ 4} - d_2^{\ 4}\right)} + \dfrac{L_B}{\left(d_2^{\ 4} - d_1^{\ 4}\right)}\right]}{}$

$\phi_{C,max} = \tau_{t,allow}\left(\dfrac{2\left(d_3^{\ 4} - d_2^{\ 4}\right)}{G d_3}\right)$

$\dfrac{\left[\dfrac{L_A}{\left(d_3^{\ 4} - d_2^{\ 4}\right)} + \dfrac{L_B}{\left(d_2^{\ 4} - d_1^{\ 4}\right)}\right]}{}$

$\phi_{C,max} = \tau_{t,allow}\left(\dfrac{2\left(d_2^{\ 4} - d_1^{\ 4}\right)}{G d_2}\right)$

$\dfrac{\left[\dfrac{L_A}{\left(d_3^{\ 4} - d_2^{\ 4}\right)} + \dfrac{L_B}{\left(d_2^{\ 4} - d_1^{\ 4}\right)}\right]}{}$

3.5-1 (a) $\sigma_{max} = 6280$ psi; (b) $T = 74{,}000$ lb-in.

3.5-2 (a) $\epsilon_{max} = 320 \times 10^{-6}$; (b) $\sigma_{max} = 51.2$ MPa;
(c) $T = 20.0$ kN·m

3.5-3 (a) $d_1 = 2.40$ in.; (b) $\phi = 2.20°$;
(c) $\gamma_{max} = 1600 \times 10^{-6}$ rad

3.5-4 $G = 30.0$ GPa

3.5-5 $T = 4200$ lb-in.

3.5-6 $d_{min} = 37.7$ mm

3.5-7 $d_1 = 0.60$ in.

3.5-8 $d_2 = 79.3$ mm

3.5-9 (a) $\tau_{max} = 5090$ psi; (b) $\gamma_{max} = 432 \times 10^{-6}$ rad

3.5-10 (a) $\tau_{max} = 23.9$ MPa; (b) $\gamma_{max} = 884 \times 10^{-6}$ rad

3.7-1 (a) $\tau_{max} = 4950$ psi; (b) $d_{min} = 3.22$ in.

3.7-2 (a) $\tau_{max} = 50.0$ MPa; (b) $d_{min} = 32.3$ mm

3.7-3 (a) $H = 6560$ hp; (b) Shear stress is halved

3.7-4 (a) $\tau_{max} = 16.8$ MPa; (b) $P_{max} = 267$ kW

3.7-5 $d_{min} = 4.28$ in.

3.7-6 $d_{min} = 110$ mm

3.7-7 Minimum $d_1 = 1.221d$

3.7-8 $P_{max} = 91.0$ kW

3.7-9 $d = 2.75$ in.

3.7-10 $d = 53.4$ mm

3.8-1 $\phi_{max} = 3T_0 L/5 G I_P$

3.8-2 (a) $x = L/4$; (b) $\phi_{max} = T_0 L/8 G I_P$

3.8-3 $\phi_{max} = 2b\tau_{allow}/Gd$

3.8-4 $P_{allow} = 2710$ N

3.8-5 $(T_0)_{max} = 3680$ lb-in.

3.8-6 $(T_0)_{max} = 150$ N·m

3.8-7 (a) $a/L = d_A/(d_A + d_B)$; (b) $a/L = d_A^4/(d_A^4 + d_B^4)$

3.8-8 $T_A = t_0 L/6,\ T_B = t_0 L/3$

3.8-9 $x = 30.12$ in.

3.8-10 (a) $\tau_1 = 32.7$ MPa, $\tau_2 = 49.0$ MPa;
(b) $\phi = 1.030°$; (c) $k_T = 22.3$ kN·m

3.8-11 (a) $\tau_1 = 1790$ psi, $\tau_2 = 2690$ psi;
(b) $\phi = 0.354°$; (c) $k_T = 809$ k-in.

3.8-12 $T_{max} = 1520$ N·m

3.8-13 $T_{max} = 9.13$ k-in.

3.8-14 (a) $T_{1,allow} = 9.51$ kN·m; (b) $T_{2,allow} = 6.35$ kN·m;
(c) $T_{3,allow} = 7.41$ kN·m; (d) $T_{max} = 6.35$ kN·m

3.8-15 (a) $T_A = 15{,}292$ in.-lb, $T_B = 24{,}708$ in.-lb
(b) $T_A = 8{,}734$ in.-lb, $T_B = 31{,}266$ in.-lb

3.8-16 (a) $T_B = \dfrac{G\beta}{L}\left(\dfrac{I_{PA} I_{PB}}{I_{PA} + I_{PB}}\right)\quad T_A = -T_B$

(b) $\beta_{max} = \tau_{p,allow}\dfrac{L}{4G}\left[\left(\dfrac{I_{PB} + I_{PA}}{I_{PA} I_{PB}}\right) \cdot d_B \pi d_p^{\ 2}\right]$

(c) $\beta_{max} = \tau_{t,allow}\left(\dfrac{2L}{G d_A}\right)\left(\dfrac{I_{PA} + I_{PB}}{I_{PB}}\right)$

$$\beta_{max} = \tau_{t,allow}\left(\frac{2L}{Gd_B}\right)\left(\frac{I_{PA} + I_{PB}}{I_{PA}}\right)$$

$$(d)\ \beta_{max} = \sigma_{b,allow}\frac{L}{G}\left[\frac{(I_{PB} + I_{PA})(d_A - t_A)\cdot d_P t_A}{I_{PA}I_{PB}}\right]$$

$$\beta_{max} = \sigma_{b,allow}\frac{L}{G}\left[\frac{(I_{PB} + I_{PA})(d_B - t_B)\cdot d_P t_B}{I_{PA}I_{PB}}\right]$$

CHAPTER 4

4.3-1 $V = 333$ lb, $M = 50667$ lb-in

4.3-2 $V = -0.93$ kN, $M = 4.12$ kN·m

4.3-3 $V = 0, M = 0$

4.3-4 $V = 7.0$ kN, $M = -9.5$ kN·m

4.3-5 $V = -1810$ lb, $M = -12580$ lb-ft

4.3-6 $V = -1.0$ kN, $M = -7.0$ kN·m

4.3-7 $b/L = 1/2$

4.3-8 $M = 108$ N·m

4.3-9 $N = P \sin\theta, V = P \cos\theta, M = Pr \sin\theta$

4.3-10 $V = -6.04$ kN, $M = 15.45$ kN·m

4.3-11 $P = 1200$ lb

4.3-12 $V = -4.17$ kN, $M = 75$ kN·m

4.3-13 (a) $V_B = 6,000$ lb, $M_B = 9,000$ lb-ft;
(b) $V_m = 0, M_m = 21,000$ lb-ft

4.3-14 $N = 21.6$ kN (compression), $V = 7.2$ kN,
$M = 50.4$ kN·m

4.3-15 $V_{max} = 91wL^2\alpha/30g, M_{max} = 229wL^3\alpha/75g$

4.5-1 $V_{max} = P, M_{max} = Pa$

4.5-2 $V_{max} = M_0/L, M_{max} = M_0 a/L$

4.5-3 $V_{max} = qL/2, M_{max} = -3qL^2/8$

4.5-4 $V_{max} = P, M_{max} = PL/4$

4.5-5 $V_{max} = -2P/3, M_{max} = 2PL/9$

4.5-6 $V_{max} = 2M_1/L, M_{max} = 7\ M_1/3$

4.5-7 $V_{max} = P/2, M_{max} = 3PL/8$

4.5-8 $V_{max} = P, M_{max} = -Pa$

4.5-9 $V_{max} = qL/2, M_{max} = 5qL^2/72$

4.5-10 $V_{max} = -q_0L/2, M_{max} = -q_0L^2/6$

4.5-11 $R_B = 207$ lb, $R_A = 73.3$ lb
$V_{max} = -207$ lb, $M_{max} = 2933$ lb-in

4.5-12 $V_{max} = 1200$ N, $M_{max} = 960$ N·m

4.5-13 $V_{max} = 200$ lb, $M_{max} = -1600$ lb-ft

4.5-14 $V_{max} = 4.5$ kN, $M_{max} = -11.33$ kN·m

4.5-15 $V_{max} = -1300$ lb, $M_{max} = -28,800$ lb-in.

4.5-16 $V_{max} = 15.34$ kN, $M_{max} = 9.80$ kN·m

4.5-17 The first case has the larger maximum moment
$$\left(\frac{6}{5}PL\right)$$

4.5-18 The third case has the larger maximum moment
$$\left(\frac{6}{5}PL\right)$$

4.5-19 $V_{max} = 900$ lb, $M_{max} = -900$ lb-ft

4.5-20 $V_{max} = -10.0$ kN, $M_{max} = 16.0$ kN·m

4.5-21 Two cases have the same maximum
moment (PL)

4.5-22 $V_{max} = 33.0$ kN, $M_{max} = -61.2$ kN·m

4.5-23 $V_{max} = -800$ lb, $M_{max} = 4800$ lb-ft

4.5-24 $M_{Az} = -PL$ (clockwise), $A_x = 0, A_y = 0$

$$C_y = \frac{1}{12}P\ (\text{upward}),\ D_y = \frac{1}{6}P\ (\text{upward})$$

$V_{max} = P/6, M_{max} = PL$

4.5-25 $V_{max} = -13.75$ kip, $M_{max} = 47.3$ ft-k

4.5-26 $V_{max} = 4.6$ kN, $M_{max} = -6.24$ kN·m

4.5-27 $V_{max} = -433$ lb, $M_{max} = 776.47$ lb-ft

4.5-28 $V_{max} = -2.8$ kN, $M_{max} = 1.450$ kN·m

4.5-29 $a = 0.5858L, V_{max} = 0.2929qL,$
$M_{max} = 0.02145qL^2$

4.5-30 $V_{max} = 2.5$ kN, $M_{max} = 5.0$ kN·m

4.5-31 $M_A = -q_0L^2/6$ (clockwise),
$A_x = 0, B_y = q_0L/2$ (upward)
$V_{max} = -q_0L/2, M_{max} = q_0L^2/6$

4.5-32 $M_{max} = 12$ kN·m

4.5-33 $M_{max} = M_{pos} = 2448$ lb-ft,
$M_{neg} = -2160$ lb-ft

4.5-34 $V_{max} = -w_0L/3, M_{max} = -w_0L^2/12$

4.5-35 $M_A = -w_0L^2/30$ in FBD; moment at A in moment
diagram $= +w_0L^2/30$
$A_x = -3\ w_0L/10$ (leftward)
$A_y = -3\ w_0L/20$ (downward)
$C_y = w_0L/12$ (upward)
$D_y = w_0L/6$ (upward)
$V_{max} = w_0L/4, M_{max} = -w_0L^2/24$ at B

4.5-36 (a) $x = 9.6$ m, $V_{max} = 28$ kN;
(b) $x = 4.0$ m, $M_{max} = 78.4$ kN·m

4.5-37 $A_x = 50.38$ lb (right)
$A_y = 210$ lb (upward)
$B_x = -50.38$ lb (left)
$N_{max} = -214.8$ lb, $V_{max} = -47.5$ lb,
$M_{max} = 270$ lb-ft

4.5-38 (a) $A_x = -q_0L/2$ (leftward)
$A_y = 17q_0L/18$ (upward)
$D_x = -q_0L/2$ (leftward)
$D_y = -4q_0L/9$ (downward)
$M_D = 0$
$N_{max} = q_0L/2$, $V_{max} = 17q_0L/18$,
$M_{max} = q_0L^2$
(b) $B_x = q_0L/2$ (rightward)
$B_y = -q_0L/2 + 5q_0L/3 = 7q_0L/6$ (upward)
$D_x = q_0L/2$ (rightward)
$D_y = -5q_0L/3$ (downward)
$M_D = 0$
$N_{max} = 5q_0L/3$, $V_{max} = 5q_0L/3$, $M_{max} = q_0L^2$

4.5-39 $M_A = 0$
$R_{Ay} = q_0L/6$ (upward)
$R_{Cy} = q_0L/3$ (upward)
$R_{Ax} = 0$
$N_{max} = -3w_0L/20$, $V_{max} = -w_0L/3$,
$M_{max} = 8w_0L^2/125$

4.5-40 $M_A = 0$, $A_x = 0$
$A_y = -18.41$ kN (downward)
$M_D = 0$
$D_x = -63.0$ kN (leftward)
$D_y = -61.2$ kN (upward)
$N_{max} = -61.2$ kN, $V_{max} = 63.0$ kN,
$M_{max} = 756$ kN·m

CHAPTER 5

5.4-1 $\epsilon_{max} = 1300 \times 10^{-6}$
5.4-2 $L_{min} = 3.93$ m
5.4-3 $\epsilon_{max} = 6400 \times 10^{-6}$
5.4-4 $\rho = 68.8$ m; $\kappa = 1.455 \times 10^{-5}$ m^{-1}; $\delta = 29.1$ mm
5.4-5 $\epsilon = 255 \times 10^{-6}$
5.4-6 $\epsilon = 640 \times 10^{-6}$
5.5-1 (a) $\sigma_{max} = 52.4$ ksi; (b) σ_{max} increases 33%
5.5-2 (a) $\sigma_{max} = 250$ MPa; (b) σ_{max} decreases 20%
5.5-3 (a) $\sigma_{max} = 38.2$ ksi; (b) σ_{max} increases 10%
5.5-4 (a) $\sigma_{max} = 8.63$ MPa; (b) $\sigma_{max} = 6.49$ MPa
5.5-5 $\sigma_{max} = 21.6$ ksi
5.5-6 $\sigma_{max} = 203$ MPa
5.5-7 $\sigma_{max} = 3420$ psi
5.5-8 $\sigma_{max} = 101$ MPa
5.5-9 $\sigma_{max} = 9.53$ ksi
5.5-10 $\sigma_{max} = 7.0$ MPa
5.5-11 $\sigma_{max} = 432$ psi
5.5-12 $\sigma_{max} = 2.10$ MPa
5.5-13 (a) $\sigma_t = 30.93$ M/d^3; (b) $\sigma_t = 360M/(73bh^2)$;
(c) $\sigma_t = 85.24$ M/d^3

5.5-14 $\sigma_{max} = 10.965M/d^3$
5.5-15 $\sigma_{max} = 21.4$ ksi
5.5-16 $\sigma_c = 61.0$ MPa; $\sigma_t = 35.4$ MPa
5.5-17 $\sigma_c = 15,964$ psi; $\sigma_t = 4341$ psi
5.5-18 (a) $\sigma_c = 1.456$ MPa; $\sigma_t = 1.514$ MPa;
(b) $\sigma_c = 1.666$ MPa ($+14\%$); $\sigma_t = 1.381$ MPa (-9%);
(c) $\sigma_c = 0.728$ MPa (-50%); $\sigma_t = 0.757$ MPa (-50%)
5.5-19 $\sigma_t = 7810$ psi; $\sigma_c = 13,885$ psi
5.5-20 $\sigma_{max} = 3\rho L^2 a_0/t$
5.5-21 $\sigma_t = 18,509$ psi; $\sigma_c = 12,494$ psi
5.5-22 $\sigma = 25.1$ MPa, 17.8 MPa, -23.5 MPa
5.5-23 $d = 3$ ft, $\sigma_{max} = 171$ psi; $d = 6$ ft,
$\sigma_{max} = 830$ psi
5.5-24 $\sigma_t = -\sigma_c = 23$ q_0 L^2 $r/(27$ $I)$
5.5-25 (a) $F = 104.8$ lb;
(b) $\sigma_{max} = 36$ ksi
5.6-1 $d_{min} = 4.00$ in.
5.6-2 $d_{min} = 11.47$ mm
5.6-3 W 14 $\times$ 26
5.6-4 W 200 $\times$ 41.7
5.6-5 S 10 $\times$ 25.4
5.6-6 $b_{min} = 150$ mm
5.6-7 $S = 19.6$ in.^{3}; use 2 $\times$ 10 in. joists
5.6-8 $s_{max} = 450$ mm
5.6-9 $q_{0,allow} = 628$ lb/ft
5.6-10 $h_{min} = 30.6$ mm
5.6-11 (a) S_reqd $= 15.37$ in.^{3}; (b)S 8 $\times$ 23
5.6-12 $d_{min} = 31.6$ mm
5.6-13 (a) $q_{allow} = 1055$ lb/ft; (b) $q_{allow} = 282$ lb/ft
5.6-14 $b = 152$ mm, $h = 202$ mm
5.6-15 $b = 10.25$ in.
5.6-16 $t = 13.61$ mm
5.6-17 1 : 1.260 : 1.408
5.6-18 $q_{max} = 10.28$ kN/m
5.6-19 6.57%
5.6-20 (a) $b_{min} = 11.91$ mm; (b) $b_{min} = 11.92$ mm
5.6-21 $s_{max} = 72.0$ in.
5.6-22 (a) $\beta = 1/9$; (b) 5.35%
5.6-23 Increase when $d/h > 0.6861$; decrease when
$d/h < 0.6861$
5.7-2 (a) $\tau_{max} = 731$ kPa, $\sigma_{max} = 4.75$ MPa
(b) $\tau_{max} = 1462$ kPa, $\sigma_{max} = 19.01$ MPa
5.7-3 $M_{max} = 25.4$ k-ft
5.7-4 $\tau_{max} = 500$ kPa
5.7-5 $\tau_{max} = 2400$ psi
5.7-6 (a) $L_0 = h(\sigma_{allow}/\tau_{allow})$;
(b) $L_0 = (h/2)(\sigma_{allow}/\tau_{allow})$
5.7-7 $P_{allow} = 2.03$ k

5.7-8 (a) $M_{max} = 72.2$ N·m
(b) $M_{max} = 9.01$ N·m

5.7-9 (a) 8×12 in. beam
(b) 8×12 in. beam

5.7-10 (a) $P = 38.0$ kN; (b) $P = 35.6$ kN

5.7-11 (a) $w_1 = 121$ lb/ft^2; (b) $w_2 = 324$ lb/ft^2;
(c) $w_{allow} = 121$ lb/ft^2

5.7-12 (a) $b = 89.3$ mm (b) $b = 87.8$ mm

5.8-1 $d_{min} = 5.70$ in

5.8-2 (a) $W = 28.6$ kN; (b) $W = 38.7$ kN

5.8-3 (a) $d = 10.52$ in.; (b) $d = 2.56$ in.

5.8-4 (a) $d = 266$ mm; (b) $d = 64$ mm

5.9-1 (a) $\tau_{max} = 5795$ psi; (b) $\tau_{min} = 4555$ psi;
(c) $\tau_{aver} = 5714$ psi; (d) $V_{web} = 28.25$ k

5.9-2 (a) $\tau_{max} = 28.43$ MPa; (b) $\tau_{min} = 21.86$ MPa;
(c) $\tau_{aver} = 27.41$ MPa; (d) $V_{web} = 119.7$ kN

5.9-3 (a) $\tau_{max} = 4861$ psi; (b) $\tau_{min} = 4202$ psi;
(c) $\tau_{aver} = 4921$ psi; (d) $V_{web} = 9.432$ k

5.9-4 (a) $\tau_{max} = 32.28$ MPa; (b) $\tau_{min} = 21.45$ MPa;
(c) $\tau_{aver} = 29.24$ MPa; (d) $V_{web} = 196.1$ kN

5.9-5 (a) $\tau_{max} = 2634$ psi; (b) $\tau_{min} = 1993$ psi;
(c) $\tau_{aver} = 2518$ psi; (d) $V_{web} = 20.19$ k

5.9-6 (a) $\tau_{max} = 28.40$ MPa; (b) $\tau_{min} = 19.35$ MPa;
(c) $\tau_{aver} = 25.97$ MPa; (d) $V_{web} = 58.63$ kN

5.9-7 $q_{max} = 1270$ lb/ft

5.9-8 (a) $q_{max} = 184.7$ kN/m; (b) $q_{max} = 247$ kN/m

5.9-9 S 8×23

5.9-10 $V = 273$ kN

5.9-11 $\tau_{max} = 1.42$ ksi, $\tau_{min} = 1.03$ ksi

5.9-12 $\tau_{max} = 19.7$ MPa

5.9-13 $\tau_{max} = 2221$ psi

5.10-1 $\sigma_{face} = \pm1980$ psi, $\sigma_{core} = \pm531$ psi

5.10-2 (a) $M_{max} = 58.7$ kN·m; (b) $M_{max} = 90.9$ kN·m

5.10-3 (a) $M_{max} = 172$ k-in; (b) $M_{max} = 96$ k-in

5.10-4 $M_{allow} = \dfrac{\pi d^3 \sigma_s}{2592}\left(65 + 16\dfrac{E_b}{E_s}\right)$

5.10-5 (a) $\sigma_w = 666$ psi, $\sigma_s = 13897$ psi
(b) $q_{max} = 665$ lb/ft
(c) $M_{0,max} = 486$ lb-ft

5.10-6 $M_{allow} = 768$ N·m

5.10-7 (a) $\sigma_{face} = 3610$ psi, $\sigma_{core} = 4$ psi;
(b) $\sigma_{face} = 3630$ psi, $\sigma_{core} = 0$

5.10-8 (a) $\sigma_{face} = 14.1$ MPa, $\sigma_{core} = 0.21$ MPa;
(b) $\sigma_{face} = 14.9$ MPa, $\sigma_{core} = 0$

5.10-9 $\sigma_a = 4120$ psi, $\sigma_c = 5230$ psi

5.10-10 $\sigma_w = 5.1$ MPa (comp.), $\sigma_s = 37.6$ MPa (tens.)

5.10-11 (a) $\sigma_{plywood} = 1131$ psi, $\sigma_{pine} = 969$ psi
(b) $q_{max} = 95.5$ lb/ft

5.10-12 $Q_{0,max} = 15.53$ kN/m

5.10-13 (a) $M_{max} = 442$ k-in (b) $M_{max} = 189$ k-in

5.10-14 $t_{min} = 15.0$ mm

5.10-15 (a) $q_{allow} = 454$ lb/ft
(b) $\sigma_{wood} = 277$ psi, $\sigma_{steel} = 11782$ psi

5.10-16 $\sigma_s = 49.9$ MPa, $\sigma_w = 1.9$ MPa

5.10-17 $\sigma_a = 1860$ psi, $\sigma_p = 72$ psi

5.10-18 $\sigma_a = 12.14$ MPa, $\sigma_p = 0.47$ MPa

5.10-19 (a) $q_{allow} = 264$ lb/ft
(b) $q_{allow} = 280$ lb/ft

5.10-20 $\sigma_s = 93.5$ MPa

5.10-21 $M_{max} = 81.1$ k-in.

5.10-22 $S_A = 50.6$ mm^3; Metal A

5.10-23 $\sigma_s = 13,400$ psi (tens.),
$\sigma_c = 812$ psi (comp.)

5.10-24 $M_{allow} = 16.2$ kN·m

CHAPTER 6

6.2-1 For $\theta = 60°$: $\sigma_{x1} = 2910$ psi, $\tau_{x1y1} = -2012$ psi

6.2-2 For $\theta = 30°$: $\sigma_{x1} = 119.2$ MPa, $\tau_{x1y1} = 5.30$ MPa

6.2-3 For $\theta = 50°$: $\sigma_{x1} = -1243$ psi,
$\tau_{x1y1} = 1240$ psi

6.2-4 For $\theta = 52°$: $\sigma_{x1} = -136.6$ MPa,
$\tau_{x1y1} = -840$, $\sigma_{y1} = 16.6$ MPa

6.2-5 For $\theta = 30°$: $\sigma_{x1} = -3041$ psi, $\tau_{x1y1} = -12725$ psi

6.2-6 For $\theta = -35°$: $\sigma_{x1} = -6.4$ MPa,
$\tau_{x1y1} = -18.9$ MPa

6.2-7 For $\theta = 40°$: $\sigma_{x1} = -13032$ psi, $\tau_{x1y1} = 4954$ psi

6.2-8 For $\theta = -42.5°$: $\sigma_{x1} = -51.9$ MPa,
$\tau_{x1y1} = -14.6$ MPa

6.2-9 Normal stress on seam, 187 psi tension. Shear
stress, 163 psi clockwise.

6.2-10 Normal stress on seam, 1440 kPa tension. Shear
stress, 1030 kPa clockwise.

6.2-11 $\sigma_w = -125$ psi, $\tau_w = 375$ psi

6.2-12 $\sigma_w = 10.0$ MPa, $\tau_w = -5.0$ MPa

6.2-13 $\theta = 56.31°$

6.2-14 $\theta = 38.66°$

6.2-15 $\sigma_x = -12813$ psi, $\sigma_y = -6037$ psi,
$\tau_{xy} = -4962$ psi

6.2-16 $\sigma_x = 56.5$ MPa, $\sigma_y = -18.3$ MPa,
$\tau_{xy} = -32.6$ MPa

6.2-17 $\sigma_y = 3805$ psi, $\tau_{xy} = 2205$ psi

6.2-18 $\sigma_y = -60.7$ MPa, $\tau_{xy} = -27.9$ MPa

6.2-19 $\sigma_b = -3700$ psi, $\tau_b = 3282$ psi, $\theta_1 = 43.7°$

6.3-1 $\sigma_1 = 4988$ psi, $\theta_{p1} = 14.08°$

6.3-2 $\sigma_1 = 120$ MPa, $\theta_{p1} = 35.2°$

6.3-3 $\sigma_1 = -977$ psi, $\theta_{p1} = 62.1°$

6.3-4 $\sigma_1 = 53.6$ MPa, $\theta_{p1} = -14.2°$

6.3-5 $\tau_{max} = 13065$ psi, $\theta_{s1} = -53.4°$

6.3-6 $\tau_{max} = 19.3$ MPa, $\theta_{s1} = 61.4°$

6.3-7 $\tau_{max} = 6851$ psi, $\theta_{s1} = 61.8°$

6.3-8 $\tau_{max} = 26.7$ MPa, $\theta_{s1} = 19.08°$

6.3-9 (a) $\sigma_1 = 180$ psi, $\theta_{p_1} = -20.56°$;
(b) $\tau_{max} = 730$ psi, $\theta_{s_1} = -65.56°$

6.3-10 (a) $\sigma_1 = 27.8$ MPa, $\theta_{p1} = 116.4°$;
(b) $\tau_{max} = 70.3$ MPa, $\theta_{s1} = 71.4°$

6.3-11 (a) $\sigma_1 = 2925$ psi, $\theta_{p1} = -25.3°$;
(b) $\tau_{max} = 1165$ psi, $\theta_{s1} = -70.3°$

6.3-12 (a) $\sigma_1 = 2262$ kPa, $\theta_{p1} = -13.70°$;
(b) $\tau_{max} = 1000$ kPa, $\theta_{s1} = -58.7°$

6.3-13 (a) $\sigma_1 = 14764$ psi, $\theta_{p1} = 7.90°$;
(b) $\tau_{max} = 6979$ psi, $\theta_{s1} = -37.1°$

6.3-14 (a) $\sigma_1 = 29.2$ MPa, $\theta_{p1} = -17.98°$;
(b) $\tau_{max} = 66.4$ MPa, $\theta_{s1} = -63.0°$

6.3-15 (a) $\sigma_1 = -1228$ psi, $\theta_{p1} = 24.7°$;
(b) $\tau_{max} = 5922$ psi, $\theta_{s1} = -20.3°$

6.3-16 (a) $\sigma_1 = 76.3$ MPa, $\theta_{p1} = 107.5°$;
(b) $\tau_{max} = 101.3$ MPa, $\theta_{s1} = 62.5°$

6.3-17 2771 psi $\leq \sigma_y \leq 9029$ psi

6.3-18 18.7 MPa $\leq \sigma_y \leq 65.3$ MPa

6.3-19 (a) $\sigma_y = 1410$ psi;
(b) $\sigma_1 = 6700$ psi, $\theta_{p1} = -23.5°$

6.3-20 (a) $\sigma_y = 11.7$ MPa;
(b) $\sigma_1 = 33.0$ MPa, $\theta_{p1} = 63.2°$

6.4-1 (a) For $\theta = 24°$: $\sigma_{x1} = 9493$ psi,
$\tau_{x1y1} = -4227$ psi;
(b) $\tau_{max} = 5688$ psi, $\theta_{s1} = -45.0°$

6.4-2 (a) For $\theta = -27°$: $\sigma_{x1} = 38.9$ MPa,
$\tau_{x1y1} = 19.8$ MPa;
(b) $\tau_{max} = 24.5$ MPa, $\theta_{s1} = -45.0°$

6.4-3 (a) For $\theta = 26.57°$: $\sigma_{x1} = -4880$ psi,
$\tau_{x1y1} = 2440$ psi;
(b) $\tau_{max} = 3050$ psi, $\theta_{s1} = 45.0°$

6.4-4 (a) For $\theta = 25°$: $\sigma_{x1} = -36.0$ MPa,
$\tau_{x1y1} = 25.7$ MPa;
(b) $\tau_{max} = 33.5$ MPa, $\theta_{s1} = 45.0°$

6.4-5 (a) For $\theta = 55°$: $\sigma_{x1} = 250$ psi,
$\tau_{x1y1} = -3464$ psi;
(b) $\tau_{max} = 4000$ psi, $\theta_{s1} = -45.0°$

6.4-6 (a) For $\theta = 21.80°$:
$\sigma_{x1} = -17.1$ MPa,
$\tau_{x1y1} = 29.7$ MPa;
(b) $\tau_{max} = 43.0$ MPa, $\theta_{s1} = 45.0°$

6.4-7 (a) For $\theta = 52°$: $\sigma_{x1} = 2620$ psi,
$\tau_{x1y1} = -653$ psi;
(b) $\sigma_1 = 2700$ psi, $\theta_{p1} = 45.0°$

6.4-8 (a) For $\theta = 22.5°$: $\sigma_{x1} = -10.25$ MPa,
$\tau_{x1y1} = -10.25$ MPa;
(b) $\sigma_1 = 14.50$ MPa, $\theta_{p1} = 135.0°$

6.4-9 (a) For $\theta = 36.87°$: $\sigma_{x1} = 3600$ psi,
$\tau_{x1y1} = 1050$ psi;
(b) $\sigma_1 = 3750$ psi, $\theta_{p1} = 45.0°$

6.4-10 For $\theta = 40°$: $\sigma_{x1} = 27.5$ MPa,
$\tau_{x1y1} = -5.36$ MPa

6.4-11 For $\theta = -51°$: $\sigma_{x1} = 11982$ psi,
$\tau_{x1y1} = -3569$ psi

6.4-12 For $\theta = -33°$: $\sigma_{x1} = -61.7$ MPa,
$\tau_{x1y1} = -51.7$ MPa, $\sigma_{y1} = -171.3$ MPa

6.4-13 For $\theta = 14°$: $\sigma_{x1} = -1509$ psi,
$\tau_{x1y1} = 527$ psi

6.4-14 For $\theta = 35°$: $\sigma_{x1} = 46.4$ MPa,
$\tau_{x1y1} = -9.81$ MPa

6.4-15 For $\theta = 65°$: $\sigma_{x1} = -1846$ psi, $\tau_{x1y1} = 3897$ psi

6.4-16 (a) $\sigma_1 = 40.0$ MPa, $\theta_{p1} = 68.8°$;
(b) $\tau_{max} = 40.0$ MPa, $\theta_{s1} = 23.8°$

6.4-17 (a) $\sigma_1 = 7525$ psi, $\theta_{p1} = 9.80°$;
(b) $\tau_{max} = 3875$ psi, $\theta_{s1} = -35.2°$

6.4-18 (a) $\sigma_1 = 3.43$ MPa, $\theta_{p1} = -19.68°$;
(b) $\tau_{max} = 15.13$ MPa, $\theta_{s1} = -64.7°$

6.4-19 (a) $\sigma_1 = 7490$ psi, $\theta_{p1} = 63.2°$;
(b) $\tau_{max} = 3415$ psi, $\theta_{s1} = -18.20°$

6.4-20 (a) $\sigma_1 = 10865$ kPa, $\theta_{p1} = 115.2°$;
(b) $\tau_{max} = 4865$ kPa, $\theta_{s1} = 70.2°$

6.4-21 (a) $\sigma_1 = -6923$ psi, $\theta_{p1} = -32.4°$;
(b) $\tau_{max} = 7952$ psi, $\theta_{s1} = 102.6°$

6.4-22 (a) $\sigma_1 = 18.2$ MPa, $\theta_{p1} = 123.3°$;
(b) $\tau_{max} = 15.4$ MPa, $\theta_{s1} = 78.3°$

6.4-23 (a) $\sigma_1 = 2565$ psi, $\theta_{p1} = 31.3°$;
(b) $\tau_{max} = 3265$ psi, $\theta_{s1} = -13.70°$

6.5-1 $\sigma_x = 26,040$ psi, $\sigma_y = -13,190$ psi,
$\Delta t = -32.1 \times 10^{-6}$ in. (decrease)

6.5-2 $\sigma_x = 114.1$ MPa, $\sigma_y = 60.2$ MPa,
$\Delta t = -2610 \times 10^{-6}$ mm (decrease)

6.5-3 (a) $\epsilon_z = -\nu(\epsilon_x + \epsilon_y)/(1 - \nu)$;
(b) $e = (1 - 2\nu)(\epsilon_x + \epsilon_y)/(1 - \nu)$

6.5-4 $\nu = 0.35$, $E = 45$ GPa

6.5-5 $\nu = 1/3$, $E = 30 \times 10^6$ psi

6.5-6 (a) $\gamma_{max} = 715 \times 10^{-6}$;
(b) $\Delta t = -2100 \times 10^{-6}$ mm (decrease);
(c) $\Delta V = 896$ mm^3 (increase)

6.5-7 (a) $\gamma_{max} = 1900 \times 10^{-6}$;
(b) $\Delta t = -141 \times 10^{-6}$ in. (decrease);
(c) $\Delta V = 0.0874$ in.^{3} (increase)

6.5-8 $\Delta V = -56$ mm^3 (decrease)

6.5-9 $\Delta V = -0.0603$ in.^{3} (decrease)

6.5-10 $\Delta V = 2640$ mm^3 (increase)

6.5-11 $\Delta V = 0.0423$ in.^{3} (increase)

6.5-12 (a) $\Delta ac = 0.0745$ mm (increase);
(b) $\Delta bd = -0.000560$ mm (decrease);
(c) $\Delta t = -0.00381$ mm (decrease);
(d) $\Delta V = 573$ mm^3 (increase);

6.6-1 (a) $\tau_{max} = 8000$ psi;
(b) $\Delta a = 0.0079$ in. (increase),
$\Delta b = -0.0029$ in. (decrease),
$\Delta c = -0.0011$ in. (decrease);
(c) $\Delta V = 0.0165$ in.^{3} (increase);

6.6-2 (a) $\tau_{max} = 10.0$ MPa;
(b) $\Delta a = -0.0540$ mm (decrease),
$\Delta b = -0.0075$ mm (decrease),
$\Delta c = -0.0075$ mm (decrease);
(c) $\Delta V = -1890$ mm^3 (decrease);

6.6-3 (a) $\sigma_x = -4200$ psi, $\sigma_y = \sigma_z = -2100$ psi;
(b) $\tau_{max} = 1050$ psi;
(c) $\Delta V = -0.0192$ in.^{3} (decrease);

6.6-4 (a) $\sigma_x = -64.8$ MPa, $\sigma_y = \sigma_z = -43.2$ MPa;
(b) $\tau_{max} = 10.8$ MPa;
(c) $\Delta V = -532$ mm^3 (decrease);

6.6-5 $K = 10.0 \times 10^6$ psi

6.6-6 $K = 5.0$ GPa

6.6-7 (a) $p = \nu F/[A(1 - \nu)]$;
(b) $\delta = FL(1 + \nu)(1 - 2\nu)/[EA(1 - \nu)]$

6.6-8 (a) $p = \nu p_0$; (b) $e = -p_0(1 + \nu)(1 - 2\nu)/E$;

6.6-9 $\Delta d = 0.00104$ in. (decrease);
$\Delta V = 0.198$ in.^{3} (decrease)

6.6-10 (a) $p = 700$ MPa; (b) $K = 175$ GPa;

6.6-11 $\epsilon_0 = 276 \times 10^{-6}$, $e = 828 \times 10^{-6}$

CHAPTER 7

7.2-1 $t = 2.48$ in, $t_{min} = 2.5$ in

7.2-2 $t = 93.8$ mm, $t_{min} = 94$ mm

7.2-3 $F = 684$ lb, $\sigma = 280$ psi

7.2-4 $\sigma_{max} = 2.88$ MPa, $\epsilon_{max} = 0.452$

7.2-5 $\sigma_{max} = 405$ psi, $\epsilon_{max} = 0.446$

7.2-6 $p = 2.93$ MPa

7.2-7 (a) $f = 26.4$ k/in
(b) $\tau_{max} = 7543$ psi
(c) $\epsilon_{max} = 3.57 \times 10^{-4}$

7.2-8 (a) $f = 5.5$ MN/m
(b) $\tau_{max} = 57.3$ MPa
(c) $\epsilon_{max} = 3.87 \times 10^{-4}$

7.2-9 $t_{min} = 0.294$ in

7.2-10 $t_{min} = 6.69$ mm

7.2-11 $D_0 = 90$ ft

7.3-1 $t_{min} = 0.350$ in.

7.3-2 (a) $h = 22.2$ m
(b) zero

7.3-3 $n = 2.25$

7.3-4 $F = 3\pi pr^2$

7.3-5 $p = 50$ psi

7.3-6 $\epsilon_{max} = 6.56 \times 10^{-5}$

7.3-7 $t_{min} = 0.113$ in.

7.3-8 $t_{min} = 3.71$ mm

7.3-9 (a) $h = 25$ ft; (b) $\sigma_1 \approx 125$ psi

7.3-10 (a) $\sigma_h = 24.9$ MPa
(b) $\sigma_c = 49.7$ MPa
(c) $\sigma_w = 24.9$ MPa
(d) $\tau_h = 12.43$ MPa
(e) $\tau_c = 24.9$ MPa

7.3-11 (a) $t_{min} = 0.675$ in (b) $t_{min} = 0.338$ in

7.3-12 (a) $\sigma_1 = 93.3$ MPa, $\sigma_2 = 46.7$ MPa
(b) $\tau_1 = 23.2$ MPa, $\tau_2 = 46.7$ MPa
(c) $\epsilon_1 = 3.97 \times 10^{-4}$, $\epsilon_2 = 9.33 \times 10^{-5}$
(d) $\theta = 35°$, $\sigma_{x_1} = 62.0$ MPa,
$\sigma_{y_1} = 77.0$ MPa, $\tau_{x_1 y_1} = 21.9$ MPa

7.3-13 (a) $\sigma_1 = 7015$ psi, $\sigma_2 = 3508$ psi
(b) $\tau_1 = 1754$ psi, $\tau_2 = 3508$ psi
(c) $\epsilon_1 = 1.988 \times 10^{-4}$, $\epsilon_2 = 4.68 \times 10^{-5}$
(d) $\theta = 28°$, $\sigma_{x1} = 4281$ psi, $\sigma_{y1} = 6242$ psi,
$\tau_{x_1 y_1} = 1454$ psi

7.4-1 $\sigma_t = 5100$ psi, $\sigma_c = -5456$ psi, $\tau_{max} = 2728$ psi

7.4-2 $d_{min} = 48.4$ mm

7.4-3 $\sigma_t = 3963$ psi, $\sigma_c = -8791$ psi, $\tau_{max} = 6377$ psi

7.4-4 $\sigma_t = 16.93$ MPa, $\sigma_c = -41.4$ MPa,
$\tau_{max} = 28.9$ MPa

7.4-5 $P = 194.2$ k

7.4-6 $p_{max} = 9.60$ MPa

7.4-7 $t_{min} = 0.125$ in.

7.4-8 $\phi_{max} = 0.552$ rad $= 31.6°$

7.4-9 $\sigma_t = 39,950$ psi, $\sigma_c = -2226$ psi,
$\tau_{max} = 21,090$ psi

7.4-10 (a) $\sigma_{max} = 56.4$ MPa, $\tau_{max} = 18.9$ MPa;
(b) $T_{max} = 231$ kN·m

7.4-11 $\sigma_t = 4320$ psi, $\sigma_c = -1870$ psi, $\tau_{max} = 3100$ psi

7.4-12 $\sigma_t = 29.15\ qR^2/d^3$, $\sigma_c = -8.78\ qR^2/d^3$,
$\tau_{max} = 18.97\ qR^2/d^3$

7.4-13 $d = 1.50$ in.

7.4-14 $P = 34.1$ kN

7.4-15 (a) $\sigma_{max} = 4534$ psi, $\tau_{max} = 2289$ psi
(b) $P_{allow} = 629$ lb

7.4-16 $\tau_A = 76.0$ MPa, $\tau_B = 19.94$ MPa,
$\tau_C = 23.7$ MPa

7.4-17 (a) $\sigma_x = 0$ psi, $\sigma_y = 6145$ psi, $\tau_{xy} = 345$ psi
(b) $\sigma_1 = 6164$ psi, $\sigma_2 = -19.30$ psi,
$\tau_{max} = 3092$ psi

7.4-18 Pure shear $\tau_{max} = 0.804$ MPa

7.4-19 $\sigma_t = 10{,}680$ psi; No compressive stresses;
$\tau_{max} = 5{,}340$ psi

7.4-20 (a) $\sigma_1 = 31.2$ MPa, $\sigma_2 = -187.2$ MPa,
$\tau_{max} = 109.2$ MPa
(b) $\sigma_1 = 178.7$ MPa, $\sigma_2 = -29.1$ MPa,
$\tau_{max} = 103.9$ MPa

7.4-21 (a) $\sigma_1 = 0$ psi, $\sigma_2 = -20{,}730$ psi, $\tau_{max} = 10{,}365$ psi
(b) $\sigma_1 = 988$ psi, $\sigma_2 = -21{,}719$ psi,
$\tau_{max} = 11{,}354$ psi

7.4-22 Maximum $\sigma_t = 18.35$ MPa, $\sigma_c = -18.35$ MPa,
$\tau_{max} = 9.42$ MPa

7.4-23 Top of beam
$\sigma_1 = 8591$ psi, $\sigma_2 = 0$ psi,
$\tau_{max} = 4295$ psi

CHAPTER 8

8.2-1 $q = q_0 x/L$; Triangular load, acting downward

8.2-2 (a) $q = q_0 \sin \pi x/L$, Sinusoidal load;
(b) $R_A = R_B = q_0 L/\pi$; (c) $M_{max} = q_0 L^2/\pi^2$

8.2-3 $q = q_0(1 - x/L)$; Triangular load, acting
downward

8.2-4 (a) $q = q_0(L^2 - x^2)/L^2$; Parabolic load, acting
downward;
(b) $R_A = 2q_0 L/3$; $M_A = -q_0 L^2/4$

8.3-1 $\delta_{max} = 0.182$ in., $\theta = 0.199°$

8.3-2 $h = 96$ mm

8.3-3 $L = 120$ in. $= 10$ ft

8.3-4 $\delta_{max} = 15.4$ mm

8.3-5 $\delta/L = 1/400$

8.3-6 $E_g = 80.0$ GPa

8.3-7 Let $\beta = a/L$: $\dfrac{\delta_C}{\delta_{max}} = \dfrac{3\sqrt{3}(-1 + 8\beta - 4\beta^2)}{16(2\beta - \beta^2)^{3/2}}$
The deflection at the midpoint is close to the
maximum deflection. The maximum difference
is only 2.6%.

8.3-11 $v = -mx^2(3L - x)/6EI$, $\delta_B = mL^3/3EI$,
$\theta_B = mL^2/2EI$

8.3-12 $v(x) = -\dfrac{q}{48EI}\left(2x^4 - 12x^2 L^2 + 11L^4\right)$
$\delta_B = \dfrac{qL^4}{48EI}$

8.3-15 $v(x) = \dfrac{q_0 L}{24EI}\left(x^3 - 2Lx^2\right)$ for $0 \le x \le \dfrac{L}{2}$
$v(x) = \dfrac{-q_0}{960LEI}$
$\left(-160L^2 x^3 + 160L^3 x^2 + 80Lx^4 - 16x^5\right.$
$\left. - 25L^4 x + 3L^5\right)$ for $\dfrac{L}{2} \le x \le L$
$\delta_B = \dfrac{7}{160}\dfrac{q_0 L^4}{EI}$; $\delta_C = \dfrac{1}{64}\dfrac{q_0 L^4}{EI}$

8.3-16 $v(x) = \dfrac{q_0 x}{5760LEI}\left(200x^2 L^2 - 240x^3 L\right.$
$\left. + 96x^4 - 53L^4\right)$ for $0 \le x \le \dfrac{L}{2}$
$v(x) = \dfrac{-q_0 L}{5760EI}\left(40x^3 - 120Lx^2 + 83L^2 x - 3L^3\right)$
for $\dfrac{L}{2} \le x \le L$
$\delta_C = \dfrac{3q_0 L^4}{1280EI}$

8.3-17 $v(x) = -\dfrac{PL}{10368EI}\left(-4104x^2 + 3565L^2\right)$
for $0 \le x \le \dfrac{L}{3}$
$v(x) = -\dfrac{P}{1152EI}\left(-648Lx^2 + 192x^3\right.$
$\left. + 64L^2 x + 389L^3\right)$ for $\dfrac{L}{3} \le x \le \dfrac{L}{2}$
$v(x) = -\dfrac{P}{144EIL}\left(-72L^2 x^2 + 12Lx^3 + 6x^4\right.$
$\left. + 5L^3 x + 49L^4\right)$ for $\dfrac{L}{2} \le x \le L$
$\delta_A = \dfrac{3565PL^3}{10368EI}$; $\delta_C = \dfrac{3109PL^3}{10368EI}$

8.4-3 $v = -M_0 x(L - x)^2/2LEI$;
$\delta_{max} = 2M_0 L^2/27EI$ (downward)

8.4-4 $v(x) = -\dfrac{q}{48EI}\left(2x^4 - 12x^2L^2 + 11L^4\right)$

$\theta_B = -\dfrac{qL^3}{3EI}$

8.4-6 $v = -q_0x^2(45L^4 - 40L^3x + 15L^2x^2 - x^4)/360L^2EI$;

$\delta_B = 19q_0L^4/360EI;\ \theta_B = q_0L^3/15EI$

8.4-7 $v = -q_0x(3L^5 - 5L^3x^2 + 3Lx^4 - x^5)/90L^2EI$;

$\delta_{max} = 61q_0L^4/5760EI$

8.4-8 $v(x) = \dfrac{q_0}{120EIL}\left(x^5 - 5Lx^4 + 20L^3x^2 - 16L^5\right)$

$\delta_{max} = \dfrac{2q_0L^4}{15EI}$

8.4-9 $v(x) = -\dfrac{qL^2}{16EI}\left(x^2 - L^2\right)$ for $0 \le x \le L$

$v(x) = -\dfrac{q}{48EI}\left(-20L^3x + 27L^2x^2 - 12Lx^3\right.$

$\left. + 2x^4 + 3L^4\right)$ for $L \le x \le \dfrac{3L}{2}$

$\delta_C = \dfrac{9qL^4}{128EI};\ \ \theta_C = \dfrac{7qL^3}{48EI}$

8.4-10 $v(x) = -\dfrac{q_0L^2}{480EI}\left(-20x^2 + 19L^2\right)$ for $0 \le x \le \dfrac{L}{2}$

$v(x) = -\dfrac{q_0}{960EIL}\left(80Lx^4 - 16x^5 - 120L^2x^3\right.$

$\left. + 40L^3x^2 - 25L^4x + 41L^5\right)$

for $\dfrac{L}{2} \le x \le L$

$\delta_A = \dfrac{19q_0L^4}{480EI};\ \theta_B = -\dfrac{13q_0L^3}{192EI};\ \delta_C = \dfrac{7q_0L^4}{240EI}$

8.5-1 $\theta_B = 7PL^2/9EI;\ \delta_B = 5PL^3/9EI$

8.5-2 (a) $\delta_1 = 11PL^3/144EI$; (b) $\delta_2 = 25PL^3/384EI$;
(c) $\delta_1/\delta_2 = 88/75 = 1.173$

8.5-3 (a) $a/L = 2/3$; (b) $a/L = 1/2$

8.5-4 (a) $\delta_c = 6.25$ mm (upward)
(b) $\delta_c = 18.36$ mm (downward)

8.5-5 $y = Px^2(L - x)^2/3LEI$

8.5-6 $\theta_B = 7qL^3/162EI;\ \delta_B = 23qL^4/648EI$

8.5-7 $\delta_C = 0.0905$ in., $\delta_B = 0.293$ in.

8.5-8 (a) $\delta_A = PL^2(10L - 9a)/324EI$ (positive upward);
(b) Upward when $a/L < 10/9$, downward when $a/L > 10/9$

8.5-9 (a) $\delta_C = PH^2(L + H)/3EI$;
(b) $\delta_{max} = PHL^2/9\sqrt{3}EI$

8.5-10 $\delta_C = 3.5$ mm

8.5-11 $\theta_B = q_0L^3/10EI,\ \delta_B = 13q_0L^4/180EI$

8.5-12 $\theta_A = q(L^3 - 6La^2 + 4a^3)/24EI$;
$\delta_{max} = q(5L^4 - 24L^2a^2 + 16a^4)/384EI$

8.5-13 (a) $P/Q = 9a/4L$
(b) $P/Q = 8a(3L + a)/9L^2$
(c) $P/qa = 9a/8L$ for $\delta_B = 0$;
$P/qa = a(4L + a)/3L^2$ for $\delta_D = 0$

8.5-14 $\delta = 19WL^3/31,104EI$

8.5-15 $k = 3.33$ lb/in

8.5-16 $M_1 = 7800$ N·m, $M_2 = 4200$ N·m

8.5-17 $= \dfrac{6Pb^3}{EI}$

8.5-18 $\delta_E = \dfrac{47Pb^3}{12EI}$

8.5-19 $\delta_C = 0.120$ in.

8.5-20 $q = 16cEI/7L^4$

8.5-21 $\delta_h = Pcb^2/2EI,\ \delta_v = Pc^2(c + 3b)/3EI$

8.5-22 $\delta = PL^2(2L + 3a)/3EI$

8.5-23 (a) $b/L = 0.403$; (b) $\delta_C = 0.00287qL^4/EI$

8.5-24 $\alpha = 22.5°,\ 112.5°,\ -67.5°,$ or $-157.5°$

CHAPTER 9

9.2-1 $P_{cr} = \beta_R/L$

9.2-2 (a) $P_{cr} = \dfrac{\beta a^2 + \beta_R}{L}$ (b) $P_{cr} = \dfrac{\beta a^2 + 2\beta_R}{L}$

9.2-3 $P_{cr} = 6\beta_R/L$

9.2-4 (a) $P_{cr} = \dfrac{(L - a)(\beta a^2 + \beta_R)}{aL}$ (b) $P_{cr} = \dfrac{\beta L^2 + 20\beta_R}{4L}$

9.2-5 $P_{cr} = \dfrac{3\beta_R}{L}$

9.2-6 $P_{cr} = \dfrac{3}{5}\beta L$

9.2-7 $P_{cr} = \dfrac{7}{4}\beta L$

9.3-1 (a) $P_{cr} = 453$ k; (b) $P_{cr} = 152$ k

9.3-2 (a) $P_{cr} = 2803$ kN; (b) $P_{cr} = 953$ kN

9.3-3 (a) $P_{cr} = 650$ k; (b) $P_{cr} = 140$ k

9.3-4 (a) $M_{allow} = 1143$ kN·m

9.3-5 (a) $Q_{allow} = 23.8$ k

9.3-6 (a) $Q_{cr} = \dfrac{\pi^2EI}{L^2}$

(b) $Q_{cr} = \dfrac{2\pi^2EI}{9L^2}$

9.3-7 (a) $Q_{cr} = \dfrac{2\pi^2 EI}{L^2}$

(b) $M_{cr} = \dfrac{3d\pi^2 EI}{L^2}$

9.3-8 $\Delta T = \pi^2 I/\alpha A L^2$

9.3-9 $h/b = 2$

9.3-10 (a) $P_{cr} = 3\pi^3 E r^4/4L^2$; (b) $P_{cr} = 11\pi^3 E r^4/4L^2$

9.3-11 $P_1 : P_2 : P_3 = 1.000 : 1.047 : 1.209$

9.3-12 $P_{allow} = 604$ kN

9.3-13 $F_{allow} = 54.4$ k

9.3-14 $W_{max} = 124$ kN

9.3-15 $t_{min} = 0.165$ in

9.3-16 $P_{cr} = 497$ kN

9.3-17 $W_{cr} = 51.9$ k

9.3-18 $\theta = \arctan 0.5 = 26.57°$

9.3-19 (a) $q_{max} = 142.4$ lb/ft; (b) $I_{b\,min} = 38.5$ in^4

(c) $s = 0.264$ ft, 2.42 ft

9.4-1 $P_{cr} = 235$ k, 58.7 k, 480 k, 939 k

9.4-2 $P_{cr} = 62.2$ kN, 15.6 kN, 127 kN, 249 kN

9.4-3 $P_{allow} = 253$ k, 63.2 k, 517 k, 1011 k

9.4-4 $P_{allow} = 678$ kN, 169.5 kN, 1387 kN, 2712 kN

9.4-5 $P_{cr} = 229$ k

9.4-6 $T_{allow} = 18.1$ kN

9.4-7 (a) $Q_{cr} = 4575$ lb; (b) $Q_{cr} = 10065$ lb, $a = 0$ in

9.4-8 $P_{cr} = 447$ kN, 875 kN, 54.7 kN, 219 kN

9.4-9 $P_{cr} = 4\pi^2 EI/L^2$, $v = \delta(1 - \cos 2\pi x/L)/2$

9.4-10 $t_{min} = 10.0$ mm

9.4-11 (b) $P_{cr} = 13.89 EI/L^2$

APPENDIX A

A.1-1 A
A.1-2 B
A.1-3 A
A.1-4 A
A.1-5 D
A.1-6 A
A.1-7 C
A.1-8 D
A.1-9 D
A.1-10 D
A.1-11 A
A.1-12 B
A.1-13 C
A.2-1 D
A.2-2 B
A.2-3 A
A.2-4 A

A.2-5 D
A.2-6 A
A.2-7 B
A.2-8 C
A.2-9 A
A.2-10 D
A.2-11 D
A.2-12 C
A.2-13 A
A.2-14 C
A.2-15 D
A.3-1 D
A.3-2 A
A.3-3 C
A.3-4 A
A.3-5 D
A.3-6 D
A.3-7 B
A.3-8 B
A.3-9 C
A.3-10 B
A.3-11 D
A.3-12 B
A.3-13 B
A.3-14 D
A.4-1 D
A.4-2 C
A.4-3 D
A.4-4 A
A.4-5 A
A.4-6 C
A.4-7 B
A.5-1 A
A.5-2 C
A.5-3 D
A.5-4 D
A.5-5 A
A.5-6 B
A.5-7 B
A.5-8 C
A.5-9 C
A.5-10 A
A.5-11 B
A.5-12 B
A.5-13 C
A.5-14 B
A.5-15 B
A.5-16 D

A.6-1	C		**A.7-14**	B
A.6-2	C		**A.7-15**	C
A.6-3	D		**A.8-1**	C
A.6-4	A		**A.8-2**	C
A.6-5	B		**A.8-3**	B
A.6-6	A		**A.8-4**	C
A.6-7	C		**A.8-5**	B
A.6-8	D		**A.8-6**	C
A.7-1	A		**A.8-7**	D
A.7-2	C		**A.9-1**	D
A.7-3	D		**A.9-2**	B
A.7-4	B		**A.9-3**	D
A.7-5	C		**A.9-4**	A
A.7-6	A		**A.9-5**	D
A.7-7	D		**A.9-6**	A
A.7-8	D		**A.9-7**	B
A.7-9	A		**A.9-8**	B
A.7-10	D		**A.9-9**	B
A.7-11	D		**A.9-10**	C
A.7-12	A		**A.9-11**	D
A.7-13	C			

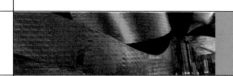

Index

PRINCIPAL UNITS USED IN MECHANICS

Quantity	International System (SI)			U.S. Customary System (USCS)		
	Unit	Symbol	Formula	Unit	Symbol	Formula
Acceleration (angular)	radian per second squared		rad/s^2	radian per second squared		rad/s^2
Acceleration (linear)	meter per second squared		m/s^2	foot per second squared		ft/s^2
Area	square meter		m^2	square foot		ft^2
Density (mass) (Specific mass)	kilogram per cubic meter		kg/m^3	slug per cubic foot		slug/ft^3
Density (weight) (Specific weight)	newton per cubic meter		N/m^3	pound per cubic foot	pcf	lb/ft^3
Energy; work	joule	J	N·m	foot-pound		ft-lb
Force	newton	N	kg·m/s^2	pound	lb	(base unit)
Force per unit length (Intensity of force)	newton per meter		N/m	pound per foot		lb/ft
Frequency	hertz	Hz	s^{-1}	hertz	Hz	s^{-1}
Length	meter	m	(base unit)	foot	ft	(base unit)
Mass	kilogram	kg	(base unit)	slug		$\text{lb-s}^2\text{/ft}$
Moment of a force; torque	newton meter		N·m	pound-foot		lb-ft
Moment of inertia (area)	meter to fourth power		m^4	inch to fourth power		in.^4
Moment of inertia (mass)	kilogram meter squared		kg·m^2	slug foot squared		slug-ft^2
Power	watt	W	J/s (N·m/s)	foot-pound per second		ft-lb/s
Pressure	pascal	Pa	N/m^2	pound per square foot	psf	lb/ft^2
Section modulus	meter to third power		m^3	inch to third power		in.^3
Stress	pascal	Pa	N/m^2	pound per square inch	psi	lb/in.^2
Time	second	s	(base unit)	second	s	(base unit)
Velocity (angular)	radian per second		rad/s	radian per second		rad/s
Velocity (linear)	meter per second		m/s	foot per second	fps	ft/s
Volume (liquids)	liter	L	10^{-3} m^3	gallon	gal.	231 in.^3
Volume (solids)	cubic meter		m^3	cubic foot	cf	ft^3